THE PHYSICAL BASIS FOR HETEROGENEOUS CATALYSIS

BATTELLE INSTITUTE
MATERIALS SCIENCE COLLOQUIA

Published by Plenum Press

1972: Interatomic Potentials and Simulation of Lattice Defects
Edited by Pierre C. Gehlen, Joe R. Beeler, Jr., and Robert I. Jaffee

1973: Deformation and Fracture of High Polymers
Edited by H. Henning Kausch, John A. Hassell, and Robert I. Jaffee

1974: Defects and Transport in Oxides
Edited by Martin S. Seltzer and Robert I. Jaffee

1975: The Physical Basis for Heterogeneous Catalysis
Edited by Edmund Drauglis and Robert I. Jaffee

In preparation:

1976: Fundamental Aspects of Structural Alloy Design

THE PHYSICAL BASIS FOR HETEROGENEOUS CATALYSIS

Edited by

EDMUND DRAUGLIS

Physics, Electronics and Nuclear Technology Department
Battelle Memorial Institute, Columbus Laboratories

ROBERT I. JAFFEE

Materials Science Department
Battelle Memorial Institute, Columbus Laboratories

BATTELLE INSTITUTE MATERIALS SCIENCE COLLOQUIA

Gstaad, Switzerland
September 2-6, 1974

Robert I. Jaffee, Chairman

PLENUM PRESS • New York — London

Library of Congress Cataloging in Publication Data

Battelle Institute Materials Science Colloquia, 9th, Gstaad, Switzerland, 1974.
   The physical basis for heterogeneous catalysis.

   Includes indexes.
   1. Catalysis—Congresses. I. Jaffee, Robert Isaac, 1917-   II. Drauglis, Ed-
mund, 1933-   III. Battelle Memorial Institute, Columbus, Ohio. IV. Title.
QD505.B37   1974   541'.395   75-41427
ISBN-13: 978-1-4615-8761-3   e-ISBN-13: 978-1-4615-8759-0
DOI: 10.1007/978-1-4615-8759-0

©1975 Plenum Press, New York
Softcover reprint of the hardcover 1st edition 1975
A Division of Plenum Publishing Corporation
227 West 17th Street, New York, N.Y. 10011

United Kingdom edition published by Plenum Press, London
A Division of Plenum Publishing Company, Ltd.
Davis House (4th Floor), 8 Scrubs Lane, Harlesden, London, NW10 6SE, England

To **PROFESSOR PAUL H. EMMETT**

for his contributions to heterogeneous catalysis over fifty years.

S. ANDERSSON *Chalmers University of Technology, Göteborg, Sweden*
J. R. ANTONINI *Battelle, Geneva Research Centre, Geneva, Switzerland*
A. BAGCHI *University of Maryland, College Park, Maryland, U.S.A.*
D. W. BASSETT *Imperial College of Science and Technology, London, England*
J. H. BLOCK *Fritz-Haber-Institut der Max-Planck-Gesellschaft, Berlin, Dahlem, West Germany*
G. C. BOND *Brunel University, Uxbridge, Middlesex, England*
M. BOUDART *Stanford University, Stanford, California, U.S.A.*
R. A. CRAIG *Battelle, Columbus Laboratories, Columbus, Ohio, U.S.A.*
E. DEROUANE *Facultés Universitaires N.-D., Bruxelles, Belgium*
E. DRAUGLIS *Battelle, Columbus Laboratories, Columbus, Ohio, U.S.A.*
H. EHRENREICH *Harvard University, Cambridge, Massachusetts, U.S.A.*
G. EHRLICH *University of Illinois at Urbana-Champaign, Urbana, Illinois, U.S.A.*
R. P. EISCHENS *Texaco, Inc., Beacon, New York, U.S.A.*
P. H. EMMETT *Portland State University, Portland, Oregon, U.S.A.*
G. ERTL *Universität München, München, West Germany*
J. FIGAR *Battelle, Geneva Research Centre, Geneva, Switzerland*
T. E. FISCHER *Esso Research and Engineering Company, Linden, New Jersey, U.S.A.*
J. J. FRIPIAT *Universite Catholique de Louvain, Heverleee, Belgium*
T. B. GRIMLEY *University of Liverpool, Liverpool, England*
H. D. HAGSTRUM *Bell Laboratories, Murray Hill, New Jersey, U.S.A.*
W. HAIDINGER *Technicum Cantonal, Bienne, Switzerland*
G. HOCHSTRASSER *Battelle, Geneva Research Centre, Geneva, Switzerland*

B. IMELIK *Institute de Recherches sur la Catalysis, Villeurbanne, France*

R. I. JAFFEE *Battelle, Columbus Laboratories, Columbus, Ohio, U.S.A.*

L. JANSEN *Battelle, Advanced Studies Center, Geneva, Switzerland*

K. H. JOHNSON *Massachusetts Institute of Technology, Cambridge, Massachusetts, U.S.A.*

C. KEMBALL *University of Edinburgh, Edinburgh, Scotland*

C. W. KERN *Battelle Institute, Columbus Center, Columbus, Ohio, U.S.A.*

W. KOHN *University of California, San Diego, La Jolla, California, U.S.A.*

D. H. LESTER *Battelle, Pacific Northwest Laboratories, Richland, Washington, U.S.A.*

R. MASON *University of Sussex, Falmer, Brighton, England*

D. MENZEL *Technische Universität München, München, West Germany*

R. P. MESSMER *General Electric Company, Schenectady, New York, U.S.A.*

E. W. MONTROLL *University of Rochester, Rochester, New York, U.S.A.*

D. NEWNS *Imperial College of Science and Technology, London, England*

J. OUDAR *Universite de Paris, Appliquée, Paris, France*

E. W. PLUMMER *University of Pennsylvania, Philadelphia, Pennsylvania, U.S.A.*

T. N. RHODIN *Cornell University, Ithaca, New York, U.S.A.*

N. RÖSCH *Universität München, München, West Germany*

L. D. SCHMIDT *University of Minnesota, Minneapolis, Minnesota, U.S.A.*

J. R. SCHRIEFFER *University of Pennsylvania, Philadelphia, Pennsylvania, U.S.A.*

G. M. SCHWAB *Universität München, München, West Germany*

J. R. SMITH *General Motors Corporation, Warren, Michigan, U.S.A.*

G. A. SOMORJAI *University of California, Berkeley, California, U.S.A.*

H. SUHL *University of California, San Diego, La Jolla, California, U.S.A.*

K. TAMARU *The University of Tokyo, Bunkyo-Ku, Tokyo, Japan*

J. T. WABER *Northwestern University, Evanston, Illinois, U.S.A.*

C. WAGNER *Max-Planck-Institut für Biophysikalische Chemie, Postfach, Germany*

G. WALTER *Battelle-Institut e.V., Frankfurt, Germany*

M. B. WEBB *University of Wisconsin, Madison, Wisconsin*

G. WOLKEN, JR., *Battelle Institute Columbus Center, Columbus, Ohio, U.S.A.*

P. WYNBLATT *Ford Motor Company, Dearborn, Michigan, U.S.A.*

J. T. YATES *National Bureau of Standards, Washington, D.C., U.S.A.*

PREFACE

THE PHYSICAL BASIS FOR HETEROGENEOUS CATALYSIS is the proceedings of the ninth Battelle Colloquium in the Materials Sciences, held in Gstaad, Switzerland, September 2–6, 1974. It took as its theme the application of modern theoretical and experimental surface physics to heterogeneous catalysis. Progress in the field by classical chemical methods seemed to have slowed down, at a time when the need for better catalysts was particularly great. The Organizing Committee thought it might be possible to accelerate progress by the application of the powerful techniques evolved in recent years for studying atomically clean surfaces. However, the translation of ideas derived from clean single crystal surfaces with well-characterized chemisorbed layers to real catalysts with high ratios of surface to mass on which reactions were taking place and requiring transport of mass and energy is a giant step, raising many questions and requiring thorough discussion by surface physicists on the one hand and catalytic chemists on the other. The 1974 Battelle Colloquium provided a forum for this exchange.

As its usual custom, the Colloquium started the first day of introductory lectures by three distinguished scientists who have contributed importantly over many years to this field. On following days, the overall field was divided into the following sessions:

Experiments on Clean Metal Surfaces with Adsorbates
Theory of Chemisorption

ix

The Effect of Small Particles and Porous Carriers
Kinetics and Transport
Applications to Catalysis

Each of the first four topics was introduced by a session of research papers, following which a general discussion of critical issues was held. Finally, a concluding agenda discussion covering issues in the overall field of heterogeneous catalysis was led by T. E. Fischer (experimental) and T. B. Grimley (theoretical). The reader is commended to these discussions for an up-to-date statement of the status of research in the field.

Each Colloquium customarily honors one of the distinguished participants by inscribing the proceedings in his name. At Gstaad we honored Professor Paul Emmett for his pioneering work in the field of heterogeneous catalysis. At the Farewell Banquet, Professor Emmett presented informal remarks about his career in catalysis, which are included in the front matter of these proceedings.

We also wish to thank our Battelle colleagues who have supported the Colloquia. In particular, we wish to thank Dr. Sherwood L. Fawcett, President of Battelle Memorial Institute, and Dr. C. W. Kern, Director of Research in the Physical Sciences of Battelle Institute, under whose auspices these Colloquia are a continuing activity. We are grateful to Dr. V. Stingelin, Associate Director of the Battelle Geneva Research Centre, for his participation in the opening events of the Colloquium, and for the excellent support provided by members of the Battelle Geneva staff: Anne-Marie Bonino in charge of local arrangements and head of the secretariat, including Julie Swor and Darlene Weaver, who provided excellent secretarial assistance; Pierre Fontaine, lighting and sound arrangements. Edna Jaffee was in charge of the Ladies Program and hospitality. To all these, and many others, we wish to extend our sincere thanks.

The Organizing Committee is grateful to J. R. Block, G. C. Bond, M. Boudart, R. Gomer, C. Herring, T. R. Hughes, T. N. Rhodin, J. H. Sinfelt, and M. B. Webb for their advice and choice of subject and participants for this Colloquium.

The Organizing Committee

R. I. Jaffee, Chairman
E. Drauglis
H. Ehrenreich
L. Jansen
W. Kohn
F. J. Milford
J. R. Schrieffer
G. Walter

AUTOBIOGRAPHICAL REMARKS—PAUL H. EMMETT

INTRODUCTION

Any accounting of the career of a scientist should, I believe, relate the circumstances that led to his choosing science as a career, the people and events that influenced his work, and then, briefly, the chronology of his working career and some recounting of a few of the things that may have been accomplished.

CHOOSING A SCIENTIFIC CAREER

During my early life, including my first 3 years in high school, I had no interest in or leaning toward science. My father was a self-made engineer and railroad man, who, together with my mother, was open minded about my future, but both were determined that I take my education seriously and push on to college, and perhaps even further. It was not until my senior year in high school, after a session with a vocational advisory teacher (a Miss Workman at Washington High School in Portland, Oregon), that I was encouraged to take both physics and chemistry in my senior year and to consider seriously one of these or some other branch of science as a career. I soon found both chemistry and physics to intrigue and interest me by the challenges that they presented. My choice of chemistry probably resulted from the superior and enlightened teaching by Dr. William (Willie) Green, who taught chemistry many years at Washington High School and who was not only an inspiration to me, but also played an important role in the life of my friend, the illustrious Linus Pauling. Linus was a year ahead of me in high school, but was, I believe, equally influenced by the teaching of Willie Green. As an aside, it may be pointed out that Linus was a "drop out" at Washington High. I suspect his dropping out resulted from his wish to finish up with more chemistry from Willie Green, and fewer of the required courses in other fields. At any rate, he had to wait until he had been awarded two Nobel prizes before being given an honorary diploma from Washington High School.

Upon graduation from high school I elected to attend Oregon Agricultural College, possibly because it was the closest College or University that was a State Institution and therefore within reach of one who planned to be at least partially self-supporting in school. At that time, the only two full-fledged chemistry programs available at OAC were Agricultural Chemistry and Chemical Engineering. I chose the latter without giving much thought as to whether I preferred the engineering aspects of chemistry or a more basic scientific approach that would be involved in a regular chemistry curriculum.

At OAC I was greeted by the military influence that existed in all schools at that time but which was particularly emphasized at schools that were land grant colleges and were obliged to offer military training in connection with the Reserve Officer's Training Corps. I was 10 days too young for the last military draft in the fall of 1918, but nevertheless continued 4 years of military training and was eventually given a commission in the Reserves. I even went to the summer camps. However, after 3 years, I resigned my commission with the conviction that if war ever came, I would be of more use as a chemist than as a field artilleryman.

The Chemical Engineering course consisted of regular chemistry courses, industrial chemistry, some mechanical engineering, some electrical engineering, and even a course in "blacksmithing" or foundry work. Linus Pauling had also elected to go to OAC. He entered without a high school diploma, proceeded in school for 2 years, and then joined the faculty for a year, in time to teach quantitative analysis to my class. For the last 2 years we were classmates, and became fast friends.

There may be something prophetic in the fact that we both took the "blacksmithing" or foundry course, in which we were supposed to learn the characteristics of steel and methods of working it and welding it. Linus, in later years, became a world authority on metals and alloys, and I was to spend a lifetime working on the preparation and properties of iron catalysts for synthesizing ammonia. As you know, in our meeting this week, I have summarized 50 years of research on iron synthetic ammonia catalysts.

Another important aspect of our training at OAC was the freshman chemistry teacher, Professor John F. G. Hicks. His son, John Hicks, is, I am sure, known to many of you, since he at one time was a Vice President of Battelle Institute. Professor Hicks was an inspiring teacher. His philosophy of teaching could be summed up by one of his favorite expressions: "In the book it is so-but *why?*" We were taught to understand the whys and wherefores of all that we studied and read. This philosophy and emphasis probably had something to do with my choosing physical chemistry rather than organic. It should be noted that catalysis as a subject for my career had not yet put in an appearance. We were all encouraged to go on to graduate work by our Professor of Chemical Engineering, but choices of research fields were yet to be made.

After 3 years of varsity debating, I was urged by some of my father's lawyer friends to take up a career in patent law. The lure of chemistry, however, was too strong to permit my seeking a career in patent law, so I turned my attention to selecting a graduate school for work toward a Ph.D. degree, and a career of research and teaching in Physical Chemistry.

My reasons for choosing the California Institute of Technology are not too clear. It was a new school headed by a group of brilliant scientists, including Professors A. A. Noyes in Chemistry, Professor R. A Millikan in Physics, and Professor R. C. Tolman in Physical Chemistry. Possibly, I was influenced by the higher teaching grants ($750, compared with $350 at some of the other schools) and probably by the fact that it was Linus Pauling's choice. At any rate, each of us applied and was accepted on one condition, namely that we work out and hand in all the problems in the first nine chapters of the book *Chemical Principles* by Noyes and Sherill. By working these problems, we practically completed an extra course in Physical Chemistry. The summer required to complete these problems was one of the most intense of my early scientific training.

My first contact with catalysis came mostly on the advice of Professor Noyes, who suggested that I work with Dr. Arthur F. Benton, who had just completed work for his Ph.D. under Professor Hugh S. Taylor at Princeton University, and was to be at Cal Tech on a National Research Fellowship for 2 years. I accepted the advice of Dr. Noyes and so cast my lot in the field of catalysis. The more I learned about it, the more enthusiastic I became. I profited greatly from the 2 years with Dr. Benton. He was an excellent laboratory technician and a profound student in the field of catalysis. Such technique as I have acquired, I learned from Benton. I also acquired from him the habit of reading and thinking critically.

At the end of my 2 years with Dr. Benton, I had to decide whether to continue my studies in the field of catalysis, or to change the trend of my work for a final year to broaden my background. When my request for a transfer was turned down by a University prominent in the field of catalysis, I decided to finish my work for my Ph.D. at Cal Tech. For my final year, I elected to work with my fellow student, Linus Pauling, who by this time had become an expert in X-ray crystal-structure work. I remember we worked out and published one paper as a correction to the crystal structure of barium sulfate. I have always valued this year spent with Linus Pauling, not only because crystal-structure work was germane to my chosen field of catalysis, but also because it was an inspiration to work with him.

A LIFETIME OF CATALYTIC RESEARCH

The Fixed Nitrogen Research Laboratory, 1926 to 1937

After finishing at Cal Tech and teaching general chemistry for a year at Oregon State College (my alma mater with a name change), I chose to continue my interest in catalysis by pursuing a career in that field. I was accepted as an Assistant Chemist at the Fixed Nitrogen Research Laboratory in Washington, D. C. This laboratory had been established at the end of World War I to work out the details of an American version of a process for synthesizing ammonia from a hydrogen-nitrogen mixture, and I arrived at the time when this version had been developed. The next obvious step was the one assigned to me, namely, to establish and pursue a line of work designed to better understand the mechanism by which the iron synthetic ammonia catalysts operated.

Dr. F. G. Cottrell, of Cottrell Precipitator fame, was director of the laboratory and proved to be one who adopted a policy of letting his key people plan and pursue their respective goals with very little interference. It was a very happy and productive 11-year period that I spent at the laboratory.

This is not the time or place to review in detail all of the research that we performed in pursuing our goal of learning more about the mode of operation and mechanism of operation of the iron synthetic ammonia catalysts. But I shall assume the prerogative of describing two of our projects that I consider as being among the chief accomplishments. The first was the development of a method for measuring the surface area of catalysts or other finely divided or porous solids; the second was concerned with the discovery of the errors introduced into the measurement of the water-gas-shift reaction by the phenomenon known as thermal diffusion.

It occurred to us at the start that if one were to try to judge why some of the iron catalysts were active and some were inactive, one had to know something about the accessible area on which the catalytic reaction might occur. Benton, my former boss at Cal Tech, had run some adsorption isotherms of nitrogen at a temperature corresponding approximately to the boiling point of nitrogen, −195°C. He had obtained a kink in the adsorption isotherm at about 1/2 atmosphere pressure and suggested that this might mark the completion of a monolayer of adsorbed gas. It later turned out that his particular "kink" was due to his failure to take into consideration the deviation of nitrogen from being a perfect gas. When the proper corrections were made for this gas imperfection, the "kinked line" turned into a straight line, but the idea of finding the point on the isotherm corresponding to a monolayer was a good one. Further investigations culminated in the conclusion that the beginning of this straight-line portion of the isotherm probably marked the true completion of a monolayer. It was at about one-tenth of the liquefaction pressure of the nitrogen gas rather than at about one-half the pressure as had been suggested by Benton. The method of selecting a monolayer of adsorbed gas as the volume of gas at the beginning of the long linear portion of the isotherm, called the "point B method", was used for estimating the volume of various gases required to form a monolayer. The multiplication of the number of molecules in such a monolayer by the estimated area occupied by each molecule permitted the calculation of the surface area of the catalyst.

Theoretical confirmation of our interpretation of the adsorption isotherms seemed desirable. Several years after the completion of the point B method, a young Hungarian scientist, Dr. Edward Teller, joined the faculty of George Washington University. Soon my energetic assistant, Dr. Stephen Brunauer, was working with him on the theory of multimolecular adsorption. Out of their derivation came the equation

$$\frac{x}{v(1-x)} = \frac{1}{V_m C} + \frac{(C-1)x}{V_m C}$$

where v is the volume adsorbed at the relative pressure x, V_m is the volume of gas required to form a monolayer, and C is a constant. The monolayer value chosen by the point B method proved to be in good agreement with V_m.

After a critical discussion, the equation was incorporated into a summary paper on the advantages of the gas-adsorption method for measuring surface area. The paper was authored by Brunauer, Emmett, and Teller; the equation became known as the BET equation and has, for 36 years, been used throughout the world for measuring the surface area of catalysts.

The second accomplishment had to do with the value for the equilibrium constant for the reaction

$$H_2O + CO = H_2 + CO_2$$

The directly measured values differed by about 40 percent from the indirect values obtained by combining the equilibria for the two separate reactions

$$Fe + H_2O = H_2 + FeO$$

and

$$FeO + CO = CO_2 + Fe$$

Floyd Shultz, my efficient assistant, set up an apparatus that we hoped would tell us why about half the workers who had published on the subject obtained one value for the equilibrium constant $(H_2)/(H_2O)$ for the iron-iron oxide system, whereas the other half obtained a ratio 40 percent higher. Luck entered into our finding a solution to the problem. We didn't have a piece of quartz tubing 2 cm in diameter for our reactor so we made a reactor by sealing a 2-mm-diameter quartz tube to a 2-cm diameter tube. We placed a boat of Fe-FeO in the middle of the furnace at the juncture of the 2-mm and 2-cm tubes. We arranged a pumping system that would enable us to circulate a water vapor-hydrogen mixture through the furnace in such a way that the exit gas could pass out either through the 2-mm capillary or through the 2-cm capillary. To our surprise, when the gas mixture came out through the small capillary, the water vapor/hydrogen ratio seemed to be the one that when combined with the constant for the Fe-CO-FeO-CO_2 system yielded the correct equilibrium constant, whereas the circulation with the gas existing from the 2-cm tube yielded a result that seemed to be 40 percent in error. But, why? The answer came by another stroke of luck. One of the scientists in our after-work journal seminar gave a paper on the thermal-diffusion separation of a helium-carbon dioxide mixture. In simple terms, a mixture of He-CO_2 in a quartz tube, one end of which was in a furnace at 1000°C and the other at room temperature, showed a large difference in the composition of the gas in the hot zone as compared with the gas in the cold zone. This difference was interpreted on the basis of a theory worked out by Enskog in 1913 and by S. Chapman in 1916 as a phenomenon called thermal diffusion. According to the theory, the light and small molecules tend to collect in the hot part of the tube and the heavy or large molecules in the cold part. In the traditional apparatus for the water vapor-hydrogen equilibration with Fe-FeO, a quartz tube 2 cm in diameter at both ends had been employed and, hence, with such apparatus, an error was always obtained which varied with temperature but was about 40 percent at 1000°C.

Time does not permit the discussion of other work at the Fixed Nitrogen Laboratory. I would, however, like to point out that during this time I had my first graduate student, Rowland Hansford, who worked at George Washington University where I taught a course in catalysis. He did a thesis on the activity of iron catalysts for the hydrogenation of ethylene. His distinguished career has always been a real joy to me.

The Department of Chemical Engineering at the Johns Hopkins University, 1937/1944

In the fall of 1937 I accepted the chairmanship of the Department of Chemical Engineering at the Johns Hopkins University and promptly hired Dr. Carlos Bonilla and Dr. Ralph Witt to help organize the department. Between them, they taught most of the actual chemical engineering, whereas I limited myself to some of the industrial chemical calculations that were, in fact, applied physical chemistry.

The department prospered and, due largely to the activity of Drs. Bonilla and Witt, was accredited in 1940, 3 years after its inception. My research during this period was concerned mostly with the study of surface areas of various materials (see dissertation by W. B. Harris) and the hydrogenating activity of iron catalysts (dissertation by John Gray) and Ni-Cu alloys (dissertation by Nis Skau). Time does not permit a detailed discussion. One thing I do want to mention—the stay at Hopkins brought me in contact with Dr. E. Emmet Reid, who had retired from the chemistry department in 1935. Back in 1922, he rendered catalytic chemists a big service by translating into English the book *Organic Catalysis* by Paul Sabatier. In 1935 he got the urge to republish this book. The publishers agreed, provided someone (and that someone turned out to be me) would summarize all the new work in catalysis since 1922. The product of our combined efforts was published under the title *Catalysis Then and Now* and is, I believe, still available. The thing I like to remember about Emmet Reid was the way he lived up to his saying: "The way to be active is to be active". He died just short of his 102nd birthday and was actively writing articles and books to the last.

After the attack on Pearl Harbor in 1941, I became engaged in Chemical Warfare work, and then in 1943, I joined the Manhattan Project to work under Urey on the development of a barrier material suitable for the diffusional separation of uranium-235 and uranium-238.

The Manhattan Project 1943–1944

Little can be said about the year and a half spent on the Manhattan Project, partly because the work done is still under secrecy orders and partly because my position was largely administrative, without much chance of contributing scientific ideas. In a few instances I may have been helpful in promoting a continuation of certain projects that some were in favor of phasing out. One such project was the preparation of the fluorocarbon plastics needed for the construction of parts of the diffusion plant. At the time the decision had to be made, Dr. William Miller, who was in charge of the fluorocarbon work, had produced a tiny piece of product by using very high pressure. The gamble was whether he could produce by a reasonable process the desired material in 6 months. The decision was made to continue and paid off handsomely.

These were thrilling times. There was an ever present tension, heightened always by the question as to whether Germany might beat us to the atom bomb with its devastating possibilities. It was a project that I am proud to have been associated with.

Mellon Institute of Industrial Research 1944–1955

By late 1944, the work at Columbia University had advanced to the point of development at which it could be transposed into the construction of the actual diffusion plant. I accordingly considered that my assignment had been completed. I accepted a position at Mellon Institute in Pittsburgh on the Multiple Petroleum Fellowship financed by the Gulf Oil Company. Thus began another 11 years of my scientific career. I might add that I continued as a consultant on the Manhattan Project and have remained a consultant with the Atomic Energy Laboratory at Oak Ridge up to the present time.

The work at Mellon Institute was concerned with various catalytic problems of interest to the Petroleum industry. Principal among these was the study of the mechanism by which the Fischer-Tropsch process operated to produce hydrocarbons from a mixture of carbon monoxide and hydrogen. It is interesting that immediately after World War II there was some apprehension over our dwindling petroleum supplies and a feeling that with an effective submarine blockade we might be cut off in an emergency from essential petroleum supplies. So considerable effort was spent both by petroleum companies and by the Bureau of Mines in studying the various ways by which hydrocarbon products, including gasoline, could be obtained from coal. The evidence for such a possibility was plentiful in the form of processes that had operated in Germany successfully during the war to yield a supply of gasoline and other essential hydrocarbons. Thus, attention at the Bureau of Mines turned to comparing the Fischer-Tropsch process for producing synthetic hydrocarbons with the process for liquefaction of coal by high-pressure hydrogenation. It is interesting to note that in the early 1950's, the same problem was worked on that faces the nation today—energy from coal. However, at that time petroleum was so cheap and easily available that the whole project was dropped except for some continuing research on the mechanism by which the catalysts for the Fischer-Tropsch process operate. During this period of 11 years, my colleagues and I elected to study the Fischer-Tropsch process by using iron catalysts and by employing radio-active tracer compounds containing radioactive carbon.

We started our carbon-14 tracer work rather early in 1945 as I recall. Our first supply of carbon-14 came from Dr. John Dunning of Columbia University. This is not the proper place for going into details of our work except to point out that it marked the beginning of about 30 years of research that my colleagues and I were to pursue in using isotopes as tracers to study catalysts and catalytic processes. We succeeded in showing that the reaction mechanism involved the formation of oxygen-carbon-hydrogen complexes on the surface as intermediates rather than the formation of intermediate metallic carbides on the catalyst. The work was carried on successfully because of the participation of a very skilled and capable group of colleagues. These included Drs. R. J. Kokes, J. T. Kummer, and Thomas De Witt as postdoctorates; Drs. Luther Browning, Keith Hall, and Harry Padgurski, who obtained Ph.D. degrees while engaged in this work; and W. B. Spencer, who obtained a master's degree.

It was in this period that isotope work on silica-alumina cracking catalysts received considerable attention. The chemisorption of the hydrocarbons on cracking catalysts as measured by radioactive tracers (work of Drs. Donald MacIver and Robert Zabor) and the rate of exchange of deuterium oxide with the light hydrogen of the cracking catalysts (work of Dr. Robert Haldeman) were prominent among our accomplishments at this time.

And so the work at Mellon proceeded at a merry pace until the summer of 1955. At that time I was invited to accept appointment to the W. R. Grace Professorship that had been set up in the Chemistry department at Johns Hopkins University. It involved being a Professor at Johns Hopkins University and at the same time serving as a consultant for the catalytic work at W. R. Grace and its subsidiary, Davison Chemical Company.

Johns Hopkins University—W. R. Grace Professor 1955–1971

At Hopkins, we continued studies on the use of tracers both for Fischer-Tropsch synthesis of hydrocarbons and for catalytic cracking. In addition, we undertook work on catalytic hydrogenation over Raney nickel, copper-nickel alloys, and synthetic ammonia catalysts. Catalytic measurements on the newly discovered xenon fluorides and equilibium data on the formation of Mo_2C completed the list of work carried on at Johns Hopkins during the closing years of my active scientific career.

In Fischer-Tropsch synthesis studies we concentrated mostly on the use of ketene (CH_2CO) as a tracer. Dr. Blyholder, who did the work, took advantage of the newly developed gas-liquid partition chromatography to analyze the products of reaction and to measure the radioactivity of each hydrocarbon product. Without going into all the details of the work, I can summarize by saying that we showed that ketene dissociates into an adsorbed CH_2 groups and gaseous CO. These CH_2 groups are capable of initiating the formation of the long-chain hydrocarbons that make up the products of the Fischer-Tropsch synthesis.

The tracer work on catalytic cracking comprised adding a number of radioactive paraffins, olefins, and aromatics to a stream of cetane being cracked over silica-alumina catalysts. (For details, the reader is referred to dissertations at the Johns Hopkins University over the period 1961 to 1972 by W. A. Van Hook, Joseph Hightower, Kun-ichi Matsushita, Barton Barclay, Robert Garten, and John Bordley.) The extent to which secondary reactions among the primary products or between the primary products and the reactants took place could then be determined by analysis of the radioactivity of the products. In general, the paraffins underwent no secondary reactions, olefins reacted extensively to form higher molecular weight hydrocarbons and aromatics were also relatively inactive, though radioactive benzene and toluene alkylated extensively with propylene to form very radioactive cumene and para cymene, respectively. In a word, this tracer work on cracking, together with our earlier work, gives a good picture of many of the working details by which catalysts operate to give our modern supply of cracked gasoline.

Catalytic hydrogenation rounded out our surface studies. We showed that iron, contrary to current theories, is a good catalyst for hydrogenating benzene to cyclohexane, but that it rapidly loses its activity, presumably because of its being converted to iron carbide by its reaction with benzene or cyclohexane (see dissertation of Miss Jane Philips). Copper and iron were shown to enhance the activity of metallic nickel, though quantitative comparisons were made difficult by the fact that the surface of the copper-nickel alloys probably assumed a different composition than that of the bulk of the catalyst (dissertation by Gervin Harkins). We also studied and learned much about the performance of the well-known Ni-Al alloy known as Raney nickel, and this brings me to the high spot of my fond recollection of the work at Hopkins, in that Raney nickel work was done by Dr. Richard Kokes. Dick joined our group and was introduced to catalytic work in 1952 at the Mellon Institute. After 3 years of valuable service, he followed me to Baltimore as a part-time assistant while he served on the faculty at Loyola College. In 1958 he joined the Hopkins faculty, and for the next 15 years proceeded to establish himself firmly at the top of the ladder in the field of catalysis. He endeared himself to all of us and was a particular inspiration to me and to his students. When I retired in 1971, I was glad to find in his hands a strong catalytic program that gave promise of continuing the tradition of catalysis that had been established 50 years earlier by the work of Frazier and Patrick. Unhappily for us all, Dick died of a heart attack 2 years later at the age of 46. We miss him.

And so, in 1971, I "retired". I use quotation marks because I am busier than ever and have not yet found the traditional rocking chair. Back in

Portland, Oregon, where I was born, and spent my early years, I keep busy writing papers, giving lectures, consulting, and occasionally visiting such beautiful spots as the one here in Switzerland we are now enjoying.

CONTENTS

Part One

INTRODUCTORY LECTURES

FIFTY YEARS OF PROGRESS IN THE STUDY OF THE CATALYTIC SYNTHESIS OF AMMONIA

P. H. Emmett

Portland State University
Portland, Oregon

ABSTRACT

The catalytic synthesis of ammonia developed in Germany in the 1908 to 1913 period has probably been studied more extensively during the past 50 years than any other catalytic reaction. Research endeavors include the development of a method for measuring the surface area of the catalysts; a study of the adsorption of both hydrogen and nitrogen by the iron catalysts; a phase-rule study of the formation of nitrides of iron from ammonia-hydrogen mixtures; a study of the distribution of promoters on the catalyst surface; methods for estimating the fraction of adsorbed nitrogen present in the molecular form; and estimates from infrared spectroscopy, from LEED measurements, and from ion-emission spectroscopy as to the path followed during synthesis and the catalyst planes important in the synthesis. Recently, revolutionary nonferrous catalysts severalfold more active than standard iron catalysts have been prepared by Ozaki. These throw new light on the mechanism of the catalytic synthesis of ammonia.

1 INTRODUCTION

The catalytic synthesis of ammonia developed in the period 1908–1913 in Germany[1] has probably been studied more intensively during the past 50 years than any other catalytic reaction. It is therefore a topic particularly well suited to provide a perspective for papers appearing in the present symposium. It has been suggested that this review should emphasize and summarize work with which my colleagues and I have been personally connected. This is especially convenient in view of the fact that a complete review of all the work that has been done on this catalytic system in the past 50 years would require a book rather than a short paper. Indeed several such fine reviews have already been written.[2,3,4] For convenience the work will be discussed under six headings: adsorption, state of adsorbed nitrogen on iron, kinetics, composition of the catalyst surface, miscellaneous recent surface mechanism studies, and new nonferrous catalyst developments.

2 ADSORPTION

Both physical and chemical adsorption proved to be important in the study of iron catalysts. Physical adsorption of nitrogen, argon, and other similar gases near their boiling points led to the BET method for measuring catalyst surface areas. It has also been useful in calculating pore size distribution on reduced iron catalysts. Chemical adsorption of both hydrogen and nitrogen were studied with a view to ascertaining whether the rate of adsorption of either might be rate determining in ammonia synthesis.

2.1 Physical Adsorption

In the early days of ammonia research it was ascertained that the activity of an iron catalyst varied with methods of preparation and with promoter content. It became evident, however, that conclusions as to the nature of changes resulting in improved activity could be made only if one knew something about the surface area of the porous solids used as catalysts. No method for measuring such areas then existed. It seemed reasonable that the adsorption isotherms for inert gases, such as nitrogen, near their boiling points might exhibit characteristics that would enable one to select the volume of adsorbate required to form a monolayer and hence furnish a basis for estimating the catalyst surface area. Accordingly, adsorption isotherms for nitrogen, oxygen, argon, carbon monoxide, carbon dioxide, methane, butane, ammonia, nitrous oxide, and nitric oxide were made at temperatures near their respective boiling points.[5] A group of such isotherms is shown in Fig. 1. The combined data seemed to indicate that by

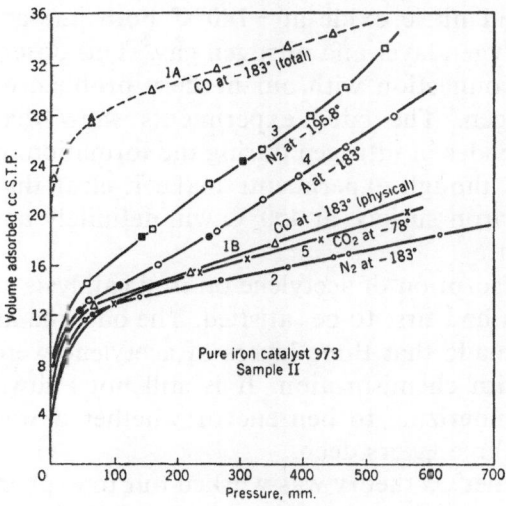

Fig. 1. Adsorption isotherms on a pure iron synthetic ammonia catalyst for various gases near their boiling points: curve 1A is for total adsorption of CO (physical plus chemical adsorption); curve 1B is for physical adsorption occurring at −183°C, after the evacuation of the sample from curve 1A at −78°C for 1 hour; solid symbols represent desorption.[5]

using reasonable estimates for the area covered by each adsorbed molecule, one could calculate surface-area values from the various adsorbates that agreed with each other, if one assumed that the beginning of the long linear part of the adsorption isotherms (called "point B") constituted a statistical monolayer. Of course, in making such measurements one had to first saturate the catalyst with any chemisorption or reaction that might be occurring. Only when this is done can reliable surface areas be obtained. In retrospect, four observations in regard to this initial chemisorption or reaction should be mentioned.

1. Carbon monoxide rapidly forms a complete chemisorbed monolayer on a pure iron catalyst at −183°C. The volume of the chemisorbed layer was about 1.25 times[6] as great as the volume of physically adsorbed nitrogen estimated by the point B method to form a monolayer. Hence, if the chemisorbed CO occupied about 12 to 13 $Å^2$ per molecule compared with 16.2 $Å^2$ per molecule for the physically adsorbed nitrogen molecule, the chemisorption of CO afforded a good confirmation that point B did indeed correspond closely to a statistical monolayer.

2. The adsorption of oxygen at −183°C or even −195°C rapidly formed not just one but five to ten layers of chemically bound oxygen.[5,6] Presumably, the outer five to ten layers of the catalyst are susceptible to low-temperature oxidation. After the completion of the formation of such chemically bound oxygen, a normal reversible isotherm for physically adsorbed oxygen can be made. This observation suggests that all chemisorption experiments for oxygen on metals, even at low temperatures, must be watched carefully for the possible formation of several layers of surface oxide.

3. Nitrous oxide at −78°C and nitric oxide at −140°C both interact with the iron surface to form an oxygen layer and nitrogen gas.[5] This observation is especially interesting in connection with our modern problem of decomposing the oxides of nitrogen. The cited experiments show that nitrogen can be released from the oxides of nitrogen during the formation of one or more layers of oxide, even though experiments make it clear that chemisorbed nitrogen put onto an iron surface at 450°C will definitely not desorb at or below room temperature.[7]

4. Attempts to measure the adsorption of acetylene on iron catalysts at −100°C showed that chemisorption had first to be satisfied. The odd and as yet unpublished[8] observation was made that three layers of acetylene were required to saturate the surface with chemisorption. It is still not known whether the acetylene might be trimerizing to benzene or whether it was merely forming a surface polymer three layers deep.

Later, in collaboration with Teller[9], a theory was worked out to explain the S-shaped character of the adsorption isotherms such as shown in Fig. 1. Specifically a BET equation was developed in the form

$$\frac{X}{V(1-X)} = \frac{1}{V_m C} - \frac{(C-1)x}{V_m C} \tag{1}$$

where x is the relative pressure p/p_0 of the adsorbate, V is the volume of adsorbate on the surface at relative pressure x, V_m is the volume of gas required to form a monolayer, and C is a constant related to the heat of adsorption. A linear plot of the left side of this equation against x yields straight lines of the type shown in Fig. 2. The slope and intercept of these

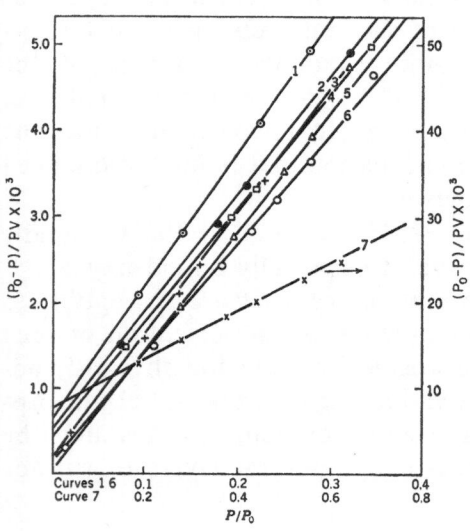

Fig. 2. BET plots for the adsorption on 0.606 g of silica gel of CO_2 at −78°C (curve 1); argon at −183°C (curve 2); nitrogen at −183°C (curve 3); oxygen at −183°C (curve 4); CO at −183°C (curve 5); nitrogen at −195°C (curve 6); and C_4H_{10} at 0°C (curve 7).[9]

straight lines enable one to calculate the volume of gas required for a monolayer and to estimate the heat of adsorption. Fortunately, the agreement between the values of V_m deduced by the point B method and by the BET method is good. These means of arriving at surface-area values for various catalysts thus for the first time made it possible to tell whether catalysts differed from each other due to the quality or to the quantity of surface available. Fortunately, these early surface-area measuring procedures have withstood the test of time and are still used today on an international scale.

Physical adsorption also proved to be useful in measuring pore size and pore size distributions. Barrett, Joyner, and Halenda[10] showed how it was possible to calculate pore distributions from nitrogen desorption isotherms. Recently their method has been revised and improved.[11,12] It also has been applied to adsorption as well as desorption isotherms.[13] It will suffice to point out here that a fairly good approximation to pore distributions can be made by methods involving physical adsorption-desorption isotherms. Figure 3 shows the pore distribution on a bone char obtained by Barrett, Joyner, and Halenda. On the same curve are the points obtained by a mercury porosimeter method[14] which show good agreement with the nitrogen desorption calculations. The data obtained by Nielsen[15] for the pore distribution of an iron synthetic ammonia catalyst are presented in Fig. 4.

2.2 Chemical Adsorption

It has long been recognized that the rate and amounts of chemisorption of reactants on a catalyst might give valuable clues to the slow step involved in a particular catalytic reaction. It was natural therefore to study on iron synthetic ammonia catalysts the chemisorption of both hydrogen and nitrogen. Chemisorption data for H_2 on a doubly promoted iron catalyst[16] (catalyst 931 containing 1.59 percent K_2O and 1.3 percent Al_2O_3 as promoters) are shown in Fig. 5. The low-temperature curve in this isobar is presumably physically adsorbed hydrogen. The parts of the isobar between -100 and $0°C$ and between 100 and $400°C$ are called types A and B chemisorption, respectively. Later it was discovered that catalysts promoted with Al_2O_3 but no potassium oxide showed chemisorption (type C) in the region -195 to $-140°C$.[17] In line with this, the singly promoted catalyst is active for the H_2-D_2 exchange at $-195°C$, whereas the doubly promoted catalyst with no type C adsorption is inactive for this exchange at $-195°C$. At any rate, the miscellaneous hydrogen-adsorption experiments showed clearly that all three types of hydrogen chemisorption occurred at temperatures well below those at which ammonia synthesis will occur. Presumably, therefore, the slow step in synthesis is not the rate of hydrogen chemisorption.

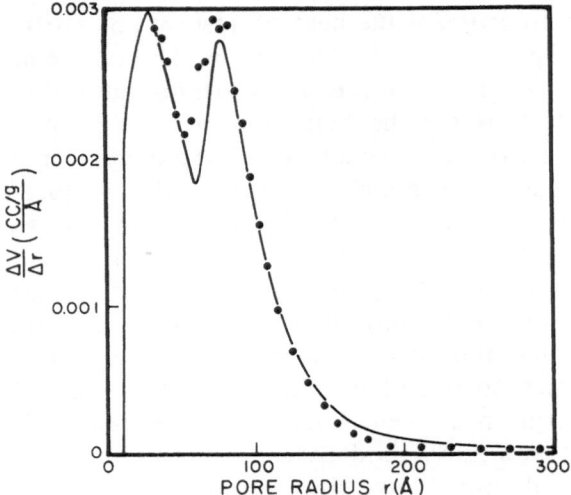

Fig. 3. Pore distribution in a sample of bone char as calculated by Barrett et al.[10] from a nitrogen desorption isotherm. Also shown are experimental points obtained for the same bone char by a mercury porosimeter.[14]

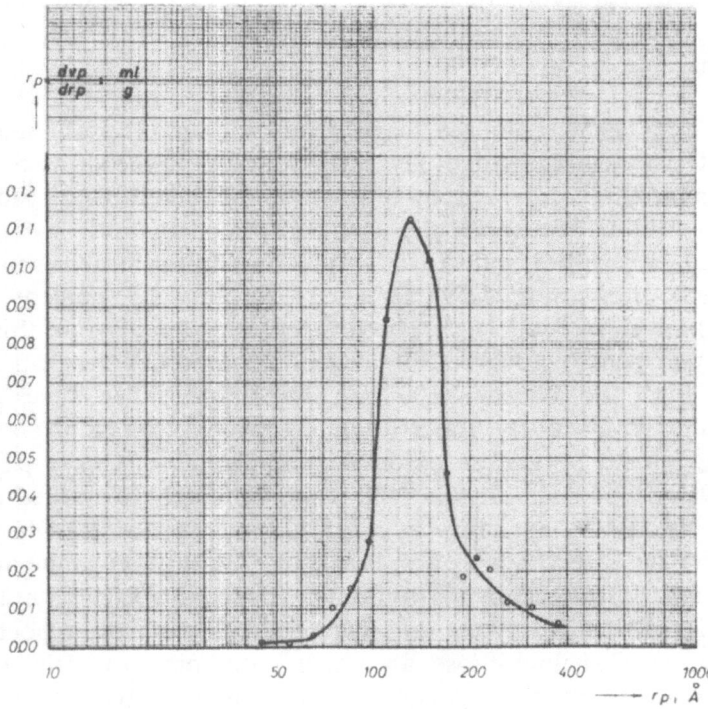

Fig. 4. Distribution of micropores in a reduced promoted iron synthetic ammonia catalyst.[2]

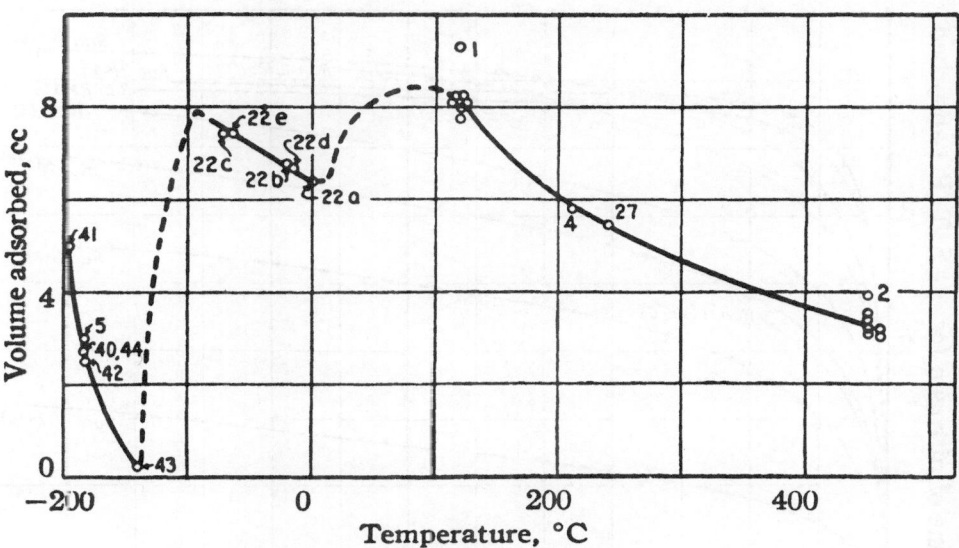

Fig. 5. A hydrogen adsorption isobar at 1 atm pressure on doubly promoted iron catalyst 931 between −195° and 450°C. The adsorption below −140° is considered to be physical adsorption; that occurring at higher temperatures includes two types of chemical adsorption, type A between −100° and 0°C and type B above 100°C.[16]

Nitrogen chemisorption characteristics can perhaps best be illustrated by the rate curves shown in Fig. 6.[7] The rate of adsorption is clearly temperature sensitive. Actually it appears to involve an energy of activation of about 16 kcal/mole of nitrogen. Later measurements by Scholten and Zwietering[18] showed that the energy of activation varied from a low value of about 10 kcal up to 30 kcal, depending on the fractional covering of the surface with adsorbed nitrogen. In connection with the rate curves shown in Fig. 6 it was pointed out that the rate of nitrogen adsorption at low coverages seemed to be equal to the rate of ammonia synthesis. In other words, the rate of synthesis appears from these rate curves to be controlled by the rate of chemisorption of nitrogen. The way in which nitrogen isotopes have entered into this argument in connection with Horiuti's "stoichiometric number" hypothesis is described below.

One rather surprising and relatively unknown observation on chemisorption was made in connection with our study of iron catalysts.[19] We wished to ascertain whether chemisorbed CO on iron catalysts at −78°C could be displaced by the adsorption of oxygen at the same temperature. To investigate this matter, a layer of 27 cc of radioactive (13,000 counts/min/per cc) CO was chemisorbed onto the surface of a singly promoted iron catalyst. Then, after a brief evacuation of the sample to remove gaseous CO, 272 cc of oxygen was added—the temperature still be-

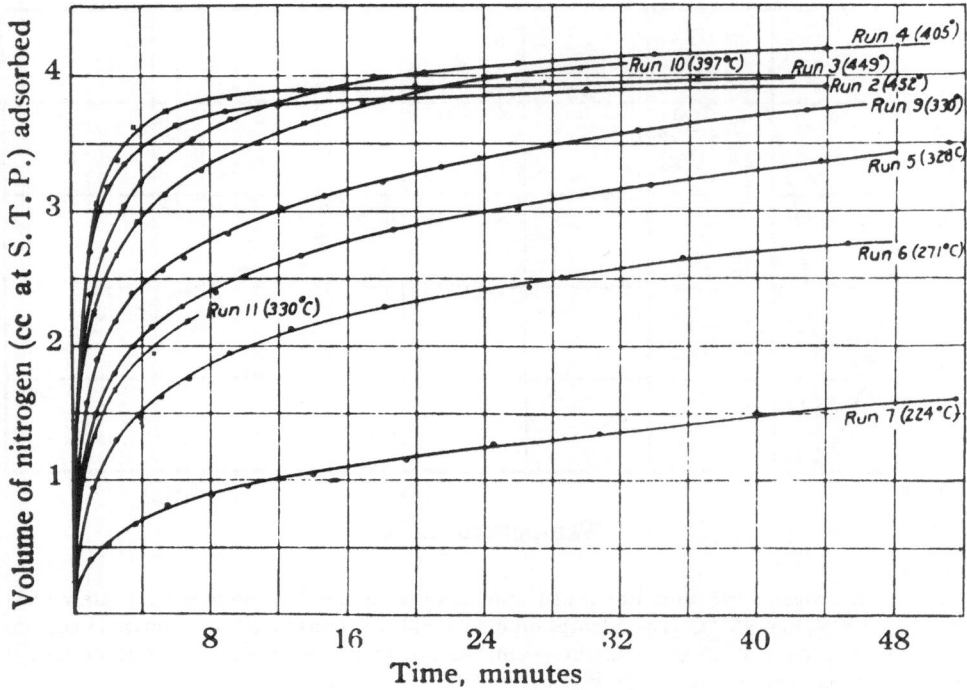

Fig. 6. Rates of nitrogen chemisorption at 1 atm pressure on a doubly promoted iron catalyst as a function of temperature; 16.46 g of catalyst 931 was used.[7]

ing −78°C. The oxygen was substantially all picked up in a very short time. The 0.12 cc of residual gas had an activity of only 1900 counts/min/cc. It can therefore be concluded that only 0.02 cc of the 27 cc of CO was removed by the added oxygen. Apparently the oxygen has penetrated the CO layer (probably through occasional holes left by the desorption of a few CO molecules during the initial evacuation) and spread out over the first ten layers of iron. It is very striking that the oxygen did not displace the adjacent CO molecules even though the heat of adsorption (or oxidation) of iron by oxygen is about 122 kcal/mole of oxygen and the heat of adsorption of CO is about 35 kcal/mole.

In summary, then, both hydrogen and nitrogen are extensively adsorbed on iron catalysts; the rate of the nitrogen adsorption appears to determine the rate of synthesis. Much additional work on the adsorption of nitrogen on iron catalysts has been reported in the literature. Some of it is reviewed in connection with the discussion below as to whether the adsorbed nitrogen is in the form of molecules or atoms on the various catalysts.

3 STATE OF THE NITROGEN ADSORBED ON IRON

The discussion of the chemisorption of nitrogen on iron synthetic ammonia catalysts leads naturally to the question of whether the nitrogen is adsorbed as atoms or molecules. Two approaches have been used in estimating the nature of chemisorbed nitrogen: (1) the study of the influence of adsorbed nitrogen on CO chemisorption and (2) the recent measurements using $^{30}N_2$ molecules as tracers.

3.1 The Influence of Chemisorbed
Nitrogen on Chemisorbed CO

If nitrogen is present on an iron catalyst as molecules, it would be expected to inhibit the adsorption of CO at −78 or −195°C for the simple reason that steric interference would make such inhibition inevitable. On the other hand, if nitrogen is present as atoms it might not cause inhibition because of its size or because the nitrogen conceivably could adsorb on the iron atom at the center of a body-centered unit cell and leave the surface free to chemisorb CO. Such measurements were made[20], and they are summarized in Table I. They show the surprising result that there is no inhibition by the nitrogen adsorbed at 450°C, on either doubly promoted or singly promoted catalysts. On pure iron catalysts, there is a suggestion that there might be a slight inhibition. The conclusion from these results is clearly that the nitrogen is adsorbed as atoms rather than as molecules at 450°C.

Table I. Effect of N_2 Chemisorption on CO Chemisorption[b]

Catalyst	N_2 Chemisorption[a], cc	CO Chemisorption, cc		Decrease in CO Chemisorption, cc
		True	On N_2 Chemisorption	
973	3.6	25.2	24.4	0.8
954	13.3	14.2	14.2	0.0
931	8.3	45.8	45.8	0.0

(a) N_2 chemisorption 450°C.
(b) CO chemisorption at −183°C.

One other observation is consistent with nitrogen being present as atoms. The crystal structure of Fe_4N shows nitrogen atoms at the center of unit cells containing iron atoms at the center of the faces and at the corners. The nitrogens are about 3.6 Å apart. If the pressure of nitrogen is boosted to about 4,300 atm at 450°C, Fe_4N will be formed by the reaction of nitrogen with iron.[21] It seems logical to infer that at pressures below the 4,300-atm dissociation pressure, the nitrogen atoms will still dissociate even though they are not striking the surface with sufficient frequency to form Fe_4N.

Later measurements by Takezawa and the author[22] have been extended down to lower temperatures. They confirm that, as judged by the inhibitive effect on CO chemisorption, the singly and doubly promoted iron catalysts showed that most of the chemisorbed nitrogen was present as atoms and not molecules at 450°C. However, at lower temperatures, inhibition definitely occurs; thus nitrogen adsorbed at 130°C on the promoted catalysts showed as much as 70 percent of the nitrogen could be present as molecules. Finally, at 300 to 350°C the measurements showed 10 to 15 percent of the nitrogen on the promoted catalysts could be present as molecules and as much as 50 percent on the pure iron catalysts.

3.2 Measurements Using $^{30}N_2$ as Tracer

Morikawa and Ozaki[23] conceived the idea of detecting the presence of nitrogen molecules in the adsorbed phase by comparing the rate of displacement of the adsorbed isotopic nitrogen with the rate of isotopic mixing according to the reaction

$$^{30}N_2 + {}^{28}N_2 = 2\ {}^{29}N_2 \tag{2}$$

By equilibrating a sample of pure iron catalyst with a nitrogen gas enriched with ^{15}N in the form of $^{30}N_2$ and $^{29}N_2$ at a given pressure, and then at time t_o circulating $^{28}N_2$ at the same pressure, he was able to compare the rate of displacement of the ^{15}N from the surface with the rate of exchange. The rate of decrease of $^{30}N_2$ and increase of $^{29}N_2$ resulting from combined displacement and exchange is shown in Fig. 7. These curves are interpreted as an indication of displacement as well as exchange. The change in the fraction of ^{15}N in the gas phase with time is shown in Fig. 8. Clearly about 80 percent of the change from f_o to f_∞ occurs in the first hour. From the observed data it can be calculated[24] that the rate of displacement V is about 6.5×10^{18} molecules/min in the early stages of the experiment and 1.1×10^{18} in the time after about 1 hour. The value of R, the exchange rate, calculates out to be 0.94×10^{18}. The authors conclude that over their pure iron catalyst at 380°C the initial displacement takes place through the presence of a different state of nitrogen (presumably molecular) than in the later stages where presumably adsorbed atoms are involved.

The authors rationalize their observations by postulating two steps in the formation of atomically adsorbed nitrogen as represented by the equation

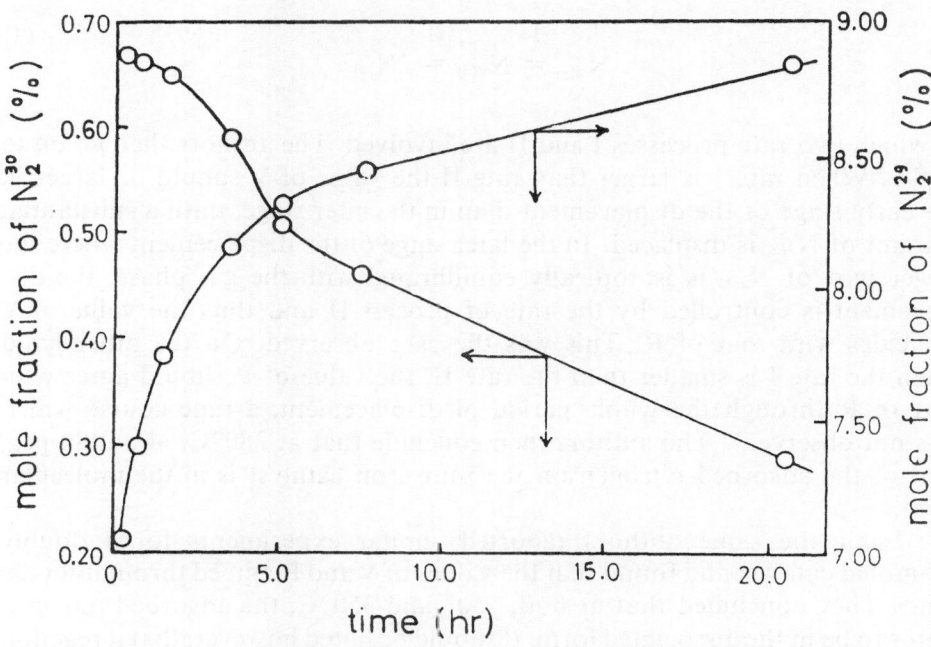

Fig. 7. Time course of mole fraction of $^{30}N_2$ and $^{29}N_2$ over a pure iron catalyst.[23]

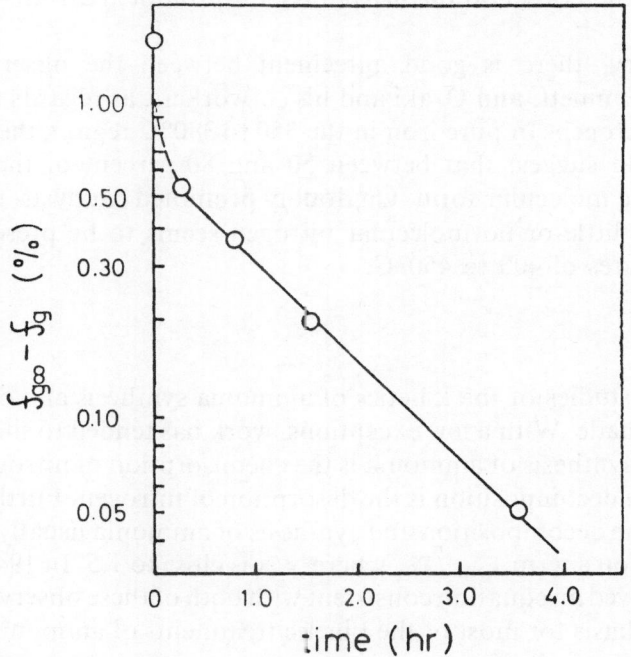

Fig. 8. Plot of log $(f_{g\infty} - f_g)$ as a function of time over unpromoted iron catalyst.[23]

$$\overset{\text{I}\qquad\quad\text{II}}{N_{2(g)} = N_{2(a)} = 2N_{(a)}} \tag{3}$$

in which two rate processes I and II are involved. The authors then go on to say[23]: "When rate I is larger than rate II the value of V should be larger in the early stage of the displacement than in the later stage, until a substantial amount of $N_{2(a)}$ is displaced. In the later stage of the displacement where the larger part of $N_{2(a)}$ is isotopically equilibrated with the gas phase, the displacement is controlled by the rate of process II and thus the value of V coincides with that of R. This was the case observed. On the other hand when the rate I is smaller than the rate II, the value of V should agree with that of R through the whole period of displacement, a time course which was not observed". The authors then conclude that at 380°C, about 80 percent of the adsorbed nitrogen on the pure iron catalyst is in the molecular form.

Later the same authors reported similar experiments for a doubly promoted catalyst and found that the values of V and R agreed throughout the run.[24] They concluded that at both 250° and 350°C, the adsorbed nitrogen seems to be in the dissociated form. It should be noted however, that if reaction II is faster than reaction I, one would expect V and R to agree throughout the run even if some molecules were present. Urabe, Aika, and Ozaki[25] point this out in regard to their observation on a potassium-promoted carbon-ruthenium catalyst.

In summary, there is good agreement between the observations of Takezawa and Emmett, and Ozaki and his co-workers, as regards the state of the adsorbed nitrogen. In pure iron in the 350 to 380°C region, the combined prediction would suggest that between 50 and 80 percent of the adsorbed nitrogen is in the molecular form. On doubly promoted catalysts, they would both agree that little or no molecular nitrogen seems to be present at synthesis temperatures of 400 to 450°C.

4 KINETICS

Numerous studies of the kinetics of ammonia synthesis and decomposition have been made. With a few exceptions, work has tended to show that the slow step in the synthesis of ammonia is the chemisorption of nitrogen and the slow step for the decomposition is the desorption of nitrogen. Furthermore, it turns out that the decomposition and synthesis of ammonia usually involve in the rate expression a term $P_{NH_3}^x / P_{H_2}^y$ where y/x is close to 1.5. In 1940, Temkin and Pyzhev derived an equation consistent with both of these observations.[26] It has formed the basis for most of the kinetic treatments of ammonia synthesis and decomposition in recent years.

Temkin assumed a heterogeneous surface and set up equations for the adsorption equilibrium of nitrogen on iron, for the rate of adsorption, and for the rate of desorption. Specifically his three pertinent equations are:

$$\theta = \frac{1}{f} \ln a_o P' \tag{4}$$

$$v = k_a P e^{-g\theta} \tag{5}$$

$$w = k_d e^{h\theta} \tag{6}$$

where θ is the fraction of the surface covered, P' is the equilibrium pressure or the "virtual" pressure of nitrogen, v is the rate of adsorption, w is the rate of desorption, and f, a_o, k_a, k_d, g, and h are constants. These equations are constructed to conform to the idea that the rates of adsorption and desorption of nitrogen depend exponentially on the fraction of the surface covered with nitrogen. At high coverage, adsorption is slow and desorption fast. Incidentally, it may be noted that measurements by Emmett and Brunauer[7] showed that up to 50-atm partial pressure, the adsorption of nitrogen increased as $(P_{N_2})^{1/6}$ regardless of whether the nitrogen was by itself or equilibrated with a 3:1 H_2:N_2 mixture.

In applying these equations, the authors assumed that the adsorption of nitrogen on the iron catalyst in the presence of an ammonia-hydrogen mixture is the same as it would be when at a nitrogen pressure equivalent to the existing partial pressure of ammonia and hydrogen in the gas mixture. Thus, since the equilibrium constant for ammonia synthesis is

$$K = \left(P_{NH_3}\right)^2 / \left(P_{H_2}\right)^3 \left(P_{N_2}\right) \tag{7}$$

the value of P' can be represented by $\left(P_{NH_3}\right)^2 / K P_{H_2}^3$, and the first of the Temkin equations becomes

$$\theta = (1/f) \ln \frac{a_o \left(P_{NH_3}\right)^2}{\left(P_{H_2}\right)^3 K} \tag{8}$$

As an illustration, the application of these equations to the decomposition of ammonia would take the form

$$w = k_d \exp\left[(h/f) \ln a_o \frac{\left(P_{NH_3}\right)^2}{\left(P_{H_2}\right)^3 K}\right] = k_d \left[\frac{\left(P_{NH_3}\right)^2 a_o}{\left(P_{H_2}\right)^3 K}\right]^{h/f} = k_2 \left[\frac{\left(P_{NH_2}\right)^2}{\left(P_{H_2}\right)^3}\right]^{\beta} \tag{9}$$

Love and Emmett[27] found experimentally that over a doubly promoted catalyst the rate of decomposition is proportional to $\left(P_{NH_3}\right)^{0.6} / \left(P_{H_2}\right)^{0.9}$. This would correspond to β having a value of 0.3.

Brunauer, Love, and Keenan[28] derived the general equations for the rate of adsorption and desorption of nitrogen on iron and for the variation of the amount adsorbed with pressure. They also showed that Eqs. (4), (5), and (6) of Temkin and Pyzhev are approximations holding true over intermediate partial pressures of nitrogen. For the first time they succeeded in calculating an adsorption isotherm from data for the rate of adsorption. The general rate equation they used is obviously of the same general form as would be expressed by a combination of Eqs. (5) and (6) of Temkin and Pyzhev. It is

$$dv/dt = k_a P_{N_2} V_m \exp\left[\frac{-Jv}{V_m RT}\right] - k_d V_m \exp\left[\frac{Bv}{V_m Rt}\right] \qquad (10)$$

Where V_m is the volume of adsorbate in a monolayer, and J and B are constants. Obviously v/V_m is the θ term of the Temkin-Pyzhev equations. At any rate, the authors evaluated the constants $k_a V_m$, $k_d V_m$, J/V_m and B/V_m from the rate curve and found them to be 0.02, 0.000957, 2100 calories, and 800 calories, respectively. If the above equation is used to evaluate equilibrium, one obtains the relation

$$\ln P = \ln\left[\frac{k_d}{k_a}\right] + \frac{B + J}{V_m RT} v \qquad (11)$$

By inserting the numerical values for the constants in this equation, they obtained values for the equilibrium adsorption of nitrogen as a function of pressure which were in good agreement with experimental values, as shown in Table II. The authors also showed that $B/(B+J)$ is equal to β of Eq. (9). Accordingly, using their rate equations for calculating β, the equation for ammonia decomposition rate would become

$$-d\left(P_{NH_3}\right)/dt = k \qquad \text{Thus, } P_{NH_3}^{0.55}/P_{H_2}^{0.85} \qquad (12)$$

which is very close to the observed data of Love and Emmett.[27] Beta turns out to be 0.276 for the doubly promoted catalyst used. Thus, from experimental measurements on the rate of adsorption of nitrogen on a doubly promoted iron catalyst, the authors were able to calculate the correct expression for the dependence of the rate of ammonia decomposition on the partial pressures of ammonia and hydrogen.

Nielsen, Kjaer, and Hansen[29] give an excellent account of their testing, experimentally, the agreement between the synthesis rate and a derived equation essentially the same as that of Temkin and Pyzhev. Nielsen and his co-workers combined four equations by Ozaki, Taylor, and Boudart[30] and added a term for the back reaction to give

Table II. Calculated and Observed Adsorption Isotherms[28] for Nitrogen on a Double Promoted Iron Synthetic Ammonia Catalyst

P, mm of Hg	V_{obs}, cc STP	V_{calc}, cc STP
25	2.83	2.88
53	3.22	3.22
150	3.69	3.70
397	4.14	4.15
768	4.55	4.45

$$\text{Rate} = \frac{k_2 \left(a_{N_2} K_a^2 - a_{NH_3}^2 / a_{H_2}^3\right)}{\left(1 + K_3 a_{NH_3} / a_{H_2}^w\right)^{2\alpha}} \tag{13}$$

where pressures have been replaced by activities and w is introduced to indicate whether the surface is covered with NH (w = 1) or with N(w = 1.5). In careful kinetic experiments at pressures up to 300 atm and over the temperature range 375 to 495° C, the authors established that w = 1.5 and alpha = 0.75. ($\beta = 0.25$ since $\alpha + \beta = 1$.) These data fit the Temkin-Pyzkev equation exactly because the above equation becomes identical to the Temkin-Pyzhev equation when, as in actual practice, the "1" in the denominator can be neglected.

As can be judged from the present discussion, the treatment of the kinetics of ammonia synthesis is difficult and full of many sources of error. A single illustration will perhaps suffice to show the errors that have been committed in the name of kinetics. Anderson and Toor[31] pointed out that for the usual differential reaction rate, a general equation can be written in the form

$$\text{Rate} = (1/273R(1+a)^n \, da/d(1/S_0)$$

The term "n" in the original Temkin-Pyzhev equation has a value of 1. Emmett and Kummer[32] corrected it to a value of 3. Anderson and Toor[31] made a further correction to give 2 which appears to be the correct value. Incidentally the analysis by Uchida and Kuraishi[33] earlier had contained the correct value of 2.

In summary, without delving further into details, one can say that the kinetics for ammonia synthesis and decomposition over a doubly promoted catalyst conform to a picture in which the surface is covered with adsorbed N atoms rather than NH groups and in which α, as defined in Eq. (13), has a value between 0.7 and 0.75. Further implications of the kinetic findings are discussed below in the section dealing with the reaction mechanism.

5 COMPOSITION OF THE CATALYST SURFACE

Two questions naturally arise as to the composition of the surface of an iron catalyst. The first has to do with whether or not the catalyst is partially converted to the nitride during synthesis. The second question concerns the surface concentrations of the promoters such as K_2O and Al_2O_3 that are commonly added to the iron and the function of these promoters in actual high-pressure synthesis.

For sometime it has been known that ammonia will interact with iron to form compounds that correspond approximately to Fe_4N, Fe_3N, and Fe_2N.[34] Early work with Brunauer and Hendricks[21] showed that at 444°C, 30 percent ammonia in hydrogen was necessary to form Fe_4N at 1 atm pressure. Calculations showed that this indicated a dissociation pressure of about 4300 atm for the Fe_4N at 444°C. It is obvious therefore that even Fe_4N will not be formed during synthesis at pressures and temperatures now used commercially. Incidentally it should be pointed out[34] that 70 and 100 percent NH_3, respectively, is required to form the higher nitrides Fe_3N and Fe_2N at this temperature.

It is well known that certain compounds such as Al_2O_3 or a combination of K_2O and Al_2O_3 act as promoters when added to iron catalysts. Usually only 1 or 2 percent of the promoter compounds is added. Two questions in regard to promoters have naturally arisen: (1) What fraction of the catalyst surface is covered by the promoters? (2) What is the mechanism by which they function in improving and maintaining the activity of an iron catalyst in actual commercial use?

One of the earliest clues as to the fraction of the surface of a reduced iron catalyst covered by promoter arose from observations as to the chemisorption of carbon monoxide and carbon dioxide. It was found that whereas carbon monoxide covered the entire surface of a pure unpromoted reduced iron catalyst at −183°C, it covered only about 40 percent of a catalyst containing 1.59 percent K_2O and 1.3 percent Al_2O_3. The results of such measurements are shown in Fig. 9. Fortunately carbon dioxide was found to adsorb on the promoter content and not on the iron, as shown in Fig. 10. Since the sum of the carbon monoxide chemisorption and the carbon dioxide chemisorption was equal approximately to the volume of physically adsorbed nitrogen in a monolayer, it was concluded that the promoter covered about 60 percent of the surface and free iron the remaining 40 percent. For catalysts containing only Al_2O_3 as promoter, there is no carbon dioxide chemisorption. The carbon monoxide chemisorption indicates that about 50 percent of the singly promoted (Al_2O_3 the only promoter) catalyst is usually covered with promoter.

Confirmation of the tendency of promoters to concentrate on the surface of iron catalysts has been obtained by the use of isotopic tracers. The first attempt at using tracers on this problem was made by Scholten.[36] He measured

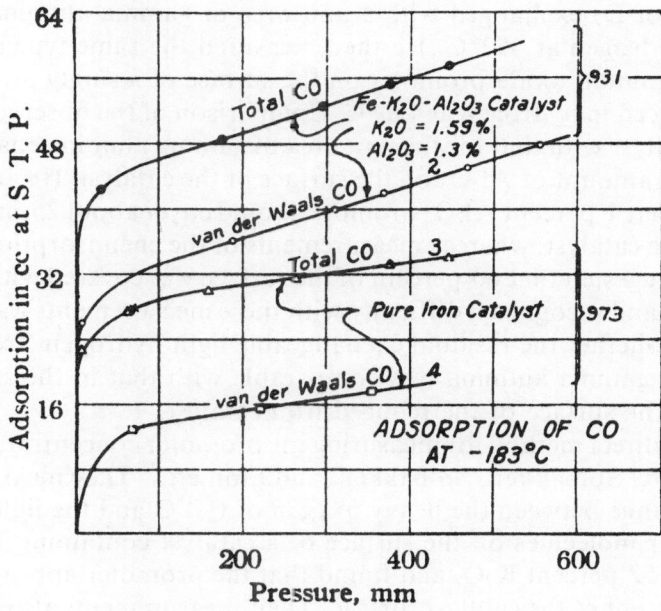

Fig. 9. Comparison of the total carbon monoxide adsorption for doubly promoted catalyst 931 and for pure iron catalyst 973.[35]

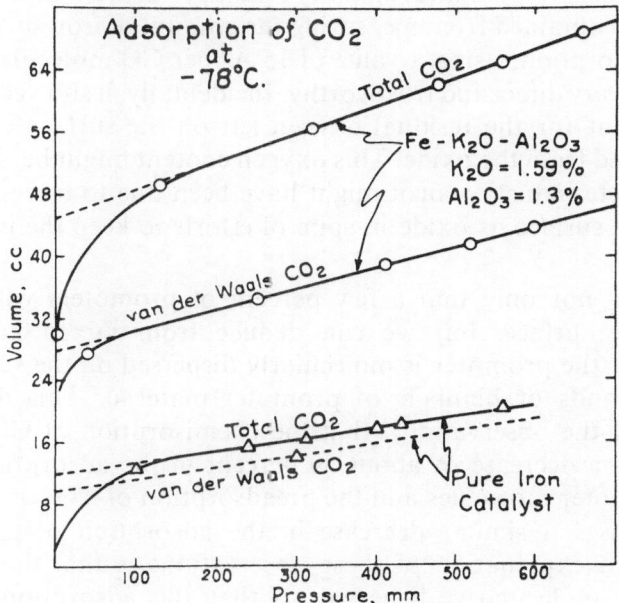

Fig. 10. Adsorption of carbon dioxide at −78°C on a doubly promoted and a pure iron catalyst. Emmett, P.H., *Advances in Colloid Science*, **I**, 1−36 (1942).

the amount of D_2 exchanged with a unit area of gamma alumina dried in a stream of hydrogen at 450°C. He then measured the same type of exchange with the aluminum oxide promoter on the surface of a singly promoted iron catalyst reduced in hydrogen at 450°C. Comparison of the observed exchange over the catalyst with that for the unit area of gamma alumina enabled him to calculate the amount of Al_2O_3 on the surface of the catalyst. He arrived at the conclusion that 1 percent Al_2O_3 promoter would cover about 25 percent of the surface of the catalyst, whereas measurements of the chemisorption of carbon monoxide suggested that 60 percent of the surface was covered with promoter. The obvious and recognized difficulty with these measurements was the uncertainty as to whether the residual exchangeable light hydrogen (from OH and H_2O) on the gamma alumina was comparable with that in the same area of alumina on the surface of the reduced iron catalyst.

A more direct method for measuring the promoter concentration was used successfully by Solbakken, Solbakken, and Emmett.[37] They measured the extent of exchange between the heavy oxygen of $H_2{}^{18}O$ and the light oxygen of the promoter molecules on the surface of a catalyst containing 1.06 percent Al_2O_3 and 0.52 percent K_2O, and found that the promoter appeared to cover about 60 percent of the catalyst surface. Their measurements also showed that the promoter was present substantially as a fraction of a monolayer and not in thick layers. The value of 60 percent was of course obtained only after assuming a cross section for the oxygen ions. A value of 10.7 Å^2 per oxygen ion was chosen on the basis of estimates made by Pauling. Furthermore, the 60 percent figure was also obtained from measuring the amount of iron surface as judged by CO chemisorption, using a value of 15 Å^2 per CO molecule. The method appears to be very direct and trustworthy. Incidentally, it showed a coverage of only 1.2 percent for the residual oxygen left on the surface of a pure iron catalyst reduced from the oxide. This oxygen content might be that remaining from incomplete reduction or it might have been due to traces of impurities present on the surface as oxide in spite of efforts to keep the iron as pure as possible.

We know not only that a few percent of promoters will cover more than half the surface, but we can deduce from mixed adsorption experiments that the promoter is molecularly dispersed on the surface and not present as islands or blankets of promoter material. This distribution is made clear by the observation[20] that the chemisorption of CO on the iron surface causes a decrease of about 50 percent in the adsorption of CO_2 on adjacent promoter molecules and the preadsorption of CO_2 on the promoter molecules causes a similar decrease in the adsorption of CO. The most straightforward explanation of these observations is that the molecules of promoter are molecularly dispersed so that the adsorption of CO_2 has enough overhang to partially block the adsorption of CO on the adjacent iron and, similarly, the chemisorbed CO molecules will, by their size, partially block the chemisorption of CO_2 on the adjacent promoter molecules.

In agreement with this picture of molecularly dispersed promoter molecules is the fact that atomically or ionically adsorbed species[20] on the iron (nitrogen or oxygen) have substantially no effect on the adsorption of CO_2. Furthermore, type A adsorption of hydrogen is slowed down by the preadsorption of CO_2 but eventually attains the same adsorption values as are obtained in the absence of carbon dioxide.

The specific function of the promoters is much more a matter of dispute than the question of the extent of coverage of the surface by the promoters and the state of subdivision of the promoters. It is generally agreed that the alumina (or similar oxides such as SiO_2 or ZrO_2) has as its chief function the prevention of crystal growth or sintering of the iron catalyst. This is shown most clearly by the fact that the surface area of a pure iron catalyst continues to decrease as it is heated to higher and higher temperature. Heating pure iron catalyst to 500°C will reduce the area to about 0.5 m^2/g; whereas singly promoted catalyst will retain about 12 m^2/g; a doubly promoted catalyst will retain an area in the 4 to 5-m^2/g range even after being heated to 575°C.[5] The crystal growth in pure iron as compared with promoted iron catalysts has been confirmed[38] by noting the relative changes in the X-ray patterns on heating samples of pure and promoted iron to 600°C for 4 hours. There may be other specific effects produced by adding Al_2O_3 as a promoter, but these are hard to establish because of the difficulty of making extensive measurements on pure iron catalysts for comparison with the singly promoted catalysts. The major uncertainty as to the function of the promoters in iron catalysts has to do with the function of and part played by the K_2O in a doubly promoted (Fe-Al_2O_3-K_2O) iron catalyst.

There is again good agreement[39] that the addition of the K_2O to the promoter produces a catalyst that is much more active for ammonia synthesis under commercial high-pressure operation than is a singly promoted catalyst containing only Al_2O_3. Incidentally, it is equally well established that K_2O by itself will not prevent sintering and results in a catalyst which, if anything, is less active than pure iron. Why then is a K_2O-Al_2O_3 promoter better than Al_2O_3 alone in promoting an iron synthetic ammonia catalyst?

One proposal is that singly promoted iron catalysts tend to be poisoned by ammonia in high-pressure synthesis and to be covered by inhibiting NH and NH_2 groups, whereas the doubly promoted catalyst containing a strong alkali drives off these basic groups and leaves an iron surface partially covered with adsorbed nitrogen atoms.[20] There are at least two well-substantiated bits of evidence supporting this hypothesis. In the first place, it has been shown that adsorbing nitrogen onto an iron catalyst at 450°C inhibits the chemisorption of hydrogen on the iron at 100°C if the catalyst is doubly promoted, but has no effect or even increases the adsorption of hydrogen if the catalyst has only Al_2O_3 as promoter.[20] This is equivalent to saying that the sites covered with chemisorbed nitrogen atoms can also hold these atoms (as NH?) or even two hydrogen atoms (as NH_2?) provided no alkali is present. With a doubly promoted catalyst, such groups are ejected; iron atoms covered by nitrogen atoms will not adsorb even one hydrogen atom. The experimental evidence is shown in Table III.[20] In the last column it may be noted that a value of 1.0 indicates a 1:1 inhibition of hydrogen adsorption by nitrogen chemisorption; a value of 0.0 would mean no inhibition of hydrogen by nitrogen and would be equivalent to NH formation; a value of −1 would mean two hydrogens adsorbed per nitrogen atom such as might suggest NH_2 formation. It is evident that these data are consistent with the formation of both NH and NH_2 on the singly promoted iron catalyst but not on the doubly promoted catalyst.

The second type of information that suggests NH and NH_2 as the species present on the singly promoted catalyst has to do with the behavior of such a catalyst with respect to the kinetics of ammonia decomposition on its surface. Whereas the doubly promoted catalyst yields kinetic data[27] that fit the Temkin-Pyzhev equation both for ammonia decomposition and high pressure synthesis[29], the singly promoted catalyst yields kinetic data for ammonia decomposition that do not obey the Temkin-Pyzhev equation. The temperature coefficient as seen in Fig. 11 almost vanishes in the temperature range 400 to 425°C and is abnormally small both at higher and lower temperatures. The rate of ammonia decomposition is almost independent of the partial pressures of ammonia and of hydrogen at about 390 and 450°C, whereas at 429°C it is proportional to $(P_{H_2})^{1.2}/(P_{NH_3})^{0.84}$—the inverse of the kinetic expression for the reaction on the doubly promoted catalyst. A

Table III. Effect of Chemisorbed Nitrogen on Type B Hydrogen Adsorption[6]

| Catalyst | N_2 Chemisorption, cc | H₂ Adsorption at 100°C, cc | | Decrease in Vol H_2 Adsorption per CC Chemisorbed Nitrogen |
		Clean Surface	Top or Chemisorbed Nitrogen	
973 (pure Fe, 0.15% Al₂O₃ as impurity)	2.5	3.0	1.3	0.68 (3 min)
		3.5	2.5	0.40 (1 hr)
954 (singly promoted, 10.2% Al₂O₃)	14.7	25.8	26.5	−0.05 (1 hr)
			44.8	−0.52 (21.5 hr)
931 (doubly promoted, 0.59% K₂O, 1.3% Al₂O₃)	8.0	12.7	4.1	1.07 (3 min)
		14.7	7.9	0.85 (1 hr)

Note: The nitrogen was chemisorbed at 400°C. and the hydrogen at 100°C. at a pressure of 1 atm.

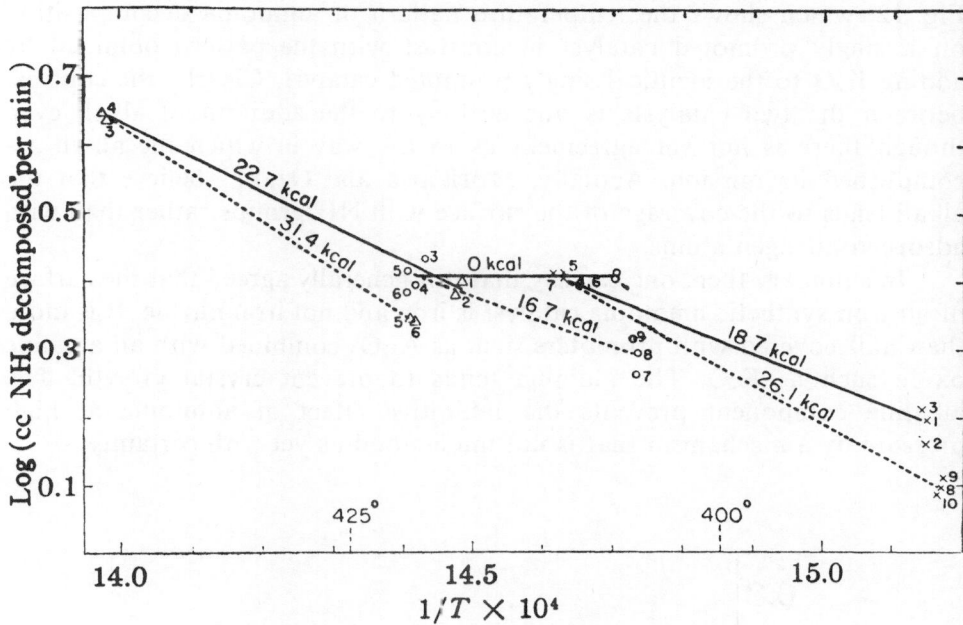

Fig. 11. Temperature dependence of rate of ammonia decomposition on a 1.55-cc sample of singly promoted iron catalyst 954 (10.2% Al$_2$O$_3$); hydrogen flow 400 cc/min; ammonia flow 100 cc/min. Numbers indicate the chronological order of data for a given temperature-dependence run; dotted lines are for data taken from higher to lower temperatures.[27]

quantitative equation for the ammonia decomposition over a singly promoted catalyst has not yet been worked out. However, it has been suggested[2] that perhaps in the low-temperature range the decomposition mechanism may involve the interaction on the surface of two adsorbed NH$_2$ or NH groups to form N$_2$. Perhaps the region of zero energy of activation is one in which the NH and NH$_2$ groups are gradually giving way to adsorbed nitrogen atoms. At any rate it is evident that an equation (Temkin-Pyzhev) designed and developed on the hypothesis that the surface is covered with nitrogen atoms applies to ammonia decomposition over the temperature range 400 to 450°C on doubly promoted catalysts but not on singly promoted catalysts.

Frankenburg[3] explained the action of added K$_2$O in the promoter by assuming that NH and NH$_2$ groups and NH$_3$ tend to be adsorbed at the iron-alumina interface because of acidic groups present in the alumina. The alkali eliminates these sites and thus prevents the accumulation of NH, NH$_2$, and NH$_3$ on the catalyst surface.

An interesting way of contrasting the properties of Al$_2$O$_3$ as a single promoter with those of K$_2$O-Al$_2$O$_3$ as a double promoter is illustrated in

Fig. 12, which shows the temperature pattern of ammonia decomposition on a singly promoted catalyst in contrast with the pattern obtained by adding K₂O to the identical singly promoted catalyst. Clearly, the contrast between the two catalysts is due entirely to the addition of alkali even though there is not yet agreement as to the way in which the alkali accomplished its mission. Actually, Morikawa and Ozaki[24] believe that the alkali leads to the coverage of the surface with NH groups rather than with adsorbed nitrogen atoms.

In summary then, one can say that it is generally agreed that the surface of an iron synthetic ammonia catalyst is iron and not iron nitride. It is more than half covered with promoters such as Al_2O_3 combined with an alkaline oxide such as K_2O. The alumina tends to prevent crystal growth. The alkaline component prevents the inhibitive effect of ammonia at high pressure by a mechanism that is not understood as yet with certainty.

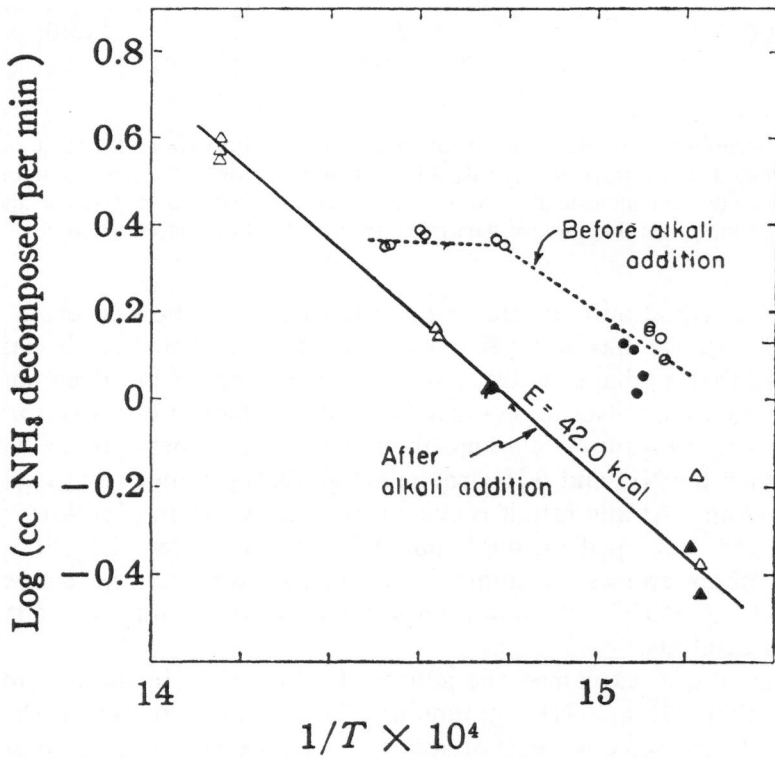

Fig. 12. Effect of the addition of K_2O to a reduced sample (1.5 cc) of singly promoted catalyst 954 upon the temperature dependence of the rate of ammonia decomposition over the catalyst; hydrogen flow, 150 cc/min; ammonia flow, 500 cc/min; no nitrogen diluent.[27]

6 MECHANISM OF AMMONIA SYNTHESIS OVER IRON CATALYSTS

It has already been stated that the slow step in ammonia synthesis involves the dissociative adsorption of nitrogen on the iron surface. This has been seriously questioned only once in the last 50 years. The challenge was from Horiuti[40] and his co-workers[41,42] and was based on his theory of stoichiometric numbers[40]. For ammonia synthesis it can be illustrated by the following series of equations:

$$N_2 + 3H_2 = 2NH_3 \tag{14}$$

Reaction	Stoichiometric Number	
$N_2 + 2Fe = 2Fe\text{-}N_{ads}$	1	(15)
$H_2 + 2Fe = 2Fe\text{-}H_{ads}$	3	(16)
$Fe\text{-}N_{ads} + Fe\text{-}H_{ads} = Fe\text{-}NH_{ads}$	2	(17)
$NH_{ads} + H_{ads} = NH_{2ads}$	2	(18)
$NH_{2ads} + H_{ads} = NH_{3ads}$	2	(19)
$NH_{3ads} = NH_{3gas}$	2	(20)

The stoichiometric number is defined as the number of times one of the steps has to occur to produce the two molecules of ammonia as product. Horiuti and his co-workers showed that by adding heavy nitrogen as $^{30}N_2$ or $^{15}NH_3$ to his sample of ammonia and comparing the rate of isotopic exchange and the rate of cracking he could evaluate the stoichiometric number. His measurements showed a stoichiometric number of 2. This was evidence against the chemisorption of nitrogen being the slow step and suggested that perhaps one of the reduction steps was rate controlling. It so happened that simultaneously in Holland, independent measurements of the stoichiometric number, being made by Bokhoven, Gorgels, and Mars[43,44], yielded a value of 1 for the stoichiometric number, thus reaffirming the long held belief that the chemisorption of nitrogen was the rate-determining step. Recently Tanaka[45] has published results agreeing with those obtained by Bokhoven et al.[44] in showing that the stoichiometric number is 1 and not 2. No explanation of the cause of the discrepancy has been given, though one striking difference in the technique used by Horiuti and that used by Tanaka and by Bokhoven et al. has been pointed out.[2] Horiuti measured the rate of synthesis and the rate of isotopic exchange in two separate experiments. The others carried out both measurements simultaneously. With separate experiments there is always a chance that the poison content could be sufficiently different in the isotopic measurement from that in the rate measurement to account for a factor of two. At any rate, two groups of workers have now found a stoichiometric number of 1, so that the chemisorption of nitrogen as a slow step still remains a respectable conclusion.

Even though the stoichiometric number is 1, it does not follow that the active species on the surface consists of nitrogen atoms. Temkin and his coworkers have recently suggested that the synthesis might take place through the successive formation of adsorbed N_2, N_2H_2, N_2H_x, and finally ammonia.[46] This mechanism would still lead to a stoichiometric number of 1 if the nitrogen chemisorption is the slow step. However, it would lead to a stoichiometric number of 1 even if the reduction steps to form adsorbed N_2H_2 or N_2H_x were the slow ones.

It might be imagined that infrared absorption measurements could differentiate between ammonia synthesis possibly taking place through the steps represented by the formation of N, NH, and NH_2 and that possibly taking place through formation of adsorbed N_2, N_2H_2, and, in general, N_2H_x. Unfortunately, one group of workers[47] has found infrared evidence for NH and NH_2 formation, whereas another group has located evidence for N_2H_x[48] formation. The infrared data are accordingly inconclusive up to the present time but should eventually find the conditions critical to the formation of NH and NH_2 intermediates as well as the formation of N_2H_x intermediates.

The two remaining questions relative to reaction mechanism are as to whether there is a preferred action of certain surface planes in the iron catalysts and whether under synthesis conditions the surface is covered mostly with adsorbed nitrogen atoms or adsorbed NH groups.

Brill, Richter, and Ruch[49] have made some spectacular observations in regard to the planes involved in ammonia synthesis and decomposition. In a field emission experiment using an iron tip they not only ascertained that nitrogen was preferentially adsorbed on the (111) face as compared with the (110) and (100) faces, but also showed that exposure of that tip to nitrogen caused a rapid transformation of (100) and (110) planes to (111) faces. They also report that quantum calculations[50] suggest that the molecular orbital arrangements in nitrogen molecules are especially compatible with the iron orbitals on the (111) faces. Furthermore, they quote observations which indicate that ammonia synthesis takes place on the (111) face but not on (100) or (110) faces and that according to LEED work, the (111) but not the (100) or (110) planes chemisorb nitrogen. Accordingly, proof is very strong that regardless of the detailed steps of synthesis the reaction occurs predominantly on the (111) faces of the iron catalysts.

As regards the question concerning whether the surface of the iron catalyst is covered with adsorbed nitrogen atoms or adsorbed NH groups, the evidence seems to favor the presence, at synthesis temperature, of adsorbed nitrogen on the doubly promoted catalysts and possibly NH or NH_2 groups on the catalysts that are singly promoted with Al_2O_3. The evidence in favor of the steady-state coverage of the surface with nitrogen groups is summarized above. Two more items should be mentioned.

The kinetics of ammonia decomposition and synthesis show a dependence on the partial pressure of ammonia and hydrogen of the form $(P_{NH_3})^x/(P_{H_2})^y$, where x/y equals 1/1.5 for ammonia decomposition both in the work of Takezawa and Toyoshima[51] and Love and Emmett[27], and in the synthesis data of Nielsen, Kjaer, and Hansen[29]. These results are consistent with the postulate of Temkin and Pyzhev[26] to the effect that the surface is covered with chemisorbed nitrogen atoms.

Furthermore, the α value for the Temkin equation [Eq. (13)] turns out to be 0.67 to 0.75, which is in good agreement with the beta values (alpha plus beta must equal unity, according to the Temkin concept) of 0.22 and 0.3 obtained by Takezawa and Toyoshima[51] and by Emmett and Love[27], respectively, for temperatures of 400 to 450°C.

The one bit of kinetic evidence that has been interpreted as favoring the existence of a layer of NH rather than a layer of nitrogen atoms at steady state is the series of experiments by Ozaki, Boudart, and Taylor.[30] They interpret their results to indicate an x/y ratio of 1/1, and this would be consistent with an NH surface coverage and not a coverage by nitrogen atoms. Logan and Philp[52] have reanalyzed the results of Ozaki, Boudart, and Taylor and claim that they would be better fit by an x/y ratio of 1/1.5 corresponding to a coverage with nitrogen atoms than by the ratio of 1:1. Aika and Ozaki[53] disputed this interpretation. In any event, the experiments of Ozaki, Boudart, and Taylor were carried out at temperatures of 300°C and lower and hence cannot be construed as necessarily representing conditions at synthesis temperatures of about 450°C. Also, it should perhaps be pointed out that the existence of an NH layer on the doubly promoted catalyst used by Ozaki, Boudart, and Taylor was originally attributed by them to the fact that the catalyst was not thoroughly reduced, whereas, more recently, they have been inclined to cite it as evidence of the influence of potassium oxide on the surface coverage.

In summary then, the experimental work of the past 50 years leads to the conclusion that the rate-determining step in ammonia synthesis over double iron catalysts is the chemisorption of nitrogen. The reaction appears to take place primarily on the (111) faces of the catalyst. The question as to whether the nitrogen species involved on the surface is molecular or atomic is still not conclusively resolved, though in the writer's opinion, the direct participation of nitrogen in an atomic form seems more likely than that in a molecular form under the conditions of high-pressure synthesis at about 450°C. On singly promoted Fe-Al$_2$O$_3$ catalysts, the Temkin-Pyzhev concepts and equations do not apply. The kinetics and mechanism are complicated over such catalysts and may well involve a coverage of the surface with NH and NH$_2$ groups. The Al$_2$O$_3$ promoter definitely functions in part by preventing crystal growth in the iron catalysts and hence preserving a high iron surface area at operating temperatures. The function of the K$_2$O promoter is not a matter of general agreement among catalytic workers, though the suggestion seems reasonable that the addition of K$_2$O to an Al$_2$O$_3$ promoter effectively increases the high-pressure synthesis activity by keeping the iron surface free of inhibiting NH, NH$_2$, or NH$_3$ groups.

It should be realized that the discussion presented here is subject to space limitations. For a more detailed and comprehensive survey of all aspects of the ammonia synthesis process, reference should be made to the book by Anders Nielsen[2] and to the book *Catalysis*, Vol. III, the Chapters 6 and 7, by Frankenburg[3] and Bokhaven, van Heerden, Westrik, and Zwietering[4], respectively, on ammonia synthesis mechanism and concepts. The kinetics of nitrogen adsorption and desorption and of the physical characteristics of iron catalysts are treated in great detail in the thesis by Scholten.[36]

7 NEW CATALYSTS FOR AMMONIA SYNTHESIS

Until a few years ago, we had settled rather smugly into the conviction that iron was the preferred metal from which to fabricate ammonia catalysts. Perhaps we overlooked some of the early observations of Haber to the effect that osmium was very active as an ammonia catalyst. Indeed even the use of potassium atoms seems to have been recognized in the early work. For example Mittasch[1] states: "Alkali metals can act favorably on molybdenum but oxides of the alkali metals are harmful". At any rate, it came as very much of a surprise to most of us when Ozaki, Aika, and Hori[54] announced a few years ago that they had prepared a potassium-charcoal-ruthenium catalyst that was severalfold as active as a doubly promoted iron catalyst for the synthesis of ammonia at 1 atm pressure.

The initial observations of Ozaki and co-workers can be illustrated by Fig. 13. All of the activities shown are for K-C-metal catalysts, except the ruthenium on celite and the standard Fe-Al$_2$O$_3$-K$_2$O catalysts. The activities of these two catalysts are shown without the benefit of added potassium atoms. Clearly K-C-Ru catalyst is several times as active as the iron synthetic ammonia catalyst. Furthermore it should be pointed out that K-C has no activity and Ru-C has a negligible activity.

The metals shown in Fig. 13 were, as pointed out by Ozaki et al., all recognized in the early German work as active elements. He extended the list to include rhodium, iridium, and platinum as K-C-metal catalysts. The composite results of all these tests are shown[55] in Fig. 14, in which the abscissa is the value of the ΔH of formation of the highest metal oxide per metal atom.

Nitrogen isotope experiments[56] seem to show the nitrogen chemisorbed on the K-C-Ru catalyst is all exchangeable in a few hours with $^{30}N_2$. Apparently the nitrogen is present as atoms rather than as molecules. The mechanism believed by Ozaki to represent the action of the potassium atoms is the donation by them of electrons, which in turn increase the electron concentration of the ruthenium surface. It is postulated that the nitrogen may be adsorbed on the ruthenium as N^- particles and that the formation of such ions becomes easier the higher the electron concentration of the ruthenium. In line with this hypothesis, cesium with its lower ionization potential than potassium is even more effective[57] than the latter as a promoter for Ru-C catalysts.

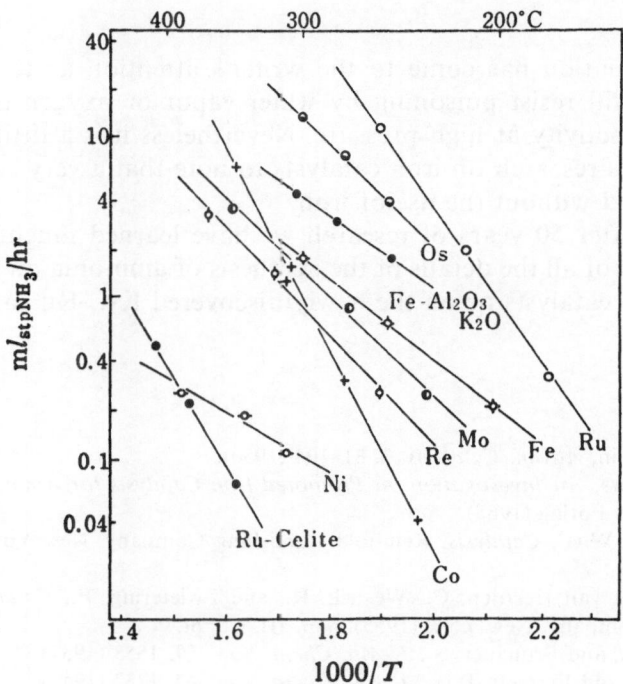

Fig. 13. Arrhenius plots of synthesis rates at a flow rate of 4.5 l/hr at 600 mm Hg pressure over catalysts of 2.5 g each except Fe-Al$_2$O$_3$-K$_2$O (0.77 g) and Ru-celite (1.5 g); osmium catalyst was prepared from OsO$_4$.[54]

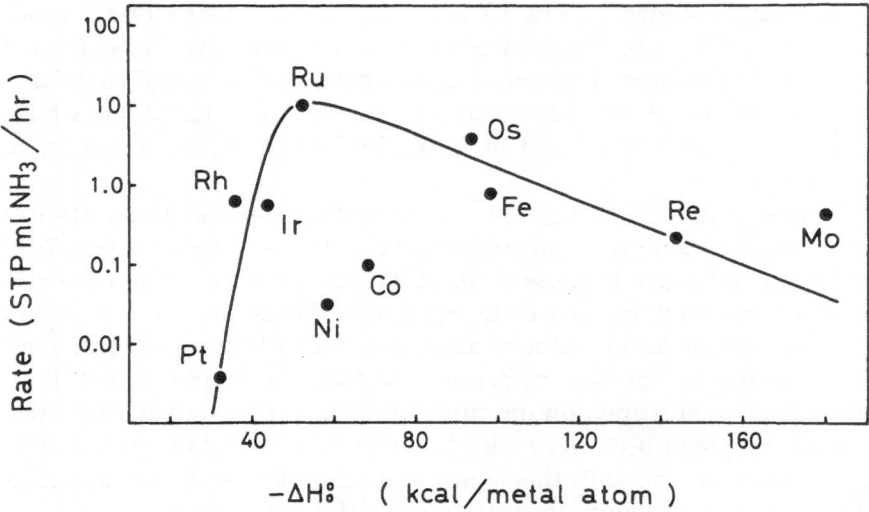

Fig. 14. Rate of ammonia synthesis over 2.5 g of 5 wt % transition metal-AC-K at 250°C, 600 Torr pressure, and 4.5 l/hr flow rate as a function of the heat of formation of the highest metal oxide per gram atom of metal, -ΔH°.[55]

No information has come to the writer's attention as to whether the new catalyst will resist poisoning by water vapor or oxygen or whether it will retain its activity at high pressure. Nevertheless it is a fitting climax to our many years research on iron catalysts to note that a very active catalyst can be prepared without the use of iron.

Clearly, after 50 years of research we have learned much, but we still cannot be sure of all the details of the synthesis of ammonia on the standard promoted iron catalysts or on the newly discovered K-C-Ru catalysts.

REFERENCES

1. Mittasch, Alwin, *Advan. Catalysis*, **2**, 81–104 (1950).
2. Nielsen, Anders, *An Investigation on Promoted Iron Catalysts for Ammonia Synthesis*, Jul. Gjellerups Forlag (1968).
3. Frankenburg, W.G., *Catalysis*, Reinhold Publishing Company, New York (1955), Vol. III, Chapt. 6.
4. Bokhoven, C., van Heerden, C., Westrik, R., and Zwietering, P., *Catalysis*, Reinhold Publishing Company, New York (1955), Vol. III, Chapt. 7.
5. Emmett, P.H., and Brunauer, S., *J. Am. Chem. Soc.*, **59**, 1553 (1937).
6. Brunauer, S., and Emmett, P.H., *J. Am. Chem. Soc.*, **62**, 1732 (1940).
7. Emmett, P.H., and Brunauer, S., *J. Am. Chem. Soc.*, **56**, 35 (1934).
8. Brunauer, S., and Emmett, P.H., unpublished results.
9. Brunauer, S., Emmett, P.H., and Teller, E., *J. Am. Chem. Soc.*, **60**, 309 (1938).
10. Barrett, E.D., Joyner, L.G., and Halenda, P.C., *J. Am. Chem. Soc.*, **73**, 373 (1951).

11. Roberts, B.F., *J. Colloid Interface Sci.*, **23**, 266 (1967).
12. Brunauer, S., Mikhail, R.S., and Bodor, E.E., *J. Colloid Interface Sci.*, **25**, 358 (1967).
13. Cranston, R. W., and Inkley, F.A., *Advan. Catalysis*, **9**, 143 (1957).
14. Joyner, L.G., Barrett, E.P., and Skold, *J. Am. Chem. Soc.*, **73**, 3155 (1951).
15. Nielsen, Anders, *An Investigation on Promoted Iron Catalysts for Ammonia Synthesis*, Jul. Gjellerups Forlag (1968), p. 206.
16. Emmett, P.H., and Harkness, R.W., *J. Am. Chem. Soc.*, **57**, 1631 (1935).
17. Kummer, J.T., and Emmett, P.H., *J. Phys. Chem.*, **56**, 258 (1952).
18. Scholten, J.J.F., and Zwietering, P., *Trans. Faraday Soc.*, **53**, 1363-70 (1957).
19. Kummer, J.T., and Emmett, P.H., *J. Am. Chem. Soc.*, **73**, 2886 (1951).
20. Brunauer, S., and Emmett, P.H., *J. Am. Chem. Soc.*, **62**, 1732 (1940).
21. Emmett, P.H., Hendricks, S.B., and Brunauer, S., *J. Am. Chem. Soc.*, **52**, 1456 (1930).
22. Takezawa, N., and Emmett, P.H., *J. Catalysis*, **11**, 131 (1968).
23. Morikawa, Y., and Ozaki, A., *J. Catalysis*, **12**, 145 (1968).
24. Morikawa, Y., and Ozaki, A., *J. Catalysis*, **23**, 97 (1971).
25. Urabe, K., Aika, K., and Ozaki, A., *J. Catalysis*, **32**, 108 (1974).
26. Temkin, M.I., and Pyzhev, V., *Acta Physiochim., U.R.S.S.*, **12**, 327 (1940).
27. Love, K.S., and Emmett, P.H., *J. Am. Chem. Soc.*, **63**, 3297 (1941).
28. Brunauer, S., Love, K.S., and Keenar, R.J., *J. Am. Chem. Soc.*, **64**, 751 (1942).
29. Nielsen, A., Kjaer, J., and Hansen, B., *J. Catalysis*, **5**, 68 (1964).
30. Ozaki, A., Taylor, H.S., and Boudart, M., *Proc. Roy. Soc. (London)*, **A258**, 47 (1960).
31. Anderson, R.B., and Toor, H.L., *J. Phys. Chem.*, **63**, 1982 (1959).
32. Emmett, P.H., and Kummer, J. T., *Ind. Eng. Chem.*, **35**, 677 (1943).
33. Uchida, H., and Kuraishi, M., *Bull. Chem. Soc. Japan*, **28**, 106 (1955).
34. Brunauer, S., Jefferson, M.E., Emmett, P.H., and Hendricks, S.B., *J. Am. Chem. Soc.*, **53**, 1778 (1931).
35. Emmett, P.H., and Brunauer, S., *J. Am. Chem. Soc.*, **59**, 310 (1937).
36. Scholten, J.J.F., Thesis, Delft, Holland (1959), p. 65.
37. Solbakken, V., Solbakken, A., and Emmett, P.H., *J. Catalysis*, **15**, 90 (1969).
38. Wyckoff, R.W.G., and Crittenden, E.D., *J. Am. Chem. Soc.*, **47**, 2866 (1925).
39. Krabetz, R., and Peters, Cl., *Angew. Chem.*, **77**, 333 (1965).
40. Horiuti, J., *J. Res. Inst. Catalysis, Hokkaido Univ.*, **1**, 8 (1948).
41. Enomoto, S., and Horiuti, J., *J. Res. Inst. Catalysis, Hokkaido Univ.*, **2**, 87 (1953).
42. Enomoto, S., Horiuti, J., and Kobayashi, H., *J. Res. Inst. Catalysis, Hokkaido Univ.*, **3**, 185 (1955).
43. Bokhoven, C., Gorgels, M.J., and Mars, P., *Trans. Faraday Soc.*, **55**, 315 (1959).
44. Bokhoven, C., Gorgels, M.J., and Mars, P., *J. Res. Inst. Catalysis, Hokkaido Univ.*, **9**, 287 (1961).
45. Tanaka, K., *J. Res. Inst. Catalysis, Hokkaido Univ.*, **13**, 119 (1965).
46. Temkin, M.I., Morozev, N.M., and Snapatina, E.N., *Kinetika i Kataliz*, **4**, 565 (1963).
47. Nakata, T., and Matsushita, S., *J. Phys. Chem.*, **72**, 458 (1968).
48. Brill, R., Jiru, P., and Schulz, G., *Z. Physik. Chem. Neue Folge*, **64**, 215 (1969).
49. Brill, R., Richter, E.L., and Ruch, E., *Angew. Chem.* (International Edition in English), **6**, 882 (1967).
50. Ruch, E., *Zehn Jahre Fonds der Chem. Ind.*, Verband der Chem. Ind. e.V. Fonds der Chem. Ind., Dusseldorf (1960), p. 163.
51. Takezawa, N., and Toyoshima, I., *J. Phys. Chem.*, **70**, 594 (1966).
52. Logan, S.R., and Philp, J., *J. Catalysis*, **11**, 1 (1968).
53. Aika, K., and Ozaki, A., *J. Catalysis*, **16**, 97 (1970).
54. Ozaki, A., Aika, K., and Hori, H., *Bull. Chem. Soc., Japan*, **44**, 3216 (1971).

55. Aika, K.M., Yamaguchi, J., and Ozaki, A., *Chem. Letters, Chem. Soc., Japan*, pp. 161–64 (1973).
56. Urabe, K., Aika, K., and Ozaki, A., *J. Catalysis*, **32**, 108 (1974).
57. Aika, K., Hori, H., and Ozaki, A., *J. Catalysis*, **27**, 424 (1972).

DISCUSSION on Paper by P. H. Emmett

BOUDART: In the case of ammonia synthesis on iron, if the mechanism discussed by Prof. Emmett is correct, it follows that it is possible to predict what the rate of the reaction between N_2 and D_2 will be, providing the rate of the reaction between N_2 and H_2 is determined under identical conditions. This prediction, which we made a few years ago, has now been verified by Temkin and co-workers quite recently. There are very few such satisfying examples of quantitative prediction in heterogeneous catalysis, and thus it is the more worthwhile to mention this one.

EMMETT: Thanks for reminding us of the deuterium work on ammonia synthesis. It is indeed true that your prediction has been beautifully verified and fits in well with the widely accepted mechanism proposed.

EHRLICH: Could you elaborate upon the field emission measurements of nitrogen adsorption on iron, which were interpreted by Brill et al. as indicating that interaction with nitrogen caused an enlargement of the (111) planes? It appears to me that it is difficult to arrive at unequivocal conclusions about structural changes of a field emitter just from changes in the appearance of the pattern, and without carrying out ancillary measurements.

EMMETT: I am not sufficiently well acquainted with the details of the work reported by Brill and his associates to give an opinion as to the reproducibility of the results he reported in regard to the influence of nitrogen on the distribution of surface planes on the field emission tip. I am inclined to rely on the experience of the people making this field emission experiment that some change in the developed faces from (100) and (110) to (111) planes was observed. Certainly additional work on this point is very much to be desired. It is particularly important because there are other indications that the (111) plane is the important one in ammonia synthesis.

BLOCK: The field emission work on iron was performed in our laboratory in Berlin. The FEM images shown by Prof. Emmett are the clean field-evaporated iron emitter (110) oriented, imaged at 80°K (upper left) and the same emitter after nitrogen adsorption at 80°K (upper right). After heating this tip to 400°C for a short time without applied field, the nitrogen adsorption (FEM at 80°K at the bottom of the slide) displays a completely different image. The most striking observation is the enlargement of the (111) faces with high work function. From these experiments it was concluded that nitrogen preferentially chemisorbs at the (111) surfaces and that these surfaces enlarge geometrically at 400°C owing to a decreased surface tension after nitrogen chemisorption.

The interpretation of FEM images of course involves difficulties. In the present case a nonequilibrium surface of the bcc lattice of iron becomes the most pronounced dark plane in field emission. This is particularly astonishing since on a field-evaporated clean iron tip, this surface has only a minor size prior to adsorption. FEM images are in agreement, but this is not, however, an unequivocal proof for the enlarged (111) faces for which other arguments have been found in Brill's group. This, of course, also applies for the molecular or atomic structure of nitrogen at the iron surface.

IMELIK: It appears from your work that the surface of the catalyst is covered to 50 to 70 percent by promoters. Is it then possible that a dilution effect occurs, similar to what is observed in some alloys? This would mean that clusters may be formed, or, at least, that the particle size of iron in promoted catalysts could be significantly smaller than in the unpromoted catalyst. On the other hand, if the particles or clusters become too small, the specific catalytic activity should decrease since the probability of finding low-index crystallographic planes, which seems to be active in this reaction, decreases. Could you comment on this point?

EMMETT: I think you are quite right in being concerned about the possibility of the promoters covering such a large fraction of the surface that no room is left for iron. All I can say is that the catalysts so far produced still have a high percentage of iron on the surface, even when as much as 10 percent aluminum oxide is present as promoter. Accordingly, the condition that you predict has apparently not been obtained in practical catalyst operations. Incidentally, the crystal size of the promoted catalysts is about one-quarter as great as that for the unpromoted pure iron catalyst, but it is still about 500 Å and hence is not small enough to alter seriously the surface plane distribution.

You question as to whether promoter content might alter the nature of the crystallographic planes' is one that we cannot as yet answer. Presumably, the surface planes are the active (111) planes, although adequate proof of this has not yet been obtained.

YATES: Can you briefly describe the early history of the *discovery* of the Al_2O_3 and K_2O activity for promotion of NH_3 synthesis on iron surfaces?

EMMETT: I would not like to state definitely the exact priority in the various patents involved in specifying promoters for iron catalysts. However, it is my opinion that the early patents by Mittasch cover the use of promoters, such as aluminum oxide. Later patents were issued to Dr. Larson on the use of potassium oxide as well as aluminum oxide to form the so-called doubly promoted catalyst. Perhaps Dr. Schwab or some of the other German scientists here could give you a more accurate statement on the details of the German patents.

SCHWAB: Mittasch in his book on the history of ammonia synthesis mentions that the first German patent concerned only the single promoted catalyst containing Al_2O_3 and that the double promoted one with K_2O was submitted some years later.

THERMODYNAMICS OF ADSORPTION

Carl Wagner

*Max-Planck-Institut für biophysikalische Chemie
(Karl-Friedrich-Bonhoeffer-Institut)
Göttingen, West Germany*

ABSTRACT

Introducing appropriate operational definitions of adsorption on uniform surfaces and on solids involving nonuniform surfaces, one obtains thermodynamic relations between measurable quantities without conceptual difficulties by use of a modified Helmholtz energy defined as $F - n_2\mu_2 - n_3\mu_3 - \ldots$. In particular, one obtains

1. Relations between change of surface free energy with composition of a liquid or solid phase and excess surface concentrations of the adsorbates
2. Relations for mutual displacement of the adsorbates
3. Relations between temperature dependence of adsorption and heats of adsorption.

1 INTRODUCTION

Adsorption of gases on well-defined faces of single crystals is investigated in order to obtain an understanding of interaction between adsorbate and adsorbent from the viewpoint of molecular theory, especially in conjunction with the kinetics and the mechanism of surface reactions, e.g., in heterogeneous catalysis. In addition, investigations on solid adsorbents involving a nonuniform surface are conducted in conjunction with industrial processes, e.g., recovery of volatilized solvents and fractionation of mixtures of hydrocarbons in natural gas. Both situations are to be covered in a comprehensive treatment of the thermodynamics of adsorption. In the present paper, however, only adsorption on solids from the gas phase is considered. A more complete treatment including adsorption on solids from liquid phases is presented in a recent publication by the author.[1]

2 OPERATIONAL DEFINITION OF THE SURFACE EXCESS CONCENTRATION OF ADSORBATE i WITH RESPECT TO A REFERENCE COMPONENT

A comprehensive treatment of the phenomenon of adsorption and the pertinent thermodynamic relations was first given by Gibbs.[2] Accordingly, present textbooks and monographs refer to Gibbs, although this treatment is not readily assimilated by students and even by experts, as has been pointed out by various authors.[3-5] Difficulties arise since the "surface excess" of a component is defined by use of the concept of the "dividing surface". This concept has a clear significance if the exact positions of the molecules of the components in the vicinity of an interface at a given time are known. However, this knowledge, is in general not available. Likewise, there are difficulties if one uses the concept of a "surface phase" located between two bulk phases.

In accord with Bridgman[6], the definition of a thermodynamic quantity such as enthalpy change, specific heat, etc., is supposed to have a definite operational meaning in terms of measurable macroscopic quantities, regardless of a molecular model. For the concept of adsorption, consider the experiment shown schematically in Fig. 1. A glass tube contains a liquid and a gaseous phase of a two-component system, e.g., $Hg + HCl$, at $0°C$ where the rate of the reaction $2 Hg(\ell) + 2 HCl(g) = Hg_2Cl_2(s) + H_2(g)$ is negligible. In accord with the Gibbs phase rule, there are two independent intensive variables (= degrees of freedom). Thus, for given values of T and P, the chemical potentials of the components, the mole fractions, and likewise the concentrations per unit volume in either phase have definite values. When the axis of the tube is in the vertical position as shown in

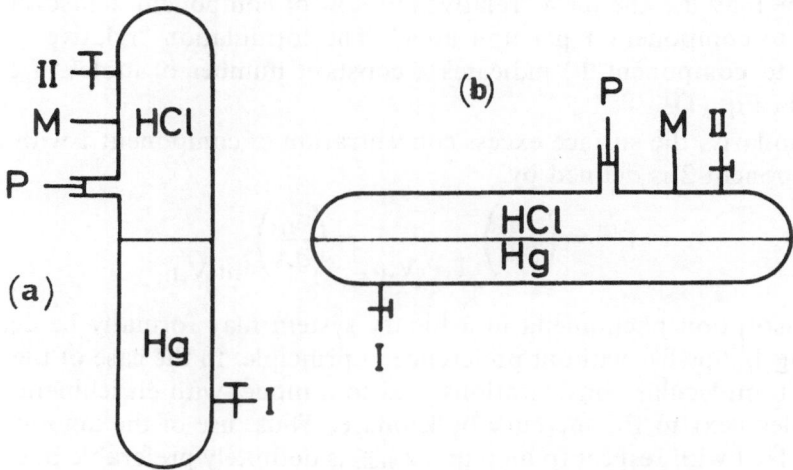

Fig. 1. Measurement of adsorption in the system Hg(ℓ)-HCl(g); I and II are valves for addition of mercury and HCl, respectively; M indicates manometer and P indicates piston for change of volume.

Fig. 1a, the interfacial area between the liquid and the gas phase is small. When the tube is turned 90 degrees as shown in Fig. 1b, the interfacial area is enlarged. If the volumes of the liquid and the gaseous phase are equal and the liquid phase does not wet the solid material of the tube, the interfacial areas between the gas phase and the solid tube and between the liquid phase and the solid tube are not changed if effects due to a finite contact angle at the three-phase boundary solid/liquid/gas are neglected. Thus, only the interfacial area between the liquid and the gas phase is increased. In this experiment, one observes a change in the pressure P indicated by manometer M of sufficient sensitivity when the tube is turned and the temperature is kept constant. Likewise, one observes a change in the concentration of one and/or the other component especially in the gas phase, e.g., with the help of optical absorption measurements. The values of the initial intensive variables may be restored by adding a certain amount of one or the other component at constant volume V by valve I or II. The number of moles of component 2 which must be added for a unit increase of the interfacial area in order to restore the initial values of the intensive variables may be used as the definition of the surface excess concentration $\Gamma_2^{(1)}$ of component 2 with respect to component 1, in accord with Seelich[7] and Pitzer and Brewer[8]:

$$\Gamma_2^{(1)} = \left(\frac{\partial n_2}{\partial A}\right)_{T,P,V,n_1} = \left(\frac{\partial n_2}{\partial A}\right)_{int,V,n_1} \tag{1}$$

where the subscript "int" indicates that all intensive variables be kept constant. Alternatively, especially in conjunction with a molecular interpretation, one may use the term "relative amount of component 2 adsorbed with respect to component 1 per unit area". The formulation "relative . . . with respect to component 1" indicates a constant number of moles of component 1 in Eq. (1).

Similarly, the surface excess concentration of component 1 with respect to component 2 is defined by

$$\Gamma_1^{(2)} = \left(\frac{\partial n_1}{\partial A}\right)_{T,P,V,n_2} = \left(\frac{\partial n_1}{\partial A}\right)_{int,V,n_2} \tag{2}$$

Adsorption phenomena in a binary system may formally be described by using $\Gamma_2^{(1)}$ or $\Gamma_1^{(2)}$ without preference in principle. In the case of the system Hg-HCl, molecular considerations lead to a model with enrichment of HCl molecules next to the mercury bulk phase. Thus, use of the amount of adsorbed HCl with respect to mercury, $\Gamma_{HCl}^{(Hg)}$, is definitely preferable in conjunction with a discussion of the structure of the interface, e.g., use of the Langmuir adsorption isotherm eventually modified in view of energetic interaction among neighboring HCl molecules, etc.

By and large, in a system liquid-gas or solid-gas with one component as the majority constituent in the denser phase, this component is best chosen as reference component 1.

The definition of surface excess concentrations may readily be generalized for multicomponent systems. In a two-phase system involving ν components one has ν independent intensive variables. Thus, all intensive variables such as the mole fractions in either phase are completely determined when T, P, and the chemical potentials of $(\nu-2)$ components or T and the chemical potentials of $(\nu-1)$ components are given. Upon increase of the interfacial area in a multicomponent system at constant volume, it is, in general, necessary to change the amounts of $(\nu-1)$ components in order to restore the initial values of the intensive variables. Thus, the surface excess concentration of component i with respect to component 1 is defined as

$$\Gamma_i^{(1)} = \left(\frac{\partial n_i}{\partial A}\right)_{int,V,n_1} \tag{3}$$

In a two-phase system involving ν components, one has $(\nu-1)$ independent values $\Gamma_i^{(1)}$.

The definitions in Eqs. (1) through (3) are analogous to the definitions of energy or enthalpy changes of chemical reactions as a basis of calorimetric measurements. Under adiabatic conditions, one observes a change in the temperature of the system. The initial temperature may be restored by exchange of energy with an auxiliary heat reservoir, electrical heating or cooling, etc. Accordingly, the enthalpy change ΔH of a chemical reaction is defined by

$$\Delta H = \left(\frac{\partial H}{\partial \xi}\right)_{T,P} \qquad (4)$$

where ξ denotes the degree of the advancement of the chemical reaction.

On the basis of the definitions in Eqs. (1) through (4), one may devise appropriate experimental procedures which permit easy operation and yield eventually a high accuracy. For instance, the value of ΔH may be obtained from the observed temperature change under adiabatic conditions when the heat capacity of the system has been determined in a separate experiment. Likewise, the surface excess of a gaseous component, e.g., HCl at the surface of finely divided silver at $0°C$, may be deduced from the observed pressure when a given amount of HCl is passed into an evacuated vessel containing solid silver as adsorbent, the solubility of HCl in the bulk of the solid is negligible, and the volumes of the empty vessel and the silver sample are known.

The general definition of $\Gamma_i^{(1)}$ in Eq. (3) is especially advantageous in conjunction with thermodynamic calculations (see Section 3). Other equivalent definitions of $\Gamma_i^{(1)}$ are possible, e.g., by use of the so-called algebraic method due to Hansen[9] and Goodrich[10] which has been adopted in the IUPAC "Manual of Definitions, Terminology and Symbols in Colloid and Surface Chemistry"[11].

In Fig. 1, adsorption of gas on the surface of a liquid is considered. Adsorption on a solid may be treated similarly. The following special problems must be noticed.

1. Adsorption depends in general on the crystallographic surface of the solid, e.g., it is different for the (100), (110), or (111) face of a cubic crystal. When adsorption is investigated with a sample involving different kinds of crystallographic surfaces, one obtains average values of $\Gamma_2^{(1)}$, etc., which are adequately defined only for samples with known fractions of the individual crystallographic faces.

2. Adsorption may be enhanced by presence of steps, kinks, and dislocations beyond adsorption on an ideal crystallographic face.

3. Diatomic gases such as H_2 or N_2 may be bound to the surface of a solid either as molecules or as atoms. Thus, adsorption data depend on whether dissociation of diatomic molecules is suppressed or equilibrium is established. Dissociation may be negligible at sufficiently low temperatures and relatively short spans of time for attainment of adsorption equilibrium, whereas at elevated temperatures, dissociation equilibrium may be established rather rapidly. Accordingly, different values of $\Gamma_i^{(1)}$, etc., may be defined, viz., (a) for suppressed dissociation and (b) for established dissociation equilibrium. An analogous situation is found in homogeneous gas phases in which chemical reactions may occur. As an example, consider the com-

pressibility of a NO_2-N_2O_4 mixture. If the gas is compressed quickly, the reaction $2NO_2 \rightleftharpoons N_2O_4$ may be suppressed. If the gas is compressed more slowly, equilibrium of the reaction $2NO_2 \rightleftharpoons N_2O_4$ may be established. Thus, one has two different values of the adiabatic, i.e., isentropic, compressibility, which determines the velocity of the propagation of sound. Accordingly, the velocity of the propagation of sound in a NO_2-N_2O_4 mixture is a function of frequency, that is, dispersion occurs.

In addition, one may, in principle, distinguish between (a) adsorption without transfer of electrons and (b) adsorption with transfer of electrons under formation of adsorbed ions, e.g., O_2^- or O^-. It is most probable that on metals, electronic equilibrium is established virtually immediately and, accordingly, adsorption with suppression of transfer of electrons is practically not observable. The situation may be different in the case of ionic crystals as adsorbents if transfer of electrons is associated with a rearrangement of ionic defects in the electrical double layer next to the surface. In this case, one may anticipate different values of $\Gamma^{(1)}$, etc., depending on whether the rearrangement of ionic defects is suppressed, or equilibrium of the distribution of the defects is established.

4. At relatively low temperatures, it may be safe to assume that atoms of a solid are virtually immobile when a foreign gas is admitted to an evacuated vessel accommodating the solid phase. Investigations with the help of LEED, however, have shown that adsorption of oxygen on a solid metal, e.g., nickel above room temperature may result in a rearrangement of the metal atoms, especially in formation of two-dimensional ordered structures consisting of metal and oxygen atoms (or ions). In addition, facetting may occur, i.e., crystallographic faces stable *in vacuo* may disappear in favor of other crystallographic faces which are more stable in the presence of a strongly adsorbed component in the gas phase.[12,13] Accordingly, one has different values of $\Gamma^{(1)}$, etc., depending on whether the constituent atoms of a solid are virtually immobile or a rearrangement of the atoms takes place.

5. So far, it has been assumed tacitly that the extension of the phases α and β normal to the interface is insignificant for the amount of adsorption. By and large, the interaction energy among atoms and/or molecules decreases very sharply with distance. Thus, in most systems, adsorption phenomena are practically independent of the thickness of the phases involved if the thickness is greater than ten monolayers. However, exceptions may occur in the case of semiconductors in which the density of the charge carriers in the bulk phase is fairly low and accordingly the Debye length is relatively large, e.g., 10^{-6} cm or even greater.

3 THERMODYNAMIC EQUATIONS

For a two-phase system, use of the Helmholtz energy F as the so-called characteristic function is appropriate if the independent variables are temperature T, total volume V, interfacial area A, and the number of moles n_1, n_2, etc. The total differential of F is

$$dF = - SdT - PdV + \sigma dA + \mu_1 dn_1 + \mu_2 dn_2 + \ldots \qquad (5)$$

where S is entropy, V is volume, T is temperature, P is pressure, σ is the work required for a reversible enlargement of the interface by unit area at constant values of the intensive variables (also called surface free energy), and μ_1, μ_2, etc., are the chemical potentials of components 1, 2, etc., respectively.

If both phases are fluid, the value of σ is readily measurable and equal to the interfacial tension. If one phase is solid, σ may be determined by use of special experimental techniques[14,15] providing the solid undergoes plastic flow within the time available for an experiment. In the case of a crystal, σ depends on the crystallographic surface. Thus, eventually, the term σdA in Eq. (5) is to be replaced by the sum of the products pertaining to different crystallographic surfaces.

In conjunction with adsorption phenomena, especially Eq. (3), the independent variables are T, V, A, n_1, μ_2, ... Thus it is appropriate to introduce a modified Helmholtz energy $(F - n_2\mu_2 - n_3\mu_3 \ldots)$ whose total differential is

$$d(F-n_2\mu_2-n_3\mu_3-\ldots) = -SdT-PdV+\sigma dA+\mu_1 dn_1-n_2 d\mu_2-n_3 d\mu_3-\ldots \qquad (6)$$

Upon differentiating $(F-n_2\mu_2-n_3\mu_3-\ldots)$, once with respect to A and once with respect to μ_i for $i \neq 1$, one obtains

$$\frac{\partial^2(F-n_2\mu_2-n_3\mu_3-\ldots)}{\partial A \partial \mu_i} = \left(\frac{\partial \sigma}{\partial \mu_i}\right)_{T,V,A,n_1,\mu_{k\neq 1,i}}$$

$$= -\left(\frac{\partial n_1}{\partial A}\right)_{T,V,n_1,\mu_{k\neq 1,i}} \qquad (7)$$

where the subscript $\mu_{k\neq 1,i}$ indicates that the chemical potential of all components except for components 1 and i be kept constant. The value of σ depends only on the value of the intensive variables. According to Eq. (3), $\partial n_i/\partial A$ is the relative surface excess concentration of component i per unit area with respect to component 1 denoted by $\Gamma_i^{(1)}$. Thus, it follows from Eq. (7) that

$$\Gamma_i^{(1)} = - \left(\frac{\partial \sigma}{\partial \mu_i}\right)_{T,\mu_{k \neq 1,i}} \tag{8}$$

which is the classical adsorption equation due to Gibbs.[2]

In general, σ is determined experimentally as a function of composition of one of the coexisting phases rather than as a function of the chemical potentials μ_2, μ_3, etc. A particularly simple situation occurs when σ has been determined as a function of the partial pressures or mole fractions in an ideal gas phase (= phase β). Then, one may rewrite Eq. (8) as

$$\Gamma_i^{(1)} = - \frac{1}{RT} \left(\frac{\partial \sigma}{\partial \ln p_i}\right)_{T,p_{k \neq 1,i}} \tag{9}$$

Buttner, Udin, and Wulff[16] have used Eq. (9) to calculate the adsorption of oxygen on solid silver in a He-O_2 atmosphere. In this case, the total pressure P was kept constant at 1 atm and, therefore, $p_{He} = P - p_{O_2}$. Equation (9) can be used rigorously only if p_{He} is kept constant. In the system Ag-O_2-He, however, it is safe to assume that adsorption of helium is negligible and, accordingly, σ is practically independent of p_{He} in a range from 0 to 1 atm. At 932°C, σ has been found to decrease linearly with log p_{O_2} between $p_{O_2} = 10^{-4}$ and 0.2 atm. Thus, according to Eq. (9), $\Gamma_0^{(Ag)}$ is in essence independent of the oxygen partial pressure corresponding to adsorption of $2 \cdot 10^{15}$ oxygen atoms/cm^2. Since the compound Ag_2O decomposes at about 190°C under atmospheric pressure, occurrence of a nearly completely filled monolayer of oxygen atoms on silver at 932°C is somewhat surprising. This situation, however, is in accord with the results of investigations on other systems. Oudar[17] has investigated adsorption of sulfur on copper from H_2S-H_2 mixtures containing radioactive sulfur (^{35}S) at 830°C and has shown that a nearly complete monolayer of sulfur atoms is formed at sulfur activities considerably lower than the sulfur activity for coexistence of copper and Cu_2S as bulk phases. According to investigations with LEED, an ordered array of sulfur atoms interspersed with metal atoms is formed. Analogous results have been obtained for other metals, e.g., silver and nickel.[19,20]

Upon further differentiation of Eq. (7), it follows that

$$\frac{\partial^3(F - n_2\mu_2 - n_3\mu_3 - \ldots)}{\partial A \, \partial \mu_i \, \partial \mu_j} = - \frac{\partial}{\partial \mu_j}\left(\frac{\partial n_i}{\partial A}\right) = - \frac{\partial}{\partial \mu_i}\left(\frac{\partial n_j}{\partial A}\right) \tag{10}$$

whereupon, in view of Eq. (3),

$$\left(\frac{\partial \Gamma_i^{(1)}}{\partial \mu_j}\right)_{T,\mu_{k \neq 1,j}} = \left(\frac{\partial \Gamma_j^{(1)}}{\partial \mu_i}\right)_{T,\mu_{k \neq 1,i}} \tag{11}$$

in accord with Cassel.[21]

Equation (11) may be used to describe competition of molecules of different adsorbates for a given number of adsorption sites on a solid adsorbent. When the concentration or partial pressure of component i in the gas phase is increased, more molecules of component i are adsorbed, the number of unoccupied sites decreases, and accordingly fewer molecules of the other components are adsorbed in the absence of mutual attraction between molecules of different adsorbates. According to Eq. (11), the displacement of molecules of kind i by molecules of kind j is related to the displacement of molecules of kind i, regardless of any special model. Equation (11) may also be applied to systems in which presence of adsorbate 2 enhances adsorption of adsorbate 3 and vice versa, because of mutual attraction of 2 and 3 resulting in "associative adsorption", e.g., at the liquid/gas interface of the system water-phenol-aniline at room temperature[22] and the system Fe-Cr-C at 1350°C[23,24].

4 ADSORPTION OF GASES ON SOLIDS INVOLVING A NONUNIFORM SURFACE

The surface of solid adsorbents is, in general, not uniform. In particular, different crystal faces may be present. The surface excess concentration of an adsorbate depends on the kind of the crystal face. For practical purposes, it is appropriate to report the surface excess of an adsorbate per unit mass of the adsorbent, especially in the case of highly disperse adsorbents, e.g., charcoal or alumina. Even under these complex conditions, a straightforward thermodynamic treatment is possible provided that thermodynamic equilibrium is attained among molecules adsorbed on different crystallographic faces or different sites on an amorphous solid. If the latter condition is not satisfied, results of adsorption measurements may depend on the sequence of addition of different adsorbates and may differ for increasing and decreasing partial pressures of the adsorbates, i.e., hysteresis may occur. In this case, no thermodynamic treatment is possible.

In accord with the foregoing treatment of adsorption on uniform surfaces, the following equations for nonuniform surfaces involve only quantities which are observable at least in principle. The scope of this approach is in accord with Hill[25], Schay[26], and others.

For adsorption from a gas phase on a solid adsorbent (= phase α) which is virtually not volatile and does not dissolve the species present in the gas phase to any significant extent, the IUPAC manual[11] defines the relative surface excess of component i with respect to the solid adsorbent (= component 1) by

$$n_i^{\sigma(1)} = n_i - c_i^{\beta}(V - V^{\alpha}) = n_i - c_i^{\beta}V^{\beta} \tag{12}$$

where n_i is the total number of moles of component i=2, 3, etc., c_i^β is the concentration of i in moles per unit volume in the gas phase β, V is the total volume of the system (= adsorbent + gas phase), V^α is the volume of the solid adsorbent, which may be set equal to the quotient of mass, m_{ads}, and bulk density, ρ_{ads}, and

$$V^\beta = V - V^\alpha = V - m_{ads}/\rho_{ads} \qquad (13)$$

is the volume of the gas phase β. Alternatively, the volume of the gas phase in a vessel with a given amount of the adsorbent may be determined by use of helium as an inert gas under the assumption that $n_{He}^{\sigma\,(1)} = 0$, whereupon $V^\beta = n_{He}^\beta/c_{He}$ according to Eq. (12) and $V^\alpha = V - V^\beta$.

The relative surface excess $\Delta_i^{(1)}$ of adsorbate i with respect to the solid adsorbent per unit mass of the adsorbent with given surface structure is defined by

$$\Delta_i^{(1)} = \frac{n_i^{\sigma\,(1)}}{m_{ads}} = \frac{n_i - c_i^\beta V^\beta}{m_{ads}} \qquad (14)$$

Solving Eq. (14) for n_i and differentiating with respect to m_{ads} at constant volume V^β of the gas phase, one obtains the alternative definition

$$\Delta_i^{(1)} = \left(\frac{\partial n_i}{\partial m_{ads}}\right)_{int,\,V^\beta} \qquad (15$$

where the subscript int indicates constant values of the intensive variables especially the partial pressures p_2, p_3, etc., in the gas phase.

If the gas phase involves several adsorbates, competition of the adsorbates for a given number of adsorption sites may be important. For a thermodynamic treatment it is expedient to use the modified Helmholtz energy $F - n_2\mu_2 - n_3\mu_3 - \ldots$ with the total differential

$$d(F - n_2\mu_2 - n_3\mu_3 - \ldots) =$$

$$- SdT - PdV + \mu_{ads}dm_{ads} - n_2d\mu_2 - n_3d\mu_3 - \ldots \qquad (16$$

where μ_{ads} is the chemical potential of the adsorbent per gram with a given surface area per gram. Thus, μ_{ads} in Eq. (16) is higher than that of the bulk material and depends on the composition of the gas phase. Since the surface area per gram of the adsorbent is assumed to be given, Eq. (16) does not involve a term for an independent change in the surface area A, in contrast to Eq. (6). The area A can be increased only by an increase of the mass m_{ads} of the adsorbent.

Substituting $V = V^\beta + m_{ads}/\rho_{ads}$ according to Eq. (13) and neglecting the dependence of ρ_{ads} on temperature and pressure, one may rewrite Eq. (16) a

$$d(F-n_2\mu_2-n_3\mu_3-\ldots) = -SdT-PdV^{\beta}+(\mu_{ads}-P/\rho_{ads})dm_{ads}$$

$$-n_2d\mu_2-n_3d\mu_3-\ldots \tag{17}$$

From Eq. (17) it follows that

$$\frac{\partial^3(F-n_2\mu_2-n_3\mu_3-\ldots)}{\partial m_{ads}\,\partial\mu_i\,\partial\mu_j} = -\frac{\partial}{\partial\mu_j}\left(\frac{\partial n_i}{\partial m_{ads}}\right) = -\frac{\partial}{\partial\mu_i}\left(\frac{\partial n_j}{\partial m_{ads}}\right) \tag{18}$$

for T, V^{β}, m_{ads}, μ_2, μ_3, etc., as independent variables. For an ideal gas phase, one has

$$\mu_i = \overset{\circ}{\mu_i} + RT \ln p_i \tag{19}$$

Substitution of Eqs. (15) and (19) in Eq. (18) yields

$$\left(\frac{\partial\Delta_j^{(1)}}{\partial\ln p_i}\right)_{T,p_{k\neq1,i}} = \left(\frac{\partial\Delta_i^{(1)}}{\partial\ln p_j}\right)_{T,p_{k\neq1,j}} \tag{20}$$

as an analogue to Eq. (11).

5 HEATS OF ADSORPTION

Calorimetric determinations of heats of adsorption are possible only with samples which have a fairly large surface area which is, in general, non-uniform. The following considerations are confined to adsorption from an ideal gas phase.

The molar differential heat of adsorption of component i from a gas phase on a solid is defined as the enthalpy change associated with the transfer of 1 mole of component i from the bulk gas phase to the surface of the adsorbent at constant T and P and constant amounts of the other adsorbates.

$$\Delta_aH_i = (\partial H/\partial n_i^{\sigma(1)})_{T,P,n_k,\Delta_{k\neq1}^{(1)}} \tag{21}$$

In view of this definition, in accord with the IUPAC manual[11], heats of adsorption are in general negative. The respective heats of desorption $(= -\Delta_aH_i^{(1)})$ also called isosteric heats of adsorption[26] $(= q_{i,st})$ are positive. Heats of adsorption for a single adsorbate are, in general, deduced from calorimetric measurements in closed systems, i.e., at constant volume rather than at constant pressure upon use of appropriate corrections.[27,28]

The determination of heats of adsorption in systems involving several adsorbates is more involved since an increase in the partial pressure p_i of adsorbate i results in changes of the surface excess of the other adsorbates. Pertinent equations have been derived elsewhere.[1]

6 THE TEMPERATURE DEPENDENCE OF ADSORPTION

If differential heats of adsorption are available, one may calculate the temperature coefficient of the relative surface excess $\Delta_i^{(1)}$ of adsorbates 2, 3, ... with respect to the solid adsorbent per unit mass of the adsorbent at given partial pressures p_2, p_3, etc. Pertinent relations are derived upon differentiating the modified Helmholtz function once with respect to T, once with respect to the mass of the adsorbent, and once with respect to μ_i. The resulting entropy changes are related to the respective enthalpy changes, i.e., heats of adsorption. Thus it follows that[1]

$$\left(\frac{\partial \Delta_i^{(1)}}{\partial T}\right)_{p_2, p_3, \ldots} = \frac{1}{RT^2} \sum_{j=2,3,\ldots} \Delta_a H_j \left(\frac{\partial \Delta_j^{(1)}}{\partial \ln p_i}\right)_{T, p_{k \neq 1,i}} \tag{22}$$

At present, data for systems with several adsorbates in conjunction with use of Eq. (22) are not available. Eventually, Eq. (22) may be used for qualitative considerations. Since $\Delta_a H_i$ values are negative, it follows that the decrease in adsorption with increasing temperature is especially high for adsorbates with high absolute values of the heat of adsorption corresponding to strong adsorption. Thus, the selectivity of adsorption decreases with increasing temperature.

For a single adsorbate $i=2$, it follows from Eq. (22) that

$$\left(\frac{\partial \Delta_2^{(1)}}{\partial T}\right)_{p_2} = \frac{\Delta_a H_2}{RT^2} \left(\frac{\partial \Delta_2^{(1)}}{\partial \ln p_2}\right)_T \tag{23}$$

Upon use of the auxiliary equation

$$d\Delta_2^{(1)} = \frac{\partial \Delta_2^{(1)}}{\partial \ln p_2} d \ln p_2 + \frac{\partial \Delta_2^{(1)}}{\partial T} dT \tag{24}$$

one obtains for the pressure p_2 as a function of T at constant surface excess $\Delta_2^{(1)}$

$$\left(\frac{\partial \ln p_2}{\partial T}\right)_{\Delta_2^{(1)}} = -\frac{\Delta_a H_2}{RT^2} \tag{25}$$

as an analogue to the Clausius-Clapeyron equation for the temperature dependence of the vapor pressure of a pure liquid. The negative sign in Eq. (25) is due to the definition of $\Delta_a H_2$ as enthalpy change for adsorption rather than for desorption, which would be the analogue to vaporization of a pure liquid.

Equations (23) and (25) can also be applied to adsorption on a uniform surface of single crystals in order to calculate the heats of adsorption from the temperature dependence of the surface excess concentration $\Gamma_2^{(1)}$. Then one has the well-known formula

$$- \Delta_a H_2 = RT^2 \left(\frac{\partial \ln p_2}{\partial T} \right)_{\Gamma_2^{(1)}} \tag{26}$$

Also, in systems involving several adsorbates, one may calculate partial molar heats of adsorption from measurements of the temperature dependence of adsorption. Instead of Eq. (26), one has[27]

$$\left(\frac{\partial \ln p_i}{\partial T} \right)_{\Delta_2^{(1)}, \Delta_3^{(1)}, \ldots} = - \frac{\Delta_a H_i}{RT^2} \tag{27}$$

Thus, in order to calculate values of $\Delta_a H_i$ for i=2,3, etc., one has to obtain values of $\partial \ln p_i / \partial T$ for constant surface excess values of all adsorbates by interpolation of data which have been obtained for given partial pressures p_2, p_3, etc.

Alternatively, one may evaluate values of $\partial \Delta_2^{(1)} / \partial T$, $\partial \Delta_3^{(1)} / \partial T$, etc., for given partial pressures of the adsorbates. In the case of two adsorbates, one obtains from Eq. (22) for i=2 and 3

$$\Delta_a H_2 = \frac{RT^2}{D} \begin{vmatrix} \partial \Delta_2^{(1)} / \partial T & \partial \Delta_3^{(1)} / \partial \ln p_2 \\ \partial \Delta_3^{(1)} / \partial T & \partial \Delta_3^{(1)} / \partial \ln p_3 \end{vmatrix} \tag{28a}$$

$$\Delta_a H_3 = \frac{RT^2}{D} \begin{vmatrix} \partial \Delta_2^{(1)} / \partial \ln p_2 & \partial \Delta_2^{(1)} / \partial T \\ \partial \Delta_2^{(1)} / \partial \ln p_3 & \partial \Delta_3^{(1)} / \partial T \end{vmatrix} \tag{28b}$$

where the determinant D is given by

$$D = \begin{vmatrix} \partial \Delta_2^{(1)} / \partial \ln p_2 & \partial \Delta_3^{(1)} / \partial \ln p_2 \\ \partial \Delta_2^{(1)} / \partial \ln p_3 & \partial \Delta_3^{(1)} / \partial \ln p_3 \end{vmatrix} \tag{29}$$

Interpolation and numerical differentiation of moderately accurate data may result in large errors in the calculation of heats of adsorption by use of Eqs. (27) through (29). This is true even for investigations involving a single adsorbate.[29,30] Thus, use of Eqs. (27) through (29) is restricted to systems for which sufficiently precise adsorption measurements at different temperatures are available.

7 THE DEPENDENCE OF HEATS OF ADSORPTION ON SURFACE COVERAGE

In general, $\Delta_a H_2$ for adsorption of a single adsorbate 2 on a uniform surface depends on the surface excess concentration $\Gamma_2^{(1)}$. This dependence is of interest in conjunction with models. For the derivation of the Langmiur adsorption isotherm it is presumed that $\Delta_a H_2$ is independent of $\Gamma_2^{(1)}$, i.e., energetic interaction among adsorbed molecules is negligible. Actually, there may be mutual attraction analogous to van der Waals attraction in a gas. Then, $\Delta_a H_2$ becomes more negative with increasing surface coverage by adsorbed molecules. On the other hand, there may be mutual repulsion of adsorbed molecules, e.g., in the case of the adsorption of molecules representing electrical dipoles with parallel orientation.[31,32] If dipole interaction exceeds van der Waals interaction, $\Delta_a H_2$ may become less negative with increasing surface coverage.

In the case of adsorbents involving a nonuniform surface, the change of $\Delta_a H_2$ with increasing surface excess $\Delta_2^{(1)}$ may be especially large. When the surface coverage is low, adsorption takes place preferentially at centers with a high binding energy. With increasing coverage, adsorption also occurs at centers with lower binding energies. Accordingly, $-\Delta_a H_2$ may decrease with increasing surface coverage in a rather spectacular manner. Experimental data of a $\Delta_a H_2$ as a function of surface excess $\Delta_2^{(1)}$ may therefore give important clues on the abundance of centers with different binding energies. For example, one may test whether the distribution of adsorption sites is in essence a linear or an exponential function of the binding energy as assumed in the derivation of adsorption isotherms by Slygin and Frumkin[33], Temkin and Pyzhev[34], and Zeldowitsch[35]. In this way, Frankenburg[36] and Halsey and Taylor[37] have evaluated experimental data for the adsorption of hydrogen on tungsten powder.

8 CONCLUDING REMARKS

Information provided by thermodynamic investigations is limited since all data are inherently complex in terms of quantities considered in molecular theory. Even in the relatively simple case of a single adsorbate and a uniform surface with two types of adsorption sites, the differential heats of adsorption comprise energy changes due to energetic interaction between adsorbent and adsorbate, energetic interaction among molecules of the adsorbate, redistribution of molecules of the adsorbates located at sites involving different bond energies, and eventually dissociation of molecules of the adsorbate and rearrangement of atoms or molecules of the adsorbent. In contrast, structural investigations, e.g., by LEED and spectroscopic measurements, in a broad sense provide more detailed information on special molecular reactions.

Both approaches supplement each other. In particular, thermodynamic investigations yield information for conditions at which a solid is used as an adsorbent or a catalyst in industrial applications, i.e., at elevated temperatures and a pressure of the gas phase of 1 atm or higher, whereas investigations by LEED or spectroscopic measurements yield mostly information on adsorption layers at low temperatures and low gas pressures.

REFERENCES

1. Wagner, C., *Nachr. Akad. Wiss. Gottingen, II. Math.-Physik.*, Kl., **1973**, p. 37.
2. Gibbs, J. W., *Collected Works*, Yale University Press, New Haven (1948), Vol. I, pp. 219ff.
3. Guggenheim, E. A., *Trans. Faraday Soc.*, **36**, 397 (1940).
4. Moelwyn-Hughes, E. A., *Physical Chemistry*, Pergamon Press, Oxford (1957), pp. 912ff.
5. Defay, R., Prigogine, I., Bellemans, A., and Everett, D. H., *Surface Tension and Adsorption*, Longmans, London (1966), pp. 22ff.
6. Bridgman, P. W., *The Logic of Modern Physics*, The Macmillan Company, New York (1960).
7. Seelich, F., *Monatsch. Chem.*, **79**, 333 (1948).
8. Lewis, G. N., and Randall, M., *Thermodynamics*, 2nd edition, revised by K. S. Pitzer and L. Brewer, McGraw-Hill Book Company, New York (1961), p. 479.
9. Hansen, R. S., *J. Phys. Chem.*, **55**, 1195 (1951).
10. Goodrich, F. C., *Trans. Faraday Soc.*, **64**, 3403 (1968); *Surface and Colloid Science*, **1**, 1 (1969).
11. "Manual of Definitions, Terminology and Symbols in Colloid and Surface Chemistry", *Pure and Appl. Chem.*, **31**, 577 (1972).
12. Chalmers, B., King, R., and Shuttleworth, R., *Proc. Roy. Soc. (London)*, **A193**, 465 (1948).
13. Moreau, J., and Benard, J., *Acta Met.*, **10**, 247 (1962).
14. Udin, H., Shaler, A. J., and Wulff, J., *Trans. AIME*, **185**, 186 (1948).
15. Hondros, E. D., in *Techniques of Metals Research* (R. F. Bunshaw, Ed.), Interscience Publishers, New York (1970), Vol. IV, Part 2, pp. 293ff.
16. Buttner, F. H., Udin, H., and Wulff, J., *J. Phys. Chem.*, **56**, 657 (1952).
17. Oudar, J., *Compt. Rend.*, **249**, 91 (1951).
18. Domauge, J. L., and Ouder, J., *Surface Sci.*, **11**, 124 (1968).
19. Benard, J., Oudar, J., and Cabane-Brouty, F., *Surface Sci.*, **3**, 359 (1965).
20. Perdereau, M., and Oudar, J., *Surface Sci.*, **20**, 80 (1970).
21. Cassel, H., *Physik. Z.*, **26**, 863 (1927).
22. Wagner, C., *Z. Physik. Chem. A*, **143**, 389 (1929).
23. Whalen, T. J., Kaufman, S. M., and Humenik, M., *Trans. Am. Soc. Metals*, **55**, 778 (1962); Kaufman, S. M., *Acta Met.*, **15**, 1089 (1967).
24. Belton, G. R., *Metallurg. Trans.*, **3**, 1465 (1973).
25. Hill, T. L., *Advances in Catalysis*, **4**, 244 (1952).
26. Schay, G., *Surface and Colloid Science*, **2**, 155 (1969); *J. Colloid Interface Sci.*, **35**, 254 (1971).
27. Hill, T. L., *J. Chem. Phys.*, **17**, 520 (1949).
28. Kingston, G. L., and Aston, J. G., *J. Phys. Chem.*, **73**, 1929 (1950).
29. Sandstede, G., *Z. Physik. Chem. N.F.*, **29**, 99 (1961).
30. Ross, S., and Oliver, J. P., *On Physical Adsorption*, Interscience Publishers, New York (1964).

31. Magnus, A., *Z. Physik. Chem. A,* **142,** 401 (1929).
32. de Boer, J. H., and Zwikker, C., *Z. physik. Chem. B,* **3,** 407 (1929).
33. Slygin, A., and Frumkin, A., *Acta Physicochim., U.R.S.S.,* **3,** 791 (1935).
34. Temkin, M., and Pyzhev, V., *Acta Physicochim., U.R.S.S.,* **12,** 327 (1940).
35. Zeldowitsch, J., *Acta Physicochim., U.R.S.S.,* **1,** 961 (1934).
36. Frankenburg, W. G., *J. Am. Chem. Soc.,* **66,** 1827, 1838 (1941).
37. Halsey, G., and Taylor, H. S., *J. Chem. Phys.,* **15,** 624 (1947).

DISCUSSION of paper by Carl Wagner

EHRLICH: The approach to the thermodynamics of surfaces outlined by Professor Wagner is extremely interesting. I would just like to insert a historical note by pointing out that the modified Helmholtz function that he has introduced so profitably is related to a function which has found considerable application in both statistical and thermodynamic problems, namely to $L = F - \Sigma n_i \mu_i$, where the summation extends over all components i. This thermodynamic potential is related to the grand partition function by $L = RT\ln'\Xi$, and as such dates back to J. W. Gibbs. Its use in thermodynamic applications, however, is more recent, and seems to have first been emphasized in Landau and Lifshitz *Statistical Physics* (Oxford University Press, 1938). Although the term grand potential has sometimes been applied to L, it may be more in keeping with current usage to call it the Landau potential. For bulk systems, $L = -pV$; when surfaces make a contribution, $L = pV + \sigma A$. The consistent development of surface thermodynamics using this function was initiated by Landau and Lifshitz in their book; it has been further extended by Cabrera (*Surface Science,* Vol. 2) and has recently received attention in connection with the nucleation of bubbles in liquids. It will certainly be interesting to compare the relative advantages of the Landau potential, and Professor Wagner's function, for the discussion of surface problems. However, it is clear already that for systems containing an interface, a characteristic function that involves the chemical potential μ as an independent variable is most useful.

WAGNER: In conjunction with adsorption, it is important to introduce a modified Helmholtz energy where the sum of the products of the amounts and the chemical potentials of the components except for the reference component 1 is subtracted. This asymmetry is due to the fact that adsorption can be defined only with respect to a reference component in accord with Gibbs.

SOMORJAI: Your treatment certainly simplifies the use of surface thermodynamic concepts that relate to experiments. It would be important to apply it to studies of multicomponent systems, for example, to alloy surfaces. In developing new catalysts, metal alloys will play an important role. Thus, determination of their surface phase diagrams (surface composition versus T) will become necessary. Using Auger Electron Spectroscopy (AES) for example, one obtains the surface composition, and from the temperature dependence of the surface composition, the surface tensions may be determined. The surface tension is a key parameter that enters into the treatment of surface thermodynamics of multicomponent systems. For solids, it is very difficult to determine the surface tension in any other way. It appears that the surface thermodynamics of metal-alloy surfaces will become an important field of surface science in the near future.

WAGNER: The experimental determination of the surface tension of solids is possible, but difficult, and limited to special cases. Thus, the development of new methods for the direct determination of the surface excess concentrations of the components of an alloy will be very valuable.

OUDAR: There are only few results concerning surface tension measurements during the adsorption of a reactive gas on metal single crystals. Precise results have been obtained by Hondros (Colloquium CNRS No. 187, Paris, 1969) concerning the variation of surface tension of a (100) copper face as a function of the partial pressure of H_2O in a H_2O/H_2 mixture. The surface tension σ first decreases in agreement with the Gibbs equation with one oxygen atom adsorbed for four copper superficial atoms. Then, at a critical partial pressure of H_2O, which is much lower than that for the formation of bulk oxide, σ increases suddenly and the Gibbs equation is no longer valid. It seems to me that we can conclude that the increase of σ is related to the crystallization of a complete monolayer of adsorbent atoms which introduces some interfacial stress which is not taken into account in the Gibbs equation. This interpretation is consistent with the fact that according to measurements of adsorption isotherms, a complete monolayer of adsorbent atoms can be formed at pressures which are much lower than the pressure which corresponds to the formation of a bulk compound. Do you agree with this interpretation?

WAGNER: The observations reported by Hondros are interesting. The increase in the surface tension of solid copper with increasing oxygen activity above a certain critical value indicates some kind of thermodynamic instability. Probably, hysteresis may occur when the oxygen activity is lowered. I cannot comment further before I have studied the paper by Hondros.

THE SPECIFICATION OF ACTIVE CENTERS IN METAL CATALYSTS

G. C. Bond

School of Chemistry, Brunel University
Uxbridge, Middlesex, England

ABSTRACT

After reviewing the classical picture of metal surfaces and modifications necessitated when very small particles are considered, Taylor's theory of active centers is translated into modern idiom expressed in terms of surface coordination number. It is shown in outline that the principal reactions undergone by hydrocarbons at metal surfaces can be understood if the number of bonds which can be formed between a carbon atom and a single metal atom is determined by coordination number and electronic structure. Consideration of compensation effect plots indicates that the fraction of surface active in some catalyzed reactions may be extremely small.

1 INTRODUCTION

Although the mechanisms of heterogeneously catalyzed reactions have been the subject of research for some three quarters of a century, there is no case in which it can be said with confidence that the mechanism is established beyond reasonable doubt. There is indeed no generally accepted definition of the information needed to describe a reaction mechanism, although I tried a decade ago to express one.[1] In this respect the situation is worse than with homogeneously catalyzed, and even enzymatically catalyzed, reactions, and a moment's reflection shows why this is so. Enzymes are highly effective in bringing about one specific molecular transformation; there is thus only one transition state to describe in a given enzymatically catalyzed reaction. The same is true, but to a lesser degree, with reactions catalyzed by organometallic complexes in solution. Here the active center is usually one, two, or at most three vacant coordination positions on the metal atom or ion; the number of possible modes of coordination of the reactants, and of their interaction, is quite limited, and knowledge of a plausible reaction mechanism is not hard to obtain.

The situation with a heterogeneous catalyst is, however, vastly more complex. The surface consists of a large number of atoms or ions, which though chemically equivalent are topographically different. Now, even if **one** surface atom were capable of acting as an active center in a catalytic reaction, the presence of the other neighboring atoms permits other operations to be performed simultaneously on a reactant molecule. It will help to quote an example. In the catalytic hydrogenation of *p*-methoxynitrobenzene, it is typically found that the product contains, besides the desired *p*-methoxyaniline, a proportion of aniline arising from loss of the methoxy group by hydrogenolysis. This occurs under conditions where anisole (phenyl methyl ether) is quite stable to hydrogenolysis. The catalyst surface must therefore contain at least two kinds of active sites, and the second must comprise the first as well as some additional atoms. It is the coexistence of a variety of active centers which differentiates heterogeneous catalysts from others in respect to difficulty of defining reaction mechanisms.

In some instances of practical interest, the desired product is the result of a long sequence of transformations all occurring on the same catalyst. Consider, for example, the mechanistic complexity of converting benzene to maleic anhydride, or n-heptane to toluene. In such multistep processes, each unit reaction needs description, though only one is rate limiting, if the total reaction mechanism is to be claimed to be established. There is thus usually a complex potential energy diagram, and in each unit step, a different combination of surface atoms, i.e., a different active center, is likely to be involved.

It will be necessary later to mention additional complicating factors, as it is essential to realize the full horror of the complexity. The purpose of this introductory section is merely to hint at it. Nevertheless, the importance to both the theory and the practice of catalysis warrants a sustained experimental attack on the specification of active centers in heterogeneous catalysts. This requires desirably a theoretical framework, and it is the object of this paper to sketch this; its construction calls for imagination and a willingness not to be confined by current concepts. Not until we can say for many catalyzed reactions that the mechanisms are known will heterogeneous catalysis finally pass from an art to a science.

2 THE NATURE OF METALLIC SURFACES

The conventional view of a metallic surface in the region of ambient temperature is that it consists of an ordered array of atoms forming one of the possible low-index crystallographic planes appropriate to its crystal structure, and that even with polycrystalline metals this order pervades some tens or hundreds of atomic diameters. LEED observations made with monocrystals have repeatedly confirmed this picture. The perfect order may, however, be disturbed by defects such as missing atoms, or a partly completed second layer, the latter generating a series of highly reactive step sites at its periphery. Interatomic distances may, however, be a few percent shorter than in the bulk. In short, the typical metal surface has been regarded, not unreasonably, as very similar to the surface created by the notional cleavage of a crystal.

The energetic description of surface metal atoms presents a more difficult problem. Such atoms are of course more mobile in two dimensions than atoms in the bulk, and sintering of evaporated films has been detected well below room temperature. Vibration amplitudes are larger than in the bulk, although frequencies may be lower, and recent studies[2] of the surface compositions of alloys show that interdiffusion of components normal to the surface occurs with surprising ease at moderate temperatures when a sufficient driving force exists to displace the normal equilibrium. There is a growing feeling that it may be a mistake to regard the surface layer as identical to the bulk: in some respects, surface atoms may be more liquidlike in their behavior.

3 VERY SMALL METAL PARTICLES

Preparation of supported metal catalysts by impregnating the support with a solution of a reducible salt, followed by drying and careful hydrogen reduction, leads to the formation of metal particles between 5 and 100 Å; average particle size can be controlled within this range by variation of metal content and reduction temperature[3,4,5], which permits investigation of the so-called mitohedrical region. Ion-exchange methods give smaller average particle sizes for the same metal content.[6] Colloidal dispersions containing particles predominantly less than 20 Å in size can also be made.[7] The preparation of minute particles containing only six atoms of platinum within a zeolite cage has also been claimed[8], while the decomposition of "cluster compounds" may be an alternative approach to making such minuscule groups of atoms.

By reason of the widespread occurrence of very small metal particles in technical catalysts, much attention has been given to the relevant theory. Direct observation of particle shape in this size range is almost impossible, although exceptionally careful electron microscopy of a Rh/SiO_2 catalyst has shown the existence of groups of five rhodium atoms in a ring. Small particles are, of course, thermodynamically unstable with respect to larger ones, and several attempts have been made to calculate the most stable form of very small particles[9]; it is usually supposed that the form which maximizes the number of interatomic bonds is the most stable, and some calculations show that the icosahedron bounded by (111) surfaces is more stable than all alternatives. Model calculations employing molecular orbital theory[10] and the scattered-wave (SCF-Xα) method[11] have also been performed. The latter method promises to be exceptionally useful in calculating surface densities of states for small groups of atoms.

Very small particles naturally contain in their surfaces a high proportion of atoms of low coordination number; these are the atoms occupying edges and corners in geometrically regular bodies, and with particles having an incomplete outer layer, are additionally those atoms at the edges of such layers. Simple[12] and highly detailed[3,13] calculations on the dependence of the number of atoms of given coordination number on particle size, both for regular bodies and those containing an incomplete outer layer, have been carried out. The mean coordination number of surface atoms increases with increasing particle size. No attempt has yet been made to estimate how the occurrence of a distribution of particle size affects this conclusion. It is thus to be expected that mobility parallel to and normal to the surface will be enhanced in comparison with extended surfaces, and indeed evidence is accumulating for the disordered or liquidlike nature of the surfaces of small particles.

4 THE CONCEPT OF ACTIVE CENTERS

The path of research workers in heterogeneous catalysis has been illuminated by a number of theoretical concepts, of which only two have really stood the test of time: Balandin's Multiplet Hypothesis[14] and Taylor's theory of "Active Centres"[15]. Both theories in their early expression tried in complementary ways to cast light on the nature of the surface atoms constituting a catalytically active site. Unfortunately, the former hypothesis has fallen somewhat into disrepute among Western scientists by reason of its emphasis in the popular mind on the sextet mechanism for benzene hydrogenation, a notion which is almost certainly incorrect; nevertheless Balandin's basic idea that each catalytic reaction has a specific requirement for one atom or a grouping of two, three, four, or whatever number of surface atoms is probably right. Taylor's perspicacious paper[16] published in 1925 gave clear expression to the concept of a geometrically and energetically heterogeneous surface, describing it as shown in Fig. 1. There is little difference, except in terminology, between his picture and that of small particles having atoms of different coordination number at their surface. Taylor believed that the most coordinatively unsaturated atoms constituted the "active centers" for certain reactions, whereas for other reactions the requirements might be less stringent:

> "The amount of surface which is catalytically active is determined by the reaction catalysed. There will be all extremes between the case in which all the atoms in the surface are active, and that in which relatively few are so active. The former will occur in the case of a readily catalysed process."

Thus, current ideas concerning facile and demanding reactions had been anticipated in outline by Taylor almost half a century ago.

There is now abundant evidence for the energetically heterogeneous nature of adsorbed states of many simple molecules, and these can now be ascribed, at least in part, to the existence of a variety of crystal planes, of

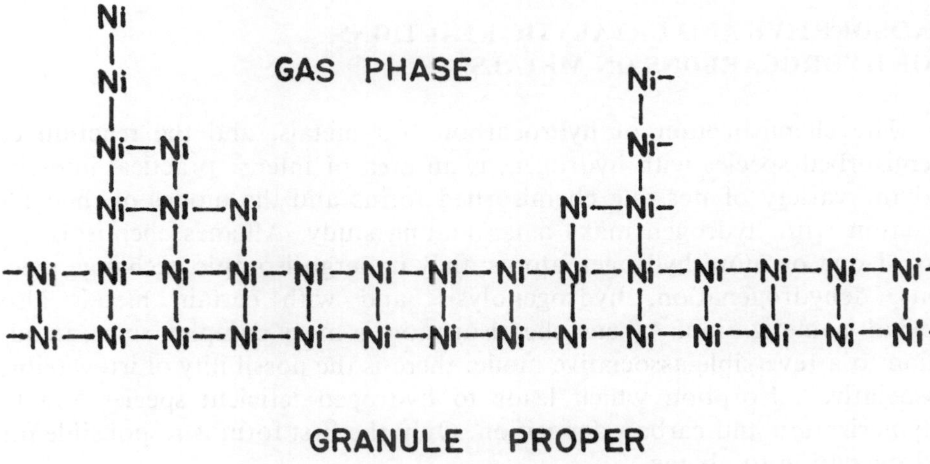

Fig. 1. Representation of active centers.[16]

atoms of low coordination number at edges and corners, and of atoms of high coordination number at the base of steps on planes.[12,13] Important evidence for the extraordinary properties of steps has come from the recent elegant work of Somorjai[17] who has shown that stepped surfaces can be obtained by cutting a single crystal at a slight angle to a low index plane. Such surfaces exhibit higher sticking coefficients for hydrogen and oxygen, and higher catalytic activity in n-heptane dehydrocyclization, than do the corresponding plane surfaces. It is of course important to realize that higher order index planes are, from the point of view of catalysis, mathematical abstractions, and even the (110) surface of a face-centered cubic crystal can be regarded as a stepped (111) surface.

In the progressive refinement and application of the theories of Balandin and of Taylor, it is necessary to seek to define for each elementary adsorptive or catalytic act

1. The number of surface metal atoms constituting the active center
2. The topographical relation of the atoms to each other (e.g., in the case of a triplet site, linear or bent on the (100) plane of a face-centered cubic metal, or triangular on the (111) plane)
3. The degree of coordinative unsaturation of each atom
4. The relevance of occupation of adjacent atoms or sites to activity of the selected site.

In the case of alloys and bimetallic clusters[18], further, more penetrating, questions arise. We may thus imagine that in the general case the active center or site for a surface reaction comprises one or more atoms suitably disposed and having in aggregate a specified (or not less than a specified) number of unsatisfied valencies.

5 ADSORPTIVE AND CATALYTIC REACTIONS OF HYDROCARBONS ON METALS

The chemisorption of hydrocarbons on metals, and the reaction of chemisorbed species with hydrogen, is an area of intense practical interest, and the variety of possible chemisorbed forms and the modes of their interaction with hydrogen make a fascinating study. Alkanes chemisorb by loss of one or more hydrogen atoms; they undergo isotopic exchange reactions, dehydrogenation, hydrogenolysis, and with certain metals also skeletal isomerization. Alkene chemisorption is more complex since, in addition to a reversible associative mode, there is the possibility of irreversible dissociative adsorption which leads to hydrogen-deficient species and to polymerization and carbon deposition. Only the first form is responsible for hydrogenation to alkane.

The trends exhibited by metals in respect of the key reactions undergone by hydrocarbons are briefly reviewed, noting the kinds of species which have been proposed in mechanistic discussions.

5.1 Irreversible Chemisorption of Hydrocarbons

It was Beeck who first showed[19] that adsorption of ethylene onto a clean metal film leads ultimately to the formation of ethane and a strongly bonded "acetylenic residue" which is not however identical to chemisorbed acetylene, but which decreases the film's activity for subsequent reactions. Methane as well as ethane is a product of the interaction of ethylene with platinum black.[20] The fraction of surface covered by irreversibly adsorbed species increases on moving to the left from Group VIII_3[19], and the amounts of ^{14}C-labelled ethylene retained by different supported metals have been measured[21], with the results shown in Fig. 2.

It is common experience that metal catalysts employed in alkene hydrogenation decline in activity with use, sometimes attaining a limiting value, and that pretreatment with alkene leads to lower subsequent rates. Partially deactivated catalysts show significant differences in reaction mechanism from fresh catalysts. Supported metals, unlike films and powders, are quite thermally stable and can have their activity regenerated by heating in hydrogen, a procedure which gives alkanes of both higher and lower carbon number than the reactant alkene. There is also much industrial experience to show that strongly bonded residues arise during dehydrogenation, and that they are precursors to coke formation.

These strongly bonded residues have never been studied in depth, but from their H/C ratios (close to unity) and their considerable stability, they

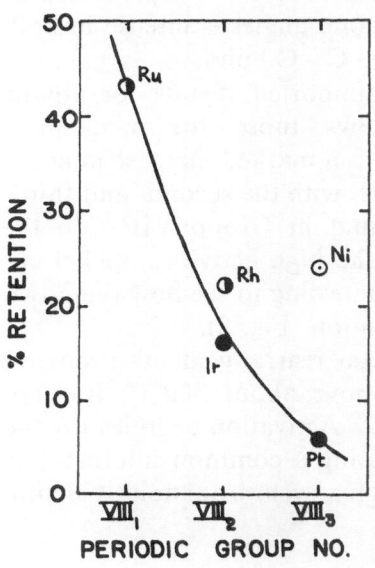

Fig. 2. Extent of retention of ^{14}C-labelled ethylene on various supported metals.[21]

(A)   (B)

Fig. 3. Intermediates proposed in hydrocarbon
reactions on metals (see text).

(C)   (D)

are very probably multiply bonded to the metal. Structures such as (A) and (B) in Fig. 3 have been suggested, and their polymerization discussed.[22]

5.2 Hydrogenolysis and Skeletal Isomerization of Alkanes

The hydrogenolysis of ethane to methane has been widely studied.[23] The reaction requires much higher temperatures than does alkene hydrogenation, one reason being the greater difficulty of alkane adsorption because of its necessarily dissociative character, and the ensuing low surface coverage by the active intermediates; these are thought to be wholly or substantially dehydrogenated versions of ethane.[23] Complete dehydrogenation to C_2 is not easily credible, but if true it would need to be represented as either (C) or (D) in Fig. 3. Distortion of the bond angles is intense in both cases, but this probably aids the splitting of the C—C bond.

Figure 4 shows the activities of silica-supported metals for ethane hydrogenolysis at $205°C$[23] and Fig. 5 shows those for neopentane hydrogenolysis at $250°C$[24]. In both cases there is a marked increase in activity on passing to the left from Group VIII_3 or IB with the second- and third-row elements, maximum activities being found in Group VIII_1. In the former case, the first-row elements exhibit quite high activities, nickel approximating to rhodium in this respect; it is interesting to see how these two elements compare with regard to ethylene retention (Fig. 2).

With C_4 and higher hydrocarbons, skeletal rearrangements occur on metals simultaneously with hydrogenolysis above about 300°C; iridium, platinum, and gold are the most active metals.[24] Activation energies for the two processes are the same or similar, betokening a common intermediate thought to be either (E) or (F) (Fig. 6), although a cyclopropanoid structure is an alternative possibility.

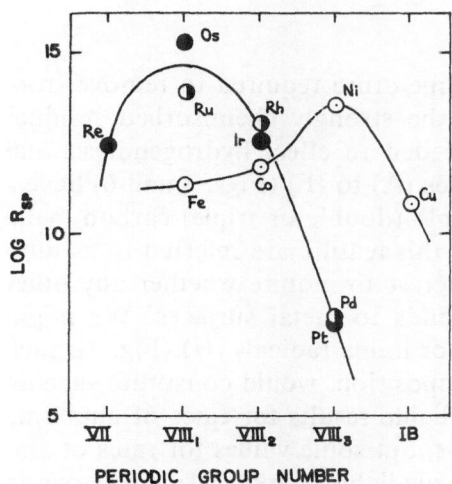

Fig. 4. Activities of metals for the hydrogenolysis of ethane at 205°C.[23]

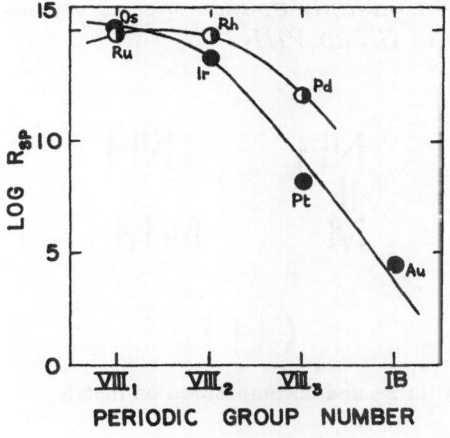

Fig. 5. Activities of metals for the hydrogenolysis of neopentane at 250°C.[24]

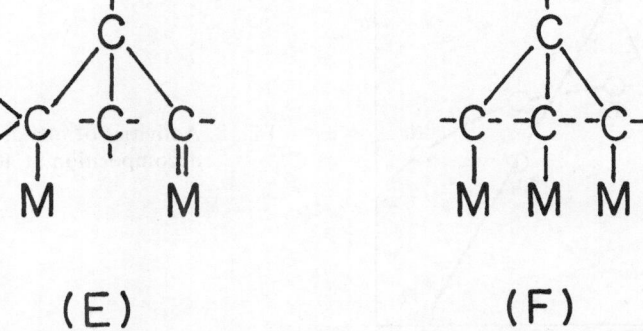

Fig. 6. Intermediates proposed in skeletal isomerization of hydrocarbons on metals.

5.3 The Role of Multiply Bonded Species

The similarity between the high temperature required to remove from metal surfaces by hydrogen treatment the strongly chemisorbed residues from alkene chemisorption and that needed to effect hydrogenolysis and skeletal isomerization suggests that species (A) to (F) (Figs. 3 and 6) have a common feature; this is clearly a multiple (double or triple) carbon-metal bond, and chemisorbed entities showing this feature are referred to as multiply bonded species. It is clearly of interest to inquire whether any other atoms besides carbon form multiple bonds to metal surfaces. We might perhaps expect that nitrogen atoms (G) or imine radicals (H) (Fig. 7), such as exist in ammonia synthesis and decomposition, would constitute such an example. Unfortunately, there are no reliable results for rates of ammonia synthesis on a number of different metals, but some values for rates of ammonia decomposition on metal films are available[25], and these are shown in Fig. 8. The similarity with Fig. 4 is especially striking, and permits the general theorem that *in reactions wherein multiply bonded species are intermediates, catalytic activity is maximal in Group VIII₃.*

Fig. 7. Intermediates proposed in ammonia synthesis and decomposition on metals.

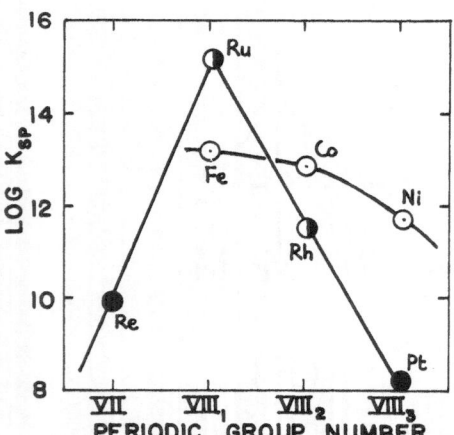

Fig. 8. Activities of metals for ammonia decomposition at 400°C.[25]

5.4 Alkene Hydrogenation and Alkane Exchange

There is much evidence to show that in the hydrogenation of ethylene and other alkenes, maximum activity is shown by rhodium and iridium in Group VIII2; nickel is much less active than palladium and platinum.[22] In alkane exchange with deuterium, high activity is shown by tungsten, molybdenum, and rhodium, but the very large differences such as are seen in hydrogenolysis are not apparent.[26] It is also interesting to see that within Group VIII there is qualitatively an inverse correlation between activities for ammonia exchange[27] and ammonia decomposition[25]. A similar antipathetic relation exists between rates of hydrogenation and hydrogenolysis of cyclopropane.[28] From this we deduce that *the sites responsible for alkane and ammonia exchange, and for alkene hydrogenation, are not those preferred by multiply bonded species.*

5.5 Specific Particle Size Effects

The present consensus is that for a surprising variety of reactions the specific rate is constant over a wide range of particle sizes, although in some instances particles smaller than 20 or 30 Å have not been examined. Such reactions have been termed[29] facile or structure insensitive. There is however a small but growing number of reactions whose specific rates do appear to increase with decreasing particle size; these have been called demanding or structure sensitive, and the clearest examples include hydrogenolysis and isomerization of alkanes.[23,24] Even benzene hydrogenation, which had been regarded as a facile reaction, has been shown to have a maximum rate when nickel particles of 10 Å in size are used as the catalyst[3]; Balandin would have seen this as a result of tremendous significance. Similarly, ethane hydrogenolysis, which appears to be a demanding reaction, can show a decrease in specific rate when very small rhodium particles are employed. Low-temperature oxidations[30] or reactions in the presence of oxygen[31] show an apparent increase in specific rate with *increasing* particle size, but this is due partly to rapid inactivation of the smaller particles. Thus, examples of nearly all the formal possibilities set down[12] some years ago have now been discovered. Further interesting developments in this area can be confidently expected.

That so many reactions are facile and so few demanding has occasioned much discussion, and a number of possible explanations have been tentatively advanced. One idea is that surface reconstitution or restructuring is a general phenomenon, so that crystal plane specificity cannot exist. Related to this, and based on the concept of a semifluid surface outlined above, is the possibility that surface metal atoms are sufficiently mobile to destroy the geometric identity of the crystal faces.

5.6 Hydrocarbon Reactions on Alloys

This is a virtually unexplored area, but initial studies have indicated that it will be a most fruitful field for future attention. The availability of techniques for preparing supported alloys[32,33] and the recognition that the normal miscibility rules do not apply to very small particles[18], opens up tremendous possibilities. The activity of nickel for hydrogenolysis is rapidly lost as copper is added in quite small amounts, while activities for hydrogenation and dehydrogenation are scarcely affected.[34,35] More subtle effects have been observed with this system using alkane-deuterium exchange as the test reaction.[36] Observations on alloy catalysts are undoubtedly complicated by the fact that chemisorption of a species strongly attracted to one component of the alloy tends to enrich the surface in that component[2]; therefore, there is a need to measure surface composition while the surface is acting catalytically.

6 THE SPECIFICATION OF ACTIVE CENTERS FOR REACTIONS OF HYDROCARBONS ON METALS

To assist in understanding and rationalizing the foregoing body of information, the following simplifying hypotheses are advanced.

1. With any metal, low-coordination-number surface atoms such as exist at edges, corners, and steps can form a greater number of covalent bonds to atoms in a chemisorbed species than can atoms in a low-index crystallographic plane; the more coordinatively unsaturated the atom, the greater is the number of bonds it can form.

2. Processes involving the formation or removal of multiply bonded intermediates (disruptive chemisorption, alkane hydrogenolysis and isomerization, ammonia synthesis, and decomposition) occur at active centers incorporating at least one atom capable of forming a multiple bond.

3. With palladium and platinum, these atoms will have low coordination numbers; and, following the first hypothesis, on moving to the left through Group VIII, atoms of progressively higher coordination number can fulfill this role. *The fraction of surface active for reactions involving multiply bonded species increases from Group IB to VIII₃, at which point it may be maximal.* Nickel atoms have more free valencies than topographically equivalent platinum atoms.

4. Those reactions involving multiply bonded intermediates are frequently demanding with metals of Group VIII₃; they are likely to be progressively less demanding on metals further to the left, and more demanding with metals of Group IB. Ammonia decomposition should prove to be a demanding reaction, especially with Group VIII₃ and IB

metals, while ammonia synthesis may be demanding on metals of Group VIII$_1$.

5. Alkene hydrogenation, alkane and ammonia exchange, and alkane dehydrogenation (which are facile reactions) can proceed on one or more classes of sites composed of atoms in flat, low-index planes; with sites incorporating low-coordination-number atoms, or more generally with atoms capable of multiple-bond formation, disruptive reactions of alkanes and especially alkenes occur which constitute self-poisoning. The lower activity of ruthenium and osmium as compared with rhodium and iridium for alkene hydrogenation may be due to a smaller fraction of the surface being active, or a larger fraction self-poisoned.

6. On alloying a Group VIII$_3$ metal with a Group IB metal, surface atoms are now believed to retain some but not all of their chemical identity. Since surface atoms of Group IB metals are almost incapable of forming multiple bonds, activity for hydrogenolysis should decline with increasing content of Group IB metal; since such atoms can readily accommodate a single bond, activity for alkene hydrogenation and related reactions should not decline.

The contemporary validity of Taylor's concept of the energetically heterogeneous surface is thereby indicated; but it can be extended to cover the situation where some sites are too active and produce strongly bonded, self-poisoning residues.

7 ESTIMATION OF THE FRACTION OF ACTIVE SURFACE

From the foregoing it seems likely that measurement of specific rates for a reaction such as ethane hydrogenolysis through an estimation of the surface area of the metal by, say, hydrogen chemisorption could be grossly misleading, since especially with Group VIII$_3$ and Group IB metals, much of the surface may be covered with hydrogen atoms and ethyl radicals which are not intermediates in the target reaction. Direct gravimetric measurements have shown that the fraction of copper surface active in formic acid decomposition is quite small.[37]

Some possible assistance in determining the fraction of active surface comes from a somewhat unexpected quarter, namely, from consideration of compensation effect plots. The phenomenon of sympathetic variation between activation energy (E) and the logarithm of the preexponential factor (log A) is well known to all workers in the field of heterogeneous catalysis, and indeed beyond, since "linear free energy relationships" are frequently encountered in homogeneous reactions. A very peculiar manifestation of the compensation effect was noted some years ago[22], when it was seen that for ethylene hydrogenation over some alloy systems, the points fell on two distinct and parallel lines (see Fig. 9 for an example). No interpretation could be offered at the time, but a possible explanation is the following.

Although there is no universal agreement concerning the origin of the compensation effect, it is most reasonable and plausible to suppose that it arises from a correlation between the enthalpy required to form the transition state and the entropy of activation $\Delta S^{\ddagger}$. Thus the tighter the transition state, the greater will be the enthalpy input necessary to achieve it and the more favorable also will be the entropy change. A more trivial explanation of the compensation effect applies when apparent rather than true activation energies are used. Such is frequently the case, and the reactions to be considered below may well fall into this category. Whatever the true interpretation, if we find a series of points obeying the equation

$$\log A = mE + c$$

(where m and c are constants), we may suppose that the transition states are similar in composition, and differ only in their energy. Another term besides $\exp(\Delta S^{\ddagger}/R)$ which is contained in the experimentally observed log A is of course the concentration of active sites. If, as in Fig. 9, some points lie on a lower line, we may imagine it is because this term has a lower value by a factor of about 10^6. The abrupt drop in activity occurring at about 70 percent copper is thus attributed entirely to a fall in the fraction of active surface at this point, while the mechanism (as defined by the nature of the transition state in the slowest step) remains unaltered. This is in harmony with the remark above concerning Group IB metals.

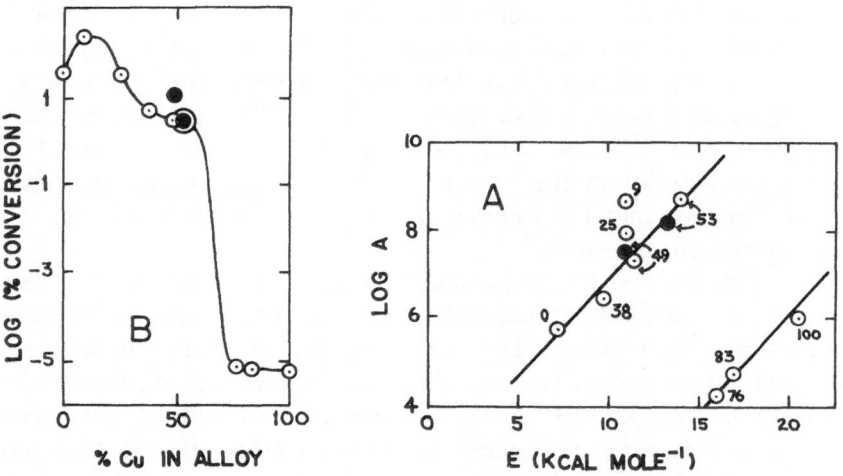

Fig. 9. Ethylene hydrogenation over palladium-copper foils[22]: A, specific activity at 100° C as a function of composition; B, compensation effect (figures give percentage of copper in alloy). Open points denote disordered alloys; filled points denote ordered alloys.

We should now examine the compensation-effect plots corresponding to Figs. 4, 5, and 8; these are shown in Figs. 10, 11, and 12. Considering first the results for ethane hydrogenolysis[23] (Fig. 10), we see that the points for six metals (Cu, Co, Re, Ir, Rh, and Ni) lie close to a "main sequence" line, while the points for ruthenium and osmium (which are more active) lie above this line and those for palladium and platinum (which are less active) lie below. The broken lines, drawn with the same slope as the main sequence line, should not be taken too seriously. A similar situation exists with neopentane hydrogenolysis[24] (Fig. 11), but it is less clearly defined because there are fewer results. Gold now joins platinum as a metal having a very low density of active sites. The results for ammonia decomposition are shown in Fig. 12. Again the situation is not absolutely clear, but with some imagination a main sequence comprising the points for nickel, rhenium, and rhodium may be seen; the points for iron, cobalt, and ruthenium lie above and that for platinum lies below this line.

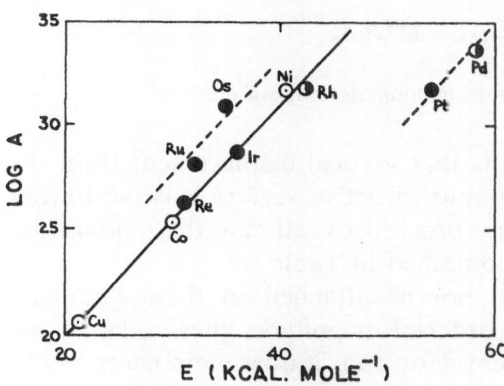

Fig. 10. Compensation effect for ethane hydrogenolysis.[23]

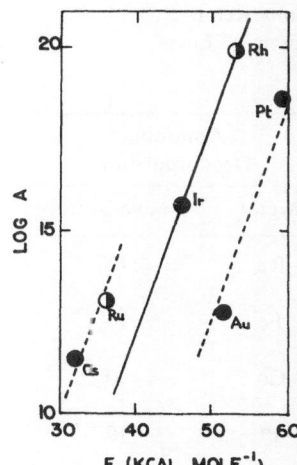

Fig. 11. Compensation effect for neopentane hydrogenolysis.[24]

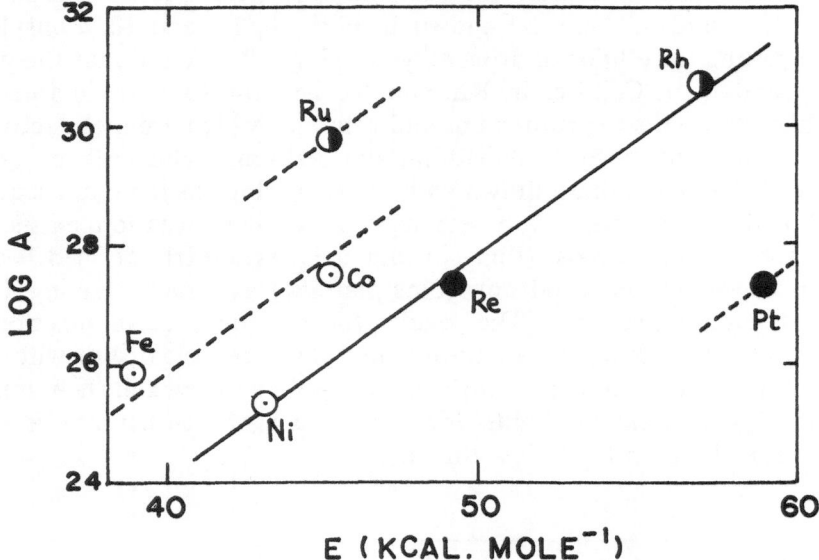

Fig. 12. Compensation effect for ammonia decomposition.[25]

Accepting the working hypothesis that vertical displacement from the main sequence lines measures the amount of active surface relative to that used by metals on these lines, we may proceed to estimate these quantities from Figs. 10 to 12. The results are contained in Table I.

Quantitative significance should not be attached to these numbers, which should be regarded only as order-of-magnitude guesses as to the amount of active surface in each system. However, it does seem necessary to

Table I. Amount of Active Surface in Several Reactions Relative to That for Metals Lying on the "Main Sequences" Lines (Taken to be Unity)

Ethane Hydrogenolysis		Neopentane Hydrogenolysis		Ammonia Decomposition	
Metal	Active Surface	Metal	Active Surface	Metal	Active Surface
Os	10^3	Os	10^4	Ru	10^4
Ru	40	Ru	10^3	Fe	10^2
Pd	10^{-7}	Pt	10^{-5}	Co	30
Pt	10^{-7}	Au	10^{-6}	Pt	10^{-4}

postulate that the fraction of active surface exhibited by different metals in a given reaction can vary by as much as 10^{10}, and although this may seem to be an inordinately large range, it is little greater than the range of isothermal activities.

It is noteworthy that the two metals seeming to show the highest fractions of active surface are hexagonal close packed in structure; they may therefore present an intrinsically higher density of appropriate configurations of atoms than metals having other structures.

8 CONCLUSION

Recent ideas concerning the possible semifluid or at least highly mobile character of metal surfaces have not yet been integrated with classical notions of crystalline surfaces and rigid active centers. The disordered nature of real surfaces may yet emerge to be an important cause of catalytic activity. One fruitful way of simplifying the problem of the multisite surface is to eliminate some sites by selective poisoning. There are hints that partially sulfided surfaces may be easier to study for this reason. Surfaces of alloys should contain simpler active centers for demanding reactions.

REFERENCES

1. Bond, G. C., and Wells, P. B., *Advan. Catalysis,* 15, 92 (1964).
2. Bouwman, R., Lippits, G.J.M., and Sachtler, W.M.H., *J. Catalysis,* 25, 350 (1972).
3. Coenen, J.W.E., van Meerten, R.Z.C., and Rijnten, H. Th., *Proc. 5th International Congress on Catalysis,* North Holland (1973), p. 671.
4. Dorling, T. A., and Moss, R. L., *J. Catalysis,* 7, 378 (1967).
5. Poltorak, O. M., and Boronin, V. S., *Russ. J. Phys. Chem.,* 40, 1436 (1966).
6. Dorling, T. A., Lynch, B.W.J., and Moss, R. L., *J. Catalysis,* 20, 190 (1971).
7. Bond, G. C., *Trans. Faraday Soc.,* 52, 1235 (1956).
8. Dalla Betta, R. A., and Boudart, M., *Proc. 5th International Congress on Catalysis,* North Holland (1973), p. 1329.
9. Romanowski, W., *Surface Science,* 18, 373 (1969).
10. Baetzold, R. C., *Solid State Physics,* 4, 62 (1972).
11. Slater, J. C., and Johnson, K. H., *Phys. Rev.,* B5, 844 (1972).
12. Bond, G. C., *4th International Congress on Catalysis,* Akadémiai Kaidó, Budapest (1971), Paper 67.
13. van Hardeveld, R., and Hartog, F., *Advan. Catalysis,* 22, 75 (1972).
14. Schlosser, E.-G., *Ber. Bunsenges. Physikal. Chem.,* 73, 358 (1969).
15. Trapnell, B.M.W., *Advan. Catalysis,* 3, 1 (1951).
16. Taylor, H. S., *Proc. Roy. Soc.,* A108, 105 (1925).
17. Somorjai, G. A., *Catalysis Reviews,* 7, 87 (1973).
18. Sinfelt, J. H., *J. Catalysis,* 29, 308 (1973).
19. Beeck, O., *Faraday Soc. Discussions,* 8, 118 (1950).

20. Baird, T., Paál, Z., and Thomson, S. J., *J. Chem. Soc. Faraday Trans. I*, **69**, 50 (1973).
21. Taylor, G. F., Thomson, S. J., and Webb, G., *J. Catalysis*, **12**, 191 (1968).
22. Bond, G. C., *Catalysis by Metals*, Academic Press, London and New York (1962), Chap. 11.
23. Sinfelt, J. H., *Catalysis Reviews*, **3**, 175 (1969); *Advan. Catalysis*, **23**, 91 (1973).
24. Boudart, M., and Ptak, L. D., *J. Catalysis*, **16**, 90 (1970).
25. Logan, S. K., and Kemball, C., *Trans. Faraday Soc.*, **56**, 144 (1960).
26. Anderson, J. R., and Kemball, C., *Proc. Roy. Soc.*, **A233**, 361 (1954).
27. Kemball, C., *Proc. Roy. Soc.*, **A214**, 413 (1952).
28. Merta, R., and Poneč, V., *4th International Congress on Catalysis*, Akadémiai Kaidó, Budapest (1971), Paper 50.
29. Boudart, M., *Advan. Catalysis*, **20**, 153 (1969).
30. Ostermaier, J. J., Katzer, J. R., and Manogue, W. H., *J. Catalysis*, **33**, 457 (1974).
31. Pusateri, R. J., Katzer, J. R., and Manogue, W. H., *AIChE Journal*, **20**, 219 (1974).
32. Allison, E. G., and Bond, G. C., *Catalysis Reviews*, **7**, 233 (1972).
33. Cormack, D., Thomas, D. H., and Moss, R. L., *J. Catalysis*, **32**, 492 (1974).
34. Sinfelt, J. H., Carter, J. L., and Yates, D.J.C., *J. Catalysis*, **24**, 283 (1972).
35. Beelen, J. M., Poneč, V., and Sachtler, W.M.H., *J. Catalysis*, **28**, 376 (1973).
36. Roberti, A., Poneč, V., and Sachtler, W.M.H., *J. Catalysis*, **28**, 381 (1973).
37. Tamaru, K., *Advan. Catalysis*, **15**, 64 (1964).

DISCUSSION on Paper by G. C. Bond

KEMBALL: I believe that it is an oversimplification to suggest that the small rates of hydrogenolysis of ethane on platinum are due to the requirement of special sites which amount to only 10^{-9} of the total sites present. If this were the case, the activity of platinum catalysts would show tremendous dependence on their method of preparation, and such dependence is not found. The low rate is more likely to be associated with a substantial loss of entropy on forming the transition state which may require a more critical arrangement of the various atoms to lead to carbon-carbon bond rupture than on metals which are better than platinum at forming multiply bonded adsorbed species.

BOND: I know of no evidence that the specific activity of platinum catalysts is *not* strongly dependent on method of preparation *for demanding reactions.* Kemball's point concerning entropy of activation may well be valid, but it does not assist in understanding why with ethane hydrogenolysis over platinum and palladium, the smaller preexponential factor is not compensated by a smaller activation energy as is the case with those metals (including copper) lying on the "main sequence" line.

KEMBALL: I am glad that Prof. Bond has joined those of us who have been emphasizing the importance of multiply bonded hydrocarbon species for catalytic hydrogenolysis on metals. This idea was supported by a correlation which I gave in a plenary lecture in 1968 at the 4th International Congress on Catalysis in Moscow (the lecture was subsequently reprinted in *Catalysis Reviews* in 1971). The multiple exchange of methane with deuterium, which involves as a rate-determining step the formation of adsorbed methylene species, is a good measure of the ability of metals to form multiply bonded hydrocarbon species. The correlation showed that metals which are active for this exchange are also active for the hydrogenolysis of butane and the rates of the two reactions correlate well on a series of metal films.

BOND: It was not possible within the compass of my paper to mention all instances wherein multiply bonded species are likely to be of importance. I am grateful to Kemball for drawing attention to the relevance of Type II methane exchange to this discussion.

SCHRIEFFER: I would like to note that the relation, discussed in chemical terms by Prof. Bond, between the coordinative unsaturation of active substrate atoms and the strength of chemisorption bonds with these atoms, is supported in general terms by recent work from the solid-state-physics point of view. However, the simple view of single, double, and triple bonds is replaced by bonds of varying strength and symmetry, depending on the appropriately projected local density of states on the surface. We will hear more about this later in the colloquium.

Part Two

EXPERIMENTS ON CLEAN METAL SURFACES WITH ADSORBATES

THE USE OF X-RAY PHOTOELECTRON SPECTROSCOPY (ESCA) FOR STUDYING ADSORBED MOLECULES

*John T. Yates, Jr., Nils E. Erickson,
S. D. Worley*, Theodore E. Madey*

*Surface Chemistry Section
National Bureau of Standards
Washington, D.C.*

ABSTRACT

An ultrahigh vacuum X-ray photoelectron spectrometer has been used to study a number of cases of adsorption on tungsten single crystals. The choice of adsorbates spans a wide range from dissociative chemisorption to nondissociative chemisorption to physisorption.

A method for estimating absolute surface coverages from ESCA data has been verified by comparison with absolute molecular-beam measurements of monolayer coverage for the system oxygen + W(100).

*Guest Worker, National Bureau of Standards, 1973—74. Now affiliated with Department of Chemistry, Auburn University, Auburn, Alabama.

The ESCA technique for measuring chemical shifts has been found to be useful in discriminating various modes of surface bonding in all adsorption systems studied. For oxygen chemisorbed on W(100), two distinct oxygen states are seen. In the case of CO chemisorbed on W(100), a direct correlation between ESCA and thermal desorption behavior has been observed for four states of chemisorbed CO. For H_2CO interaction with W(100), dissociative adsorption occurs initially, followed by multilayer condensation of H_2CO. For xenon physisorption on W(111), high-energy sites (defects?) are covered first by mobile xenon; this is followed by adsorption on the uniform surface.

Coverage-dependent shifts in binding energy have been observed in several cases. The magnitudes of these shifts are independent of changes in average work function, and are best explained by considering local dipole interactions.

In all cases studied, it appears that final-state extra-atomic relaxation effects dominate in determining the magnitude of the chemical shift for adsorbed species. Core-level binding energies have been found to decrease upon adsorption in all cases so far studied, and the magnitude of the decrease seems to correlate in a crude way with increasing strength of adsorption.

1 INTRODUCTION

Examples of the application of X-ray photoelectron spectroscopy (XPS) or electron spectroscopy for chemical analysis[1,2] (ESCA) to the study of adsorbed species on atomically clean single crystals of tungsten are discussed herein. In X-ray photoelectron spectroscopy one measures the binding energy E_B of electrons photoemitted from deep-lying electronic core levels of substrate and adsorbate atoms. Small changes in binding energy of the adsorbate core levels (chemical shifts) are caused by various modes of bonding in the adsorbed layer, and it is the purpose of this paper to delineate the major physical factors responsible for these chemical shifts. Studies of this type, on well-defined single-crystal substrates, form a baseline for the eventual application of ESCA to the study of practical surfaces of interest to workers in areas such as surface catalysis, electronic materials, and corrosion. The technique has previously been applied to the study of adsorbed species on polycrystalline films[3-5], polycrystalline ribbons[6-9], and single crystals[10,11].

There are two major factors that determine chemical shifts in the binding energy of core electrons in atoms bound in molecules or to surfaces. The first has to do with the initial charge state of the adsorbed atom and its surroundings.[1] If we assume that an atom, when bonded, is partially ionic and has associated with it an excess or a deficiency of charge qe located in a shell of radius r, then the change in binding energy associated with the atom's charge will be approximately $\Delta E_B = qe^2/r$. For a partially ionic atom associated with other charged atoms in a crystal, molecule, or on a surface, one must additionally sum over all of the neighbor charges, obtaining for the new binding energy

$$E_B' = E_B \text{ (free atom)} + \frac{qe^2}{r} + \sum_j \frac{q_j e^2}{r_j} \qquad (1)$$

where the sum is over all neighbors, j, of charge $q_j e$ at distance r_j. The application of these ideas to alkali halide crystals has been discussed by Citrin and Thomas.[12]

The second factor involved in determining the chemical shifts in binding energy has to do with the final state of the core-hole ion produced by photoemission. Consider the change in state for atom A upon photoejection:

$$h\upsilon + A \rightarrow (A^+)^* + e^- \qquad (2)$$

where $(A^+)^*$ is the symbol for the core-hole ion having total energy $E_{(A+)}^*$.

The photoelectron is ejected with a measured energy E_K, and from the definition $E_B \equiv h\upsilon - E_K$, one can calculate from Eq. (2) that

$$E_B = E_{(A+)}^* - E_A = \text{(final-state total energy)} \qquad (3)$$
$$- \text{(initial-state total energy)}$$

Historically, it has usually been assumed that variation in E_B for different modes of bonding in molecules was due only to changes in initial-state energy, as for example in Eq. (1). That is, the assumption was made that an atom, no matter what its bonding, would acquire the same value of $E_{(A+)}^*$ upon photoejection of a particular core electron. It has recently been shown in papers by Citrin[12,13] and by Shirley[14-16] that variations occur in $E_{(A+)}^*$ for different surroundings, or for different modes of bonding of A. These effects are termed extra atomic relaxation (XR) effects and may be thought of as polarization effects in the neighborhood of the core-hole ion leading to a decrease in binding energy because of the flow of negative charge to the core-hole ion. Thus, building on Eq. (1), the true binding energy is given by the expression:

$$E_B = E_B' - E_{XR} \qquad (4)$$

In practice it is usually not possible to determine exactly the charge distribution associated with either an adsorbed atom or an adsorbed molecule, so that the initial-state binding energy, E_B', cannot be calculated exactly, and the relaxation energy cannot be determined exactly by experiment using Eq. (4). What can be done at present is to determine the *total* chemical shift $[E_B - E_B \text{ (free atom or molecule)}] = \Delta E_B$, and to relate this qualitatively to the bonding character of the adsorbed species as characterized by other methods.

2 EXPERIMENTAL

The experimental apparatus used in this work has been described previously.[10] An AEI ES 200 ESCA spectrometer has been modified to achieve a base pressure of $\sim 3 \times 10^{-10}$ Torr. Various single-crystal disks of tungsten may be mounted at the sample position as shown in Fig. 1. For

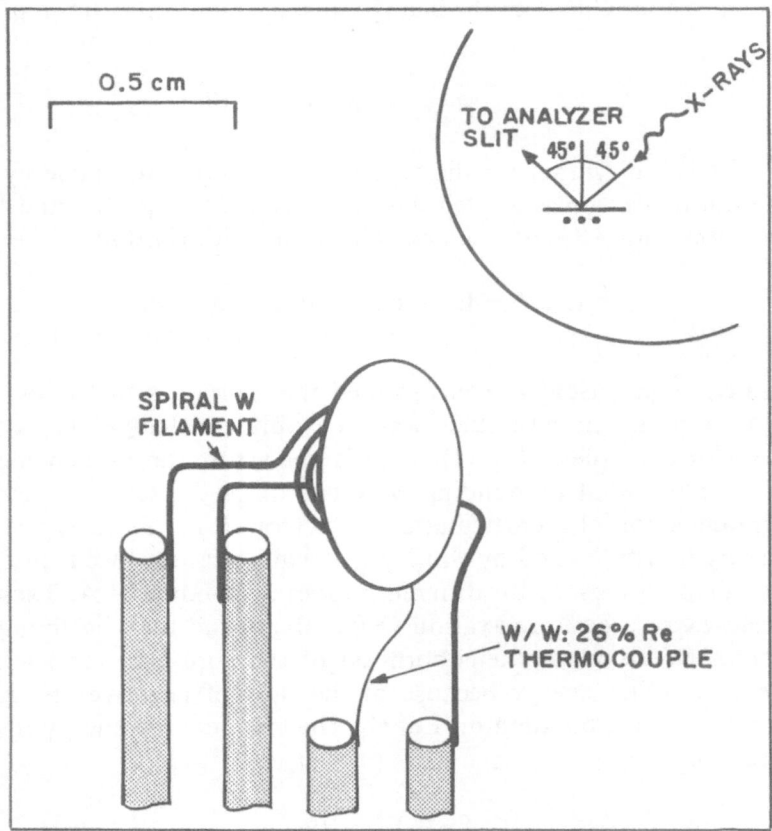

Fig. 1. Support assembly for tungsten single crystal disk in ultrahigh-vacuum ESCA apparatus. Crystal is cleaned at 2600°K by electron bombardment from spiral tungsten filament.

cleaning, a tungsten single crystal can be heated to 2600°K by electron bombardment from the spiral tungsten filament, mounted behind. The crystal can also be heated to ~1100°K by radiation from the spiral filament for careful flash desorption measurements, *in situ*. The crystal is initially heat treated in O_2 for removal of impurity carbon, and following several flashes at 2600°K in vacuum, the surface cleanliness is verified by XPS. The crystal may be cooled to either 300°K or ~80°K prior to adsorption experiments, and its degree of cleanliness again verified by XPS. All gas-adsorption and -desorption measurements are made using a quadrupole mass spectrometer which is frequently calibrated against a Bayard-Alpert gauge of specified sensitivity. To avoid gas-phase contamination processes, the Bayard-Alpert gauge is not operated during adsorption experiments.

$MgK_{\alpha 1,2}$ radiation (1253.6 eV) filtered through a thin aluminum window, is used to excite photoemission at an incidence angle of 45 degrees as shown in the upper right corner of Fig. 1. The small daily variation in intensity of the X-ray source is followed by observation of the W(4f) emission intensity from the clean substrate.

All photoelectron energies are referenced to the Fermi energy of the substrate crystal which is directly measured. The variation of the electron energy scale (i.e., spectrometer work function) was less than 0.05 eV over the course of these experiments, as judged by repeated measurement of the W(4f) lines.

3 RESULTS

3.1 Chemisorption of Oxygen by W(100) and W(111)

The integrated O(1s) intensity versus O_2 exposure at 300°K has been measured for W(100) as shown in Fig. 2. Also, in this figure, the rise in work function of W(100) during O_2 adsorption is plotted from the data of Madey.[17] It is seen that the rate of change of coverage with exposure falls off as exposure increases. We arbitrarily define the monolayer ($\theta_0 = 1.0$) at an O_2 exposure of 3 x 10^{15} O_2 molecules/cm^2 incident. It is well known that slow oxygen adsorption occurs at higher O_2 exposures[18,19], presumably as rearrangement of the surface layer occurs, so that any definition of a monolayer is necesarily arbitrary. Beyond the coverage defined as $\theta_0 = 1.0$, a 10-fold increase in exposure produces only a 20 percent increase in O(1s) intensity (approximate coverage) and a work function fractional change of only approximately 6 percent of the total change, indicating that our definition of $\theta_0 = 1.0$ is probably valid to within 10 to 20 percent.

By comparing the monolayer O(1s) photoyield to the total no-loss photoyield from the clean tungsten crystal, it is possible to estimate the

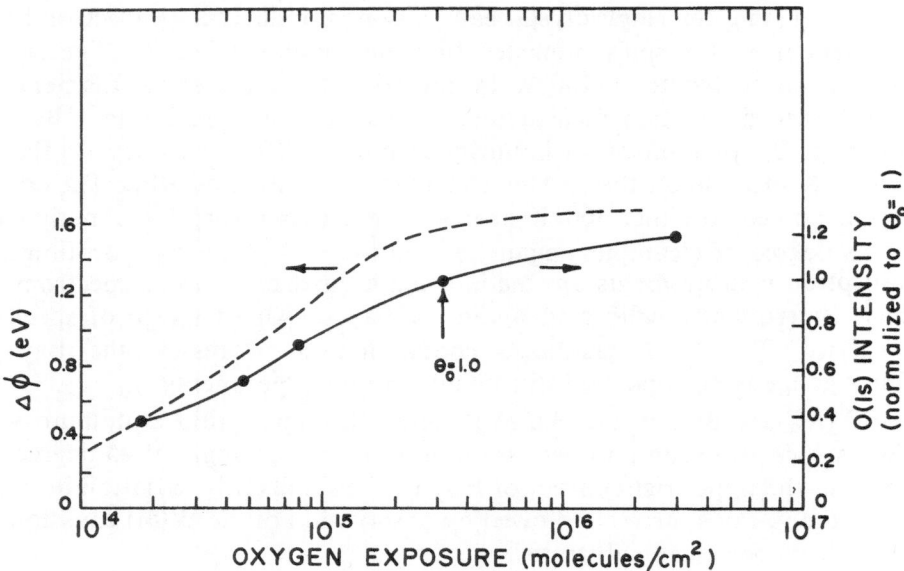

Fig. 2. Integrated O(1s) intensity and work function change[17] for oxygen chemisorption on W(100). The point at 3×10^{15} O_2 molecules/cm^2 exposure is defined as one monolayer ($\theta_0 = 1.0$).

coverage of oxygen on the crystal in the monolayer. An expression for calculating the coverage has been previously derived[6]; viz.,

$$\sigma = \frac{Y_m}{Y_s} \frac{N \mu_s \rho \lambda \cos \theta}{\mu_m M} \tag{5}$$

where σ = monolayer coverage (atoms/cm^2); N = Avogadro's number; μ_s = substrate (W) X-ray mass adsorption coefficient for MgK$_\alpha$ radiation [2330 cm^2/g][20]*; λ = electron attenuation length in substrate [13 x 10^{-8} cm for electrons of energy 800 to 1250 eV][21]; ρ = density of substrate W [19.3 g/cm^3]; θ = 45 degrees = angle of incidence of X-rays; μ_m = adsorbate mass absorption coefficient [2530 cm^2/g(O)][20]; M = atomic weight of adsorbate atom [16 g/g atom]; Y_m = no-loss photoyield from the monolayer; Y_s = no-loss photoyield in all photolines from the clean substrate.

For the system O_2 + W(100), Madey[17] has previously made an absolute measurement of the monolayer coverage using a calibrated molecular beam having a known angular intensity distribution, so that this system provides an excellent opportunity to check on the validity of the photoemission method for measuring surface coverage. The results obtained by the two methods are given in Table I, and it is seen that there is excellent agreement. The W(100) surface contains 1.0×10^{15} tungsten atoms/cm^2 in its uppermost

*The values in Ref. 20 are in very good agreement with recent theoretical calculations of X-ray absorption coefficients by J. H. Scofield (private communication).

Table I. Monolayer Coverage of Oxygen on W(100)

Method	Coverage
Molecular beam[17]	$(1.14 \pm 0.17) \times 10^{15}$ oxygen atoms/cm^2
ESCA—no-loss photoyield ratio	1.08×10^{15} oxygen atoms/cm^2

layer of atoms, and to within the accuracy of these two measurements, this same number of oxygen atoms are present on the surface at $\theta_0 = 1$.

The O(1s) photoemission spectra for increasing O_2 exposure are shown in Fig. 3. Spectrum (a) is a measure of the initial contamination of the crystal by oxygen-containing species during initial crystal cooling in vacuum. Spectrum (d) corresponds to one monolayer of oxygen. It can be seen that two O(1s) peaks are resolved at all coverages. These have been fitted to Gaussian curves having a full width at half maximum (fwhm) of 1.30 eV, as shown. This peak shape best fits the high-kinetic-energy edge of the O(1s) spectra. The existence of two peaks in the O(1s) spectrum for O(ads) + W(100) is in qualitative agreement with the data of Bradshaw et al.[11], but there are differences in the details of the spectral development versus coverage.

A tail on the low-kinetic-energy side of the peaks is present at all coverages and may be due to inelastic processes in the layer, or perhaps to small amounts of oxygen in states other than the two represented by the Gaussian peaks. The entire O(1s) peak area has been used for the coverage calculation previously discussed.

It may be seen that the Gaussian associated with state 2 shifts very little with oxygen coverage, whereas the state 1 curve shifts to lower binding energy as oxygen coverage increases. These shifts, along with the work-function change, for W(100) are plotted versus oxygen coverage in Fig. 4. The work-function dependence on coverage was determined from comparison of the curves in Fig. 2, assuming that the O(1s) integrated intensity is proportional to oxygen coverage, θ_0. It can be seen that both oxygen states shift with coverage, but by different amounts, and that neither state shifts directly with the increase in ϕ. If the photoemission process sensed the average work function of the surface, ϕ, then, as the diagram at the right in Fig. 4 indicates, E_B^f should decrease directly as ϕ increases, and both peaks should shift together with the slope of the $\Delta\phi$ versus θ_0 curve. This behavior is *not* observed, which indicates that the application of the coverage-dependent average work function is not appropriate for determining binding energy shifts as coverage changes. A possible double-dipole sheet model to explain the behavior of the O + W(100) two-state system is discussed later.

It is well known that extensive rearrangement of an oxygen-covered

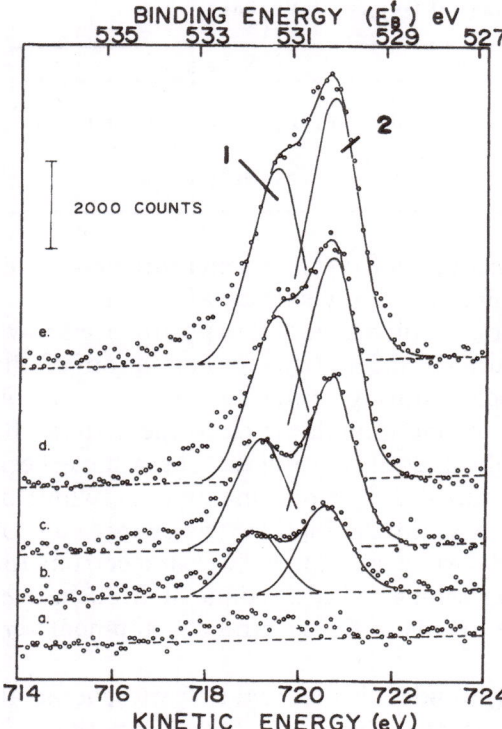

Fig. 3. O(1s) ESCA spectra for oxygen chemisorption on W(100) at 300°K: (a) background spectrum; (b) $\theta_o = 0.4$; (c) $\theta_o = 0.7$; (d) $\theta_o = 1.0$; (e) $\theta_o = 1.2$. All spectra, 6 scans, 1 sec/channel/scan, 0.1 v/channel. The two states are fitted with Gaussian curves having fwhm = 1.30 eV.

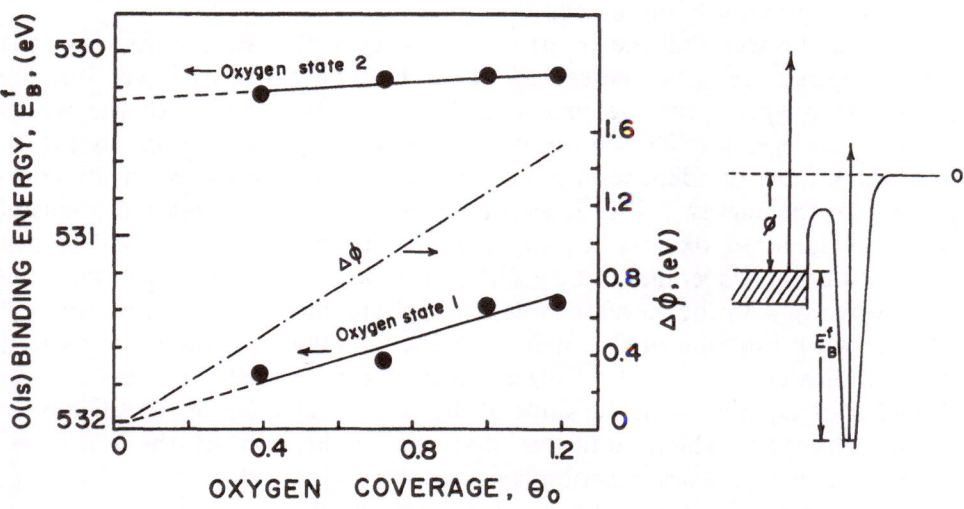

Fig. 4. O(1s) binding energy shifts (relative to Fermi edge) compared to average work-function change during oxygen chemisorption by W(100). At the right is a schematic of the method used to measure the O(1s) binding energy relative to the Fermi edge, E_B^f

W(100) surface may be achieved by heating to near $1000°K$.[22] This is accompanied by a substantial drop in work function[23], and has been interpreted as being due to oxygen penetration into the tungsten lattice giving dipole inversion. In Fig. 5, the effect of heating on the O(1s) spectrum is observed at two oxygen coverages. The open circles define the spectrum prior to heating. Upon heating to $\sim900°K$, the doublet structure disappears and a single broad O(1s) peak is produced. It appears that the two peaks tend to coalesce by means of shifts in opposite directions, as judged from comparison of points on the wings of each peak. This would imply that the two states tend to become equivalent as might be expected for extensive oxygen penetration into the tungsten lattice.

As a further example of the influence of structural factors on the O(1s) spectra for oxygen chemisorbed on tungsten single-crystal surfaces, several spectra for oxygen chemisorption on W(111) are shown in Fig. 6. Spectrum (a) is a measure of the small contamination by oxygen-containing substances on the crystal during cooling in vacuum prior to O_2 adsorption. Spectra (b) through (e) are produced by increasing exposure to O_2 (g), and the O(1s) peak grows in intensity with very little shift in peak-center energy. The O(1s) spectra are not resolved into multiple states; however, by using two 1.30-eV fwhm Gaussian peaks, as for W(100), the spectra can be fitted as shown, although it should be emphasized that the fit to two peaks is shown only to illustrate that a two-state model is possible for O + W(111) also. The splitting of the two states shown here is ~0.9 eV, compared with

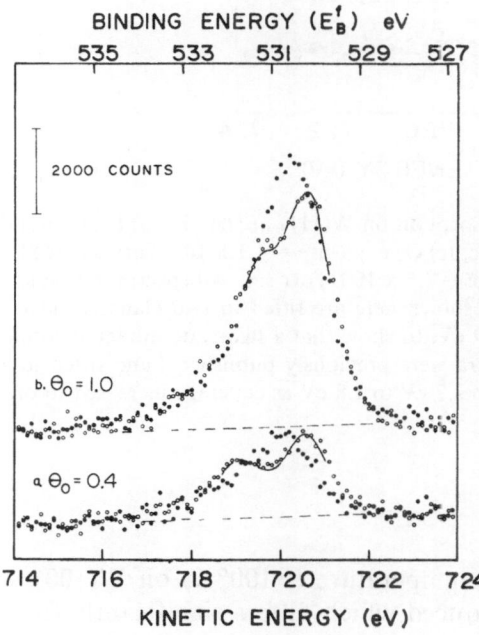

Fig. 5. Effect of heating on O(1s) spectra for oxygen chemisorbed on W(100) at two coverages: (a) heated to $915°K$; (b) heated to $> 1000°K$. Open circles—before heating; filled circles—after heating.

1.2 to 1.7 eV for W(100). The spectra for oxygen adsorption on W(111) resemble that achieved on W(100) by heating and rearranging the surface layer.

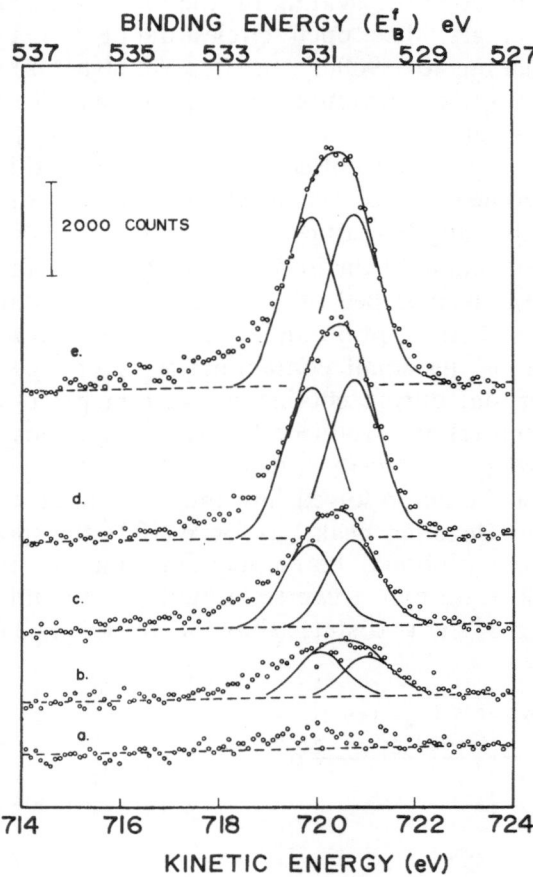

Fig. 6. O(1s) ESCA spectra for oxygen chemisorption on W(111) at ~80°K: (a) background spectrum; (b) O_2 exposure = 5.2 x 10^{-7} Torr sec; (c) O_2 exposure = 1.1 x 10^{-6} Torr sec; (d) O_2 exposure = 2.2 x 10^{-6} Torr sec; (e) O_2 exposure = 4.5 x 10^{-6} Torr sec. All spectra, 6 scans, 1 sec/channel/scan, 0.1 v/channel. The spectra shown here are fitted to two Gaussian peaks having fwhm = 1.30 eV and separated by ~0.9 eV, to show that a two-state model as found on W(100) is possible on W(111). These spectra were previously published[10] and fitted to a single Gaussian of variable fwhm ranging from 2.7 eV to 1.8 eV as coverage increased to one monolayer.

3.2 Chemisorption of CO by W(100)

When CO is chemisorbed at low temperature (~100°K) on W(100), a combination of adsorbed states is produced which differs significantly from those produced by adsorption at 300°K.[24-31] The low-temperature layer was originally named "virgin" CO because of its unique property to convert to strongly bound β-CO upon heating to ~300°K.[27] The virgin-CO state exhibits a negative dipole outward as judged from the work-function increase associated with this state. Early views that low-temperature adsorption produced a single virgin-CO state have been modified, and it is now postulated from a number of studies[8,28,31] that the low-temperature CO layer contains a small quantity of β-CO, as well as α-CO, in addition to the virgin CO which is the precursor to the thermally activated formation of additional β-CO. Although structures for the virgin- and β-CO states are not known, the α-CO state has been studied by reflection infrared spectroscopy[32,33] and electron-energy-loss spectroscopy[34] and is spectroscopically similar to the sp-hybridized CO ligands in metal carbonyls. In Fig. 7 a simplified picture of the three CO states is given for purposes of acquainting the reader with the general character of these states; the structure of virgin CO and β-CO shown is speculative but consistent with present XPS information, as will be seen. Gomer[35] has recently suggested that virgin CO passes through a β-CO precursor state upon thermal conversion to β-CO. Also, β-CO exhibits a number of desorption substates on W(100).

We first examine the O(1s) spectrum for CO adsorption at ~80°K on W(100), as seen in spectrum (a) of Fig. 8. A broad band of photoemission is seen. Upon heating to 275°K [spectrum (b)], a dramatic change in band shape occurs and new emission is seen at higher kinetic energy above ~720 eV. This is a result of virgin CO to β-CO conversion. A residual peak, centered near 717.8 eV is due to α-CO. On further heating to 550°K [spectrum (c)], α-CO desorbs and β-CO O(1s) emission is enhanced at ~720.2 eV. The O(1s) spectra shown here have been fitted with a number of Gaussian peaks (fwhm = 1.50 eV) which are consistent with desorption experiments discriminating α-CO and various β-CO substates (see Fig. 11 and accompanying discussion). Errors of several tenths of an electron volt may be present in these assignments. Note that virgin CO and $β_1$-CO have very similar O(1s) energies. The same experiment was repeated while following C(1s) spectral changes, as shown in Fig. 9. Spectrum (a) is a background spectrum for clean W(100) in the C(1s) region and is curved and steeply sloped because of its overlap with energy-loss peaks originating from W(4d) emission. Spectrum (b) is the 80°K virgin-CO layer and exhibits a broad band of emission intensity. On heating to 275°K, a sharp C(1s) peak at 967.8 eV is produced due to virgin CO to β-CO conversion. Further heating to 550°K is accompanied by a slight loss of intensity near 964.6 eV and by further enhancement of the sharp C(1s) peak at 967.8 eV. Thus, both the O(1s) and C(1s) behavior of chemisorbed CO indicate that extensive chemical changes occur when a low-temperature layer of CO is activated by heating, as was seen previously on polycrystalline tungsten using ESCA.[8]

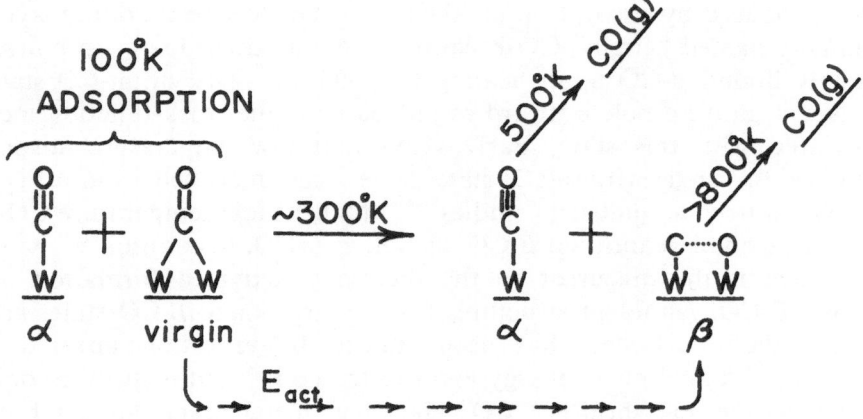

Fig. 7. Simplified picture of possible bonding structures for chemisorbed CO on Tungsten. The α-CO state has been studied spectroscopically and is postulated to involve sp-hybridized carbon. The structure of virgin-CO is speculative, and a number of β-CO species exist together with possible large variations in the strength of the C ... O bond.

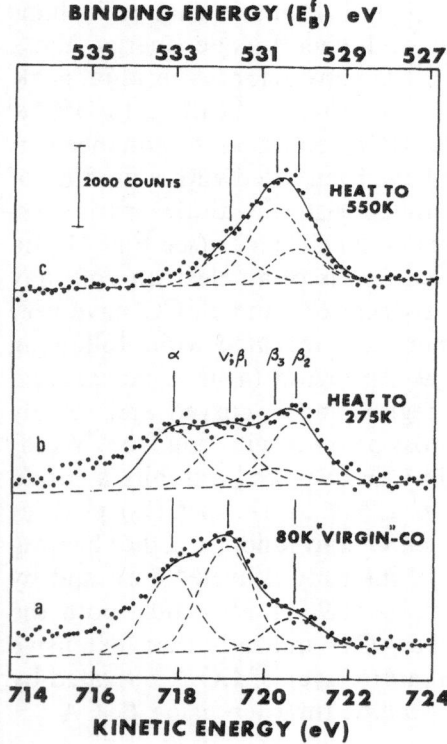

Fig. 8. O(1s) spectra for monolayer CO chemisorption on W(100) at ~80°K (exposure = 2.4 x 10^{-4} Torr sec); (b) virgin-CO → β-CO conversion by heating to 275°K; (c) desorption of α-CO by heating to 550°K. (Further conversion to β-CO states is also observed.) All spectra, 6 scans, 1 sec/channel/scan, 0.1 v/channel. All Gaussian curves shown have fwhm = 1.50 eV.

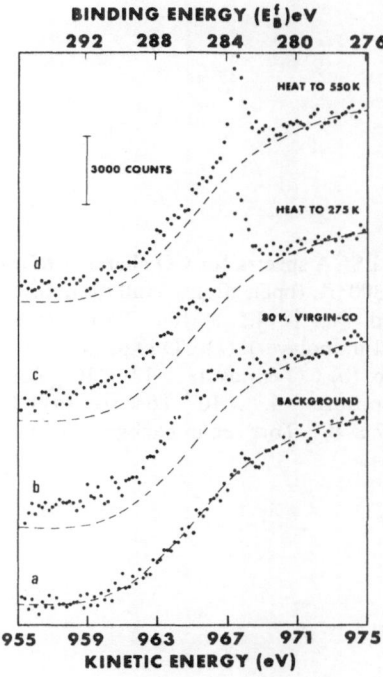

BINDING ENERGY (E_B^f)eV

292   288   284   280   276

HEAT TO 550 K

3000 COUNTS

HEAT TO 275 K

d

80 K, VIRGIN-CO

c

BACKGROUND

b

a

955   959   963   967   971   975

KINETIC ENERGY (eV)

Fig. 9. C(1s) spectra for monolayer CO chemisorption on W(100) at 80°K, and for various heat treatments: (a) background spectrum; (b) CO monolayer adsorption at ~80°K (exposure = 2.4 x 10^{-4} Torr sec); (c) partial virgin-CO → β-CO conversion by heating to 275°K; (d) desorption of α-CO by heating to 550°K. (Further conversion to β-CO is also observed.) All spectra, 12 scans, 1.07 sec/channel/scan, 0.2 v/channel.

When CO is adsorbed on W(100) at ~300°K, a different distribution of binding states occurs than when the adsorption temperature is at 80°K. The 300°K-state distribution consists of α-CO and β-CO, and displays a significantly different O(1s) photoemission spectrum from that produced by α + virgin CO at ~80°K. The development of O(1s) spectra for increasing coverage of CO at 300°K is shown in Fig. 10, where the open circles correspond to monolayer CO coverage and the spectra consisting of dark points are for partial CO coverages. In the early stages of adsorption, O(1s) photoemission intensity is concentrated in the β-CO region, centered near 720.2 eV. Only in the later stages of adsorption do O(1s) features attributable to α-CO develop below ~718 eV. This behavior is consistent with reflection infrared spectroscopic studies on polycrystalline tungsten, where α-CO was detected only during the final stages of adsorption[32,33], as α-CO fills sites left following β-CO adsorption.

It was of interest to selectively desorb various CO species from the CO monolayer by using a flash desorption procedure which is interrupted at various points following desorption of various fractions of the monolayer This is shown in Fig. 11 where displayed on the right are the various partia flash desorption experiments; various stages of CO desorption are indicated by the cross-hatched portions of the total $\alpha + \beta_1 + \beta_2 + \beta_3$ CO-desorptio traces. The full-coverage distribution of CO-desorption states shown here

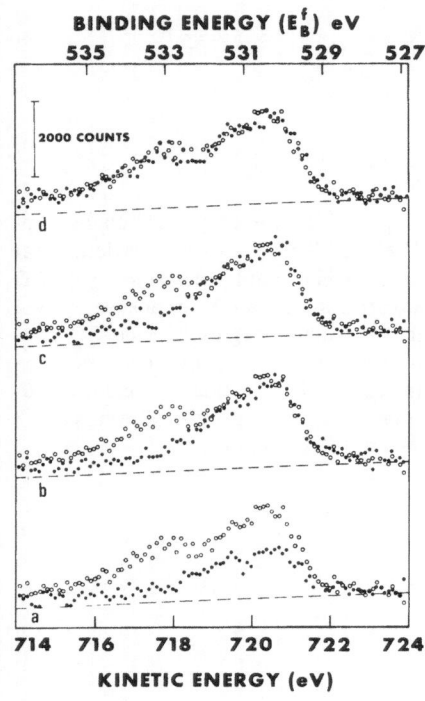

Fig. 10. O(1s) ESCA spectra for CO chemisorption on W(100) at 300°K; (open circles—full monolayer in background $P_{CO} = 4.5 \times 10^{-6}$ Torr; closed circles—partial monolayer): (a) CO exposure = 5.0 $\times 10^{-7}$ Torr sec; (b) CO exposure = 1.9 x 10^{-6} Torr sec; (c) CO exposure = 4.7 x 10^{-6} Torr sec; (d) CO exposure = 3.7 x 10^{-4} Torr sec in background, P_{CO} = 4 x 10^{-7} Torr.

in excellent agreement with the results of Clavenna and Schmidt[36] on W(100). On the left is shown a comparison of the O(1s) photoemission spectrum for monolayer CO with the spectrum obtained after each partial desorption experiment at the right. It can be seen that α-CO desorption (curve a) results in a loss of intensity in the lowest O(1s) kinetic energy regions. β_1-CO desorption causes loss of intensity on the low-kinetic-energy edge of the broad β-CO peak near 719 eV (curve b); β_2-CO desorption (curve c) causes loss of O(1s) intensity on the *high*-kinetic-energy edge of the broad β-CO peak, and in curve d we see that the remaining β_3-CO is associated with a central O(1s) peak located near 720.2 eV kinetic energy. These experiments indicate that the ESCA method is useful for distinguishing the various CO chemisorbed species, as is flash desorption. The desorption of specific binding states is accompanied by losses of photoemission intensity in various specific regions of the spectrum. The O(1s) spectrum for the CO remaining behind as β_3-CO is very similar to that obtained by heating 0.4 monolayer of oxygen to ~900°K, as shown in Fig. 5 and may indicate that CO in this β_3-CO state is dissociated[37,38] or that O ... C interactions are too weak to be detected by ESCA chemical shifts, even though this β_3-state desorbs as CO.

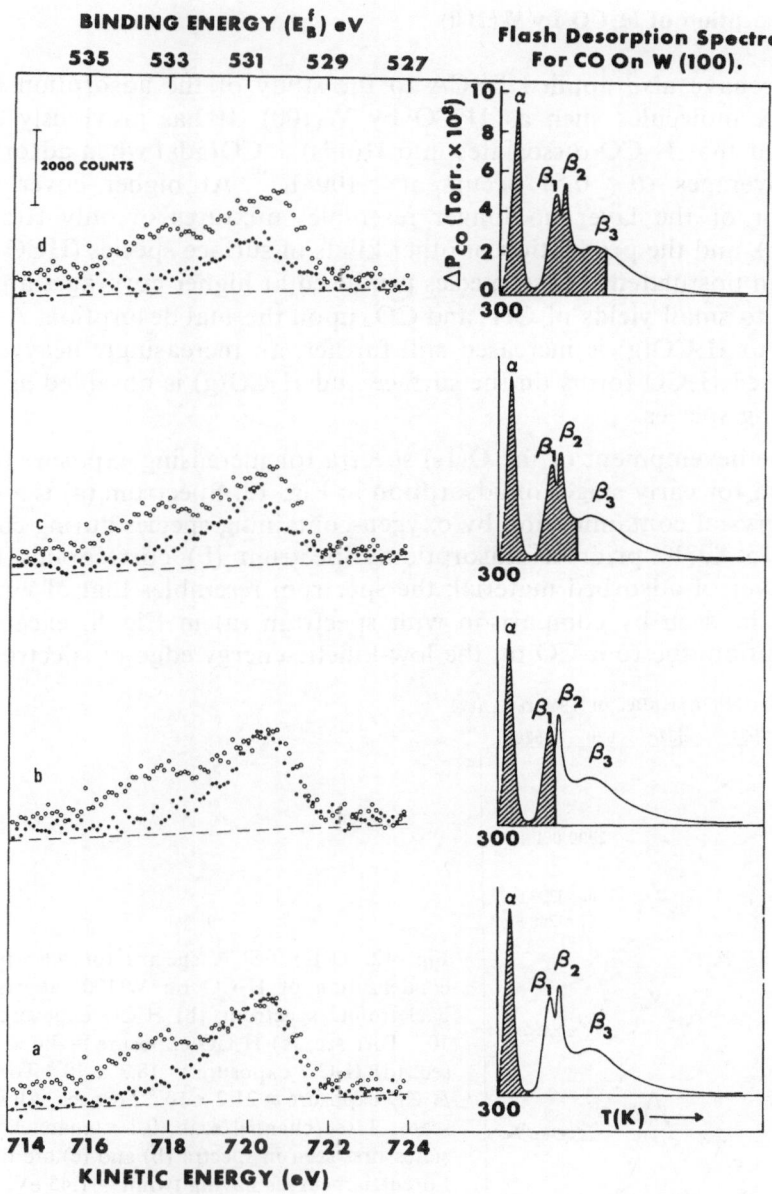

Fig. 11. Partial desorption of CO from W(100). Right portion of figure shows flash desorption spectrum for CO monolayer from W(100); cross-hatched regions depict partial desorption experiments in which thermal programming was interrupted. Left portion compares full coverage O(1s) spectrum (open circles) with spectrum following each partial desorption experiment. (a) Following α-CO desorption; (b) following $\alpha + \beta_1$-CO desorption; (c) following $\alpha + \beta_1 + \beta_2$-CO desorption; (d) following $\alpha + \beta_1 + \beta_2 +$ partial β_3-CO desorption. All spectra, 6 scans, 1 sec/channel/scan, 0.1 v/channel.

3.3 Adsorption of H₂CO by W(100)

We have also applied ESCA to the study of the adsorption of more complex molecules such as H_2CO by W(100). It has previously been established that H_2CO dissociates into H(ads) + CO(ads) when adsorption to low coverages ($\theta \lesssim 0.5$) occurs at ~100°K.[39] At higher coverages, the behavior of the layer no longer resembles mixtures of only H(ads) and CO(ads), and the production of other kinds of surface species (HCO? CH_2?) has been postulated. These species produced at higher coverages are known to lead to small yields of CH_4 and CO_2 upon thermal desorption. As the exposure to $H_2CO(g)$ is increased still further, an increasingly heavy layer of condensed H_2CO forms on the surface and $H_2CO(g)$ is observed as a major desorbing species.

The development of the O(1s) spectra for increasing exposure to H_2CO is shown for early stages of adsorption in Fig. 12. Spectrum (a) is a measure of the crystal contamination by oxygen-containing species during cooling *in vacuo* to 80°K prior to adsorption. Spectrum (b) corresponds to ~0.5 monolayer of adsorbed material; the spectrum resembles that of virgin CO, as may be seen by comparison with spectrum (a) in Fig. 8, except that a contribution due to α-CO on the low-kinetic-energy edge of spectrum (a) in

E_B^f, BINDING ENERGY RELATIVE TO FERMI LEVEL (eV)

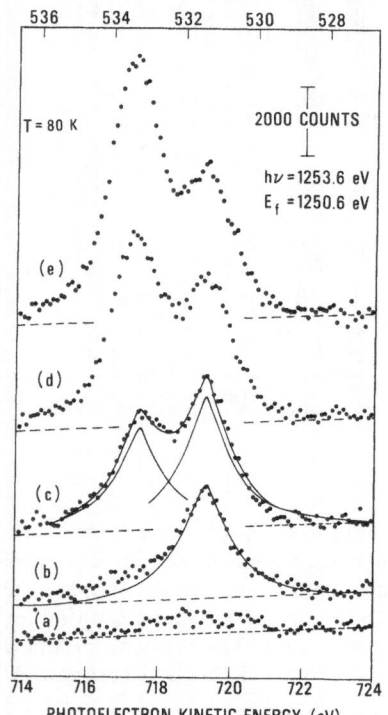

Fig. 12. O(1s) ESCA spectra for adsorption and condensation of H_2CO on W(100) at ~80°K: (a) background spectrum; (b) H_2CO exposure = 2.4 x 10^{-6} Torr sec; (c) H_2CO exposure = 4.6 x 10^{-6} Torr sec; (d) H_2CO exposure = 18.3 x 10^{-6} Torr sec; (e) H_2CO exposure = 21.2 x 10^{-6} Torr sec. All spectra, 6 scans, 1 sec/channel/scan, 0.1 v/channel. The two states first seen in spectra (b) and (c) are fitted with Lorentzian curves having fwhm = 1.45 eV.

PHOTOELECTRON KINETIC ENERGY (eV)

Fig. 8 (717.8 eV) is absent; presumably, H(ads) competes on W(100) for sites capable of α-CO adsorption when H_2CO is the adsorbate, as is found for coadsorption of hydrogen and CO on W(100).[40] Upon increasing the H_2CO exposure, a new O(1s) peak initially centered at about 717.5 eV develops and grows in intensity, while the virgin-CO peak reaches a maximum intensity. This low-kinetic energy peak may initially be due to either HCO(ads) or to H_2CO(ads). However, upon continual exposure to H_2CO(g), the peak grows extensively and shifts to lower kinetic energy while broadening, as shown in Fig. 13 [where spectrum (a) is identical to spectrum (e) in Fig. 12)]. Associated with the appearance and growth of this peak, we find that increasing amounts of H_2CO(g) are desorbed by heating from 80°K to ~600°K. In the later stages of condensation of H_2CO into a thick layer, it can be seen that O(1s) photoemission from the underlying virgin-CO layer is attenuated due to energy-loss processes in the thick H_2CO overlayer.

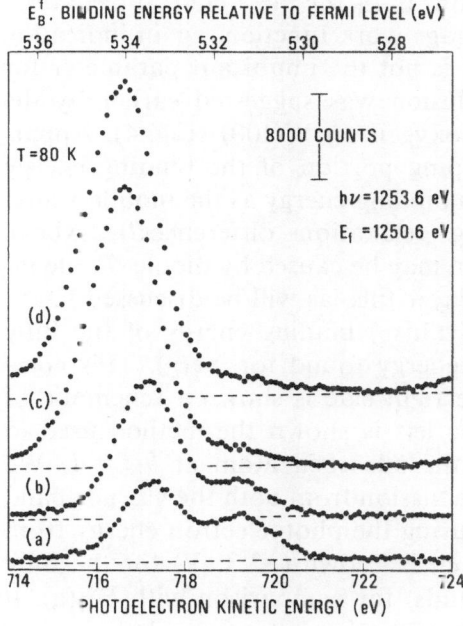

E_B^i. BINDING ENERGY RELATIVE TO FERMI LEVEL (eV)

536   534   532   530   528

8000 COUNTS

T = 80 K

$h\nu$ = 1253.6 eV

E_f = 1250.6 eV

(d)

(c)

(b)

(a)

714   716   718   720   722   724

PHOTOELECTRON KINETIC ENERGY (eV)

Fig. 13. O(1s) ESCA spectra for condensation of H_2CO on W(100) at ~80°K: (a) corresponds to spectrum e in Fig. 12; (b) H_2CO exposure = 15.4 x 10^{-6} Torr sec; (c) H_2CO exposure = 63.5 x 10^{-6} Torr sec; (d) H_2CO exposure = 91.8 x 10^{-6} Torr sec. All spectra, 6 scans, 1 sec/channel/scan, 0.1 v/channel.

3.4 Physical Adsorption of Xe by W(111)

The previous examples have been concerned with the application of ESCA to chemisorption and to mixtures of chemisorbed and condensed substances. In this section the use of ESCA to study a physically adsorbed atom is briefly discussed. The system Xe + W(111) is an inherently simple adsorption system where multiple binding states do not coexist.[41,10]

Several Xe($3d_{5/2}$) spectra for increasing xenon coverage on W(111) at 120°K are shown in Fig. 14. The $3d_{5/2}$ line is the most intense xenon line for MgK$_\alpha$ excitation. Spectrum (a) is the background spectrum, while spectra (b) through (f) are for coverages ranging from $\theta_{Xe} = 0.05$ to $\theta_{Xe} = 1.0$, where saturation of the surface occurs. The data are fitted to a Lorentzian curve having a constant full width at half maximum of 1.24 eV. The center of the xenon peak has been located for a number of coverages, and a small coverage-dependent shift has been observed, as shown in Fig. 15. Also plotted in Fig. 15 are measurements[41] of the work-function change as a function of θ_{Xe} on W(111) as well as separate measurements of desorption energy[41] for the xenon, as determined by flash desorption. It can be seen that all of these measurements indicate that below $\theta_{Xe} \simeq 0.2$, the behavior is different than for $\theta_{Xe} > 0.2$. We believe this is due to the initial adsorption of mobile xenon on extraneous high-energy sites on the W(111) crystal[10,41]; for $\theta_{Xe} > 0.2$, adsorption probably occurs on perfect (111) regions.

In Fig. 15 it can be seen that the shift in E_B^f for the Xe($3d_{5/2}$) peak does not follow the measured change in average work function, again indicating that the average work-function change is not the important parameter for calculating line shifts. The same conclusion was suggested earlier by the measured shifts of the O(1s) lines for oxygen on W(100) (Fig. 4). Apparently, below $\theta_{Xe} \simeq 0.2$, the steeply sloping portion of the binding energy curve reflects small differences in xenon binding energy as the mobile xenon atoms fill high-energy sites (causing relaxation differences?). Above $\theta_{Xe} \simeq 0.2$, the slope of the linear portion may be caused by dipole-dipole interactions which increase as the xenon layer fills, as will be discussed later.

It is of interest to compare the Xe($3d_{5/2}$) binding energy of the fully covered xenon layer with the binding energy found for Xe(g). This comparison is illustrated in Fig. 16. On the right side is shown a schematic of the $3d_{5/2}$ energy level for Xe(g). On the left is shown the method used to determine E_B^o (ads) for this level for the adsorbed atom at $\theta_{Xe} = 1$. We measure E_B^f directly by measuring photoemission from both the valence band (locating the Fermi edge) and by measuring the photoelectron energy from the adsorbed atom. Since the value of ϕ_T is known (=3.3 eV) for the fully covered surface[41], we may estimate E_B^o(ads) for comparison with E_B^o(g). It must be remembered that in this diagram, the final state core-hole ions in the two cases are *different;* relaxation effects due to the tungsten surface are present in the adsorbed case. Thus, $\Delta E_B^o = -2.1$ eV is a reflection of both initial- and final-state effects. The initial-state contribution is probably small since the charge on the physisorbed xenon atom is of order $\delta^+ = +0.03e$ or less, which would give $\Delta E_B^o \simeq +0.1$ eV.[10] The cross-hatched region represents schematically the range in E_B^o found in this study as coverage is varied.

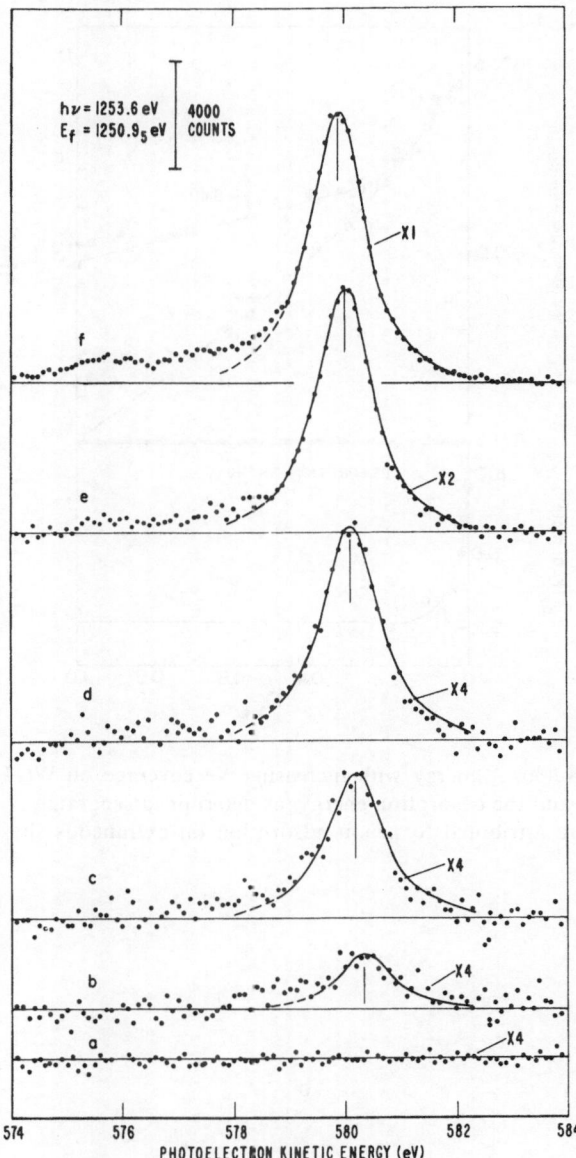

Fig. 14. Xe(3d$_{5/2}$) ESCA spectra for xenon adsorbed on W(111) at ~120°K: (a) clean W(111); (b) xenon exposure = 8.5 x 10^{-8} Torr sec; (c) xenon exposure = 4.8 x 10^{-7} Torr sec; (d) xenon exposure = 7.0 x 10^{-7} Torr sec; (e) xenon exposure = 1.3 x 10^{-6} Torr sec; (f) xenon exposure = 5.0 x 10^{-6} Torr sec. All spectra, 4 scans, 1 sec/channel/scan, 0.1 v/channel. Curves are fitted to a Lorentzian line shape having fwhm = 1.24 eV. A small satellite peak near 576 ev is due to the Xe(3d$_{3/2}$) line excited by MgK$\alpha_{3,4}$ radiation. Kinetic energies can be converted to binding energies (relative to the Fermi level) by subtracting abscissa from 1250.95 eV.

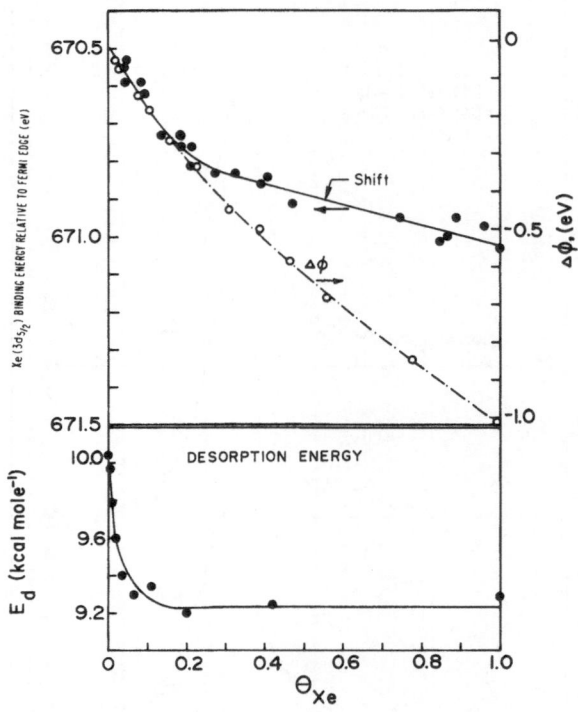

Fig. 15. Shift of Xe($3d_{5/2}$) energy with increasing Xe coverage on W(111) at 120°K, work function behavior, and the desorption energy, as determined separately.[41] The rapid changes below $\theta_{Xe} \sim 0.2$ are attributed to xenon adsorption on extraneous sites or defects on the W(111) crystal.

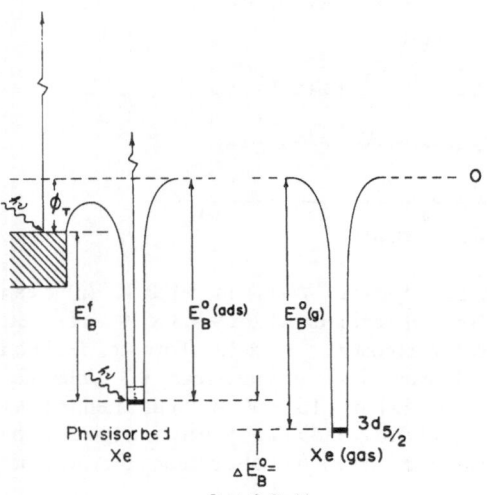

Fig. 16. Schematic of energy level for xenon physisorption on W(111).

4 DISCUSSION

4.1 Sensitivity of ESCA for Detection of Adsorbed Species

In the four examples chosen for presentation here, it is seen in each case that ESCA is sensitive to small fractions of a monolayer of adsorbed species. The limit of sensitivity for detecting an adsorbate atom is dependent upon a number of factors such as adsorbate and substrate mass absorption coefficients and the electron attenuation length, as shown in Eq. (5). In addition, the signal-to-noise ratio and the signal-to-background ratio in the digital data directly affect the detection sensitivity. The signal-to-noise ratio may, of course, be improved by increasing data-accumulation time, or by decreasing instrument resolution. However, even in ultrahigh vacuum, the possibility of extensive contamination of the surface during long data-accumulation periods must be considered as a limiting factor for sensitivity enhancement. As currently applied in our laboratory, ESCA has a working sensitivity of ~0.05 monolayer. A practical sensitivity improvement factor of 2 to 4 may be anticipated using a variety of procedures such as background vacuum improvement, increased X-ray intensity, and photoelectron takeoff angles near grazing.

4.2 Use of ESCA to Measure Absolute Surface Coverages of Adsorbed Species

At present, there are few absolute methods for determining surface coverages of adsorbed species on single-crystal adsorbent surfaces. The calibrated molecular-beam method employed by Madey[17,42] has been applied only to the systems $O_2 - W(100)$ and $H_2 + W(100)$. Various estimates of monolayer coverages are sometimes made by using LEED measurements to determine unit cell dimensions for an ordered adsorbed overlayer; these measurements detect only the coherent surface lattice, and the method requires an assumption concerning the molecular identity of the adsorbed species (atoms? molecules?). The use of conventional thermal desorption procedures for absolute coverage determinations suffers from serious uncertainties in gauge calibration, system pumping speeds, and system wall effects.

Viewed in this light, therefore, ESCA may be considered an important direct tool for absolute surface coverage determination, although incomplete knowledge of accurate electron escape depths is at present a limitation in the use of this method. In addition, the accurate measurement of the no-loss photoyields is difficult in the presence of a large inelastic background. Another problem is evaluation of the relative contribution of satellite features due to multielectron processes. It is difficult to assess the accuracy of various assumptions about these factors.

4.3 Coverage-Dependent Shifts in Core-Level Photoemission Peaks

The chemisorption system O(ads) + W(100) and the physisorption system Xe(ads) + W(111) offer examples of widely different kinds of adsorbed species where the work function-versus-coverage behavior has been thoroughly studied. As shown in Figs. 4 and 15, there is evidence in both cases that the core-hole photoemission peaks do not shift directly in accordance with changes in average work function. It is true that in the cases shown here, the *direction* of the peak shift is the same as would be expected from the sign of $\Delta\phi$; consideration of the local electrostatic effect of neighboring charged species could also produce such shifts.

The case of O(ads) + W(100) is of particular interest since the two O(1s) states discriminated by ESCA shift by different amounts as coverage increases. This may be understood qualitatively (using initial state arguments only) by considering the electrostatic potential in a double-dipole-sheet model as shown schematically in Fig. 17. Here we have approximated a possible structure for O(ads) + W(100) by placing oxygen atoms on body-centered tungsten atoms (state 2) and on uppermost tungsten atoms (state 1). This assignment is consistent with a larger amount of final-state relaxation in state 2, due to the more extensive coordination of state 2 oxygen atoms with tungsten atoms, leading to a lower O(1s) binding energy, as observed. In the bottom portion of Fig. 17, an arrangement of

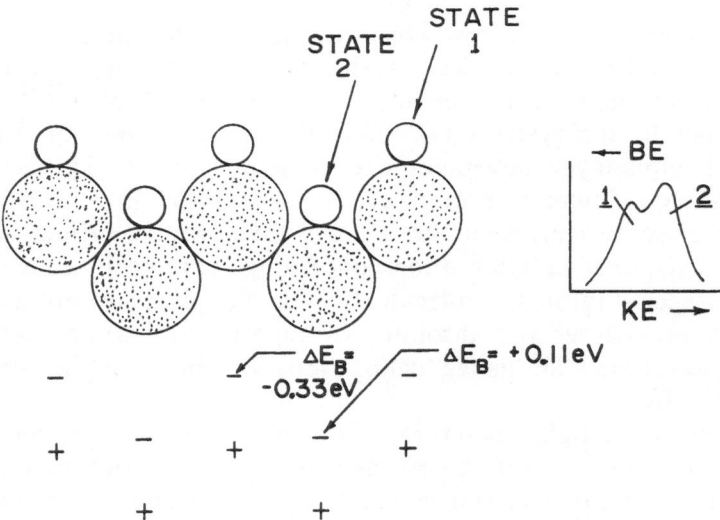

Fig. 17. Schematic of possible dipole sheet model for oxygen chemisorption on W(100). The cross section through the W(100) surface is cut along a W(110) plane. The schematic arrangement of surface dipoles is shown, along with calculated shifts in E_B due to the interaction of the different oxygen states with nearest neighbor dipoles; in the plane of the (110) cut.

$W^{\delta+}$ $O^{\delta-}$ dipoles corresponding to the structural model is shown. The dipole moment, μ, has been calculated from the relation

$$\mu = \frac{\Delta\phi \, (eV) \times 10^{18}}{2 \, \pi \, \sigma \times 300} \, \text{(Debye)} \tag{6}$$

where the assumption is made that the two states have equal dipole moment. Assuming a dipole length $= r_0 + r_w \simeq 1.87\text{Å}$, the appropriate ionic charge $\delta = qe = 0.094e$. Using the spacing shown, it may be seen that the potential at the oxygen atom in state 1 is strongly affected by two state 2 dipoles, and for this local interaction, $\Delta E_B(1) = -0.33$ eV. However, state 2 is buried inside the state 1 dipole sheet, and its two nearest neighbor dipoles cause $\Delta E_B(2) = +0.11$ eV. The sign of the state 2 shift, as well as its magnitude, is dependent on the relative location of the two dipole sheets; this calculation is given only for illustrative purposes, since we do not know definitely the location or the magnitude of the dipoles for oxygen atoms chemisorbed on W(100). If the structure and the appropriate dipole moments were known, then summation over all neighbors, as shown in Eq. (1) should be carried out to calculate ΔE_B.

The illustrative model in Fig. 17 is consistent with the tendency of the two states to become equivalent upon heating, owing to oxygen penetration into the lattice, leading to a reduction in ϕ. Recent work[43] involving the study of the angular distribution of O^+ produced by electron impact on O(ads) + W(100) has suggested bonding configurations similar to the two states suggested in Fig. 17.

Other models are also possible, however, to explain the two O(ls) states. For instance, one might invoke the mixed adsorption of O_2 molecules and oxygen atoms. This model is unlikely to be valid since additional ESCA studies at 80°K on W(100) have shown that the same two adsorption states are produced in similar ratio, whereas one might expect an enhancement of molecular adsorption at low temperatures, by analogy with other chemisorption systems. Also, thermal desorption of O_2 is not observed from tungsten.[18]

For the case of the physisorption of xenon on W(111), above $\theta_{Xe} \simeq 0.2$, a linear change in ΔE_B with coverage is observed. The slope of this portion of the curve is $+0.3$ eV/monolayer. Using Eq. (1) and summing over 42 xenon neighbor dipoles having $\delta^+ = 0.03e$ gives $\Delta E_B = +0.3$ eV/monolayer, in agreement with the slope of the linear portion of the curve in Fig. 15. The value of δ^+ was determined from work-function changes during adsorption, assuming charge transfer between adsorbate and substrate. Assumptions were also made concerning the packing in the xenon monolayer and the dipole length.[10]

4.4 Magnitude of Chemical Shift for Adsorption

As illustrated by Fig. 16, it is instructive to calculate the total magnitude of the chemical shift for an adsorbed molecule compared with that of the *same* molecule in the gas phase. This has been done for the limited range of data now available for adsorption on tungsten and the values are reported in Table II. Also, kinetic measurements of the activation energy for desorption, E_d, of the particular species are shown where available. The data are arranged in order of increasing strength of interaction with the solid.

A note of caution with respect to the values of ΔE_B^o reported in Table II is offered. The absolute binding energies for core-level electrons of adsorbate species are taken as $E_B^o = E_B^f + \phi$; these quantities are compared with values of E_B^o for gas-phase species to compute the chemical shift ΔE_B^o. There is an uncertainty in ΔE_B^o arising from the uncertainty in the choice of the appropriate value of ϕ in particular cases (clean-surface work function?; work function of covered surface?; constant (but unknown) inner potential?). The magnitude of this correction is generally agreed to be of the order of typical-surface work functions, and the values assumed in the present cases are tabulated in Table II. The fundamental quantity in measurements of the core-level binding energies for adsorbed species appears to be E_B^f, the binding energy relative to the Fermi level.

Several features of the data in Table II should be emphasized. First, the chemical shift, ΔE_B^o, is negative in each case, independent of the adsorbate charge state as deduced from work-function changes measured during adsorption. Second, the value ΔE_B^o becomes more negative as the interaction energy with the solid increases, or as E_d increases. These two general trends are not influenced by the uncertainty associated with the use of a proper work-function value for the various examples in Table II. These two observations together imply that final-state relaxation effects dominate for adsorbed species of these types and that the magnitude of the relaxation term in Eq. (4) reflects the strength of bonding in a crude sense. Furthermore, it may be seen for CO that adsorption causes similar chemical shifts to occur in *both ends* of this heteronuclear molecule; the small differences seen in Table II are within the range of error involved in estimating peak positions where overlapping states exist.

4.5 Conclusions Regarding the Use of ESCA for Studying Adsorbed Species

The results reviewed in this work indicate that ESCA can presently be used for qualitative analysis of surface layers with a practical sensitivity in

Table II. Summary of Chemical Shifts for Adsorption

Gas	State	Charge State	E_d, eV	ϕ, eV	E_B^f	Level	ΔE_B^o
Xe	Physisorbed on W(111); θ=1	$(Xe)^{\delta+}$	0.4	$3.3^{(a)}$	671.03	Xe($3d_{5/2}$)	−2.1
	Ion implanted in tungsten[45]	—	—	$4.5^{(b)}$	670.8	Xe($3d_{5/2}$)	−1.1
	Ion implanted in silver[13]	—	—	$4.0^{(c)}$	669.62	Xe($3d_{5/2}$)	−2.8
H_2CO	Condensed heavy layer on W(100)	$(H_2CO)^{\delta+}$	—	$\sim3.7^{(a)\,39}$	534.1	O(1s)	−1.6
CO	α-CO on W(100)	$(CO)^{\delta+}$	0.8	$4.6^{(d)}$	533.0	O(1s)	−5.0
					286.2	C(1s)	−5.4
	"Virgin" CO on W(100)	$(CO)^{\delta-}$	—	$4.6^{(d)}$	531.7	O(1s)	−6.3
					285.4	C(1s)	−6.2
	β-CO on W(100)	$(CO)^{\delta-}$	2.5 − 4.0	$4.6^{(d)}$	531.7 to 530.1	O(1s)	−6.3 to −7.9
					283.0	C(1s)	−8.6

(a) Using work function for fully covered surface.
(b) Using average work function for polycrystalline tungsten.
(c) Using photoelectric work function for polycrystalline silver.
(d) Using clean W(100) work function.

Xe(g): E_B [Xe($3d_{5/2}$)] = 676.4 eV.
CO(g): E_B [O(1s)] = 542.6 eV; E_B [C(1s)] = 296.2 eV.
$H_2CO(g)^*$: E_B [O(1s)] = 539.42 eV.

the range 1 to 5 percent of a monolayer. ESCA, when applied carefully, can also be used for making reasonable quantitative estimates of absolute surface coverages, and as a tool for checking surface cleanliness prior to experiments. Its sensitivity, compared with Auger spectroscopy, is poorer at present by perhaps a factor of 10 to 100. However, electron-impact-induced Auger spectroscopy itself suffers from the disadvantage of electron-beam damage[44] to the adsorbed layer, and some of the adsorbed species discussed here could not be easily characterized using Auger spectroscopy because of this factor. Also, the dependence of Auger line energy and line shape is a complex subject, not well understood at present.

In ESCA one is dealing with a spectroscopy in which *both* the initial-state total energy and the final-state total energy combine to determine the measured binding energies for core levels. Unfortunately, chemical effects cause variations in both the initial and final states. It appears that in both physisorption and chemisorption, final-state extra-atomic relaxation factors are dominant. In some cases, coverage-dependent *shifts* may be attributed to reasonable initial-state arguments involving interaction of the emitting atom with neighbor dipoles.

The combination of factors contributing to chemical shifts provides a severe limitation to the usefulness of ESCA as a primary tool for investigation of the *structure* of adsorbed species. Correlation of binding energies with adsorbate valence state is confounded by the overriding effect of the solid on the final-state core-hole ion.

However, ESCA used in conjunction with techniques such as thermal desorption mass spectrometry, work-function measurement, and other surface probes, is a useful tool for further characterization of surface species. ESCA should continue to prove valuable in model-system studies of surface catalysis by single-crystal substrates, especially if combined with other surface-measurement techniques.

REFERENCES

1. Siegbahn, K., et al., *ESCA: Atomic, Molecular, and Solid State Structure Studied by Means of Electron Spectroscopy*, Almqvist and Wiksells, Uppsala, Sweden (1967).
2. Siegbahn, K., et al., *ESCA Applied to Free Molecules*, North Holland, Amsterdam, The Netherlands (1969).
3. Atkinson, S. J., Brundle, C. R., and Roberts, M. W., *J. Electron Spectroscopy and Related Phenomena*, **2**, 105 (1973).
4. Brundle, C. R., and Roberts, M. W., *Proc. Roy. Soc. (London)*, **A331**, 383 (1972).
5. Brundle, C. R., and Roberts, M. W., *Chem. Phys. Letters*, **18**, 380 (1973).
6. Madey, T. E., Yates, J. T., Jr., and Erickson, N. E., *Chem. Phys. Letters*, **19**, 487 (1973).
7. Madey, T. E., Yates, J. T., Jr., and Erickson, N. E., *Electron Fis. Apli. (Spain)*, **17**, 190 (1974).
8. Yates, J. T., Madey, T. E., and Erickson, N. E., *Surface Sci.*, **43**, 257 (1974).
9. Madey, T. E., Yates, J. T., Jr., and Erickson, N. E., *Surface Sci.*, **43**, 526 (1974).
10. Yates, J. T., Jr., and Erickson, N. E., *Surface Sci.*, **44**, 489 (1974).
11. Bradshaw, A. M., Menzel, D., and Steinkilberg, M., *Jap. J. Appl. Phys.*, Suppl. 2, Pt. 2, 841 (1974).
12. Citrin, P. H., and Thomas, T. D., *J. Chem. Phys.*, **57**, 4446 (1972).
13. Citrin, P. H., and Hamann, D. R., *Chem. Phys. Letters*, **22**, 301 (1973).
14. Davis, D. W., and Shirley, D. A., *Chem. Phys. Letters*, **15**, 185 (1972).
15. Shirley, D. A., *Chem. Phys. Letters*, **16**, 220 (1972).
16. Shirley, D. A., *Chem. Phys. Letters*, **17**, 312 (1972).
17. Madey, T. E., *Surface Sci.*, **33**, 355 (1972).
18. King, D. A., Madey, T. E., and Yates, J. T., Jr., *J. Chem. Phys.*, **55**, 3236 (1971).
19. King, D. A., Madey, T. E., and Yates, J. T., Jr., *J. Chem. Phys.*, **55**, 3247 (1971).
20. Henke, B. L., in *Proceedings of the International Conference on Inner Shell Ionization Phenomena*, Atlanta, Georgia, 1972.
21. Tarng, M. L., and Wehner, G. K., *J. Appl. Phys.*, **44**, 1534 (1973).

22. Tracy, J. C., and Blakeley, J. M., *Surface Sci.*, **13**, 313 (1968).
23. Zingerman, Ya. P., and Ishchuk, V. A., *Fiz. Tverd. Tela.*, **8**, 2999 (1966) [*Sov. Phys. Solid State*, **8**, 2394 (1967)], as well as many earlier articles.
24. Klein, R., *J. Chem. Phys.*, **31**, 1306 (1959).
25. Ehrlich, G., and Hudda, F. G., *J. Chem. Phys.,* **35**, 1421 (1961).
26. Ehrlich, G., and Hudda, F. G., *J. Chem. Phys.*, **34**, 39 (1961).
27. Swanson, L. W., and Gomer, R., *J. Chem. Phys.*, **39**, 2813 (1963).
28. Gomer, R., *Discussions Faraday Soc.*, **41**, 14 (1966).
29. Engel, T., Gomer, R., *J. Chem. Phys.*, **50**, 2428 (1969).
30. Menzel, D., *Ber. Bunsenges. Physik. Chem.*, **41**, 591 (1968).
31. Yates, J. T., Jr., and King, D. A., *Surface Sci.*, **38**, 114 (1973).
32. Yates, J. T., Jr., and King, D. A., *Surface Sci.*, **30**, 601 (1972).
33. Yates, J. T., Jr., Greenler, R. G., Ratajczykowa, I., and King, D. A., *Surface Sci.*, **36**, 739 (1973).
34. Propst, F. M., and Piper, T. C., *J. Vacuum Sci. Technol.*, **8**, 53 (1967).
35. Gomer, R., *Jap. J. Appl. Phys.*, Suppl. 2, Pt. 2, 213 (1974).
36. Clavenna, L., and Schmidt, L. D., *Surface Sci.*, **33**, 11 (1972).
37. King, D. A., Goymour, C. G., and Yates, J. T., Jr., *Proc. Roy. Soc. (London)* A**331**, 361 (1972).
38. Goymour, C. G., and King, D. A., *J. Chem. Soc., Faraday Trans. I*, **69**, 749 (1973).
39. Yates, J. T., Jr., Madey, T. E., and Dresser, M. J., *J. Catalysis*, **30**, 260 (1973).
40. Yates, J. T., Jr., and Madey, T. E., *J. Chem. Phys.*, **54**, 4969 (1971).
41. Dresser, M. J., Madey, T. E., and Yates, J. T., Jr., *Surface Sci.*, **42**, 533 (1974).
42. Madey, T. E., *Surface Sci.*, **36**, 281 (1973).
43. Czyzewski, J. J., Madey, T. E., and Yates, J. T., Jr., *Phys. Rev. Letters*, **32**, 777 (1974); also, *Surface Sci.*, (April, 1975).
44. Madey, T. E., and Yates, J. T., Jr., *J. Vacuum Sci. Technol.*, **8**, 525 (1971).
45. Erickson, N. E., unpublished work.
46. Carroll, T. X., and Thomas, T. D., to be published.

DISCUSSION on Paper by J. T. Yates

WALTER: How does roughness of the surface affect the absolute calibration by ESCA of surface coverage by adsorbates?

YATES: We have not measured the surface roughness of our W(100) crystal, so I cannot answer this question from an experimental basis. Usually, however, polished crystals such as this have roughness factors near unity. Since, with this method for estimating surface coverage, one is measuring no energy loss-photoyield ratios from two thin layers (adsorbate layer and the outermost layers of adsorbent atoms which yield no-loss electrons), it may be that even with rough surfaces or surfaces of rather small particles, the method would still work in principle. It would be interesting to check this in the case of chemisorption on supported adsorbents as a function of particle size distribution. One must of course be sure that the average adsorptive properties of the particles do not change over the range of particle sizes studied.

WALTER: What is the physical meaning of a mass adsorption coefficient of a monolayer? It seems that the situation here is similar to that in ellipsometry, where, empirically, the dielectric constant of a monolayer is found to be very similar to that of the bulk, although the concept is theoretically unsound.

YATES: We have used Henke's tabulation of X-ray mass absorption coefficients in our calculations. These values are based on X-ray absorption experiments with bulk materials. However, very recently, J. H. Scofield has calculated X-ray absorption cross sections for many elements from first principles of atomic physics. For tungsten and oxygen, the theoretical mass absorption coefficients (units cm^2/g) obtained from Scofield's theoretical cross sections agree to within a few percent of Henke's tabulated experimental values. This suggests that the difference in cross section for a bulk atom and an isolated atom is small, and implies that a surface adsorbate atom exhibits a cross section similar to that in the bulk or in the gas phase.

SMITH: Dr. Yates, your dipole moment for chemical shifts of the two oxygen states on tungsten was quite interesting. Did you assign different dipole moments per atom for the two layers of adsorbed oxygen? We have found, for hydrogen on tungsten, that the dipole moment is rather sensitive to the location of the adatom. Our results for the variation with adatom position are not quantitatively accurate for small metal-adatom spacings. However, the sensitivity of the dipole moment to adatom location is likely a valid result. One might also look for such a variation in the oxygen-tungsten system.

YATES: The example shown in Fig. 17, which postulates an "outer" and an "inner" state of adsorbed oxygen on W(100), was presented only as an illustration of a possible surface configuration. For illustrative purposes, we assumed the simplest situation—namely that both species possessed the same dipole moment of 0.85D. This average dipole moment is consistent with the measured work-function increase (1.6 eV) and the measured monolayer coverage (1 x 10^{15} atoms-cm^2). The coverage-dependent shifts in E_B^f due to nearest-neighbor dipole interactions are calculated in this example only to show that different shifts may occur owing to differing electrostatic environments for the two kinds of oxygen. A proper calculation would consider the true dipole moment for each oxygen state, and would sum over the neighbor dipoles throughout the entire dipole layer.

MENZEL: We have done similar XPS measurements, in conjunction with UPS and LEED [D. Menzel, *J. Vacuum Sci. Technol.* (to be published), and work cited there], and in most respects obtained very good agreement with the results reported by Dr. Yates. The agreement is particularly striking for the absolute values of core-level binding energies, e.g., the O(1s) level of oxygen and CO on different tungsten faces, where we agree with Yates et al., to 0.1 eV.

A surprising result is found for the values just mentioned. For O(1s) of oxygen adsorbed on W(100), W(110), and polycrystalline tungsten, and on Ru(001), Ag, Cu, Mo, and Ni, essentially the same binding energy is found by different authors[1]; the same value is found for β-CO on W(110) and W(100) (in case of existence of several levels, this statement applies to the most strongly adsorbed species). In view of the large differences of the adsorption bonding in these cases, this seems to stress the overriding effect of relaxation as opposed to electrostatic shifts.

For more weakly bound CO [e.g., α-CO on W, CO on Ru(001)], the O(1s) energy is larger by about 1.8 eV than for oxygen or strongly bound CO. This can probably also be understood in terms of smaller relaxation for the O(1s) standing away from the surface in the—probably linearly bound—weakly adsorbed CO, while β-CO on tungsten either lies down on the surface or is even dissociated.

Dr. Yates has mentioned only the XPS measurements of adsorbate core levels. We have also observed shifts of substrate core levels upon adsorption. For example, adsorption of oxygen on W(100) causes the 4f levels of tungsten to shift in part by about 0.3 eV.

IMELIK: In your interpretation, nearly 20 percent of the monolayer of xenon is built up by adsorption on high-energy tungsten sites (defects, etc.). This proportion seems to be rather high for a good, atomically clean monocrystal, being in the range of what we may estimate for polycrystalline samples.

YATES: The 20 percent number represents an upper limit estimate, and is a number of low accuracy. In a more extensive paper concerned with ESCA studies of xenon adsorbed on W(111), we have shown that impurity adsorption may produce high energy xenon adsorption sites [*Surface Sci.,* **42**, 533 (1974)]. Since our W(111) surface contains about 0.05 monolayer of impurities adsorbed during cooling, we may also explain part of this effect away in this manner. However, HEED studies by others have shown that in some crystals extensive fractions of polished crystal surfaces may consist of extraneous planes or disordered regions.

IMELIK: Is there any connection between this defect's concentration and the state of CO chemisorption (α, virgin)?

YATES: The observation that α-CO forms only in the later stages of completion of the CO monolayer suggests that induced heterogeneity is responsible. CO is immobile on tungsten at 300°K, and if α-CO were due to adsorption on extraneous sites, it should appear on the surface at low coverages. Both ESCA and reflection infrared spectroscopy have shown that this is not the case. [Reference: *Surface Sci.,* **30**, 601 (1972); **36**, 739 (1973)].

OUDAR: In (100)W with oxygen adsorbed, did you observe some correlation between LEED pattern and the phase transformation observed by ESCA?

YATES: We are not able to study LEED behavior in conjunction with the ESCA experiments. However, Dr. Menzel and his co-workers have observed that treating an oxygen covered W(100) surface to 1250°K results in conversion from a (1 x 4, 4 x 1) quarter-order streaky pattern to a (1 x 2, 2 x 1) streaky pattern. [Reference: *Jap. J. Appl. Phys.*,, Suppl. 2, Pt. 2, 841 (1974).

EMMETT: I have two questions. In the first place, did you mention the sticking coefficient of xenon on your tungsten crystals?

YATES: We have measured the sticking coefficient for xenon on W(111) over a coverage range from $\theta_{Xe} = 10^{-2}$ to $\theta_{Xe} = 1$. The sticking coefficient is constant over the entire range and has a value of 0.5 or higher. Inaccuracy in gauge calibration prevents accurate measurement of the absolute sticking coefficient in these experiments. [Reference: *Surface Sci.,* **42**, 533 (1974)].

EMMETT: My second question is: Did you measure the amount of carbon monoxide chemisorbed per square centimeter of surface?

YATES: Comparison of the monolayer O(1s) ESCA spectra for oxygen on W(100) and CO on W(100) indicates a saturation CO coverage of $\sim 8 \times 10^{14}$ CO/cm^2 for CO adsorbed at 300°K.

PLUMMER: The objective of ESCA surface studies is to deduce the binding state of an adsorbate from the shift of the core-level binding energy relative to the gas-phase binding energy. Many workers have proceeded on the assumption that the shift will be a monatonic function of the binding energy. Your spectra for the β_1, β_2, and β_3 states of CO show that the core binding energy for the β_3 state is greater than that for the β_2 state, even though the β_3 state is bound to the surface more tightly than the β_2 state.

YATES: Yes, this was surprising to us at first. However, keeping in mind the shifts that occur when an oxygen layer on W(100) is heated, it is not surprising that similar effects are seen for β_2- and β_3-CO, particularly if surface penetration or lattice rearrangement are involved at the elevated temperatures associated with β-CO desorption.

NEWNS: It may be worth pointing out that the final-state shifts discussed by Dr. Yates may be regarded as initial-state shifts, if one does a better job on the initial state, using something like an optical model to describe the potential seen by the electron before photoexcitation. In this way the electron level is shifted upward, the ionization of sodium on tungsten revealing this effect even within the initial state.

YATES: Are the results calculated in this model of the right order of magnitude?

NEWNS: The shift calculated is the same as in the final-state approach within a certain approximation. Shifts of the order of 2 eV are explicable, but shifts of the order of 5 eV in both the carbon and oxygen core levels in α-CO on W(100) are less easily understood. For some purposes, the initial-state interpretation is the more appropriate.

THE INTERACTIONS AND REACTIONS OF ALKENES ON SINGLE-CRYSTAL SURFACES OF PLATINUM AND A COMPARISON OF SOME REACTIONS IN HOMOGENEOUS AND HETEROGENEOUS CATALYSIS

*Terence A. Clarke, Ian D. Gay, and Ronald Mason**

School of Molecular Sciences
University of Sussex
Brighton, England

ABSTRACT

A summary is given of our knowledge of the structures and reactions of organic ligands bonded to low valent metals in coordination complexes. Oxidative-addition reactions of surface metal atoms to carbon-hydrogen bonds of chemisorbed alkenes are shown to have analogies with reactions in homogeneous systems; the dissociation of alkenes on nickel and platinum surfaces, which is demonstrated by photoelectron spectroscopy, follows metallation of a carbon-hydrogen bond and the intermolecular elimination of H_2 or, in the case of halo-alkenes, HX. The stoichiometric reactions depend on the substrate geometry as evidenced by a comparative study of sorption on clean Pt(100) and Pt(111) surfaces.

*To whom correspondence should be addressed.

The relation between the concept of coordinatively unsaturated complexes and their catalytic activity is briefly examined for species based on d^8 and d^{10} metal ions and is speculatively extended to a consideration of electronic and structural changes at surfaces during chemisorption.

1 INTRODUCTION

There have been striking developments, during the past decade, in our understanding of the chemical bond between unsaturated organic molecules and transition metals, of the reactions of coordinated ligands and, closely related to that point, in the mechanisms of homogeneously catalyzed reactions such as hydrogenation, isomerization and hydroformylation. Three main themes have emerged which may relate, *mutatis mutandis*, to problems connected with our understanding of heterogeneously catalyzed reactions, or at least, to views of the equilibrium structures of metal-ligand surfaces.

First, the geometries of unsaturated molecules coordinated to metals in low or relatively low oxidation states are often substantially different from those of uncoordinated ligands.[1,2] Illustratively, the trigonal stereochemistry of uncoordinated olefinic carbon atoms becomes tetrahedral while the linear geometry of alkyne carbon atoms evolves towards trigonal symmetry; carbon disulfide becomes nonlinear; interatomic distances may change by up to 0.3 Å or so. These geometrical consequences of coordination can be rationalized[1] by the simple Dewar-Chatt bonding model[3] coupled with the kind of considerations that Walsh enunciated[4]; more quantitative theories of the organometallic bond, which are based implicitly on the second-order Jahn-Teller effect, have also been provided[5] and amount to an understanding that major stereochemical changes are consequent upon a net electron transfer from bonding ligand molecular orbitals to their antibonding molecular orbital counterparts. And, when we consider simple workfunction changes ($\triangle\phi$ = -0.5 to -1.0 V) which follow, say, chemisorption of alkenes and alkynes on platinum, we can infer that these coordinated hydrocarbons have more carbonium ion rather than carbanion character[6] and that their geometries will be close to those of their ground states and not to quasi-excited states.[1,5,7] Some implications of this remark to the reactions of coordinated ligands are discussed below.

Second, the reactivity of transition-metal complexes of simple ligands has been connected with the concept of coordinative unsaturation. Just as much heterogeneous catalytic activity is shown by transition metals with eight, nine, and ten valence electrons, so many studies over the past few years have concentrated on the homogeneous catalytic activity of complexes based on d^8 and d^{10} metal ions. Representative of these complexes are

$[(C_6H_5)_3P]_3RhCl$ and $[(C_6H_5)_3P]_3Pt$ which have received the most attention so far as the mechanisms of their catalysis of a number of reactions are concerned.[8,9] Wilkinson's early suggestion that the 16-electron structure, $(PhP)_3RhCl$, was completely dissociated in solution to the 14-electron structure, $(Ph_3P)_2RhCl$, and that the hydrogenation of olefins could then be represented by the scheme

$$RhClL_2 \quad \underset{}{\overset{H_2}{\rightleftharpoons}} \quad H_2RhClL_2$$

olefin $\updownarrow$ ↓ olefin

$$(olefin)\,RhClL_2 \quad \overset{H_2}{\longrightarrow} \quad RhClL_2 + paraffin$$

relied on the concept of the high reactivity of the coordinatively unsaturated, 14-electron complex, $RhClL_2$. This proposal has been scrutinized carefully in the very recent literature[10] and there is no need for us to further emphasize the problems other than to reinforce the now generally accepted view that, from an energetic standpoint, there is relatively facile interconversion between complexes having 14-, 16-, and 18- valence electron configurations, e.g.,

$$Pt(PPh_3)_2 \rightleftharpoons Pt(PPh_3)_3 \rightleftharpoons Pt(PPh_3)_4$$
$$Pt(PPh_3)_3 \rightleftharpoons Pt(PPh_3)_2 + PPh_3 \overset{+L}{\rightleftharpoons} Pt(PPh_3)_2L$$

where L might be an alkene or an alkyne. Within a localized bond description, these structural relationships can be understood by recollecting that d-p promotion energies, for metals of d^8 and d^{10} configurations of 0 or +1 formal oxidation state, are very small[11] so that there is easy interconversion between different hybridization states. Again, we shall want to see whether the idea of flexible coordination geometry, determined largely by the electronic nature of the coordinating ligands, has some value in thinking about chemisorption bonds at the surfaces of transition metals.

Third, and closely related to the concept of coordinative unsaturation, are those reactions of transition-metal complexes which are given the generic description of oxidative-addition reactions.[12] Illustratively, the reactions of the iridium (I) complex $(Ph_3P)_2Ir(CO)X$ (where X is a halogen) can be considered. Two general schemes of reaction can be envisaged:

(1) $(Ph_3P)_2Ir(CO)X + O_2 \rightarrow (Ph_3P)_2Ir(CO)X(O_2)$

where a simple covalent molecule is associatively coordinated to the metal, and

(2) $(Ph_3P)_2Ir(CO)X + CH_3I \rightarrow (Ph_3P)_2Ir(CO)(X)(CH_3)I$

where there is dissociation of the covalent molecule on binding. In both cases, however, we think of the iridium (I) complex as having increased its coordination number from 4 to 6, with a concomitant change of the formal oxidation state to $+3$ (d^6). Most oxidative-addition reactions can be understood in terms of the driving force of the reaction being the achievement of a stable metal configuration: for example, t_{2g}^6, e_g^0 (Co(III), Rh(III), Ir(III), Pt(IV)) in a quasi-octahedral configuration, t_{2g}^0, e_g^0 for oxidative addition reactions leading, say, to Ti(IV) and V(V) complexes or t_{2g}^3, e_g^0 for six-coordinate Cr(III) species generated from Cr(I) complexes.

The examples mentioned earlier relate to extramolecular covalent species associating themselves in the coordination sphere of the complex. A further example, of direct concern to us later in connection with the reactions of halo-alkenes on platinum surfaces, is the characterization of the reactions of vinyl halides and other halo-alkenes with the zero-valent complex mentioned earlier, $Pt(PPh_3)_4$, or the similar molecules, $Pt(PPh_2Me)_4$.[13] By summary, the reactions can be correlated with the ease of carbon-halogen bond cleaveage,

with subsequent elimination of acetylene from the vinyl platinum (II) complex to give the bisphosphine platinum dihalide.

But some of the more interesting reactions observed in phosphine complexes are the intramolecular oxidative-addition reactions[14] in which metallation of an aromatic carbon-hydrogen bond occurs,

with, in some cases, the subsequent reductive elimination of the hydride. The mechanism of this reaction is thought to be the nucleophilic attack of the metal fragment on the hydrogen of the carbon-hydrogen bond which, for steric reasons, is brought into relatively close proximity (ca. 2.5 Å) with the metal. We shall show that such reactions may have precise analogues on metal surfaces.

2 THE STOICHIOMETRIES OF ORGANOMETALLIC SURFACES

It is obvious that, before one considers in any serious way what the crystallography of a metal-ligand surface might be, one needs to know whether chemisorption of an organic molecule is associative, dissociative, or a combination of binding events. For some time Auger spectroscopy has been held to be a major technique for the determination of surface composition but it is clear that, in the case of chemisorption of organic molecules, it has little value—this form of electron spectroscopy causes major dissociation or fragmentation of the surface organometallic species, a simple example being the time dependence of the chlorine Auger spectrum following chemisorption of vinyl chloride on platinum.[15]. We have shown recently[16,17] that X-ray photoelectron spectroscopy has quantitative value in assessing the nature of the surface species that result from chemisorption. The essence of the method is obvious. Associatively chemisorbed organic molecules should have core electron-emission energies and intensities essentially identical to those of their gas-phase spectra: the escape depth of photoelectrons originating from ligands constituting up to a monolayer coverage of a substrate is effectively zero and the ratio of integrated emission intensities from chemically nonequivalent atoms will be well approximated by their respective ionization cross sections. With these points in mind, the data in Tables I through IV, together with the data we have reported for the chemisorption of fluoroalkenes on the Pt(100) surface, can be summarized as follows:

1. The theoretical fluorine-to-carbon peak-intensity ratio was obtained from the ratio of the tabulated[18] atomic cross sections. For chlorine, both the 2s and 2p electrons contribute appreciably to the atomic cross section, and the chlorine (2p)-to-carbon (1s) peak-intensity ratio cannot be obtained directly. Moreover, the differing angular distribution of 2s and 2p photoelectrons makes this ratio difficult to calculate with any degree of certainty and we have accordingly used the observed gas-phase ratio given in Table 4. The gas-phase data have not been corrected for the dependence of analyzer sensitivity on the kinetic energies of the photoelectrons since acceptable agreement for the F(1s)/C(1s) ratio is obtained without such a correction. Escape-probability considerations should be taken into account[19]; in the case of chlorine (2p) level, however, the kinetic energy correction is small and the use of the gas-phase value should not introduce any significant error.

Table I. Chemisorption of Fluoro-Alkenes and -Alkynes at a Pt(111) Surface

Ligand	Ligand Atoms' Binding Energies (eV) Referred to Pt4f$_{7/2}$ = 71.1 eV		Exposure (Langmuir)	Integrated C(1s) Peak Intensity[a]	Integrated "Other Atom" Intensity	$R_{1/C}$ (Expt)	$R_{F/C}$ (Theoret)
F$_3$C'-Ch=CH$_2$	C(1s)	= 284.4	100	8.15	62.0	15.2	13.7
	C'(1s)	= 291.4		4.1			
	F(1s)	= 687.0					
H$_2$C=CHF	C(1s)	= 284.5	100		5.4	0.49	2.27
	F(1s)	= 685.6		11.0			
H$_2$C=CF$_2$	C(1s)	= 285.2	200	6.35	24.8	3.9	4.55
	F(1s)	= 686.1					
cis-FHC=CHF	C(1s)	= 286.0	100	6.9	24.8	3.6	4.55
	F(1s)	= 685.9					
F$_2$C=CFCl	C(1s)	= 286.0	100	3.9	30.0	7.7	6.8
	F(1s)	= 685.8			Cl = 2.98		
	Cl(2p)	= 198.5					
H$_2$C=CH$_2$	C(1s)	= 284.3	200	10.7	—	—	—
CO	C(1s)	= 287.1	100	10.0	O = 29.8	3.0	2.9
	O(1s)	= 532.0					

(a) The chlorine, fluorine, and oxygen intensities have been corrected for the dependence of the analyzer sensitivity on the kinetic energies of the ejected electrons; variations in the multiplier gain have been corrected by continuous reference to the Pt4p$_{1/2}$ emission from the clean surface. Peak intensities normalized to C(1s) = 10.0 for a saturation layer of C$_2$H$_4$ on the Pt(100) surface.

Table II. Chemisorption of Chloro-Alkenes at a Pt(111) Surface

Ligand	Ligand Atoms' Binding Energies (eV) Referred to Ptrf$_{7/2}$ = 71.1 eV		Exposure (Langmuir)	Integrated C(1s) Peak Intensity	Integrated Cl(2p) Peak Intensity	$R_{Cl/C}$ (Expt)	$R_{Cl/C}$ (Theoret)
H$_2$C=CHCl	C(1s)	= 284.4	100	11.0	3.0	0.27	0.78
	Cl(2p)	= 198.4					
H$_2$C=CCl$_2$	C(1s)	= 284.3	100	7.8	13.4	1.7	1.55
	Cl(2p)	= 198.4					
trans-ClHC=CHCl	C(1s)	= 284.3	100	8.0	12.6	1.57	1.55
	Cl(2p)	= 198.6					
cis-ClHC=CHCl	C(1s)	= 284.3	100	7.6	11.2	1.47	1.55
	Cl(2p)	= 198.6					
ClHC=CCl$_2$	C(1s)	= 284.3	100	5.6	14.3	2.55	2.33
	Cl(2p)	= 198.6					
Cl$_2$C=CCl$_2$	C(1s)	= 284.4	100	4.43	14.5	3.27	3.10
	Cl(2p)	= 198.7					

Table III. Chemisorption of Chloro-Alkenes at a Pt(100) Surface

Ligand	Ligant Atoms' Binding Energies (eV) Referred to Pt4f$_{7/2}$ = 71.1 eV		Exposure (Langmuir)	Integrated C(1s) Peak Intensity	Integrated Cl(2p) Peak Intensity	$R_{Cl/C}$ (Expt)	$R_{Cl/C}$ (Theoret)
H$_2$=CHCl	C(1s) =	284.2	2	4.0	1.15	0.29	
	Cl(2p) =	198.6	3	5.2	2.6	0.50	0.78
			10	7.2	3.7	0.51	
			30	10.4	5.8	0.56	
			100	10.3	6.0	0.58	
H$_2$C=CCl$_2$	C(1s) =	284.3	100	6.25	9.1	1.46	1.55
	Cl(2p) =	198.6					
trans-ClHC=CHCl	C(1s) =	284.0	100	4.85	7.2	1.49	1.55
	Cl(2p) =	198.8					
cis-ClHC=CHCl	C(1s) =	284.2	100	7.25	11.3	1.56	1.55
	Cl(2p) =	198.8					
ClHC=CCl$_2$	C(1s) =	284.3	100	6.6	17.3	2.62	2.33
	Cl(2p) =	198.8					
Cl$_2$C=CCl$_2$	C(1s) =	284.6	100	5.1	16.4	3.21	3.10
	Cl(2p) =	199.0					

Table IV. Gas-Phase X-ray Photoelectron Spectroscopy Data for Halo-Alkenes

Molecule	Binding Energies Referenced to N$_2$(1s) = 409.9 eV		$R_{F(1s)/C(1s)}$[a]	$R_{Cl(2p)/C(1s)}$[a]	$R_{Cl(2s)/C(1s)}$[a]
CH$_2$=C'HCl	C(1s) =	290.8	–	0.72	0.52
	C'(1s) =	291.7			
	Cl(2p) =	206.6			
	Cl(2s) =	277.3			
CH$_2$=C'HF	C(1s) =	290.8			
	C'(1s) =	293.0			
	F(1s) =	692.6	2.5	–	–
CH$_2$=C'F$_2$	C(1s) =	291.1			
	C'(1s) =	295.8			
	F(1s) =	694.4	4.8	–	–

(a) The chlorine and fluorine intensities have not been corrected for the dependence of the analyzer sensitivity on the kinetic energy of the ejected electron.

2. At low coverage, the chemisorption of vinyl halides on the Pt(100) and Pt(111) surfaces is almost entirely dissociative with little halogen on the surface. As Fig. 1. shows, the emission intensities of the C(1s), F(1s), and Cl(2p) photoelectrons are coincident with theoretical expectations in that the theoretical slopes of 0.78 [Cl(2p)/C(1s)] and 2.27 [F(1s)/C(1s)] reflect the observed data, within experimental error, *but only at high coverage*. Using the C(1s) emission intensity at saturation as a measure of monolayer coverage, dissociative chemisorption of the vinyl halides is important up to 25 percent coverage; after that point, sorption is primarily associative.

3. The vinyl halides are somewhat more dissociated on the Pt(111) surface than on the Pt(100) surface; indeed, the fluoroalkenes generally show more dissociation on the close-packed surface.

4. Apart from vinyl chloride, the chloroalkenes are all associatively chemisorbed on the (111) and (100) surfaces. This feature obviates a mechanism which relies principally on cleavage of the carbon-halogen bond by the metal; elimination of halide from the surface must have quite different origins.

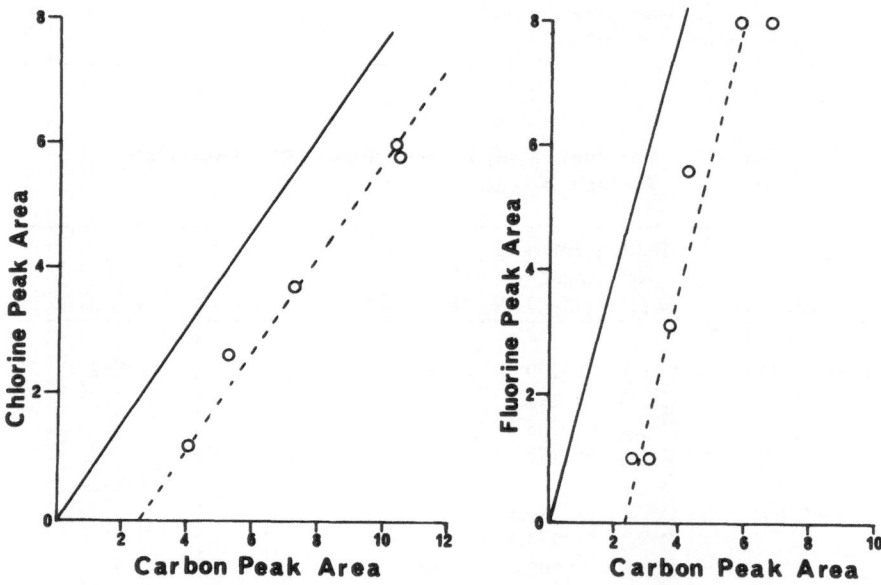

Fig. 1. Chemisorption of vinyl chloride and vinyl fluoride on the Pt(100) surface: integrated photoelectron intensities at various coverages. Solid lines refer to the ratio of intensities for associatively sorbed species. ($R_{Cl(2p/C(1s)}$ is taken from the gas-phase data (Table 4); $R_{F(1s)/C(1s)}$ is calculated from the ionization cross sections which, in turn, closely approximate the gas phase data.)

3 THE GEOMETRY OF THE PLATINUM-OLEFIN SURFACE

A discussion of these reactions must be based on a knowledge of the initial mode of attachment of the alkene or alkyne and the electron-transfer processes between the admolecule and substrate. The only technique available that could, in principle, provide details of the metal-ligand bond at an atomic resolution level, is low-energy electron diffraction, and relevant results can now be summarized.

In contrast to the (111) surface, the clean Pt(100) surface does not provide a simple (1 x 1) LEED pattern; but the (5 x 1) pattern for the clean (100) surface—which has been interpreted[20] in terms of a ruffled hexagonal top layer—transforms to a c(2x2) pattern on sorption of alkenes and alkynes. We have extended the earlier work of Somorjai et al.[21] to show that the LEED patterns for a variety of substituted alkenes, ranging from C_2H_3F to C_2Cl_4, chemisorbed on Pt(100) are all of c(2x2) symmetry although they are rather diffuse, particularly for the perchloro-substituted alkene-platinum surfaces. The diffuseness of the patterns is hardly surprising when one remembers that the areas of the van der Waals' envelopes for alkenes range from 17.7 $Å^2$ (C_2H_4) to 46.5 $Å^2$ (C_2Cl_4) and the p(2x2) substrates on (100) and (111) have areas of 31.4 $Å^2$ and 27.1 $Å^2$, respectively; one must have a statistical occupancy of metal binding sites for the associatively chemisorbed species and the c(2x2) symmetry must result from the superposition of out-of-registry p(2x2) domains since, even for ethylene, there is insufficient room for two molecules per unit mesh.

We can develop this argument further by examining the limitations placed on the structure of the metal-ligand surface by intraadlayer non-bonded interactions. Aside from adlayer-substrate nonbonded interactions, these must be the most structure sensitive, for, at least on the (100) surface, both σ- and π-bond energies are rotationally invariant.[17] Figure 2 shows the possible packing of the vinyl halides on the (100), p(2x2) substrate—it is particularly obvious for vinyl chloride that acceptable arrangements are limited to $\phi \approx 0^0$ and that in all cases, there are a number of short nonbonded H. . .X contacts; these are increased in number when one considers adlayer structures on the Pt(111) p(2x2) substrate.

Given the preferred head-to-tail arrangement,

which will be stabilized also by attractive dipolar interactions, a sensible mechanistic interpretation for the experimental data becomes available.[16,17] The surface oxidative-addition reactions cannot proceed via an initial attack on a carbon-halogen bond—this is proven by the observed associative sorp-

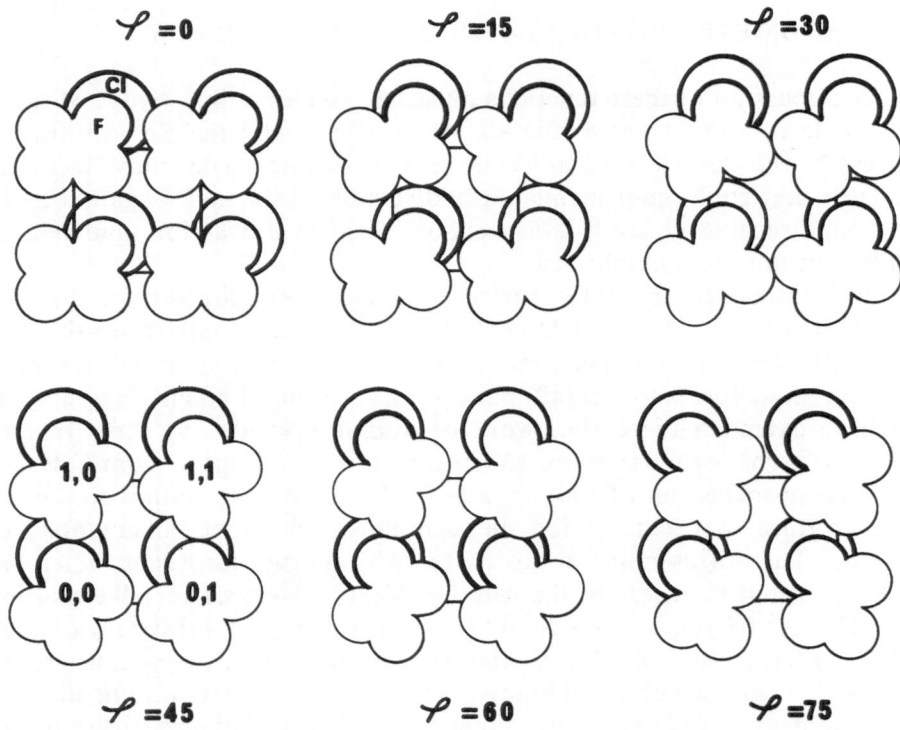

Fig. 2. Ordered adlayer arrangements of vinyl fluoride and vinyl chloride on a Pt(100) p(2x2) substrate mesh.

tion of the perchloro-alkenes and of cis-1, 2-difluoroethene; equally, the difference in behavior between, for example, 1,1-difluoroethene (dissociative sorption) and cis-1,2-difluoroethene argues against a mechanism which relies on intramolecular elimination of HF to explain the dissociative pathway. Intermolecular elimination of HX can be envisaged as follows:

1. Symmetric π-bonding of the olefin to a metal center or centers leaves a hydrogen atom of a carbon-hydrogen bond at a distance of 2.5 Å from noncoordinating metal atoms in the p(2x2) mesh—a situation exactly equivalent to that in complexes which undergo metallation of carbon-hydrogen bonds.

2. Oxidative addition of the metal to the carbon-hydrogen bond gives the metal-vinyl and metal-hydride species

3. Short nonbonded interactions in the adlayer guarantee that, subsequent to the reductive elimination of the hydride, HX is eliminated intermolecularly through nucleophilic attack on the carbon-halogen bond.

4. In the ordered head-to-tail arrangements, the nonbonded H. . .F distances for cis-1,2-difluoroethene are much larger[17] than those in 1,1-difluoroethene and HF elimination is accordingly inhibited.

5. With the exception of vinyl chloride, the chloroalkenes have van der Waals' envelopes which prevent nearest-neighbor packing on a p(2x2) substrate; the nonbonded H. . .Cl distances are, therefore, statistically greater and facile HCl elimination will be prevented.

Three other comments seem appropriate. The first relates to the observed break in the dissociation of the vinyl halides at coverages of approximately one-quarter to one-third. Alkene occupancy, at high coverage, of platinum sites separated by 5.6 A must be large, and the empty coordination sites at (1/2, 0), (0, 1/2), and (1/2, 1/2) become denied to the potential oxidative-addition reaction by virtue of nonbonded interactions in the adlayer. The second point that deserves emphasis is that we have deployed *ex post facto* arguments for the changing pattern of sorption between the difluoro-alkenes. What has been said seems right given a well-ordered structure based on a p(2x2) substrate; but both nonbonded interactions and dipolar forces would favor the following arrangement

which is a p(4x2) arrangement with two translationally nonequivalent molecules in the unit mesh. It is doubtful whether any LEED analysis could distinguish this arrangement from that of the p(2x2) mesh, but our earlier arguments would allow HF elimination from such a surface. Finally, we note that Demuth and Eastman[22] have used low-energy photoelectron spectroscopy to comment on the low-temperature chemisorption of ethylene on Ni(100), dissociation to acetylene taking place at ambient temperature. This reaction is in line with our studies on the halo-alkenes in that dihydrogen elimination from the surface will follow the initial oxidative-addition reaction, with hydride intersite transfer and reductive elimination being low-activation-energy processes. We have recently added He(I) and He(II) photoelectron-spectroscopy facilities to our combination LEED-Auger-mass spectrometer-ESCA instrument and have therefore been able to extend our high-energy photoelectron studies *(vide infra)* of the density of states of the valence band of platinum, and its modification following chemisorption of carbon monoxide[35] and alkenes, and to examine the valence orbital energies of chemisorbed alkenes. The photoelectron spectrum of ethylene valence orbitals appears very similar to that of Demuth and Eastman but, as expected, chemisorbed halo-alkenes such as C_2Cl_4 have emissions which are diffuse versions of their gas-phase spectra.

4 SOME REMARKS ON STEREOCHEMICAL AND ELECTRONIC FACTORS IN CATALYSIS

We can now return to our earlier comments on the reactions of transition-metal complexes and see whether some of the better understood features of homogeneously catalyzed reactions can be used to help us sketch in some of the gaps which presently exist in discussions of heterogeneous catalysis.

It is intuitively obvious that the concept of coordinative unsaturation—of vacant sites in the coordination sphere of a discrete complex—is somehow related to the idea of active centers in heterogeneous systems. The simplest molecular orbital theory of bonding in transition-metal complexes provides a rationale for the connection between coordination number and electronic configuration of the metal: in octahedral complexes with σ-metal-ligand bonds, for example, 18 electrons only can be accommodated in bonding, nonbonding, or weakly antibonding orbitals. π-bonding ligands such as CO and C_2H_4 stabilize those nonbonding or antibonding orbitals that are largely localized on the metal—this is the basis for the generally observed 18-electron (rare gas) rule of organometallic chemistry.[23] Coordinatively unsaturated species may be formed, therefore, when the ligands are relatively strong electron donors (tertiary phosphines and so on) and poor acceptors, with the concomitant buildup of electron density on the metal which is in a formally low oxidation state. In the more localized bond language of Nyholm[11], the facility for a complex to pass from, say, a two-coordinate 14-valence-electron configuration [Pt(PPh₃)₂] to the 16-electron structures [Pt(PPh₃)₃] [Pt(PPh₃)₂(C₂H₄)] or to the 18-electron complex, Pt(PPh₃)₄, is related to orbital promotion energies of the metal: if, by way of illustration, metal d-p promotion energies are small (zero valent d^{10} and univalent d^9 ions) orbital hybridization will allow the metal ion to react flexibly to the electronic nature and requirements of the ligands.

As an intermediate step between these considerations of simple complexes and heterogeneous surfaces, we can briefly consider the situation in cluster complexes—those polynuclear molecules whose architecture is based on metal-metal bonding. Molecular orbital calculations[24] for symmetrical species such as $(Co_6(CO)_{14})^{4-}$ suggest that the metal atoms use hybrid orbitals with predominant s and p character for forming skeletal molecular orbitals. The d-orbitals interact only weakly and, therefore, the cobalt atoms behave as pseudo main group (p block) atoms in this ion; the d-orbitals are largely associated with the metal-carbonyl bonds. We shall carry this over to ligand bonding on metal surfaces, but one other comment on cluster complexes bears directly on our remarks on variable coordination number in mononuclear complexes and the assumption[25] that metal-metal bonding within a surface is not modified on chemisorption of ligands.

For polygons and polyhedra of metal atoms with vertices of order three, the occupied delocalized skeletal orbitals can be transformed into a set of equivalent orbitals[26], each of which is localized on one of the edges (the corresponding situation for the bulk structures is that with an insulator and full bands, the Bloch functions can be transformed to a determinant of localized Wannier functions representing the bonding orbitals); molecules containing these polyhedra have molecular orbitals which reflect the polyhedral topology in a simple way and the rare-gas "rule" can be applied successfully. But in trinuclear complexes which obey the rare-gas "rule" [e.g., $Ru_3(CO)_{12}$], the highest filled molecular orbital is antibonding with respect to metal-metal bonding, and increasing electron density in the cluster through, say, substitution of CO ligands by better electron donors leads to metal-metal bond cleavage.[27] Metal-metal bonding in finite clusters thus reflects the electronic character of the coordinating ligands; there is electronic "traffic" between the metal-ligand and metal-metal bonds which is exactly analogous to those coordination-number changes in mononuclear species which can provide coordinatively unsaturated, highly reactive metal fragments. At a surface, bonding of an admolecule may lead to a variety of *local* changes in the metal-metal bond lengths which reflect whether the group, Wannier-type orbital, which is used to overlap with appropriate ligand orbitals, has predominant bonding, nonbonding, or antibonding character with respect to nearest-neighbor metal atoms.

The closest analogies, so far as extended metal structures are concerned, to the Nyholm[11] delineation of factors influencing coordination number and symmetry in metal complexes belong to the Engel-Brewer and Altmann-Coulson structural models.

Engel and Brewer correlate metallic structures with metal promotion energies in that bcc, hcp, and fcc structures relate to isolated atoms with the ground or low-lying excited-state configurations s, sp, and sp^2, respectively.[28] The successes and difficulties of the Engel-Brewer model have been examined by Hume-Rothery[29]; a conceptual difficulty is obvious in a scheme which relates atomic-energy levels to bulk structure but is noncommittal on the extent of orbital hybridization and overlap arguments. This is in contrast to the Altmann-Coulson model[30] which preempts specific orbital hybridizations to bcc (resonance between sd^3, d^4, and d^3 hybrids) to hcp (spd^4, pd^5, and sd^2) and fcc structures (p^3d^3). A comparison of these two models suggests that what is important in the Engel-Brewer scheme is not the particular number of s or p valence electrons but their total number, and this view is underpinned when we turn to the more complete pseudopotential theory of metal structures. The topic has been reviewed comprehensively by Heine and Weaire[31] for nontransition elements; the most important structure-sensitive term in the total energy is the band-structure energy which, expressed in an approximate form, shows that with the electron to atom ratio, $z = 1$, the lowest energy is the hcp structure. The hexagonal close-packed structure is also predicted for $z = 2$; $z = 3$ is predictive of the fcc arrangement. The Engel-Brewer model is successful, therefore, not on account of the orbital form of the valence electrons but on account of their total number. Pettifor[32] has developed pseudopotential theory to a treatment of the structures of transition metals. Again the emphasis—in contrast to Brewer—is on the total number, z, of valence electrons per atom. The stability of the bcc structures of the early transition metals *vis à vis* the close-packed arrangements needs an explicit consideration of hard-core contributions, but for $z = 6$ to 10, the trend of the close-packed arrangements is nicely accounted for. This model suggests that the difference in cohesive energies between the hcp and fcc structures amounts only to about 0.002 ryd/atom and is particularly small around $z = 4$, 7, and 9; the theoretical results are too insecure to be taken literally but we take them as suggestive that a similar situation may occur on transition-metal surfaces as we have shown exists for simple complexes and polynuclear species, viz., that the coordination number and symmetry of a particular metal atom or groups of metal atoms may be sensitive to local electron densities imposed by ligand characteristics and that localized or extended surface transitions may occur. Certainly, the platinum (100) surface undergoes structural transitions on sorption of carbon monoxide or alkenes. At first sight, this is a little worrying since the work-function changes[21] ($\Delta\phi_{CO} = 0.18$ v, $\Delta\phi_{C_2H_4} = -0.75$ v) would seem to imply quite different metal valence electron densities on the two surfaces. However, single-crystal ESCA studies of the density of states of the valence band show that its well-characterized[33,34] doublet structure is modified identically by the chemisorption of carbon monoxide and ethylene in that there is a reduction of the density of states of the subband centered at ca. 2 eV below the Fermi level[35] (Fig. 3). We interpret this as meaning that the states lying close to the Fermi level have much t_{2g} character and that d, π^* (metal-to-ligand) bonding is comparable for CO and C_2H_4; the extent of ligand-to-metal charge transfer is quite different with, we have argued, this "forward donation" process taking place to metal "s" states. The fact

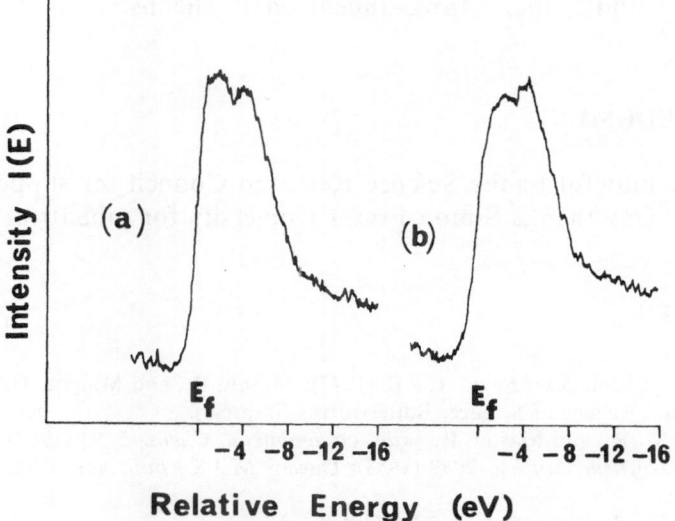

Fig. 3. The valence-band density of states for (a) the clean Pt(100) surface and (b) the Pt(100)-CO surface.

that all unsaturated hydrocarbons act as net Lewis bases (electron donors) to platinum surfaces is in significant contrast to the situation in such platinum (0) complexes as $(Ph_3P)_2Pt(C_2H_4)$ and $(Ph_3P)_2Pt(C_2H_2)$ where photoelectron-emission studies[36] show that the alkenes and alkynes are net electron acceptors; compared with these formally neutral platinum atoms, the surface atoms are electron deficient and, chemically speaking, less strong nucleophiles. We assume this is one reason for the differing pattern of reactivity towards the vinyl halides, although a more important factor must be the relative ease of intermoelcular hydrogen and halogen migration on a surface. In the homogeneous reactions of ligands such as alkynes with palladium (0) and other low-valent metals, the ligands react via carbanion mechanisms (e.g., oligomerization) as the metal fragment is a nucleophilic catalyst. Coordination of alkynes to metals in higher oxidation state leads to the situation where the ligand becomes more susceptible to nucleophilic attack.[37] That is the situation which seems to prevail for surface coordinated ligands, such as ethene and ethyne, at least as far as stoichiometric reactions are concerned. That, of course, need not be the situation for catalytic reactions as the (minority) active sites need not have electronic conditions which are at all similar to those of the remainder. But even with this caveat, it is worth emphasizing the importance of the newer physical methods in unequivocally determining, for the first time, the stoichiometry of metal-ligand surface. With comparative studies of reactions at various single-crystal and stepped surfaces, such as are partly described here, the paradox—of chemistry in two dimensions being less amenable to structural and mechanistic study than three-dimensional chemistry may be finally resolved.

ACKNOWLEDGMENT

We are grateful to the Science Research Council for support of these studies. I.D. Gay thanks Simon Fraser University for sabbatical leave.

REFERENCES

1. Mason, R., *Chem. Soc. Revs.,* **1**, 431 (1972); Mason, R., and Mingos, D.M.P., M.T.P. International Review of Science, Butterworths (in press).
2. Churchill, M. R., and Mason, R., *Adv. Organometal. Chem.,* **5**, 93 (1967).
3. Chatt, J., *J. Chem. Soc.,* p. 2939 (1953); Dewar, M.J.S., *Bull. Soc. Chim. France.,* **18**, C71 (1951).
4. Walsh, A. D., *J. Chem. Soc.,* p. 2260 (1953) (et seq.).
5. McWeeny, R., Mason, R., and Towl, A.D.C., *Disc. Faraday Soc.,* **47**, 20 (1969).
6. Mason, R., and Thomas, K. M., *Ann. N. Y. Acad. Sci.,* **239**, 225 (1974).
7. Mason, R., *Nature,* **217**, 543 (1968).
8. Osborn, J. A., Jandine, F. H., Young, J. F., and Wilkinson, G., *J. Chem. Soc. (A),* p. 1711 (1966).
9. Halpern, J., and Wong, C. S., *J. Chem. Soc. Chem. Comm.,* p. 629 (1973).
10. Tolman, C. A., Meakin, P. Z., Lindner, D. L., and Jesson, J. P., *J. Amer. Chem. Soc.,* **96**, 2762 (1974).
11. Nyholm, R. S., *Proc. Chem. Soc.,* p. 273 (1961).
12. Collman, J. P., *Accts. Chem. Res.,* **1**, 136 (1968).
13. Johnson, B.F.G., Lewis, J., Jones, J. O., and Taylor, K. A., *J. Chem. Soc. (Dalton Trans.),* p. 34 (1974).
14. Parshall, G. W., *Accts. Chem. Res.,* **3**, 139 (1970).
15. Clarke, T. A., Mason, R., and Tescari, M., *Proc. Roy. Soc. (London),* **A331**, 321 (1972).
16. Clarke, T. A., Gay, I. D., and Mason, R., *J. Chem. Soc. Chem. Comm.,* p. 331 (1974).
17. Clarke, T. A., Gay, I. D., and Mason, R., *Chem. Phys. Letters,* **27**, 562 (1974).
18. Henke, B. L., and Elgin, R. L., *Adv. X-ray Anal.,* **13**, 639 (1970).
19. Wagner, C. D., *Anal. Chem.,* **44**, 1050 (1972).
20. Clarke, T. A., Mason, R., and Tescari, M., *Surface Sci.,* **40**, 1 (1973).
21. Morgan, A. E., and Somorjai, G. A., *J. Chem. Phys.,* **51**, 3309 (1969).
22. Demuth, J. E., and Eastman, D. E., *Phys. Rev. Letters.,* **32**, 1123 (1974).
23. Mason, R., *Special Lectures, XXIII I.U.P.A.C. Congress,* **6**, 31 (1971).
24. Mingos, D.M.P., *J. Chem. Soc. (Dalton Trans.),* p. 133 (1974).
25. Bond, G. C., *Disc. Faraday Soc.,* **41**, 200 (1966).
26. Kettle, S.F.A., *J. Chem. Soc. (A),* p. 314 (1967) (and references therein).
27. Mason, R., and Zubieta, J. A., *J. Organometal. Chem.,* **66**, 279 (1974); ibid, p. 289.
28. Brewer, L., *Science,* **161**, 115 (1968).
29. Hume-Rothery, W., *Prog. Mat. Sci.,* **13**, 1 (1967).
30. Altmann, S. L., Coulson, C. A., and Hume-Rothery, W., *Proc. Roy. Soc.,* **A240**, 145 (1957).
31. Heine, V., and Weaire, D., *Solid State Physics,* **24**, 249 (1970).
32. Pettifor, D. G., *J. Physics,* **C3**, 367 (1970).

33. Baer, Y., Hedén, P. F., Hedman, J., Klasson, M., Nordling, C., and Siegbahn, K., *Phys. Scripta,* **1**, 55 (1970).
34. Kowalczyk, S., Ley, L., Pollak, R., and Shirley, D. A., *Phys. Letters,* **41A**, 455 (1972).
35. Clarke, T. A., Gay, I. D., and Mason, R., *Chem. Phys. Letters,* **31**, 29 (1975).
36. Mason, R., Mingos, D.M.P., Rucci, G., and Connor, J., *J. Chem. Soc. (Dalton Trans.),* p. 1729 (1972).
37. Chisholm, M. H., and Clark, H. C., *Accts. Chem. Res.,* **6**, 202 (1973).

DISCUSSION on Paper by Ronald Mason

BOND: You state that certain halo-alkenes are associatively chemisorbed on Pt(100). Can you distinguish between σ-diadsorption and π-adsorption (as in Zeise's salt)?

MASON: My explanation of the changing pattern of chemisorption relies exclusively on the nature of short-range interactions in the adlayer; it entirely begs the question asked by Professor Bond. I have argued in a recent *Chemical Physics Letter* article, that since the pattern of dissociation on Pt(100) and Pt(111) is similar, this may offer some evidence that it is the onefold binding site that is important. For what it is worth, the Fourier transform of the Pt(100)-alkene LEED data shows no sign of twofold bridged species. I should add also that we know of no simple "σ-diadsorbed" ethylene-cluster system in organometallic chemistry. Substituted alkenes and nonconjugated dienes form complexes with clusters such as $Os_3(CO)_{12}$, whereas ethylene itself undergoes relatively facile oxidative-addition reactions.

RHODIN: This refers to the sp-d energy-promotion effect for valence electron bonding on some surfaces of noble metals. Your discussion of the fact that certain organic complexes may change their stereo chemistries and coordination numbers in a low-energy process on these surfaces is very relevant to an explanation of the formation of the (1x5) super lattice structures on (100)Pt, (100)Ir, and (100)Au by Palmberg and Rhodin [*Journal Chemical Physics* (1966)]. The electron-level promotion corresponding to superstructure formation for these surfaces and the transformation to c(1x1) structure from a small degree of chemisorption was discussed in some detail in that publication.

MASON: I am sorry that I have overlooked the reference which Rhodin mentions, both in our paper discussing the structure of the clean Pt(100) surface (Clarke, Mason, and Tescari, *Surface Science*, **40** (1973), and in my present contribution. It is reassuring that we are using similar explanations as this may strengthen the speculations on relations between stereochemical changes in complexes and on surfaces.

ALKALI ATOM ADSORPTION ON NICKEL— STRUCTURE AND ELECTRONIC EXCITATIONS

S. Andersson

Department of Physics, Chalmers
University of Technology, Fack, S-402 20
Göteborg 5, Sweden

ABSTRACT

Adsorption of sodium and potassium on Ni(100) has been investigated by means of low-energy electron diffraction and electron-energy-loss spectroscopy. Lateral adatom distributions and tentative adatom positions are determined. Electronic excitations from occupied valence and core states and a collective excitation among the valence electrons are reported. The excitations are found to be characterized by their dependence on the surface density and charging state of the adatoms. The electron spectroscopic observations, in conjunction with work-function data, have been utilized to explore the adatom valence electron structure and charging.

1 INTRODUCTION

Alkali atom adsorption on metal substrates is characterized by a chemisorption bond of polar-metallic nature, the bond strength and charge transfer being strongly dependent on the surface density of adatoms. This qualitative picture originates primarily from experiments on work-function changes and desorption energies and is vizualized in the virtual level model.[1-3] Owing to the interaction with the substrate, the original alkali valence level is shifted and broadened into a resonance. The adatom density of states extends below the Fermi level and gives rise to a nonzero expectation value q^- of the number of electrons on the adatom. As the density of adatoms is increased, the electrostatic interaction among them will shift the virtual level downwards with respect to the Fermi level, and hence increase the occupancy q^-. Evidently, the Fermi level probes the resonance and a measurement of the charging would give information about the density of states distribution, although over a limited energy range. This has been elucidated by several authors.[1,4,5] Bennet and Falicov[1] also considered the effect of a strong external electric field in an attempt to account for experimental field emission results for the potassium on tungsten systems studied by Schmidt and Gomer.[6]

The experimental observables mentioned above are, however, implicitly related to the valence electron structures and the charge transfer, and any comparison with theoretical predictions will be tentative. For example, the charge transfer is related to the dipole moments determined from the work-function changes. Evaluation of the charge transfer thus requires knowledge about the adatom position and the dielectric properties in the substrate surface region. There is an obvious need for experiments that yield more direct information about the adatom position, the electron structure, and the charge transfer. In an attempt to obtain such information, we have utilized elastic and inelastic scattering of low-energy electrons in experiments on sodium and potassium adsorbed on Ni(100). The two-dimensional adatom distribution and the adatom position were investigated by low-energy electron diffraction (LEED). Electron excitations from occupied core and valence states have been instigated by electron-energy-loss spectroscopy in conjunction with measurements of work-function changes.

2 EXPERIMENTAL

The following experiments were carried out at different surface densities of sodium and potassium adsorbed on the clean Ni(100) surface.

(1) The lateral two-dimensional adatom distribution was determined by LEED. Tentative distances between the center of the adatom layer and the first nickel layer were deduced from effects of incoherent scattering on the electron reflectivity.

(2) Work-function changes were recorded by means of the retarding-field method. They are used primarily in conjunction with the electron spec-troscopic data [(3) below] for a determination of the local electrostatic potential.

(3) Electron-energy-loss spectra of the back-scattered electrons were obtained by means of the retarding system of the LEED optics. The re-tarding grids were modulated by a 0.3-volt peak-to-peak sinusoidal signal and the resolution was about 0.5 eV at electron energies of 5 to 30 eV. The energy-loss scale is calibrated to better than 0.1 eV.

The experiments were performed in an UHV system at an ambient pressure in the 10^{-12} Torr range. The preparation and cleaning of the Ni(100) specimen has been described previously.[7] The most effective clean-ing process after alkali metal deposition has been found to be excessive argon-ion bombardment at elevated specimen temperature (about 400°C). Sodium and potassium were then deposited onto the clean Ni(100) surface (kept at 20°C) from heated break-seal glass ampoules containing high-purity metal (>99.97 percent). The deposition rate was controlled by the source temperature and was varied in the range 1 monolayer per 1 to 25 minutes. The accuracy of the absolute-deposition-rate determinations as deduced from the source geometry and temperature is about 50 percent compared with that of average distributions observed from the LEED measurements. During deposition, the pressure rose occasionally to about 1×10^{-10} Torr but dropped to about 1×10^{-11} Torr when the vapor beam was shut off.

Deliberate exposure to the ambient at about 10^{-11} Torr for a couple of hours produced obvious changes in LEED intensities as well as in electron-energy-loss spectra, thus showing that the alkali layers are very sensitive to contamination. Since the time scale for a complete run over the coverages is of this order, separate check experiments at just one coverage were carried out.

3 ADATOM DISTRIBUTION AND POSITION

In this work we will be particularly concerned with properties of alkali layers of low densities. As discussed below, such layers do not form simple ordered two-dimensional surface structures, and a structure analysis within the conventional dynamical LEED intensity framework cannot be carried out. Tentative adatom positions will be deduced from effects of incoherent scattering on the substrate electron reflectivity.

The lateral distribution of sodium and potassium at various coverages, θ, will be considered first. The clean Ni(100) surface has ideal lateral periodicity and θ is defined such that $\theta=1$ corresponds to the density of atoms in this reference surface. The LEED observations are shown in Fig. 1. The intense outermost spots are due to the Ni $\{10\}$ beams corresponding to the Ni(100) square unit mesh edge a = 2.49 Å. At room temperature, the

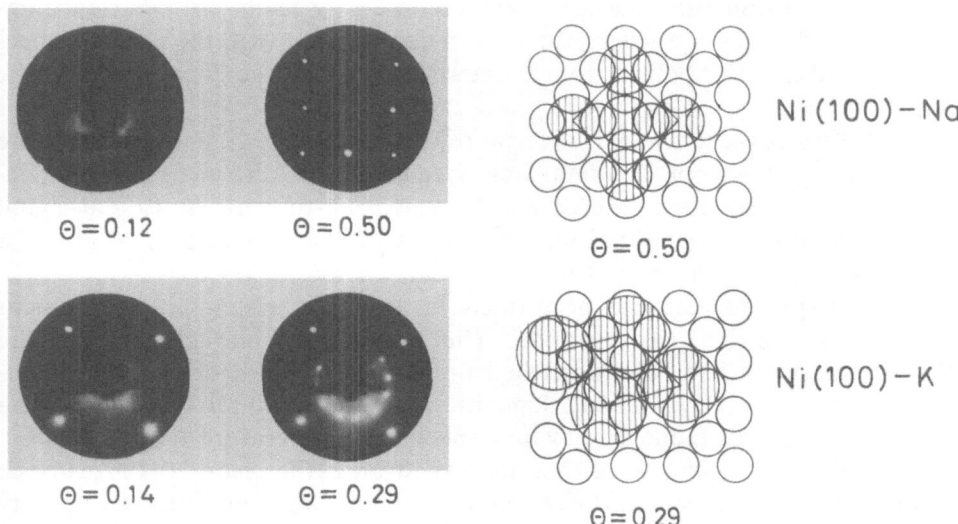

Fig.1. Diffraction patterns at various coverages θ; primary electron-beam energies were 70 and 24 eV ($\theta = 0.12$ for sodium).

sodium layer forms an ordered c(2x2) structure (square symmetry) at $\theta = 0.5$, while the potassium layer forms an ordered structure of hexagonal symmetry at $\theta = 0.29$. The ordering of the sodium layer is obviously simply related to the substrate symmetry and spacings. The ordering of the potassium layer is coincidence site on every second nickel site in either the Ni[10] or [01] direction, which gives two possible orientations for the hexagonal structure and hence 12 beams in the diffraction pattern. At lower coverages, a so-called ring pattern is observed. These patterns are interpreted to be due to uniformly distributed adatoms with a mean separation R of small spread. The coverage assignments are derived from:

$$\theta = a^2/R^2 \qquad \text{for sodium}$$

$$\theta = a^2/(\sqrt{3}/2)\, R^2 \text{ for potassium}$$

(1)

This means that even at low coverages we assume square and hexagonal symmetry for sodium and potassium, respectively. This is done so simply because the difference between the two assignments ($\approx$15 percent) is within the accuracy of the vapor rate determinations ($\approx$50 percent); thus we do not know what gradual symmetry change could occur until the formation of the ordered structures. The ultimate monolayer coverages at room temperature are $\theta = 0.5$ for sodium and $\theta = 0.38$ for potassium. According to Eq. (1) this corresponds to the mean separations R = 3.52 Å and R = 4.35 Å for sodium and potassium, respectively. This means a compression relative to the nearest-neighbor distance 3.71 Å and 4.62 Å in the respective metals. Similar observations have been reported by others.[8] In summary we conclude from the LEED observations that the sodium and potassium atoms spread uniformly with a rather well-defined mean separation when adsorbed on the Ni(100) surface.

A determination of the adatom position is a more complicated problem and can at the moment be carried out only for the simplest overlayer structures like the c (2x2)Na structure. In a previous work[9] we have reported a dynamical LEED intensity analysis of this structure. It was found that the sodium atoms occupy the fourfold symmetric hollows of the Ni(100) surface (see Fig. 1) and that the distance between the sodium layer and the first nickel layer was 2.9 Å.* The uniformly spread alkali layers at lower coverages cannot be treated in this way and a different method has been used to derive tentative layer distances. The fundamental features of the method can be simply expressed, and a more detailed description has been presented previously.[11] The electron reflectivity, R, and in particular the specular reflectivity, I_{00}, of a monochromatic electron beam normally incident on the clean Ni(100) surface is in the energy range 0 to 10 eV above the vacuum level dominated by a total energy gap $X_4' - X_1$. This is shown in Fig. 2, where R is the total electron reflectivity given by

$$R = \frac{I_o - I_c}{I_o}$$

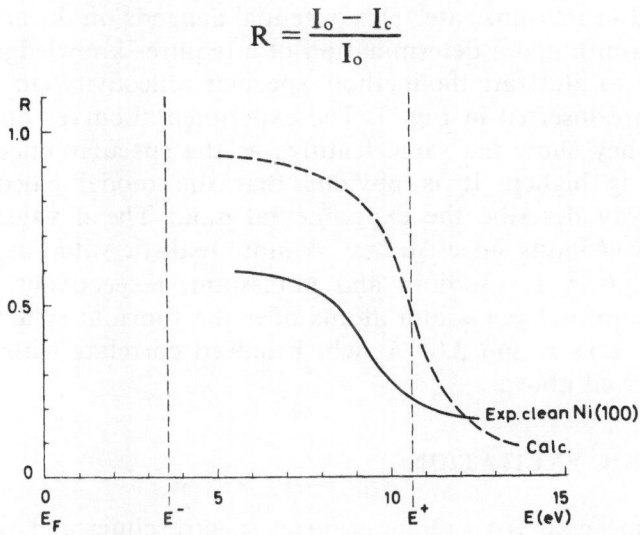

Fig. 2. Experimental electron reflectivity R (solid line) and calculated specular electron reflectivity I_{00} (dashed line) from the clean Ni(100) surface in the range of the total energy gap $X_4' - X_1$ ($E^- - E^+$). E_F denotes the Fermi level of nickel.

---

*A recent reanalysis[10] on extended intensity data still yields fourfold symmetry but gives a distance of 2.2 Å.

I_o is the primary beam current and I_c is the current trapped by the crystal. E^- and E^+ are the lower and upper edges of the gap $X_4' - X_1$ as derived from a band structure calculation. The vacuum-level is at $E = \phi_{Ni(100)} = 5.1$ eV.[12] Over the range of the energy gap, the wave function has sinusoidal form and its phase α changes continuously. The wave amplitude will then vary correspondingly in any adsorbed overlayer. Assuming that the overlayer is at a distance c from the first nickel layer and that the wave vector is $\mathbf{k}$ in the surface region, the wave function will have a *node* in the overlayer when

$$2 \, k \, c + \alpha = (2n - 1)\pi \qquad n = 1,2,3,\ldots \qquad (2)$$

If the overlayer scatters mainly incoherently, the specular reflectivity from the crystal will have a maximum for the condition above. This idea has been applied to the uniformly spread alkali layers. Thus we assume that incoherent scattering overwhelms the coherent in this case. Of particular importance is the relatively poor 2-D order that excludes strong interferences due to nonspecular beams. The specular reflectivity over the range of the energy gap is calculated in the two-beam approximation and the incoherently scattering adatoms are represented by a purely imaginary potential iV, i.e., absorbing. The average potential in the region between the overlayer and the substrate is represented by a step Δu above the bottom of the conduction band in the substrate. This potential depends on the adatom density (see next section), and a determination of c requires knowledge about it.

In order to illustrate the method, specular reflectivity curves calculated for $\Delta u = 0$ are inserted in Fig. 3. The experimental curves are total reflectivities but they show the same features as the specular ones, though the background is higher. It is obvious that the model calculations in a reasonable way describe the experimental data. The d values 2.5 Å and 3.2 Å are lower limits since $\Delta u > 0$. A more realistic value, $\Delta u = 3V$, gives 3.0 Å and 3.6 Å for sodium and potassium, respectively. Billiard-ball-packing sodium and potassium atoms over the fourfold symmetric Ni(100) hollows give 2.55 Å and 3.09 Å, which indeed correlate with the tentative distances derived above.

4 ELECTRONIC EXCITATION

Our prime goal is to relate electron spectroscopic and work-function data to the electron structure of alkali atoms adsorbed on a metal substrate. Three types of electronic excitations are discussed in connection with the experimental results: (1) excitations from occupied to unoccupied valence states; (2) excitations from a core state to unoccupied valence states; and (3) a collective excitation in the denser alkali layers. Types (1) and (2) will have particular implications with respect to the atomistic picture of adsorption,

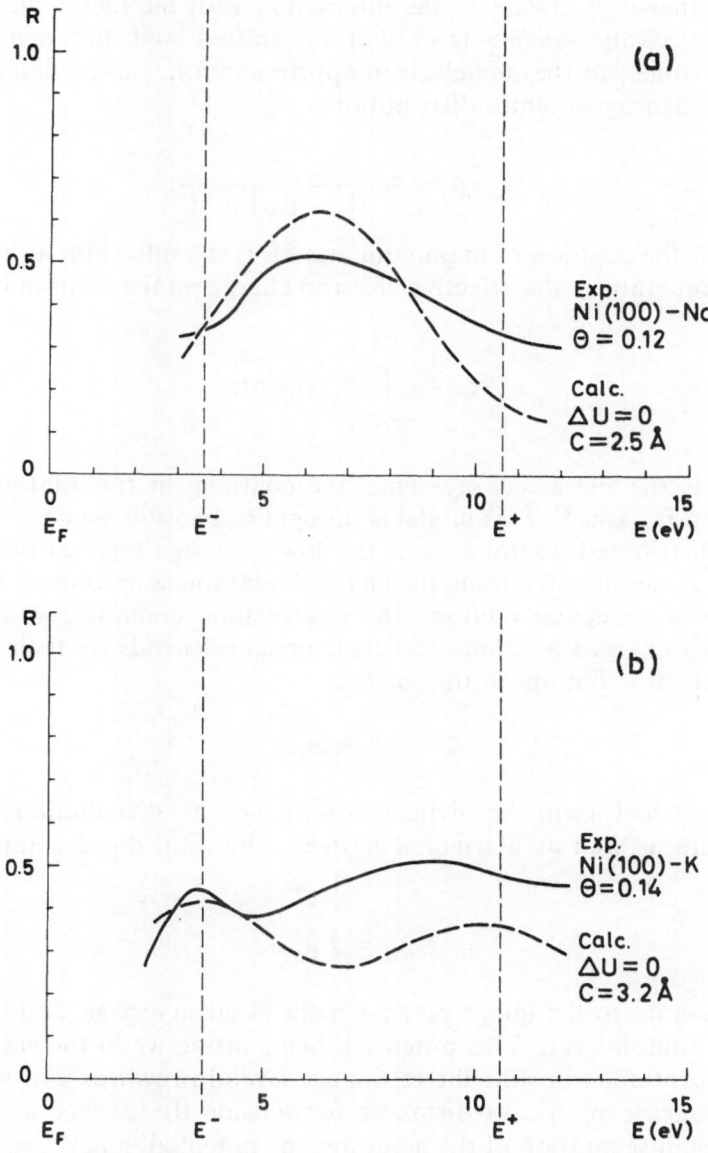

Fig. 3. Experimental electron reflectivities (solid line) versus electron energy for sodium and potassium adsorbed in Ni(100). Dashed curves represent calculated specular electron reflectivities according to the model described in the text.

with properties phrased in terms of the adatom valence electron structure. Type (3) is related to the monolayer case[13], with properties approaching those of the corresponding alkali metal.

The atomistic picture is briefly outlined here following the treatments given by others.[1,2,3] Owing to the interaction with the metal substrate, the original ns alkali valence level will be shifted and broadened into a resonance which, in the one-electron approximation, can be described by a Lorentzian density of states distribution:

$$A_\varphi(E) = \frac{1}{\pi} \frac{\Gamma}{(E - E_\varphi)^2 + \Gamma^2} \tag{3}$$

where $E\varphi$ is the position of maximum and 2Γ is the full width at half-height. At zero temperature, the effective electron charge on the adatom is given by

$$q^- = \int_{-\infty}^{E_F} A_\varphi(E)\, dE \tag{4}$$

where E_F is the Fermi energy. Thus the charging of the adatom depends critically on E_φ and Γ. This model is thought to be valid even for the case at spatially distributed adatoms (i.e., the low coverage regime) provided the electrostatic interaction among the charged adatoms is accounted for. At the position of a particular adatom, the electrostatic potential, Φ, due to the surrounding charged adatoms and their images depends on their density as well as their distribution on the surface

$$\Phi = -2\, Aed\mu n^{3/2} \tag{5}$$

where A is a coefficient that depends on the specific distributions and is ≈ 9 for a square as well as a trigonal lattice.[5] The total dipole moment 2μ is given by

$$2\mu = 2qd \tag{6}$$

d is the distance to the image plane, e is the electron charge, and the surface density of adatoms is n. This potential, being attractive in the case of alkali atom adsorption, will shift the resonance level downwards with an accompanying increase of q^-. At distances for outside the surface as compared with the mean separation of the adatoms, the potential Φ has reached its full value, giving a work-function change

$$\Delta\phi = -4\pi\mu n \tag{7}$$

A few comments on the model discussed are worthwhile. The charge distribution that builds up in the vicinity of the adatom is certainly very different from the ns valence level charge distribution in the free atom, a fact that is visualized by screening charge distributions obtained within the inhomogeneous electron gas model. [14,15] Muscat and Newns[5] approached this problem, describing the virtual level in terms of an ns − np$_z$ hybrid. Excitations from occupied to unoccupied valence states [(1) above] would in that model merely imply a transition from a bonding state to a nonbonding np$_x$, np$_y$ or antibonding ns−np$_z$ state. The density dependence of the excitation energy will plausibly be due to changes in the electrostatic potential Φ. Combining Eqs. (5) and (7) gives

$$\Phi_{if} = \frac{A}{2\pi} \text{ ed } n^{1/2} \Delta\phi \tag{8}$$

where Φ_{if} is the difference in Φ between the two states and d can be thought of as the difference between the center of gravities of the charge distributions characterizing the two states. As $\Delta\phi$ is negative, one would expect to observe a decrease in the excitation energy with increasing coverage.

This mechanism can be contrasted with that of a collective excitation in a monolayer alkali film [(3) above], a phenomenon considered by several authors. [16-19] The continuum model treated by Gadzuk[17] and the "box model" treated by Newns[18] yield essentially the same nature and dispersion for the mode. For small momentum, Q, the dispersion is linear and the frequency is, in the continuum model, given by

$$\omega(Q) \approx \omega_p (\tfrac{1}{2} - 2 Q\ell)$$
$$\omega_p = \left[\frac{4\pi n e^2}{m.\ell} \right]^{1/2} \tag{9}$$

where ℓ is the film thickness, and m and e are the effective electron mass and charge, respectively. The frequency ω_p is obviously expected to increase with increasing density as $n^{0.5}$.

Excitations of the adatoms were investigated by electron-energy-loss spectroscopy at different adatom densities. The experimental electron-energy-loss spectra, I$'_s$ (E), in the region close to the quasielastic line are shown in Fig. 4. These particular spectra were recorded at the primary electron energies 32 eV for sodium and 18 eV for potassium. At the lowest coverages, a weak loss develops that decreases in energy from about 3.5 eV to 2.3 eV for sodium in the range $0 < \theta < 0.25$ and from 3.5 eV to 1.5 eV for potassium in the range $0 < \theta < 0.2$. As the coverage is increased further, a loss of increasing energy is observed. Its intensity increases appreciably and at coverages approaching the densest monolayers ($\theta \approx 0.4$ for potassium) the double loss becomes observable. Finally, these spectra turn into the plasmon spectra for thick alkali films. This coincidence and the increase in energy are qualitatively compatible with the expected density dependence of a collective mode [Eq. (9)]. We have previously discussed this for the

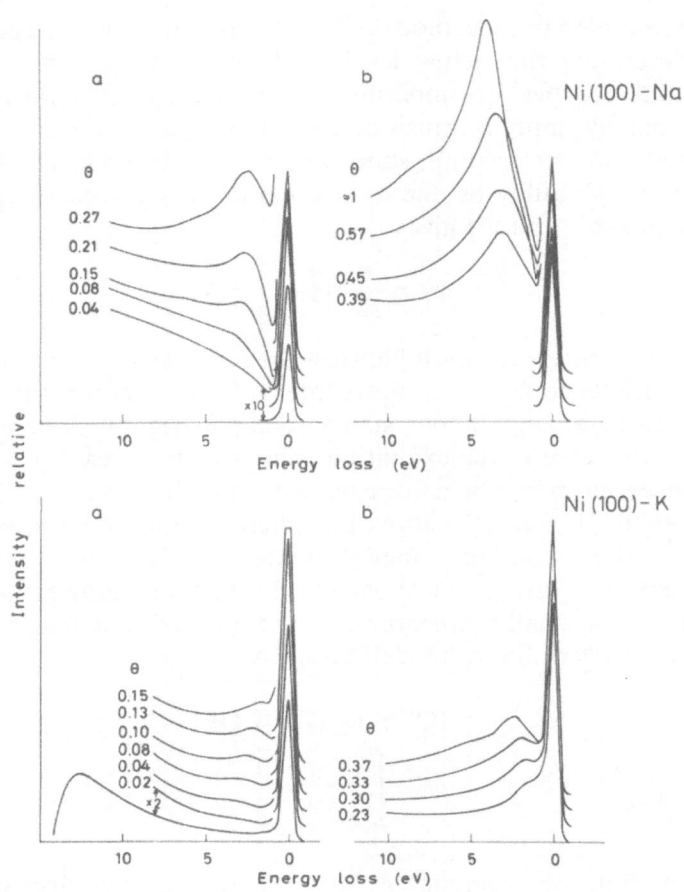

Fig. 4. Experimental electron-energy-loss spectra obtained for different coverages θ; primary electron energies were 32 eV (Na) and 18 eV (K) relative to the nickel Fermi level.

potassium-on-nickel system[20], and similar observations have been reported by others.[21,22] Concerning the nature of the mode, we notice as previously that the frequencies derived from Eq. (9) are much too large. The experimental energies are 3.0 eV at $\theta = 0.50$ for sodium and 2.3 eV at $\theta = 0.38$ for potassium, while the calculated ω_p values are 5.5 eV and 4.2 eV (using $1 = 3.71$ Å for sodium and $1 = 4.62$ A for potassium, the n.n. distances in the respective metals). Dispersion will lower the frequency as will a reduced effective charge and a larger effective mass, but we are at the moment not sure of the relative contribution of these effects. The density dependence of the loss is close to $n^{0.5}$, but, curiously enough, rather different for the two materials, $\approx n^{0.4}$ for sodium and $n^{0.8}$ for potassium ($n = \theta/a^2$).

The loss observed at low coverages can hardly be interpreted in terms of a collective mode and, in particular, we explore a possible excitation from occupied to unoccupied valence states on the adatom. Actually, the lowest excitation energies observed, 2.3 eV and 1.5 eV for sodium and potassium, are very close to the atomic sodium 3s − 3p and potassium 4s − 4p transitions at 2.1 eV and 1.6 eV, respectively. The free-atom Na 3p and K 3p levels are at −3.0 eV and −2.7 eV. If we add the low-coverage excitation energy 3.5 eV to these energies we get −6.5 eV and −6.2 eV, which are approximately in the center of the nickel d-band. The initial state, which is plausibly the bonding state, might very well be peaked there as a consequence of the bonding.

The observed loss energies are plotted versus θ in Fig. 5. The solid curves were calculated from

$$\Delta E_s = \Delta E_{so} + \Phi_{if}$$

where ΔE_{so} is the extrapolated excitation energy at $\theta = 0$ and Φ_{if} is given by Eq. (8). Φ_{if} was derived for a constant d-value from our experimental $\Delta\phi$ values shown in Fig. 6. The agreement is encouraging and certainly supports the proposed mechanism.

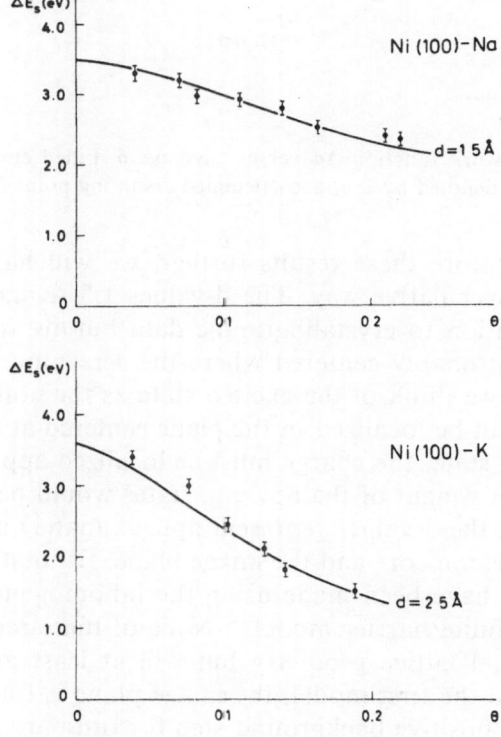

Fig. 5. Experimental energy losses ΔE_s (filled circles) versus coverage θ. Solid curves denoted by d are calculated from the electrostatic potential as described in the text.

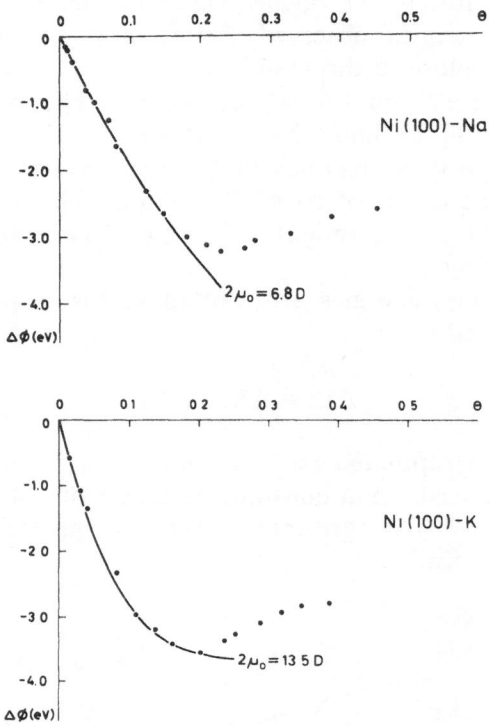

Fig. 6. The change in work function $\Delta\phi$ versus coverage θ. Filled circles are experimental values and solid curves denoted by $2\mu_0$ are calculated assuming point depolarization.

In order to explore these results further, we will have to specify the model in a rather speculative way. The d-values 1.5 Å and 2.5 Å of Fig. 5 have no simple relation to crystallographic data but are related to Φ_{if}. The bonding charge is probably centered where the screening charge is, i.e., in the image plane. If we think of the excited state as the nonbonding np_x, np_y states, the charge will be localized in the plane centered at the adatoms, and for the antibonding state, the charge must be localized approximately at the adatom. The greater weight of the np_x, np_y states would perhaps favor their contributions. Thus the d-values represent approximately an upper limit for the separation of the ion core and the image plane. Estimates of the position of the image plane have been made using the inhomogeneous background model[13] and the infinite barrier model.[23] None of these models account for the three-dimensional lattice geometry but will at least give us some idea about the matter. In the first model, the image plane is found to be located ≈0.5 Å outside the positive background step for ordinary electron gas densities. The step is thought to be located halfway between atomic planes, i.e., for Ni(100) ≈0.9 Å outside the center of the last nickel layer. We should accordingly add $0.5 + 0.9 = 1.4$ Å to our d-values to find an estimate of the separation adatom layer $-$ the last nickel layer. The values 2.9 Å and 3.9 Å derived in this way are in fact in accordance with the above-presented estimates from electron reflectivity data.

If we stick to the assumption that the d's determined above reflect the separation between the adatom and the image plane, then we are in the position to determine the charging q from the dipole moments μ according to Eq. (6). The total dipole moments 2μ derived in this way are shown in Fig. 7 and yield the extrapolated zero coverage values $2\mu_0 = 6.8$ Debye and $2\mu_0 = 13.5$ Debye for sodium and potassium, respectively. The corresponding q values are $q_0 = 0.47$ (sodium, $d = 1.5$ Å) and $q_0 = 0.56$ (potassium, $d = 2.5$ Å). These figures are considered below in connection with the shifts of the sodium 2p and potassium 3p core-level excitation thresholds. From the estimates of the adatom charge and the change in the electrostatic potential, Φ, we can make a trial investigation of the position and width of the resonance according to Eq. (3). Notice that this resonance describes the neutralization charge, q^-, and is treated as if the charge were *centered* on the adatom. We thus assume the resonance to be shifted by Φ_{if}, i.e., $E_\varphi = E_{\varphi o} + \Phi_{if}$, where $E_{\varphi o}$ is the $\theta = 0$ position. Furthermore, from Eqs. (3) and (4) we get

$$q^- = \frac{1}{\pi}\left[\tan^{-1}\left(\frac{E_F - E_\varphi}{\Gamma}\right) + \frac{\pi}{2}\right]$$

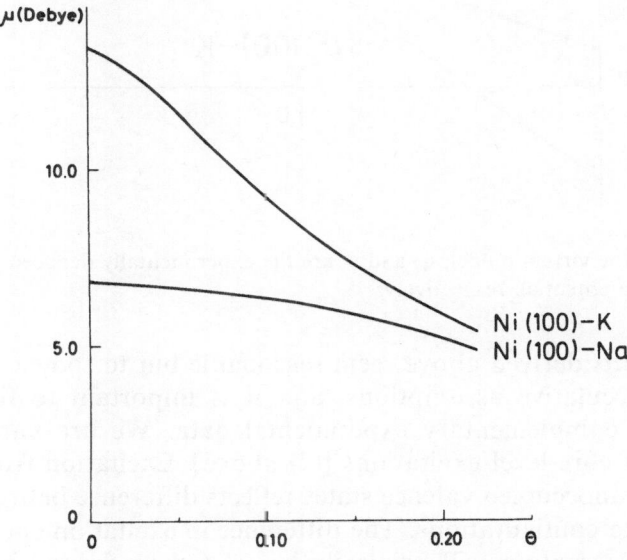

Fig. 7. Experiment total dipole moments 2μ versus coverage θ.

From the μ's and d's determined above, we can derive Φ and q⁻. The relations tan π(q⁻ − 1/2) versus Φ are shown in Fig. 8 and are indeed quite linear in the low-coverage regime (solid line). We find $E_F - E_{\varphi 0} = 0.4$ eV, $\Gamma = 4.6$ eV for sodium and $E_F - E_{\varphi 0} = 0.4$ eV, $\Gamma = 2.3$ eV for potassium. The large Γ values do not seem very meaningful. These results are, however, critically dependent on the d-values used, e.g., for $d = 1.0$ Å we get $\Gamma = 1.7$ eV and $E_F - E_{\varphi} = -1.3$ eV for sodium. If we think of the neutralization charge as a part of the screening charge distribution protruding into the adatom, it is *not centered* on the adatom and it is doubtful whether a resonance treatment of q⁻ is very useful in this case.

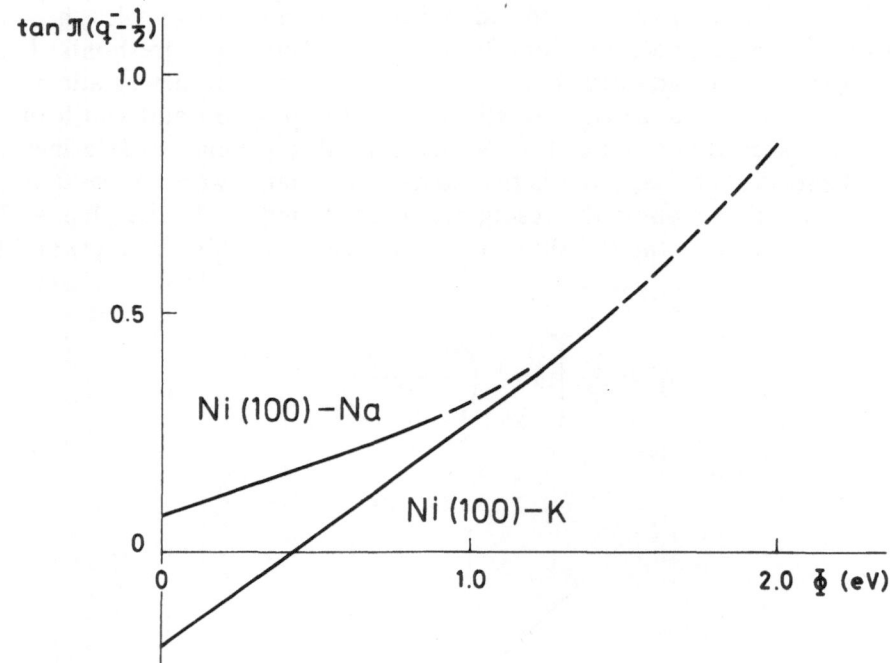

Fig. 8. Test of the virtual model; q⁻ and Φ are the experimentally deduced electronic charge and electrostatic potential, respectively.

The results derived above seem reasonable but to some extent they rely on a few speculative assumptions, and it is important to find more correlating and complementary experimental data. We are particularly concerned about core-level excitations [(2) above]. Excitation from an adatom core level to unoccupied valence states reflects difference between the initial- and final-state configurations. The difference in excitation energy relative to that for the free atom will primarily be related to the valence level occupancy q⁻ through the intra-atomic coulomb repulsions in the initial as well as the final states. A crude estimate of this difference would be

$$\Delta E_c - \Delta E_c^o = (1 - q^-)(U_{c-ns} - U_{ns-ns}) \tag{10}$$

where ΔE_c is the excitation energy for the charged atom and ΔE_c^o is that for the free atom. The coulomb repulsions among the core and valence electrons, U_{c-ns}, as well as among the valence electrons, U_{ns-ns}, are assumed to be equal in the two cases, though differences in the actual charge distributions may be important.

The energy-loss spectra around the 2p and 3p excitation thresholds for different coverages of sodium and potassium are shown in Fig. 9. The fine structure in the second derivative, $I_s''(E)$, for potassium is due to the spin-orbit splitting, 1.3 eV, of the potassium 3p level which is unresolved for the sodium 2p level. The threshold signal increases in intensity and the threshold shifts to lower energies as θ increases. The shifts are summarized in Fig. 10. For sodium we extrapolate $\Delta E_{2p} = 31.9$ eV for $\theta = 0$, which quickly decreases with θ to the value 30.7 eV. This is also the value for a thick film and agrees with the soft X-ray absorption threshold at 30.7[24] for sodium metal. The corresponding figures for ΔE_{3p} from potassium are 19.9 eV and 19.2 eV. In order to account for these shifts, we rely on soft X-ray absorption data (SXA) for free sodium atoms[25] and sodium metal[25]

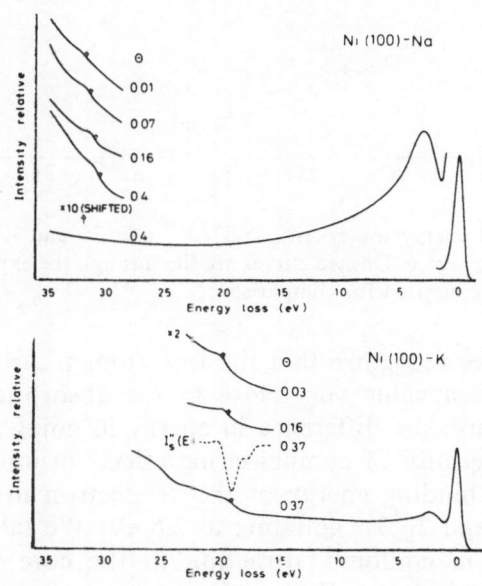

Fig. 9. Experimental energy-loss spectra around the Na 2p and K 3p core-level excitation thresholds. The spectra recorded at various coverages θ are amplified and shifted along the ordinate (intensity). $I_0''(E)$ (at $\theta = 0.37$ for potassium) is a second derivative spectrum. The primary electron beam energies were 72 eV (Na) and 42 eV (K) relative to the nickel Fermi level.

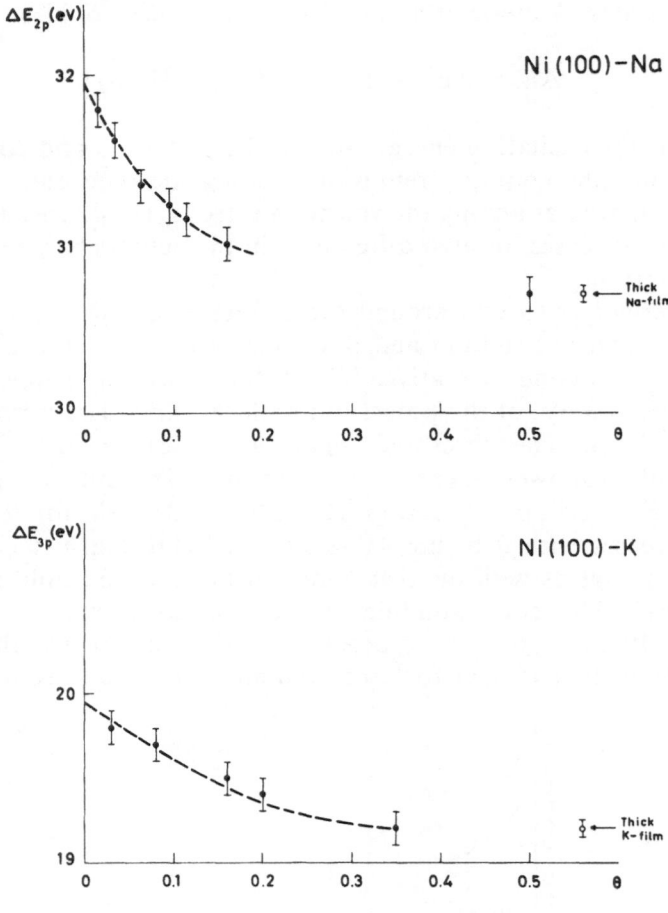

Fig. 10. Experimental energy losses ΔE_{2p} and ΔE_{3p} (Na 2p and K 3p core-level excitation thresholds) versus coverage θ. Dashed curves are fits through the experimental points. Points denoted thick films are inserted for comparison.

From such data we recognize that the free-atom transition $2p^6 3s \rightarrow 2p^5 3s^2$ occurs at 30.8 eV, a value very close to the absorption edge 30.7 eV of sodium metal; hence, the difference in energy in going to the solid state is small, plausibly because of compensating effects in initial and final states. The difference in binding energy of the 3s electron in the free-atom configurations $2p^6 3s$ and $2p^5 3s^2$ amounts to 2.6 eV. We take this difference to be the difference in coulomb attraction to the core hole and repulsion among the valence electrons. Hartree-Fock calculations[26] give $U_{2p-3s} - U_{3s-3s} = 7.8 - 5.9 = 1.9$ eV. Using these figures we can get an estimate of $(1 - q^-)$ from Eq. (10). From our experimental shift $31.9 - 30.7 = 1.2$ eV and the experimental value 2.6 eV, we get $1-q^-) = \dfrac{1.2}{2.6} = 0.46$, and from the H-F value 1.9 eV we get $(1 - q^-) = \dfrac{1.2}{1.9} = 0.63$. These figures are in accordance with the $\theta = 0$ charging state derived above. Support for this treatment is found from our experimental observation of the sodium 2p threshold at 33.1 eV for Na_2O. The shift 2.4 eV from the clean-metal value is close to the above given value 2.6 eV corresponding to complete ionization. For potassium we do not have SXA data and use the H-F value $6.0 - 4.8 = 1.2$ eV, which gives $(1 - q^-) = \dfrac{0.7}{1.2} = 0.58$. Through approximative, this exercise seems to justify our charge estimates. The θ dependence, however, as calculated from Eq. (10) and $(1 - q^-)$ (derived from $2\,\mu$) will, in particular for sodium, be smoother than in Fig. 10.

ACKNOWLEDGMENT

The author is indebted to U. Jostell and B. Kasemo for stimulating collaboration during various parts of this work. Financial support from the Swedish Natural Science Research Council is greatfully acknowledged.

REFERENCES

1. Bennet, A. J., and Falicov, L. M., *Phys. Rev.*, **151**, 512 (1966).
2. Gadzuk, J. W., *Surface Sci.*, **6**, 133 (1967); **6**, 159 (1967).
3. Newns, D. M., *Phys. Rev.*, **178**, 1123 (1969).
4. Gadzuk, J. W., *Phys. Rev.*, **154**, 662 (1967).
5. Muscat, J. P., and Newns, D. M., *J. Phys. C.*, **15**, 2630 (1974).
6. Schmidt, L. D., and Gomer, R., *J. Chem. Phys.*, **45**, 1605 (1966).
7. Andersson, S., and Kasemo, B., *Surface Sci.*, **25**, 273 (1971).
8. Gerlach, R. L., and Rhodin, T. N., *Surface Sci.*, **19**, 403 (1970).
9. Andersson, S., and Pendry, J. B., *J. Phys.*, **06**, 601 (1973).
10. Andersson, S., and Pendry, J. B., *Solid State Commun.*, **16**, 563 (1975).
11. Andersson, S., and Kasemo, B., *Surface Sci.*, **32**, 78 (1972).
12. Hagstrum, H. D., *Surface Sci.*, **30**, 5C5 (1972).
13. Lang, N. D., *Phys. Rev.*, **B4**, 4334 (1971).
14. Lang, N. D., and Kohn, W., *Phys. Rev.*, **B7**, 3541 (1973).
15. Smith, J. R., Ying, S. C., and Kohn, W., *Phys. Rev. Letters*, **30**, 610 (1973).
16. Stern, F., *Phys. Rev. Letters*, **18**, 546 (1967).
17. Ngai, K. L., Economou, E. N., and Cohen, M. H., *Phys. Rev. Letters*, **24**, 61 (1970).
18. Gadzuk, J. W., *Phys. Rev.*, **B1**, 1267 (1970).
19. Newns, D. M., *Phys. Letters*, **39A**, 341 (1972).
20. Andersson, S., and Jostell, U., *Solid State Commun.*, **13**, 833 (1973).
21. MacRae, A. U., Müller, K., Lander, J. J., Morrison, J., and Phillips, J. C., *Phys. Rev. Letters*, **22**, 1048 (1969).
22. Thomas, S., and Haas, T. W., *Solid State Commun.*, **11**, 193 (1972).
23. Newns, D. M., *Phys. Rev.*, **B1**, 3304 (1970).

24. Haensel, R., Keitel, G., Schreiber, P., Sonntag, B., and Kunz, C., *Phys. Rev. Letters*, **23**, 528 (1969).
25. Wolff, H. W., Radler, K., Sonntag, B., and Haensel, R., *Z. Physik.*, **257**, 353 (1972).
26. Mann, J. B., Los Alamos Report, LA-3690.

DISCUSSION on Paper by S. Andersson

EHRLICH: It should be pointed out that there is available a technique, quite different from LEED, to pinpoint the location of adatoms at a surface, and that is field ion microscopy. This technique is complementary to LEED, in the sense that the field ion microscope allows one to visualize individual adatoms, whereas LEED studies have been concentrated upon layers adsorbed at a surface. Using the field ion microscope, Dr. Graham at Illinois has for the first time been able to establish the binding site for a single atom specifically for tungsten self-adsorbed on the (111) plane of tungsten. It turns out that the adatoms sit in holes formed by three atoms in the first lattice layer. There are on the (111) plane two different types of such holes—a deep one with a lattice atom in the third layer, and a shallow hole for which the bottom of the depression is formed by a tungsten atom in the second layer. Direct observations indicate that only a small number of atoms, amounting to ~5 percent of the atoms incident upon a tungsten surface at a low temperature, fall into the shallow or fault sites; however, tungsten atoms held at these fault sites are surprisingly stable. Although the technique is limited to providing information about the location parallel to the surface, and does not fix the spacing perpendicular to the lattice, it is under the right conditions for yielding interesting insights concerning binding sites.

KOHN: Could you explain a little more the nature of the surface collective mode (plasmon?) which you mentioned? Over what range do you observe the $n^{1/2}$ dependence?

ANDERSSON: This mode is treated by several authors and is considered as a collective mode confined to the monolayer. Experimentally, the mode is observed for coverage larger than $\theta = 0.25$ for sodium and $\theta = 0.20$ for potassium. The density dependence of the frequency is ~0.4 for sodium in the range $0.25 < \theta < 0.50$.

KOHN: In considering the interaction of the sodium screening charge dipoles, did you allow for the screening of this interaction by the metal? I would think that this might be quite important.

ANDERSSON: This is probably taken account of by the use of experimental dipole moments in describing the electrostatic interactions.

NEWNS: Perhaps I could usefully comment on this. There are two modes. One is a cesium plasmon-like mode which at zero wave vector has a frequency approximately independent of whether screening in the substrate is included, and the frequency may be approximately proportional to the square root of electron density in the layer. The other mode is at zero wave vector dependent on screening in the substrate and decreases in frequency as the adsorbate concentration increases. Incidentally, this mode goes down to zero frequency at zero wave vector within a conducting model of the layers.

TAMARU: I should like to bring up a system, which is closely correlated with yours and which is interesting from the viewpoint of catalysis. My co-workers and I studied the system of graphite and alkali metals, which turned out to be an active catalyst, when they are in contact, for H_2-D_2 exchange reaction, hydrogenation, and isomerization of olefins, although graphite and alkali metals alone are inert for such reactions as H_2-D_2 exchange reaction and hydrogenation. It would be interesting to study sodium monolayers on graphite, as sodium does not penetrate between the graphite net plane and it has an electronic structure quite different from that of transition metals and still may catalyze the reactions mentioned above.

MASON: How well defined are the Ni(100)Na and Ni(100)K LEED patterns—do individual reflections have significant intensities to, say, 150 volts? In the comparison of observed and calculated LEED intensities, can the possibility be excluded that a proportion of the adsorbed alkali atoms occupy onefold symmetry sites?

ANDERSSON: The 2-D ordered structures are visible to 150 eV, while the ring patterns at low coverages are quite diffuse. The possibility of a smaller fraction of adatoms randomly occupying, e.g., onefold sites cannot be excluded since they will simply be defects.

LOW-ENERGY ELECTRON DIFFRACTION STUDIES OF CLEAN AND OVERLAYERED SURFACES*

*M. B. Webb, J. C. Buchholz**,*
*M. G. Lagally***, and W. N. Unertl⁺*

University of Wisconsin
Madison, Wisconsin

ABSTRACT

Low-energy electron diffraction is the most natural tool for surface crystallography. The surface unit mesh is readily and unambiguously determined. Analysis of diffracted intensities is required to determine atomic coordinates within the unit mesh. This can be done either by dynamic-theory calculations or data-reduction methods.

* Research supported in part by National Science Foundation Grant No. GH 36354 and in part by Grant No. DMR74-11971.
** Present address: Department of Chemistry and Lawrence Berkeley Laboratory, University of California, Berkeley, California.
*** Alfred P. Sloan Foundation Fellow.
⁺ Present address: Department of Materials Science and Engineering, Cornell University, Ithaca, New York.

Examples of the application of the latter type of analysis to a clean surface and to an adsorbed overlayer are given to illustrate the problems and the potential of the method. The capabilities of low-energy electron diffraction (LEED) for catalysis studies are summarized.

1 INTRODUCTION

There can be no question of the important impact that conventional X-ray crystallography has had on our understanding of solids; in fact, the atomic coordinates are generally a necessary input to theories of bulk properties. To a large extent, lack of corresponding knowledge of surface structure has impeded development of a similarly complete understanding of surfaces and processes taking place upon them. The diffraction of low-energy electrons[1,2] is the most natural technique to provide this information since their scattering takes place in the first few atomic layers. The most elementary low-energy electron diffraction (LEED) experiments, using visual display techniques and observing "spot patterns", have made a tremendous contribution to surface science in that the size and symmetry of the unit mesh, i.e., the periodicity of the structure parallel to the surface, is readily determined. This knowledge greatly limits the number of structures, i.e., atomic configurations, that need to be considered in any problem, and often with additional crystal-chemical knowledge, likely structures can be determined with considerable confidence. In addition, these simplest observations are now used rather routinely to characterize the surfaces studied in other investigations.

The interpretation of the spot patterns depends only on grating equations, which is equivalent to the conservation of the electron momentum parallel to the surface to within a reciprocal-lattice vector of the unit mesh. Because of the two-dimensional periodicity, two Laue conditions are satisfied so that

$$\mathbf{S}_{||} = \mathbf{G}_{hk} \tag{1}$$

where $\mathbf{S} = \mathbf{k} - \mathbf{k}_0$ is the diffraction vector or momentum transfer and $|\mathbf{G}_{hk}| = \frac{2\pi}{d_{hk}}$ is a reciprocal-lattice vector of the surface mesh. Here the $\mathbf{k}$'s are the propagation vectors ($|\mathbf{k}| = \frac{2\pi}{\lambda}$) of the diffracted and incident beams and d_{hk} is the interrow spacing of the surface lattice. The diffraction condition is illustrated in Fig. 1. Because the surface destroys the translational symmetry in the normal direction, the third Laue condition is relaxed and diffracted intensity is distributed continuously along the rods in reciprocal space defined by Eq. (1).

To go beyond the determination of the **G**'s, or equivalently the d's, and determine the atomic coordinates within the unit mesh, one is required to interpret the diffracted intensities. In conventional X-ray crystallography, this is relatively simple. There the intensity is given by

$$I \propto \sum_{ij} f_i f_j \, e^{i\mathbf{S}\cdot(\mathbf{r}_i - \mathbf{r}_j)} \qquad (2)$$

where the **r**'s are the atomic coordinates and the f's are known atomic scattering factors. Usually this expression is evaluated for assumed structures and compared with measured intensities. Alternatively, the Fourier transform of the intensity is calculated to give directly the pair correlation function of the structure, which is the maximum information available from a diffraction experiment alone.

For low-energy electrons, the situation is much more complex because the electrons interact strongly with the atoms in the solid. They suffer both inelastic processes which remove them from the coherent beam and they have appreciable probability of being multiply scattered elastically. Whereas single scattering alone, as included in Eq. (2), would give a series of simple broad maxima in the intensity along the reciprocal-lattice rods illustrated in Fig. 2, the actual intensity distribution is very complex because of multiple scattering. One is required to treat the attenuation of the elastic beam of

SPOT PATTERN

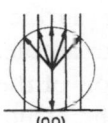

(00)

Fig. 1. Diffracted beams emerge in discrete directions satisfying $\mathbf{S}\| = \mathbf{G}\|$. This is illustrated in direct and reciprocal space on the upper and lower panels, respectively. The vertical lines in reciprocal space are the reciprocal-lattice rods of the surface two-dimensional lattice, each specified by a $\mathbf{G}_{hk}$.

Fig. 2. The single scattering would give a series of broad maxima along the reciprocal-lattice rods. Maxima would be centered at positions satisfying the third Laue condition and have a width depending on the attenuation of the electrons.

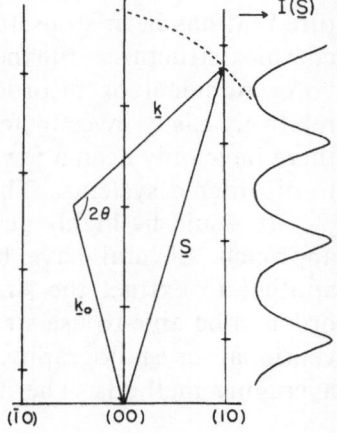

electrons and the multiple scattering in a self-consistent theory and use it to evaluate the intensity for assumed structures until one finds agreement with the observations.[2] This is a complex problem for at least two groups of reasons. First, it is a very large calculation. Inclusion of a fair number of partial-wave phase shifts is required to describe the scattering from each atomic potential since this scattering is rather strongly peaked in the forward direction. Even more partial waves are required when the thermal vibrations of the atoms are included in the calculation since then, in effect, the scattering is even more strongly peaked. Also, expansion of the wave field in the solid at the energies of interest requires keeping many beams or Bloch waves. This is particularly true for overlayer structures with their often large unit meshes. The second group of reasons stems from our incomplete knowledge of the various parameters that are required in the calculation. These include the inner potential and its variation in the immediate vicinity of the surface; the inelastic cross sections, which so far have generally been treated only as a spatially constant imaginary potential; and the thermal vibrational amplitudes, which are very different from those in the bulk. Ordinarily these parameters are considered as adjustable, which thus multiplies the number of model calculations required and often makes fits ambiguous.

These problems are more severe for overlayer structures, which require more parameters and offer a large number of plausible atomic structures to be investigated.

It is a tribute to the efforts of many investigators that realistic calculations have been done for a variety of clean surfaces and a few simple cases of overlayer structures. However such calculations remain large, expensive, and often inaccessible. There are additional limitations making dynamic calculations a less versatile and convenient basis of a surface crystallography than X-ray diffraction is for bulk structures. For example, in principle, there is no simple relation between the Fourier inversion of the multiply scattered intensity and the pair distribution function of the structure that has been so useful for X-ray crystallographers when they deal with complex structures. Further, one would like to investigate systems which involve nonideal or disordered structures. While such problems would be relatively easily investigated if single-scattering treatments were appropriate, there have only been a few treatments of the application of dynamic theories to disordered systems.[3] This is apparently an even more complex problem.

It would be highly desirable to have an alternative and complementary approach. Several have been proposed, all of which try in one way or another to extract the singly scattered intensity from the complex data in order to be able to use straightforward modifications of the analysis of conventional crystallography.[4-9] These include both Fourier inversion and averaging methods. The method of averaging data at constant momentum transfer has been most extensively tested. Presented below is a review of its status, along with a discussion of the problems, and descriptions of preliminary applications to overlayer systems.

If one generalizes Eq. (2) to include the multiple scattering, one can separate out those terms that involve only single scattering. These depend on the incident and diffracted beams only through the particular combination $S = k - k_o$. Of the other terms, which involve the multiple scattering contributions to the intensity, essentially all* depend explicitly on k and k_o, and at fixed S they oscillate as k_o is varied. Therefore, if one averages intensities measured at fixed S for a sufficient range of k_o, the desired single scattering remains but the multiple-scattering contributions to the average tend toward zero.

Experimentally, the average is accomplished with the geometries indicated in Fig. 3.

Since the range in the variation of k_o is limited, averaging out the multiple-scattering contributions to the intensity is never complete. The question is whether the available range is sufficient so that the residual multiple-scattering contributions are small compared with the desired single scattering. This is essentially an experimental question to be tested by seeing that independent averages over different ranges of data give the same result. That this can be the case has been shown on several clean surfaces: Ag(111)[4], Ni(111)[4], Ni(100)[10], W(110)[13], Zn(0001)[14], and Cu(111)[11,12]. Some results for Ni(100)[10] are presented here to demonstrate the effectiveness of the procedure and the self-consistency of the analysis and also to demonstrate the parameters that enter the interpretation. It is generally the insufficiently accurate knowledge of these parameters which limits the precision of the interpretation, but these are the same parameters or their

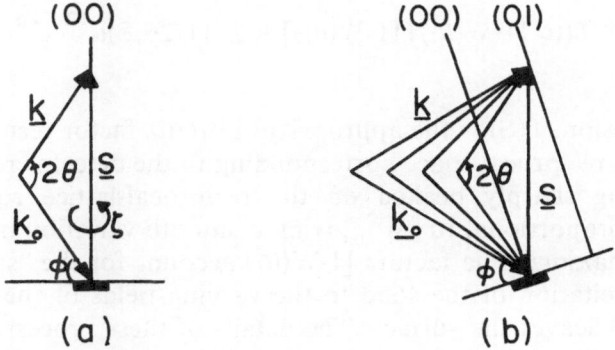

Fig. 3. Two experimental procedures to measure the diffracted intensity at fixed S while k_o is varied in reciprocal space: in (a) the crystal is rotated about S changing the azimuthal orientation of the triangle S, k_o, k; in (b) the energy and angle of incidence are changed simultaneously. In the present apparatus, the azimuthal variation is available only for the specular beam.

equivalent that must enter any analysis. They correspond to the second group of difficulties mentioned in connection with the full dynamic calculations.

2 APPLICATION TO CLEAN SURFACES: Ni(100)

Figure 4 shows an example of some measured intensity profiles (i.e., the intensity along the reciprocal-lattice rod plotted as a function of the incident energy) and Fig. 5 shows the resulting average. The average certainly shows the expected series of broadened peaks with only small additional structure attributable to the residual multiple scattering. However, these peaks are not at integral values of S_n/S_o but are shifted to lower values, corresponding to lower values of the electron energy. In general, such shifts may be due to a variety of causes but most importantly to either the inner potential or an expansion of the interplanar spacing in the vicinity of the surface. In fact, the observed shifts here are due to the inner potential, which is demonstrated by the fact that a relaxation of the interplanar spacing, while causing a shift in the positions of the intensity maxima, must also cause changes in the profile shape, which are not observed. We return to this point later. The observed shifts correspond to an energy-dependent inner potential shown in Fig. 6. The dotted curve corresponds to the calculated result of Pendry[15] who considered the effect of the forward scattering on the phase of the propagating wave. Partly for this reason Pendry would call the averaged intensity the "pseudokinematic" or "pseudosingle" scattering.

This experimentally averaged intensity is now to be compared to the expected singly scattered intensity given by

$$<I(S)> \propto \mathcal{L}(S) [1-W(\theta_1)] [1-W(\theta_2)] < \sum_i | f_i(2\theta,E)\alpha^{m_i} e^{iS \cdot (r_i - r_i)}|^2 > \quad (3)$$

In this expression, $\mathcal{L}(S)$ is the appropriate Lorentz factor resulting from the integration in reciprocal space corresponding to the detector response to the intensity being sharply peaked on the reciprocal-lattice rods. It is approximately proportional to $|S|^{-2}$, giving a smooth variation in the envelope of intensity maxima. The factors $[1-W(\theta)]$ account for the "surface losses", i.e., to the excitation of the solid to the varying fields of the electron as it approaches or leaves the surface. The details of these processes are not understood, but they can be approximately evaluated for beams of sufficiently large energy and grazing angles from known optical constants. They also are slowly varying with $|S|$. These surface losses are expected to change upon the adsorption of an overlayer. The f's are atomic scattering factors. For a single-component system, they may be factored out of the sum in Eq. (3).

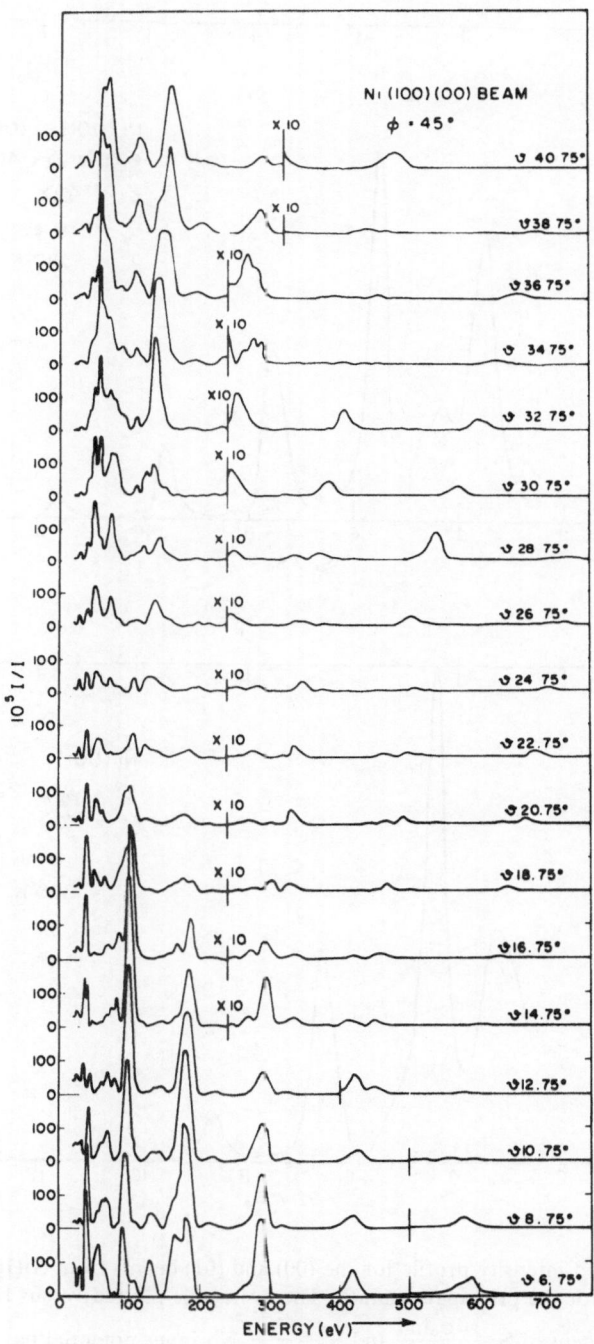

Fig. 4. Intensity profiles for the specular beam from Ni(100) for various angles of incidence and along a principal azimuth. The data correspond to the procedure illustrated in Fig. 3(b) except they are for the (00) beam; θ is the angle of incidence.

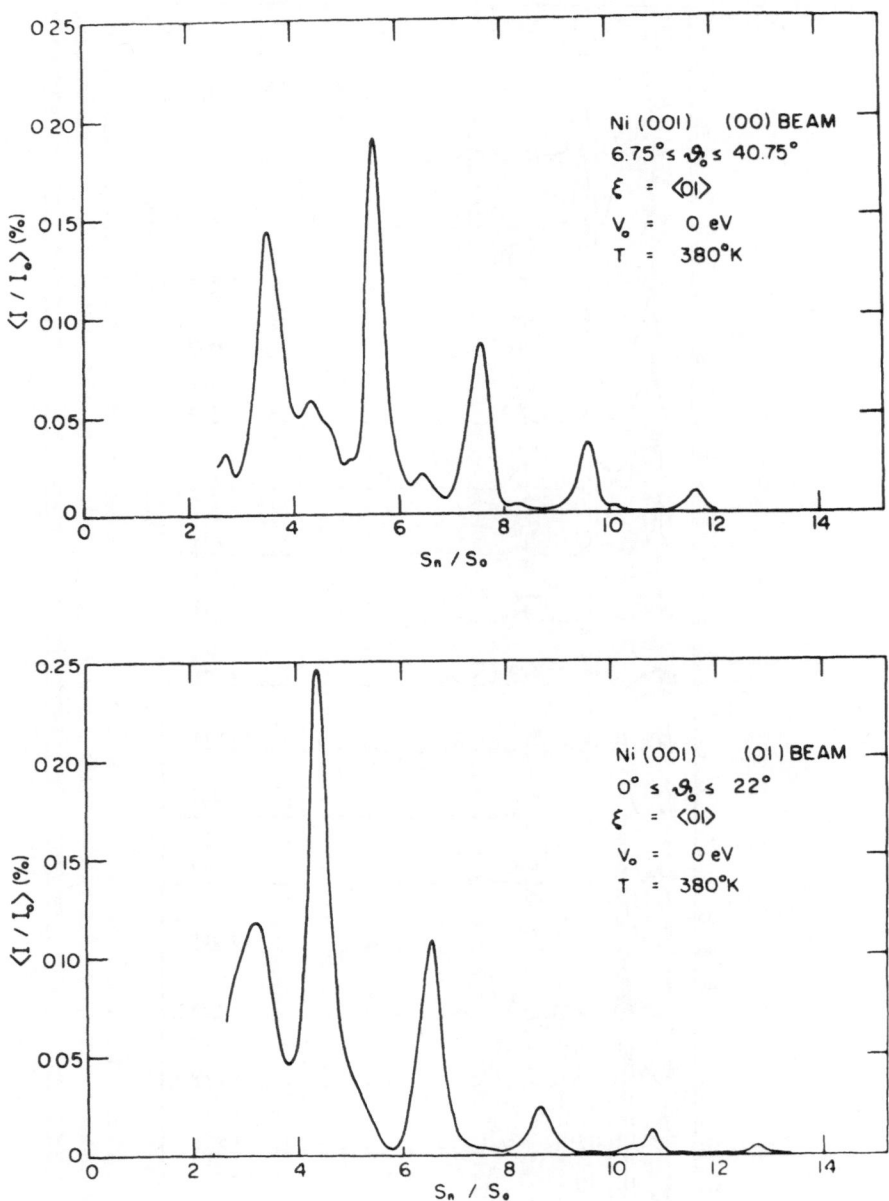

Fig. 5. The averaged intensity profile for the (00) and (01) beams from Ni(100). The averages include only data for the principal azimuth. The average is plotted versus S_n/S_o, where S_n is the normal component of $S = \dfrac{4\pi\sin\theta}{\lambda}$ and $S_o = \dfrac{2\pi}{d_{100}}$. No inner potential has been included to calculate S.

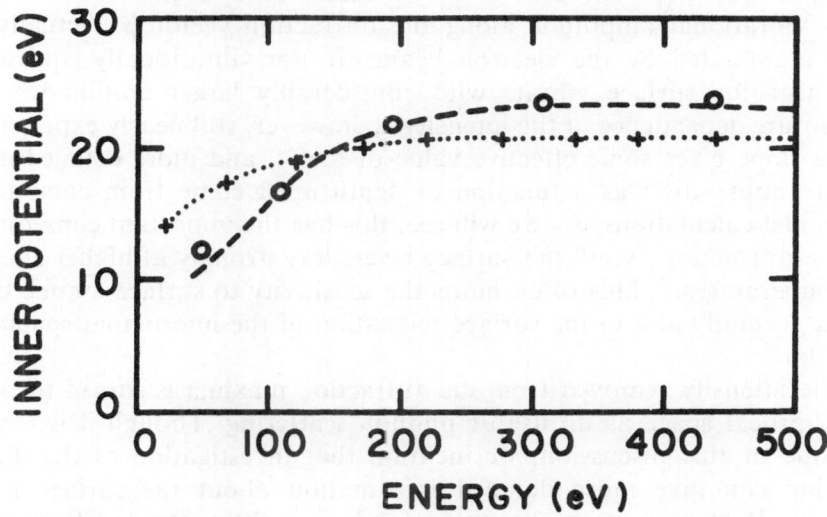

Fig. 6. The energy dependence of the inner potential derived from the displacement of maxima in the averaged intensity from integral values of S_n/S_o. The dotted line is that calculated by Pendry.[15]

The appropriately averaged $\langle|f_i(2\theta,E)|^2\rangle$ for back angles is essentially the same as for the free atom. For a several-component system the f's cannot be factored, and, further, the phases of the f's, i.e., the phase shift on scattering, must be known. This causes some additional complexity in the calculation of the single scattering, but, alternatively, f's can provide additional information since it is possible to choose regions of 2θ and E where the f's for different component atoms are very different. Doing this is the analogue of isomorphous replacement in X-ray crystallography. The $\langle|f(2\theta,E)|^2\rangle$ is somewhat modified by multiple scattering from equivalent multiplets of atoms, but these effects are small and can be calculated approximately or determined experimentally from the scattering from a disordered system.[5] The effect of $\langle|f(2\theta,E)|^2\rangle$ on the intensity is a rather slow but not necessarily monotonic modulation of relative peak intensities.

In Eq. (3) the coordinates r_i, r_j are the instantaneous positions of the atoms. If coordinates are separated into equilibrium positions and thermal displacements from equilibrium, and if all the atoms are vibrationally equivalent, the effect of thermal vibrations is to reduce the rigid-lattice diffracted intensity by a Debye-Waller factor $e^{-2M} = e^{-\langle(S\cdot u)^2\rangle}$. Since in the high-temperature limit $\langle u^2\rangle \propto T$ the diffracted intensities decrease exponentially with increasing temperature and more rapidly for larger S, i.e., for higher energies. The Debye-Waller factor is a measure of the mean square vibrational amplitude along the diffraction vector **S**. Actually the atoms illuminated by the electron beam are not vibrationally equivalent; those near the surface vibrate with considerably larger amplitudes. The temperature dependence of the intensity is, however, still nearly exponential, but the slope gives some effective value of $<u^2>$, and more detailed information about $<u^2>$ as a function of depth must come from comparison with model calculations. As we will see, this has the important consequence that the diffraction "sees" the surface layers less strongly at higher energies and temperatures.[16] This often limits the sensitivity to surface atomic coordinates generally and to the surface relaxation of the interplanar spacing in particular.

The intensity removed from the diffraction maxima is spread throughout reciprocal space as diffuse or phonon scattering. Though it is beyond the scope of this discussion, we mention that investigation of the diffuse scattering can give more detailed information about the surface lattice dynamics. It also must be accounted for before the scattering from other disordered systems can be interpreted.

The remaining parameter in Eq. (3) is α, which accounts for the attenuation of an electron beam as it moves through the material. By definition α is the ratio of the scattered amplitudes originating from successively deeper atomic layers. For the special case of uniform attenuation, α is related to the attenuation coefficient μ by

$$\alpha = \frac{A_{n+1}}{A_n} = e^{\dfrac{-\mu d}{2}\left(\dfrac{1}{\cos\theta_2} + \dfrac{1}{\cos\theta_1}\right)} \tag{4}$$

where μ includes attenuation due to both elastic and inelastic processes. α is typically between 0.3 and 0.7 in the energy range of interest and increases slowly above about 100 eV. Performing the sum in Eq. (3) assuming a constant interplanar spacing gives the intensity

$$I(\mathbf{S}) \propto \frac{1}{1+\alpha^2 - 2\alpha\cos\mathbf{S}\cdot\mathbf{d}} \tag{5}$$

which describes a series of broad maxima centered at $S_\perp/S_o$ = integers. Therefore α determines both the relative intensities and widths of the peaks as illustrated in Fig. 7; α and its energy dependence can be estimated from other experiments like the overlayer attenuation of Auger or photoelectrons or from the averaged intensity profiles. Figure 8 shows μ_T determined from the widths of peaks, from the relative intensities of peaks, from the integrated intensity, and from the relative intensities of the peaks on the (00) average profile. Such values are consistent within their uncertainties and

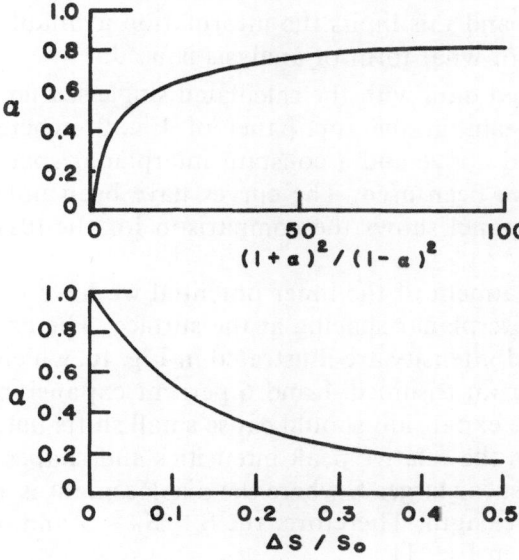

Fig. 7. The top panel gives the variation in the peak-to-valley ratio $\dfrac{(1+\alpha)^2}{(1-\alpha)^2}$ in the interference function from Eq. (5). The bottom panel gives the variation with α of the full width for peaks in the interference function.

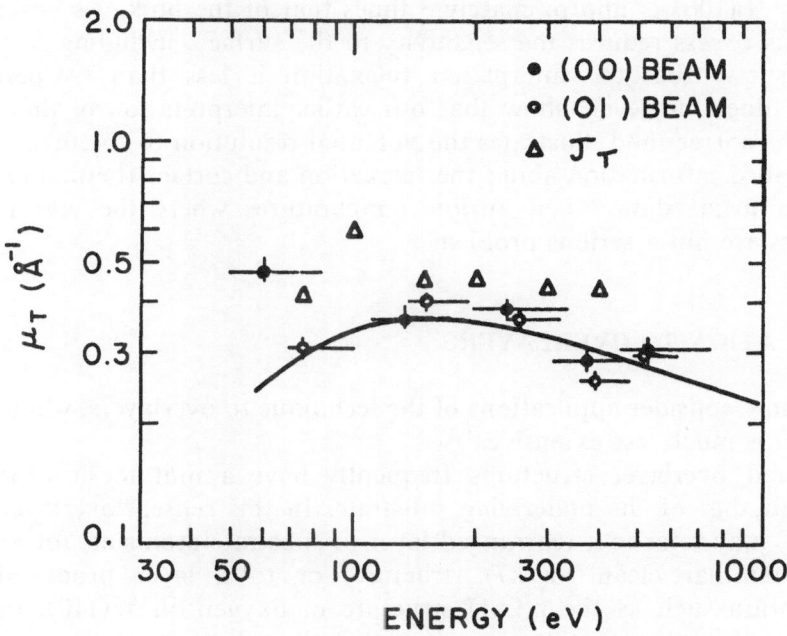

Fig. 8. Values μ_1 determined from ($\bullet$) the widths of the maxima for the specular reflection; ($\bigcirc$) the widths of the maxima for the (01) beam; ($\triangle$) the integrated intensity. The solid line shows μ_1 required to fit the relative peak heights on the (00) beam average.

with other determinations of μ_T. Less is known about the details of the attenuation in adsorbed overlayers and this limits the information available from those experiments regardless of what form of analysis is used.

The comparison of the averaged data with the calculated single scattering is shown for the specular beam in the top panel of Fig. 9, where parameters determined as described above and a constant interplanar spacing equal to that for the bulk have been used. The curves have been normalized at $S_\perp / S_o = 8$. The lower panel shows the comparison for the (01) beam using the same parameters.

In this analysis and in our treatment of the inner potential we have ignored possible relaxation of the interplanar spacing at the surface. The expected effects in the singly scattered intensity are illustrated in Fig. 10, which shows the interference function for an assumed 4 and 6 percent expansion of the outermost layer spacing. The expansion should cause small shifts but, more importantly, large changes in the relative peak intensities and shapes. As expected, these effects are largest for larger S where the displacement is a larger fraction of the electron wavelength. Therefore, the $S_\perp / S_o = 5$ and 6 peaks are examined in more detail in Fig. 11.

Unfortunately, from this comparison it is not correct to conclude that the relaxation of the outermost plane is less than ≈ 1 percent, since this particular analysis has neglected the excess vibrational amplitude at the surface, which for Ni(100) is approximately 3 times that of the bulk. As described above, this excess reduces the sensitivity to the surface; including it in the analysis shows that the interplanar relaxation is less than ≈ 4 percent. Figure 11 does, however, show that our earlier interpretation of the inner potential is correct and illustrates the potential resolution of the procedure. More detailed information about the relaxation and surface thermal expansion must await data taken at low temperatures where the vibrational amplitudes are not a serious problem.

3 APPLICATION TO OVERLAYERS

We now consider applications of the technique to overlayers, where our experience is much less extensive.

Ordered overlayer structures frequently have a unit mesh which is larger than that of the underlying substrate. In this sense, "overlayer" or "selvedge" may refer to a rearranged layer of substrate atoms as, for example, the familiar clean Si(7x7) structure, or to a layer produced by chemisorption such as the p(2x1) structure of oxygen on W(110). In the latter case, the selvedge may contain either the adsorbed species only or both the adsorbed species and those substrate atoms which may be rearranged with the overlayer periodicity. With the larger overlayer unit

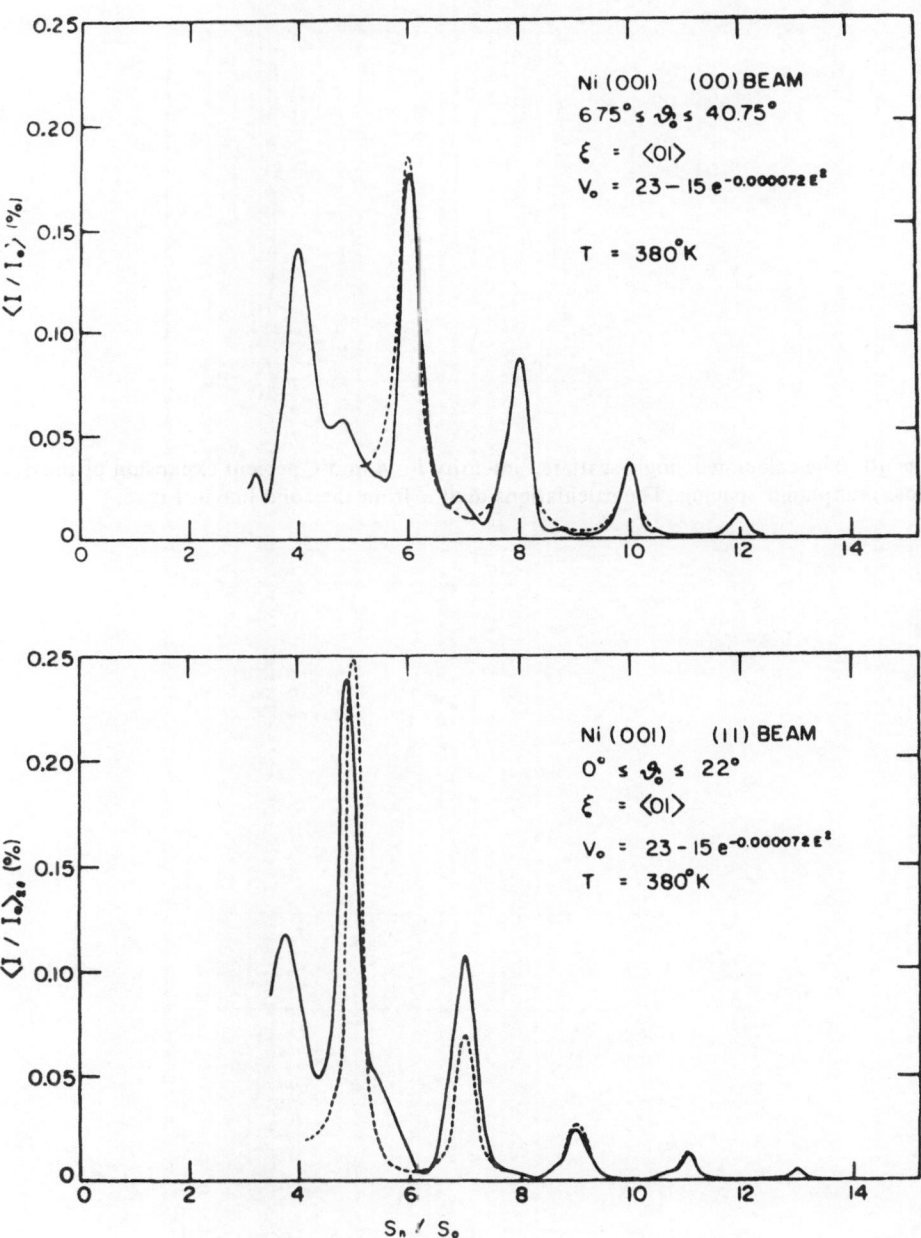

Fig. 9. Comparison of the experimental averaged intensity (solid line) and the calculated single-scattering intensity (dashed line) using the parameters discussed in the text. The calculations used the variation of α related by Eq. (4) to μ_T shown by the solid line in Fig. 8. This was chosen to fit peak heights for the (00) beam. No additional parameters were adjustable to calculate the (01) beam intensity.

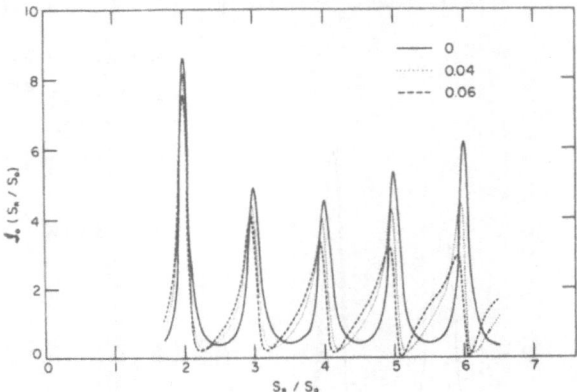

Fig. 10. The calculated single-scattered intensity for 4 and 6 percent expansion of the outer-most interplanar spacing. The calculations used α from the solid line in Fig. 8.

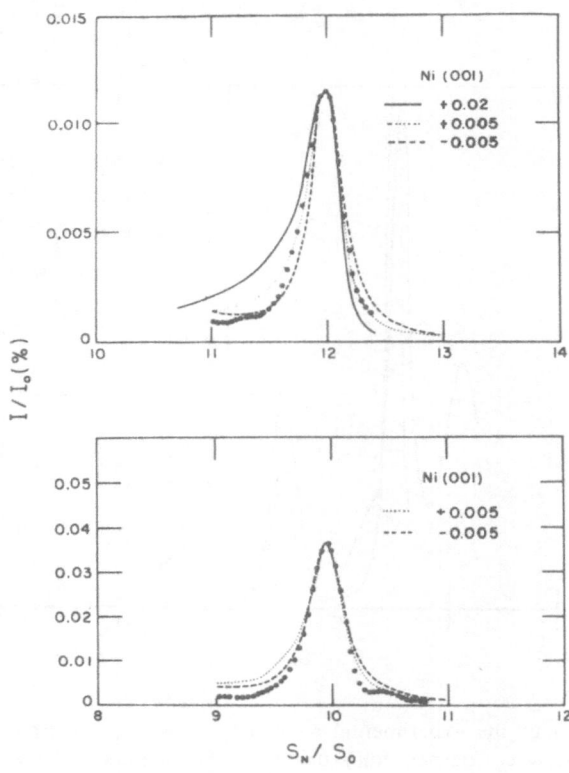

Fig. 11. Comparison of calculated intensity profiles for $\dfrac{\Delta}{d} = +2$ percent and ± 0.5 percent with the averaged data.

mesh, one observes intensity not only on the reciprocal-lattice rods characteristic of the substrate structure but also on additional and more closely spaced rods that we will refer to as "overlayer" or "fractional order" beams. These give the lateral periodicity directly.

In principle, the procedure of averaging at fixed S yields the structure (actually the pair distribution function) of adsorbed overlayers, but the problem is considerably more difficult than for clean surfaces, again for two reasons. First, since the overlayer single-scattering contribution is itself generally small, it is more difficult to extract it from the data; and second, because more parameters are required it is more difficult to calculate the singly scattered intensity for assumed structures. Ordered overlayers with a larger unit mesh than the substrate generally have fewer atoms in the plane and so give a smaller total scattered amplitude; further, the interference distributes this scattered amplitude into a larger number of diffracted beams, making the contribution to each smaller. This scattering is then to be compared to that from the substrate, which comes effectively from several atomic planes instead of one atomic plane. Each of these factors may be of the order of two, and so the overlayer single scattering may be an order of magnitude smaller than that from the substrate. Additionally, the overlayer vibrational amplitudes are often large, reducing their relative contribution still further through their Debye-Waller factor. This small overlayer intensity requires that any residue from multiple scattering due to limited data range be very small. Often one is interested in the case where the overlayer consists of light atoms on a heavy-atom substrate. Generally, the light-atom scattering factor is appreciable only for lower energies and this severely restricts the range of variation in k_0 within which the intensity is sensitive to the overlayer atoms. These features are clear in most experimental data where "overlayer beams" are generally weak compared with substrate beams, where their intensity is appreciable only at lower energies, and where "substrate beams" are very similar for the clean and overlayered surfaces.

The single scattering in the additional fractional-order beams contains information only about the relative positions of scatterers within the overlayer, since in these beams there is no single scattering amplitude from the substrate. If all the atoms whose lateral positions have the overlayer symmetry lie in a single plane, e.g., a single flat plane of adsorbed atoms, the single scattering intensity along the overlayer beams will be smooth and its energy dependence will be dominated by the atomic scattering factor, Debye-Waller factor, etc. The intensity thus contains the information about the lateral position of the atoms within the overlayer, the atomic composition, and thermal vibrations of the overlayer atoms. On the other hand, if there are atoms at several levels in each overlayer unit mesh, there is interference which modulates the intensity along the reciprocal-lattice rods. This would be the case for rearranged substrate layers, for substrate atoms which either have been moved slightly by the binding of the adsorbate or completely rearranged, or for adsorbed multilayers. This is illustrated in Fig. 12, where as an example we show the calculated single-scattered intensity along the (1/2,1/2) rod for several possible arrangements of CO in W(100) c(2x2)-CO. Apart from the overlayer periodicity, this is the easiest sort of information to retrieve since here the intensity does not compete with large substrate intensity.

The position of the overlayer relative to the substrate must be determined from the substrate beams if only single scattering is considered. In practice, this is a more difficult task than for clean surfaces, since the changes are small and often confined to low energy where there is little phase space for averaging.

We illustrate the progress to date with data for the p(2x1) structure of oxygen adsorbed on W(110).[13] The symmetry of the unit mesh and resulting diffraction patterns are illustrated in Fig. 13. The structure exists in two domains and for the data shown the domains are equally weighted, as shown by the symmetry of data. All nonequivalent beams out as far as the (20) beam have been investigated.

There have been previous diffraction studies of this lowest-coverage ordered structure of oxygen on W(110) but with contradictory interpretations.[17] Some have concluded the surface is reconstructed; others have favored a half-monolayer coverage of oxygen on top of the tungsten substrate. To date we know of no dynamic calculation for tungsten or for overlayers upon it.

The pertinent atomic scattering factors are shown in Fig. 14. The oxygen scattering is large, in fact larger than tungsten at low energy at some angles, but falls rapidly with increasing energy, while tungsten scatters appreciably over the whole energy range.

An example of the data for clean W(110) is shown in Fig. 15. The envelope of the data reflects the slowly varying factors including the atomic scattering factor, but in detail the intensity is exceedingly complex. The average of such data is compared with the calculated single scattering in Fig. 16. Some half-order structure remains in the average for this range of data. An analysis of the energy dependence of the inner potential, as for nickel, has not been made for tungsten and here a constant of 27 eV was used. Other information from the data on clean tungsten indicates that any relaxation of the interplanar spacing at the surface is less than 3 percent and that the mean square vibrational amplitude normal to the surface is between 2 and 3 times that of the bulk.[13]

Examples of intensity for some equivalent overlayer beams is shown in Fig. 17. The scale has been increased by a factor of ten from Fig. 16, so it is evident that for energies greater than 80 eV, the overlayer beams are very weak compared with substrate beams. The rapid decrease with increasing

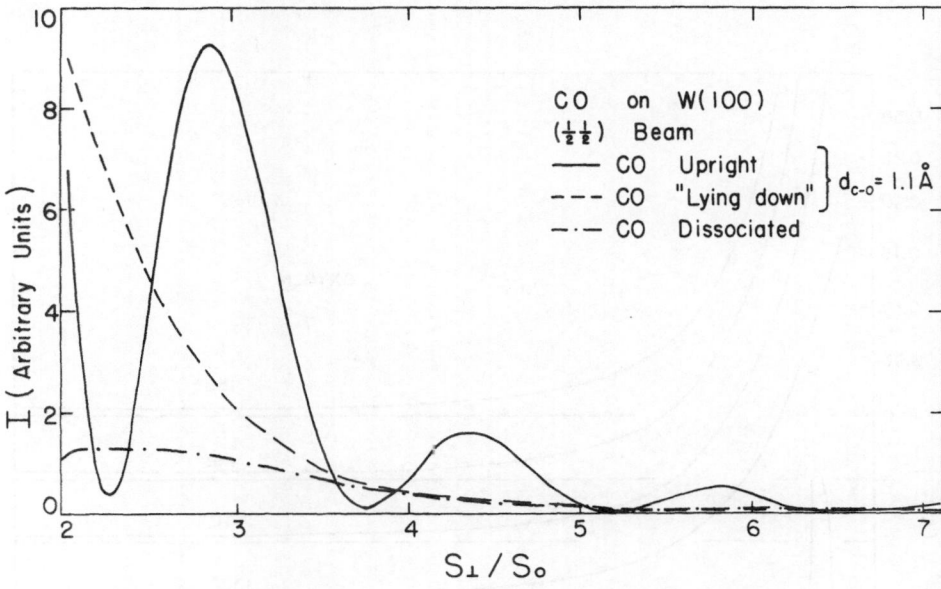

Fig. 12. The interference function along the (1/2,1/2) reciprocal-lattice rod calculated for W(110) c(2x2)-CO for three arrangements of the CO. The oscillation for the upright molecule is due to the interference between the carbon and oxygen planes.[13]

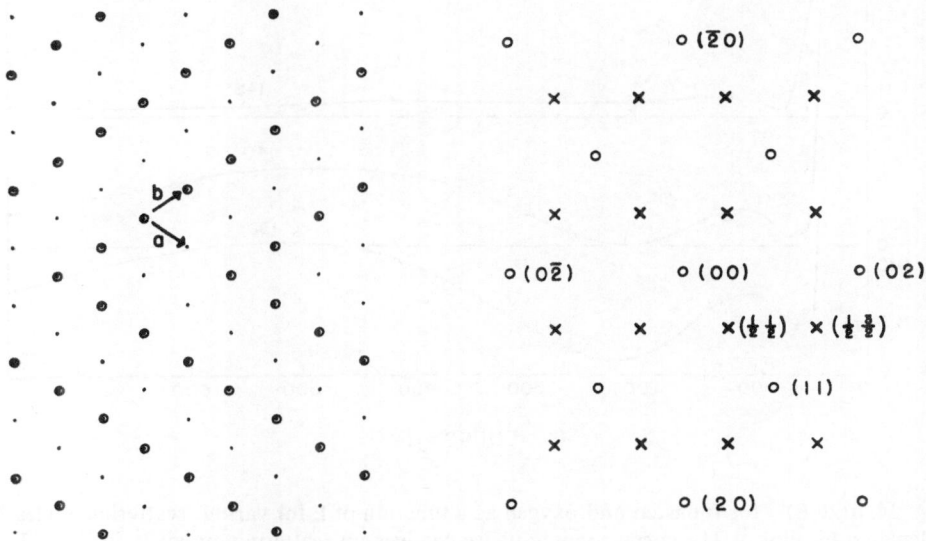

Fig. 13. The unit mesh and diffraction pattern for the p(2x1) structure on the (110) face of a bcc structure: **a** and **b** are basis vectors of the substrate mesh; open circles are the overlayer mesh.

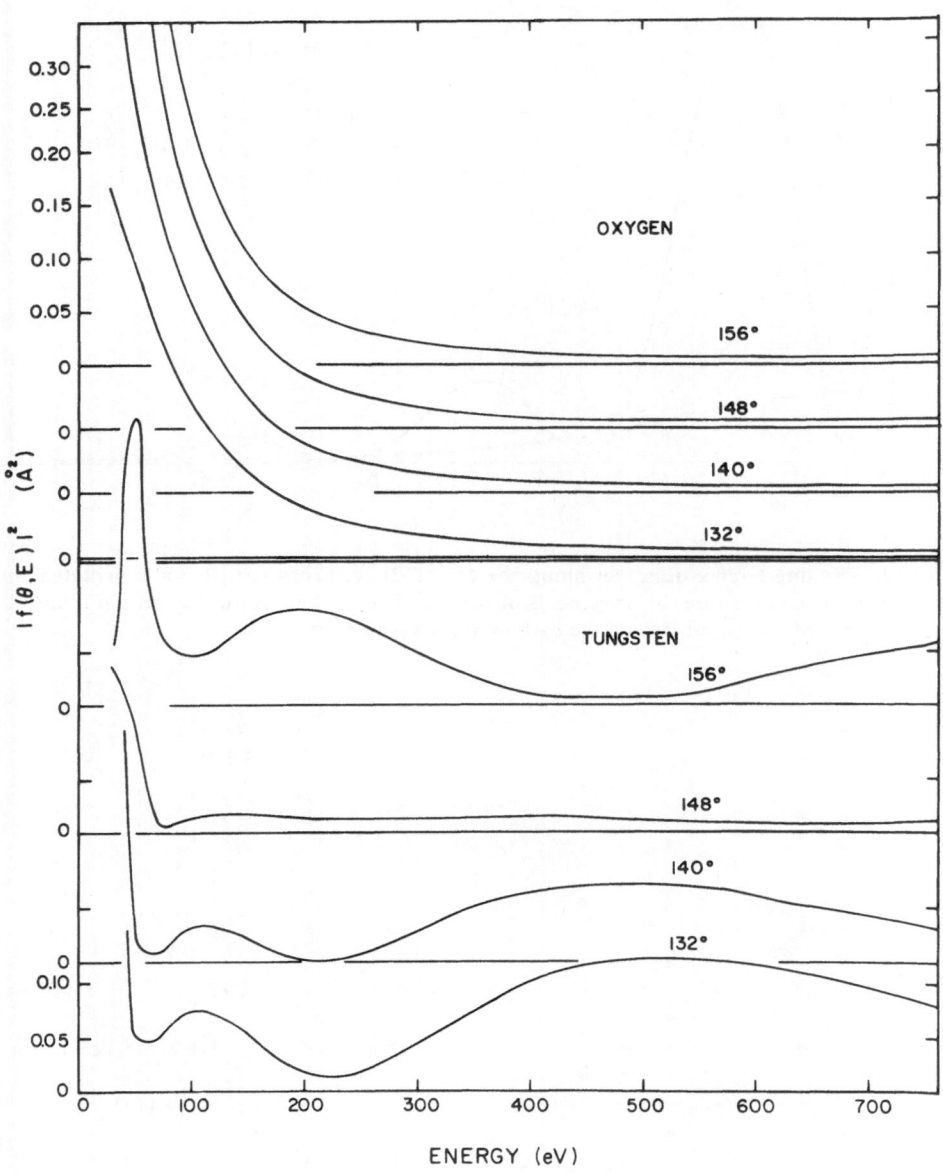

Fig. 14. $|f(2\theta,E)|^2$ for tungsten and oxygen as a function of E for various scattering angles 2θ calculated by Fink.[18] The corresponding phase changes on scattering are given in Ref. 13.

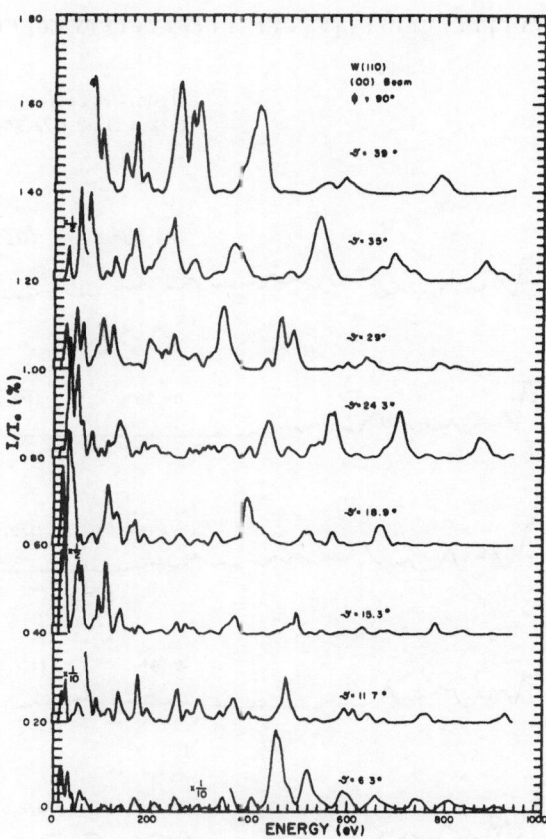

Fig. 15. The normalized specular intensity from W(110).

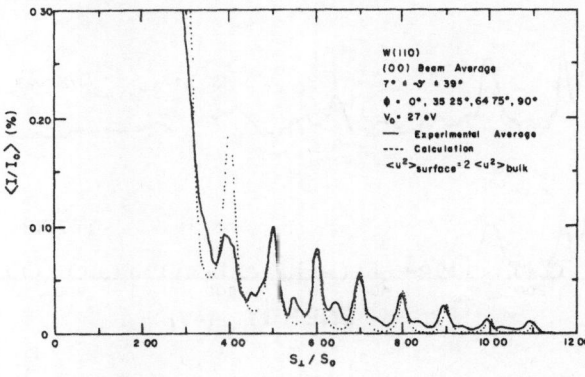

Fig. 16. The averaged intensity for the specular reflection from clean W(110). Parameters used for the calculation are given in Ref. 13.

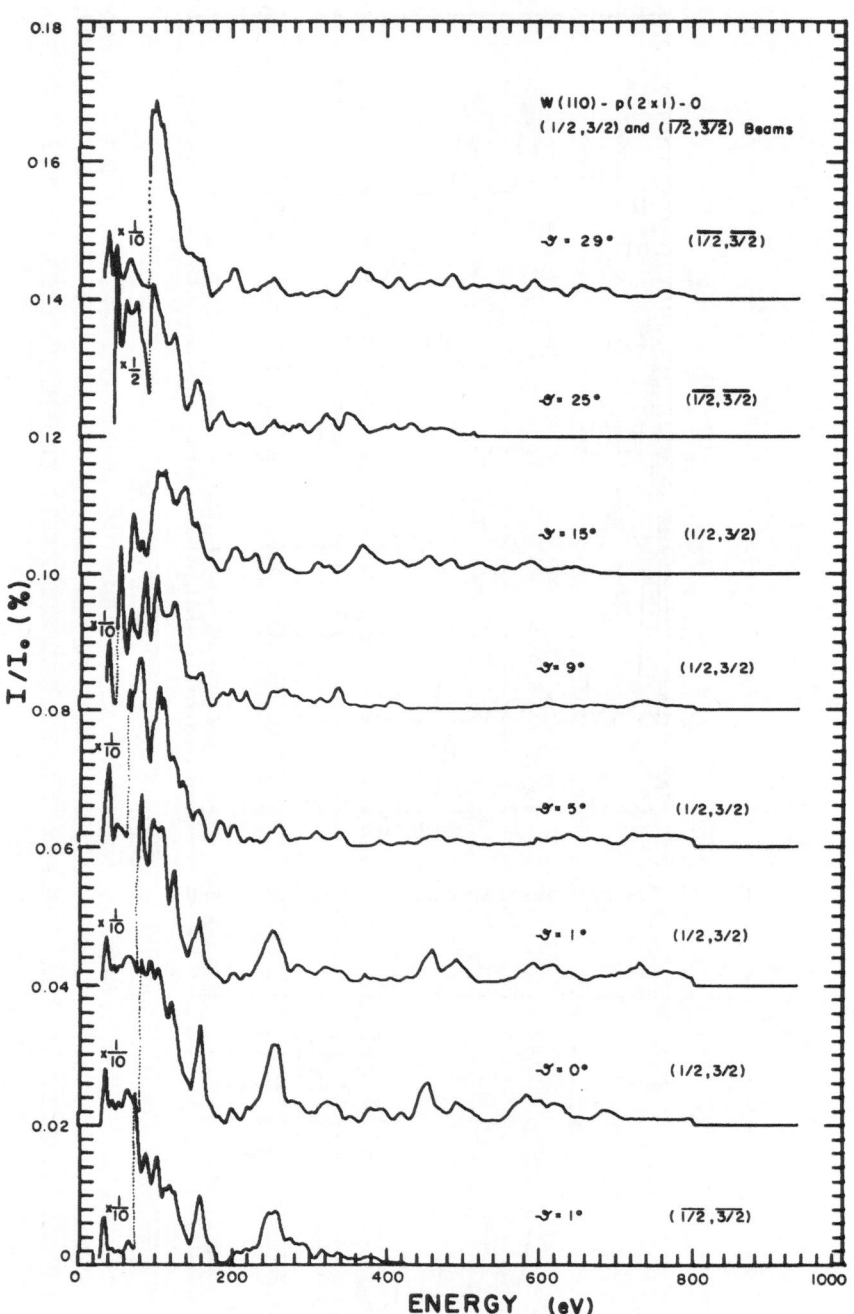

Fig. 17. The normalized intensity of the (1/2,3/2) and ($\overline{1}/2,\overline{3}/2$) overlayer beams for W(110) - p(2x1)-O; $\phi = 64.75$ degrees.

energy is suggestive of the oxygen atomic scattering factor. The diffrac-
tometer permitted nonspecular beam data only at a single azimuth, and such
a partial average for the overlayer beams is shown in Fig. 18. There is considerable multiple scattering residue as expected for the limited range of
data, but the average clearly distinguishes between a simple overlayer containing only oxygen atoms and the reconstructed layer proposed by Germer
and May[17] (Fig. 19). The general result of calculations for many models is
that any participation of the tungsten in the overlayer structure makes the
comparison worse, as is clear from the absence of overlayer beam intensity
at the higher energies. This is further verified below by data on the substrate
beams. Additional studies from the overlayer beams indicate that the layer
has a somewhat smaller vibrational amplitude that the clean tungsten and
that the overlayer has a reversible disordering near $300°C$.[13]

It remains to locate the oxygen atom relative to the tungsten substrate,
and the difficulty in doing this is shown by comparing the intensity profiles
for the clean and overlayered systems. Examples are shown in Figs. 20 and
21. As anticipated, the differences are small and only noticeable at low
energies. Averages of the data are shown in Fig. 22. The same comparison is
true also for the (00) and (11) substrate beams. These data further support
the conclusion that the tungsten does not participate in the overlayer structure. They show the severe requirements on both the precision of the data
and the analysis, whatever form that analysis takes.

A large number of calculations of the single scattering have been made
for various assumed oxygen coordinates[13]; for most, one expects differences
larger than those observed, but, unfortunately, several give expected intensities differing only very slightly from the substrate intensity. With the partial averages available, they cannot be distinguished. In particular, oxygen
in the adjacent bridge bonds along alternate rows or in regular tungsten
sites, and either of these 0.9 or 1.7 Å above the center of the tungsten layer
are consistent with the small differences seen between the clean and
overlayer averages.

4 DISCUSSION AND SUMMARY

From the experience to date it seems possible to make several
generalizations concerning the practical application of low-energy electron
diffraction techniques to problems of interest in heterogeneous catalysis
studies.

First, the technique is capable of very detailed characterization of clean
surfaces of single-crystal substrates. The simplest observation of "spot
patterns" determines the lateral periodicity and gives information about the
extent of the long-range order and surface step distributions. More detailed

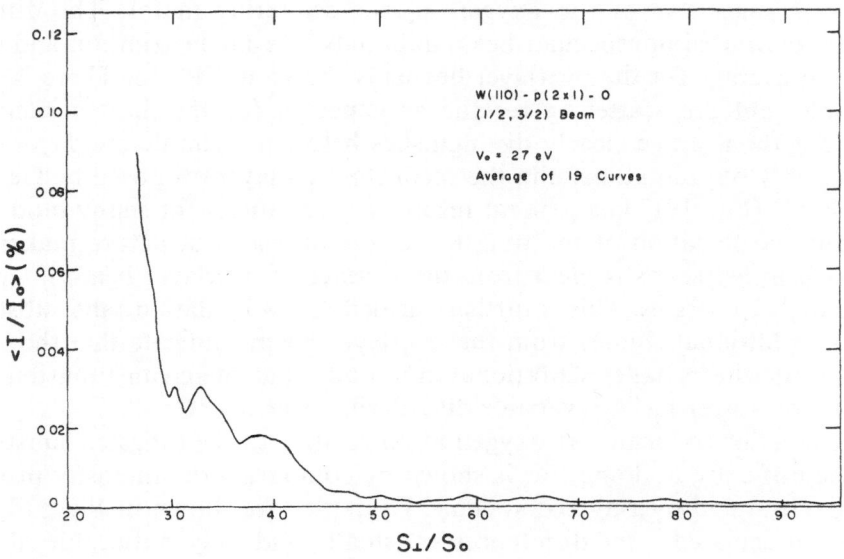

Fig. 18. The averaged intensity for the (1/2,3/2) beam at one azimuth.

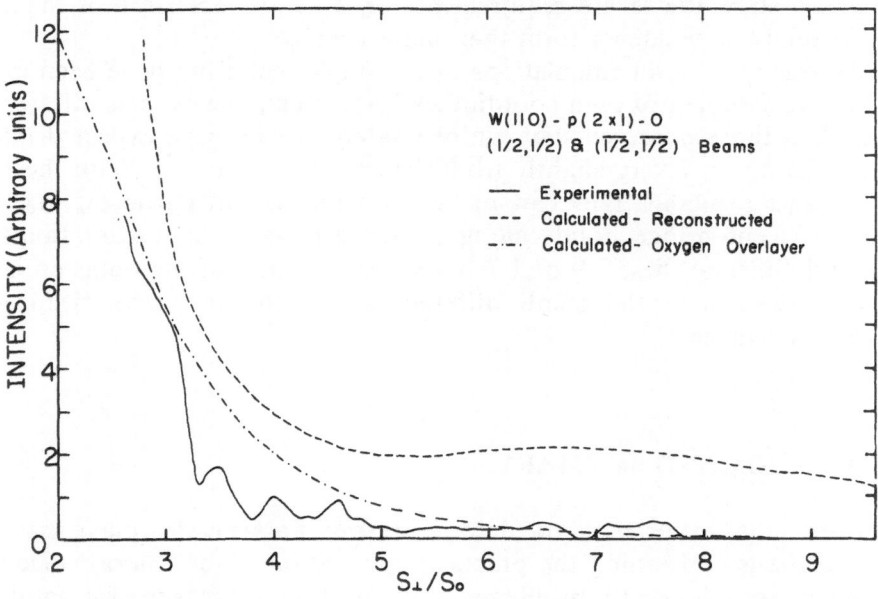

Fig. 19. Comparison of the overlayer beam partial-averaged intensity with the calculated intensity. Chain curve is calculated assuming a planar oxygen layer and the dashed curve is calculated assuming a substitutionally reconstructed layer with oxygen and tungsten occupying alternate rows in the overlayer. Debye-Waller factor, Lorentz factor, and surface loss corrections have been divided out of the data. Comparisons for a variety of other assumed models are given in Ref. 13.

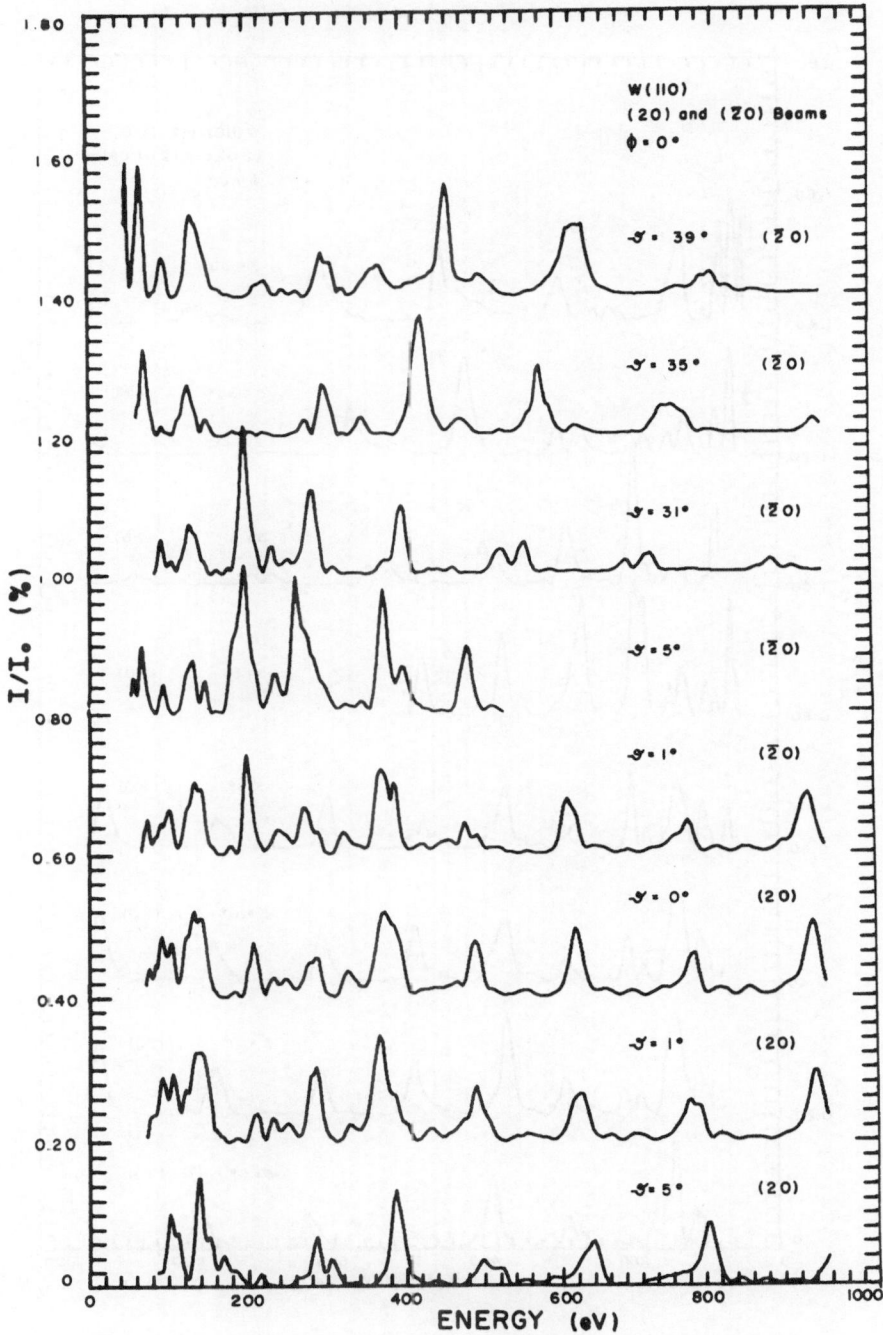

Fig. 20. The normalized intensity for the (20) and ($\overline{20}$) beams from W(110).

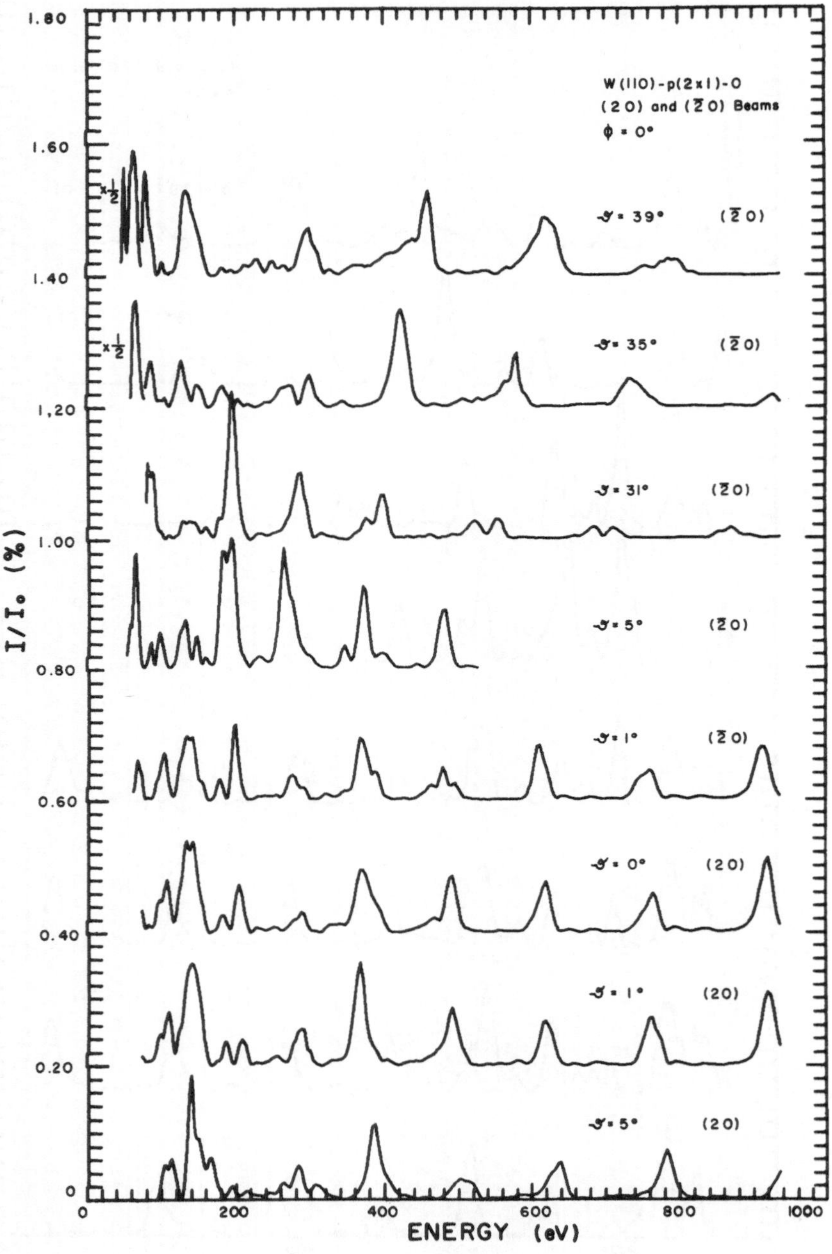

Fig. 21. The normalized intensity for the (20) and ($\overline{2}0$) beams from W(110) - p(2x1)-O.

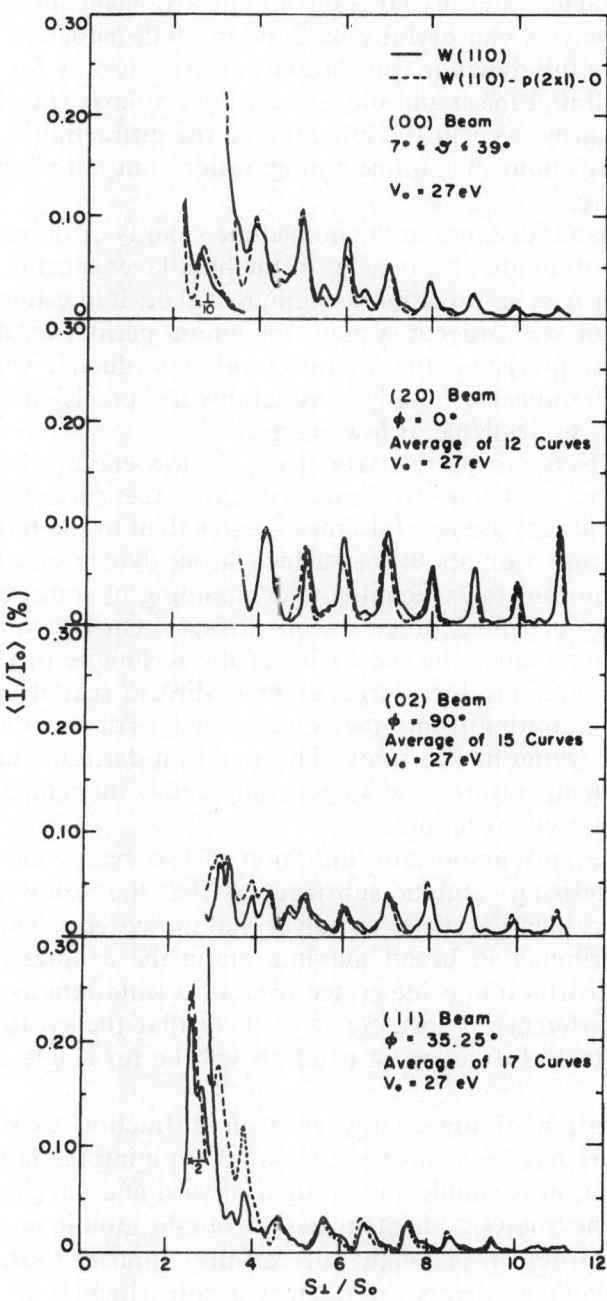

Fig. 22. Comparisons of the averaged data for various substrate beams for both clean W(110) and W(110) p(2x1)-O; the (00) beam averages include data at ϕ = 35.25, 64.75, and 90 degrees.

analysis of diffracted intensity gives the atomic arrangements within the unit mesh. These analyses can usefully be done by data-reduction schemes like averaging; or by full dynamic calculations, at present only for systems with reasonably small unit mesh and moderately light atoms. The elastic scattering occurs primarily deep in the ion core, so the diffraction contains information primarily about the atomic configurations but not about the valence electron structure.

The interplanar distance in the immediate vicinity of the surface may be determined to within about 1 percent of the bulk layer spacing by averaging techniques, which is a precision beginning to be interesting for surface studies. While for well-ordered systems the lateral periodicity may be determined to similar precision, the lateral atomic coordinates within the unit mesh can be determined with only considerably less precision because of the limited range of $S_\parallel$ available at low energies.

Thermal effects are particularly strong in low-energy electron diffraction because the electrons are scattered from the surface layer, where vibrational amplitudes are several times greater than in the bulk. This gives considerable information about the surface lattice dynamics, which must ultimately be important for a detailed understanding of transport properties and of thermally activated processes on surfaces. On the other hand, the thermal vibrations reduce the sensitivity of the technique to the outermost layer of atoms and produce large thermal diffuse scattering which can obscure diffuse scattering from other sources such as disorder and imperfections which one would like to study. This makes it desirable to work at the lowest possible temperatures and so generally below the temperatures of interest for real catalytic reactions.

Perhaps the most important limitation of low-energy electron diffraction for characterizing catalytic substrates is that the technique seems not applicable to very small particles or polycrystalline systems. The interference function is distributed in broad maxima along the reciprocal-lattice rods and when this distribution is integrated over all orientations as for a powder pattern, the interference is lost. So it is likely that the greatest impact of low-energy electron diffraction in catalysis will be for single crystal model systems.

The application of low-energy electron diffraction to studies of adsorbed overlayers has been most useful in determining the lateral periodicity. This application is simple and straightforward and has made a tremendous impact. The analysis of intensities to obtain atomic coordinates is a more difficult problem. Though successfully applied to a few simple systems, the dynamic theory of diffracted intensities is a much larger problem for overlayers than it is for clean systems where fewer beams and knowledge of fewer parameters are required. The data-reduction schemes are more difficult for overlayers because the desired single-scattering contribution to the intensity is generally small and the phase space available is often severely restricted. However, partial information, such as the arrangement within the overlayer and whether the substrate is rearranged, is available from "overlayer" beams alone. Certainly it is possible to distinguish between many of the proposed models. However, it seems yet unclear how successful low-energy electron diffraction will be in providing sufficiently simple and accurate determinations of overlayer atomic coordinates relative to the substrate to be generally useful in catalytic research. The answer must await further development of both dynamic calculations and data-reduction methods.

As for clean surfaces, the information about the thermal vibration of the adsorbed overlayers can be rather simply obtained. In addition, information about order-disorder and other surface phase transitions is available. There is enough known about atomic scattering factors and there is sufficient freedom in modern experimental arrangements of the diffraction geometry so that careful measurements of diffuse scattering should yield lateral interatomic-pair-distribution functions within dilute and disordered overlayers. In the analysis of these later types of experiments, complications from dynamic scattering are much less severe than those for determining the coordinates relative to the substrate. Though to date there have been very few quantitative investigations of dynamics of the overlayers, they undoubtedly will be important in understanding catalytic processes.

References

1. For a review of experimental aspects of low-energy electron diffraction see for example Webb, M. B., and Lagally, M. G., *Solid State Phys.,* **28**, 301 (1973).
2. For a review of theoretical aspects of low-energy electron diffraction see Pendry, J. B., *Low-Energy Electron Diffraction,* Academic Press, New York (1974).
3. Duke, C. B., and Liebsch, A., *Phys. Rev.,* **B9**, 1126, 1150 (1974).
4. Lagally, M. G., Ngoc, T. C., and Webb, M. B., *Phys. Rev. Letters,* **26**, 1557 (1971).
5. Lagally, M. G., Ngoc, T. C., and Webb, M. B., *Surface Sci.,* **55**, 117 (1973).
6. Tucker, C. W., and Duke, C. B., *Surface Sci.,* **24**, 237 (1972).
7. Clarke, T. C., Mason, R., and Tescari, M., *Proc. Roy. Soc. (London),* **A331**, 321 (1972).
8. Buchholz, J. C., Lagally, M. G., and Webb, M. B., *Surface Sci,* **41**, 248 (1974).
9. Landman, U., and Adams, D. L., *J. Vacuum Sci. Technol.,* **11**, 195 (1974).
10. Unertl, W. N., and Webb, M. B., *J. Vacuum Sci. Technol.,* **11**, 193 (1974); Unertl, W. N., Ph.D. Thesis, University of Wisconsin, 1974 (unpublished).
11. Burkstrand, J. M., Kleiman, G. G., and Arlinghaus, F. J., to be published.
12. McDonnell, L., Woodruff, D. P., and Mitchell, K.A.R., *Surface Sci.,* **45**, 1 (1974).
13. Buchholz, J. C., Wang, G.-C., and Lagally, M. G., *J. Vacuum Sci. Technol.,* **12**, 194 (1975); Buchholz, J. C., Ph.D. Thesis, University of Wisconsin, 1974 (unpublished).
14. Unertl, W. N., to be published.
15. Pendry, J. B., *J. Phys.,* **C5**, 2567 (1972).

16. McKinney, J. T., Jones, E. R., and Webb, M. B., *Phys. Rev;* **160**, 523 (1967); McKinney, J. T., Ph.D. Thesis, University of Wisconsin, 1967 (unpublished).
17. See for example: Germer, L. H., and May, J. W., *Surface Sci.,* **4**, 452 (1966); Tracy, J. C., and Blakely, J. M., *Surface Sci.,* **15** 257 (1969); and Ref. 13.
18. Fink, M., and Yates, A. C., Electronics Research Center, Technical Report No. 88, University of Texas at Austin (1970); Fink, M., and Ingram, J., *Atomic Data,* **4**, 1 (1972), and private communication.

DISCUSSION on Paper by M. B. Webb

GRIMLEY: By averaging the LEED data you lose information. Is it possible to say what sort of information is lost?

WEBB: Dynamic diffraction theory includes a band-structure calculation at the energy of the incident electron. It is not clear whether this presents a very practical way to study band structure, but one would hope to be able to learn about the variation of the potential across the surface. However, even if this is the sort of information sought, one would want the atomic arrangement first.

Another difference in general is that the multiple scattering depends on the forward scattering, whereas the single scattering observed obviously does not.

ORBITAL ENERGY SPECTRA PRODUCED BY SORBED ATOMS

H. D. Hagstrum and G. E. Becker

Bell Laboratories
Murray Hill, New Jersey

ABSTRACT

The use of two types of electron spectroscopy (ion-neutralization and ultraviolet photoemission) in studies of surface electronic structure is discussed. Observations or conclusions made on the basis of these investigations are presented.

1 INTRODUCTION

It is informative to think of the electronic structure of a solid surface as being specified by the energy-position distribution function, $\rho(\mathbf{r},E)$, also called the position-dependent density of states. The quantity $\rho(\mathbf{r},E)$ dxdydzdE is the relative probability that an electron will be found in the element of space dxdydz at position $\mathbf{r}$ with energy in the element dE at energy E. Theorists are making progress in determining this function through the outer layers of a solid and into the vacuum outside.[1,2] Adsorbed atoms will produce local electronic states or resonances in $\rho(\mathbf{r},E)$.[3-9] Our basic understanding of chemisorption certainly must be construed to include an understanding of how it modifies the electronic structure at the surface of the substrate.

A particular type of electron spectroscopy can give an angle-integrated spectroscopic function which we may, in rough approximation, take to be its own peculiar, weighted integral of $\rho(\mathbf{r},E)$ over the geometric region of the solid, as well as the energy range, from which it ejects electrons. This is one reason why it is important to view the same controlled surface with more than one type of spectroscopy. Another is that each type of spectroscopy is based on an electronic transition process which has its peculiar set of variables and its distinctive transition-probability factors. The more sophisticated we become in our understanding of these characteristics of the types of spectroscopy we employ, the more information we can extract about the systems we are studying. In this regard, too, two pairs of eyes are much better than one.

2 TYPES OF ELECTRON SPECTROSCOPY

The two types of spectroscopy employed in the work to be discussed in this paper are ion-neutralization spectroscopy (INS) and ultraviolet photoemission spectroscopy (UPS), the latter at four discrete photon energies in the range 16.8 to 40.8 eV and for variable incidence angle of the unpolarized light used. These differ in fundamental ways. UPS is one-electron spectroscopy since only one electron makes a transition when an incident photon is absorbed. INS is two-electron spectroscopy. The Auger-type process upon which it is based involves transitions in energy of both the down electron which neutralizes the ion outside the surface and the up, or excited, electron whose kinetic energy distribution is measured.

Matrix elements of INS and UPS result in quite different transition-probability characteristics. The UPS matrix element contains a factor $\mathcal{E}\cdot\text{grad}_m V$ which is the product of $\mathcal{E}$, the electric field vector of the light, and $\text{grad}_m V$, the maximum vector gradient of the unperturbed electric potential in the crystal.[10] Thus, photoemission not only requires that the electrons to be excited lie in a potential gradient but also places conditions on the geometrical relation between the $\mathcal{E}$ vector of the light and this gradient. The basic character of the INS matrix element is determined by the probability of tunneling of the down or neutralizing electron into the potential well of the incoming ion. Variation of transition probability due to the nature of wave-function overlap must also occur in both matrix elements.

The overall transition probability for either type of spectroscopy is obtained by integration of the matrix element over initial states. This requirement further contributes to the divergent characteristics of UPS and INS. In UPS, momentum conservation in the electronic transition causes the overall transition probability to depend on the joint density of states. There is evidence[11], however, that for surface orbitals the momentum selection rule is relaxed and the density-of-states factor for these orbitals involves only the product of initial- and final-state densities, as is generally true for the Auger process of INS. A UPS electron distribution is, nevertheless, a superposition of bulk and surface contributions, so one must take both types of matrix element into account. In any event, the observed kinetic energy distribution gives directly the one-electron-transition density function of UPS modified by energy straggling.

Integration over initial states in INS requires, by virtue of its two-electron nature, integration in energy over all pairs of initial electrons which can give an excited electron at a given energy.[3] It is this fact which makes the INS electron distribution depend on the convolution of the one-electron-transition density function of INS which we wish to determine.

INS and UPS differ in the degree of, and the reasons for, surface sensitivity. Electron escape depth for inappreciable energy loss determines the relative surface sensitivity in UPS. In INS, surface sensitivity is determined by the range of distances from the ion at which the up electron can be excited. This is generally smaller than the electron escape depth for electrons of comparable energy.

Another interesting and productive difference between INS and UPS arises from the circumstance that, whereas a light photon of, say, 20 eV energy may interact over an area of linear dimension 600 Å at a surface, an incident ion interacts principally with one or at most a few surface atoms. This fact combined with the requirement that two electrons must be available within something like 10^{-14} second for the Auger process to occur gives INS special characteristics in the observation of semiconductor surface states, for example, or in the detection of surface electronic inhomogeneity.[12]

3 OBSERVATIONS AND CONCLUSIONS

A number of the fundamental observations and conclusions based on investigations using INS and UPS in the same apparatus for the study of the same surface are presented below. The conclusions are quite general in nature but are illustrated by results obtained for the adsorption of tellurium and mercury on Ni(100) and Si(111) surfaces. These conclusions rest not only on the specific characteristics of each of the types of spectroscopy employed but in many instances also on the combined use of the two types. The results presented are a selection of data from more complete studies of these systems to be published in greater detail elsewhere.

1. It has been demonstrated again that the adsorption of a chalcogen on a clean nickel surface lies near the limit of adsorbate-adsorbent interaction which produces relatively narrow virtual resonances in the local density of states at the surface, the so-called surface-molecule limit.[13,14] Experimental evidence of this type of adsorption has been obtained by INS for the cases of oxygen, sulfur, and, in particular, selenium on Ni(100).[4] Other examples now exist in the literature as the result of observations by UPS[6,7] and field emission spectroscopy, FES.[8] UPS data for another such case, namely that of tellurium adsorbed on Ni(100), are presented in Fig. 1. At 1/4-monolayer and 1/2-monolayer coverages, c(2x4) and c(2x2) structures, respectively, are formed.[15] Certainly by the time the c(2x2) structure is present, the p^4 electrons of tellurium combine with nickel orbitals to form two surface molecular orbitals sufficiently localized in energy as to appear in the UPS spectrum as broadened peaks centered at the energies $E-E_F = -4.2$ and -2.1 eV. These energies differ appreciably from the energy of the p^4 electrons in the free tellurium atom. That these surface orbitals are also seen by INS is apparent in Fig. 2 where the $U(E)$ function from INS and the $N_P(E)$ function from UPS are compared for the c(2x2) and c(2x4) structures. In the terminology of the surface molecule it is concluded that the group orbital of the nickel substrate with which the tellurium interacts is restricted to relatively narrow energy ranges.

2. The INS $U(E)$ spectra of Fig. 2 demonstrate again the phenomenon that in the surface-molecule limit, different surface molecules have different molecular orbital specta.[4,5] The surface molecule for c(2x2)Te can be written Ni_2Te, whereas that for c(2x4)Te is Ni_4Te. The two orbitals at -2.1 and -4.2 eV for the former, each presumably containing two electrons of the p^4 supplied by the tellurium atom, are replaced by a single orbital peak, indicating a fundamental change in the nature of the bonding of the tellurium to the metal as is expected in the surface-molecule limit.

3. In the surface-molecule limit, the combined use of INS and UPS may in some instances be essential to a sufficient understanding of the orbital spectrum. In the orbital spectra for Ni(100) c(2x2)Te, depicted in the upper half of Fig. 2, we see that both the energy-localized orbitals are clearly visible by each type of spectroscopy, confirming their surface nature. However, the orbital spectrum in the case of the Ni(100) c(2x4)Te surface (bottom half of Fig. 2) is more clearly determined by INS than it is by UPS. The broad orbital centered at $E-E_F = -3.4$ eV in the $U(E)$ spectrum is barely visible in the $N_P(E)$ spectrum where it would certainly be unidentifiable without the INS result. The structure in the $N_P(E)$ spectrum in the range $-2.5 < E-E_F < -1.5$ eV, also seen in the spectrum for clean Ni(100) in Fig. 1, must be bulk d-band structure since it is not evident in $U(E)$.[3]

4. We have also observed a form of adsorption, distinct from that of the surface-molecule limit, in which the orbital strength of the adsorbate is

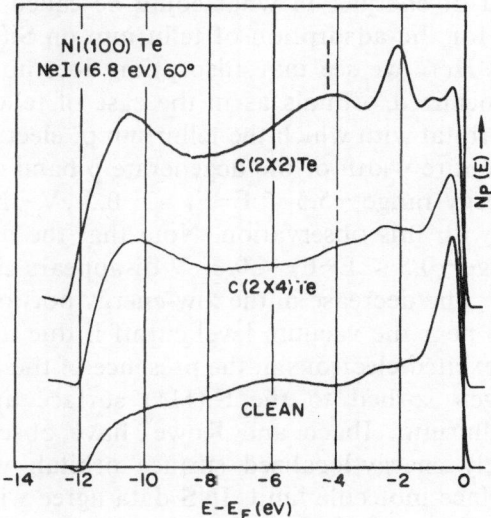

Fig. 1. Kinetic energy distributions, $N_P(E)$, of electrons photoemitted from Ni (100) by NeI radiation at 16.8 eV photon energy and at a 60 degree incidence angle. The Ni (100) surface is clean or covered with the C(2x4)Te or c(2x2)Te ordered adsorbed overlayers. The position on the E-E$_F$ energy scale of the orbital energy of the p^4 electrons of the free Te atom is shown by the vertical dashed lines. The work functions needed to plot these energies were obtained from Ref. 14.

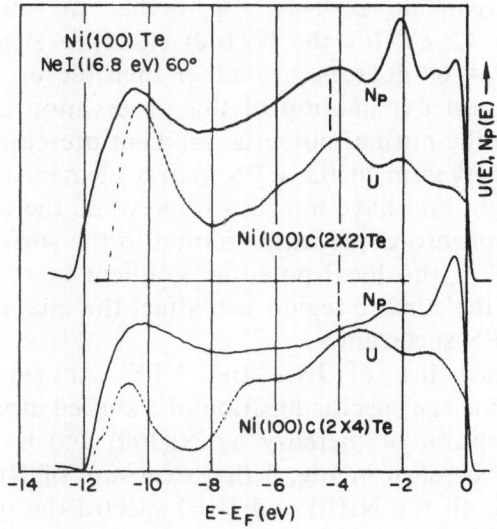

Fig. 2. UPS spectra, $N_P(E)$, and INS unfold functions, $U(E)$, for the Ni(100) c(2x4)Te and Ni(100) c(2x2)Te surfaces. As in Fig. 1, the orbital energy level of the p^4 electrons in the free Te atom is indicated by vertical dashed lines.

spread over a band of energies in what could be called the "surface-band limit". This occurs for the adsorption of tellurium on Si(111) (7x7) surface as shown in Fig. 3. Here we see that adsorption does not result in the formation of energy-localized orbitals as in the case of tellurium on Ni(100). Rather the group orbital with which the tellurium p^4 electrons bond appears to extend over the entire width of the degenerate p band of the silicon band structure in the energy range $-5.5 < E-E_F < -0.5$ eV. INS gives results in general agreement with this observation. Note that the dangling-bond surface state in the range $-0.5 < E-E_F < 0.0$ eV disappears as the tellurium adsorption progresses. The decrease in the low-energy portion of these kinetic-energy distributions near the vacuum level cutoff is due to decreased energy straggling of photoexcited electrons in the presence of the adsorbate. It is interesting that oxygen sorbed to the Si(111) surface appears to behave differently from tellurium. Ibach and Rowe[9] have observed that oxygen produces a relatively energy-localized surface orbital, which is taken to characterize the surface-molecule limit. INS data agree with this result. The type of orbital-intensity variation with incidence angle seen in Fig. 8 is apparently characteristic of adsorption in the surface-molecule limit but not in the surface-band types of adsorption of Fig. 3. For tellurium on Si(111), a relatively very much smaller dependence of the form of the UPS spectrum on the $\mathbf{\mathcal{E}} \cdot \text{grad}_m V$ term in the matrix element is observed.

5. Variation of incidence angle θ_i of the light at the sample surface in UPS provides evidence concerning the spatial position of a surface orbital in a surface molecule with respect to the gradient of the electric potential in the surface region. It can be seen in Fig. 4 that the surface orbitals at $E-E_F = -2.1$ and -4.2 eV for the Ni(100) c(2x2)Te structure, which are clearly evident at $\theta_i = 60$ degrees, are either invisible or almost so at $\theta_i = 0$ degrees. The simplest explanation of this observation is that the surface orbitals are lying in the normal potential gradient present at the surface. In this case the $\mathbf{\mathcal{E}} \cdot \text{grad}_m V$ term in the UPS matrix element would be zero for normal incidence light and have nonzero value when the $\mathbf{\mathcal{E}}$ vector of the incident light has a nonzero component normal to the surface. It is also true that the magnitude of the local potential gradient at the position of any localized orbital in the surface region will affect the intensity of the surface orbital peak in a UPS spectrum.

6. The combined use of INS and UPS can, in the proper circumstances, determine the specific location of a sorbed atom. An example is afforded by the sorption of mercury by Ni(100) and by Si(111) at room temperature. It can be seen in Fig. 5 that exposure of Ni(100) to mercury vapor produces in both the $N_P(E)$ and $U(E)$ spectra the orbital resonances corresponding to the $5d_{5/2}$ and $5d_{3/2}$ orbitals of mercury. However, the same procedure applied to Si(111) (Fig. 6) produces these orbitals only in the $N_p(E)$ spectrum and not at all in the $U(E)$ spectrum. The readiest explana-

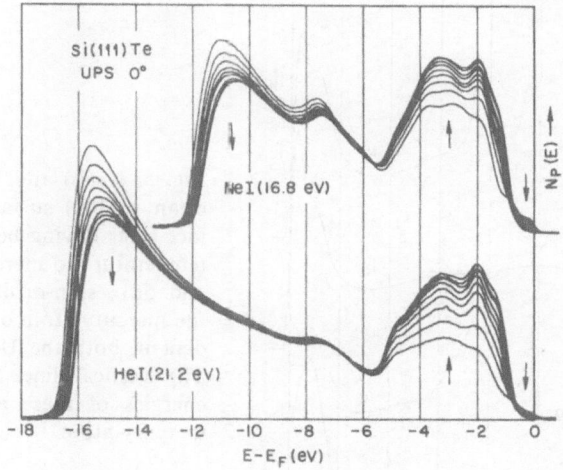

Fig. 3. UPS spectra, $N_P(E)$, obtained with HeI and NeI radiation obtained at various times as a tellurium beam interacts with the Si(111) (7x7) surface. Incidence angle of the radiation is 0 degrees. Curves were taken at equal time intervals up to a coverage corresponding to about one-half the saturation amount. The arrows indicate the directions in which the sequence progresses with increasing coverage. The starting curve is that for the clean Si(111) surface.

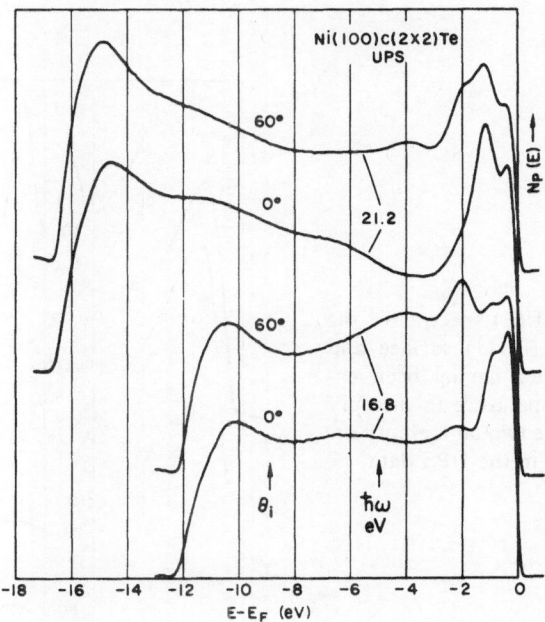

Fig. 4. UPS spectra for the Ni(100) c(2x2)Te surface at two different photon energies ($\hbar\omega$) and two different incidence angles (θ_i).

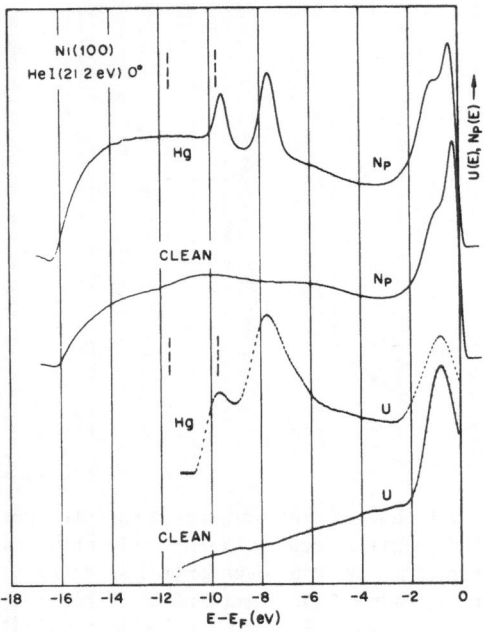

Fig. 5. $N_P(E)$ and $U(E)$ spectra for the clean Ni(100) surface and for this surface after having been exposed at room temperature to mercury vapor. The $5d_{5/2}$ and $5d_{3/2}$ spin-orbit-split resonances of the mercury atom on the surface are evident in both the UPS and INS results. The vertical lines indicate the orbital energies of these electrons in the free mercury atom.

Fig. 6. $N_P(E)$ and $U(E)$ spectra for the clean and relaxed Si(111) surface and for this surface after having been exposed at room temperature to mercury vapor. Note that the two 5d mercury orbitals are seen only in the UPS data.

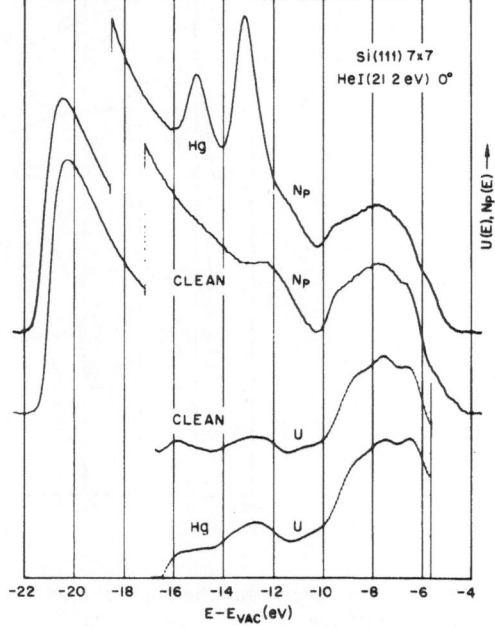

tion of this surprising result is that mercury remains adsorbed on the Ni(100) surface but is absorbed into the relatively open silicon lattice. Use of INS alone might lead to the erroneous conclusion that mercury does not interact with Si(111). Use of UPS alone might lead one to conclude erroneously that for the specific surface studied, mercury is adsorbed on the Si(111) surface as it is on the Ni(100) surface.

7. We believe that it will eventually be possible to determine certain characteristics of localized orbitals in surface molecules from intensity differences observed in the $U(E)$ and $N_P(E)$ spectra. It can be noted in Fig. 2 that the relative intensities of the two surface orbitals for the Ni(100) c(2x2)Te structure as seen by the two types of spectroscopy differ. The orbital at $E-E_F = -2.1$ eV is seen less strongly by INS than by UPS even though its form and energy position relative to the free-atom orbital energy suggest that it is relatively nonbonding. Another example of differences in orbital peak intensity in the spectra by INS and UPS is seen in Fig. 7 for the Ni(100) surface with an ordered structure upon it consisting of about two layers of tellurium. Here it can be seen that INS sees the -3.8 eV orbital strongly but sees the -6.3 eV orbital, which is strongly evident in the N_P spectrum, weakly if at all. The contribution of INS to these differences must result from the fact that the magnitude of the evanescent wave function outside the solid at the position of the probing ion when it is neutralized is smaller for the orbital seen weakly. This could result from any one, or possi-

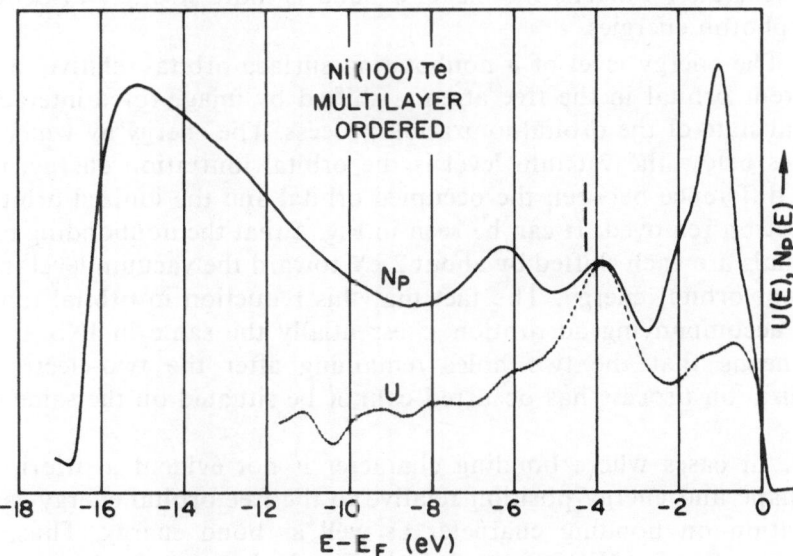

Fig. 7. $N_P(E)$ and $U(E)$ spectra for the Ni(100) surface having an ordered two-layer structure of tellurium upon it. The UPS data were taken at $h\omega = 21.2$ eV and zero incidence angle. The energy of the p^4 electrons in free tttellurium is indicated by the vertical dashed line.

ble combinations of, three effects: (a) the wave function of the orbital seen with less intensity by INS is more atomically localized (d- versus p-type, for example) than is that of the orbital seen with greater intensity; (b) the wave function of the less intense orbital has a larger value of $k_\parallel$ (momentum parallel to the surface) than does that of the more intense orbital; (c) the orbital seen weakly in INS is localized beneath the surface. It is also possible that the greater intensity of an orbital seen more strongly by UPS results, in part at least, from a matrix-element effect of UPS. As an example, a possible interpretation of the c(2x2)Te data of Fig. 2 is that the −2.1 eV orbital has more d character and the −4.2 eV orbital has more p character than its partner. Then INS will see the −2.1 eV orbital less strongly because it has less intensity in its evanescent tail outside the solid [item (a) above]. UPS, on the other hand, will see the −2.1 eV orbital more strongly because it is drawn closer to a nucleus where the $\text{grad}_m V$ term in the matrix element is larger. It is clear that more experiment and more theory are necessary before definite assignment of these possibilities can be made.

8. Symmetry character of surface orbitals may be judged from their intensity variation with photon energy in UPS in an experiment for surface orbitals analogous to that of Eastman and Kuznietz[15] for bulk orbitals. We expect the photoemission transition probability to decay with increasing $\hbar\omega$ less and less strongly through the sequence, s, p, d, f of orbital symmetry character. This can be seen in the UPS data of Fig. 8 where the p-derived surface orbitals at −2.1 and −4.2 eV are most strongly evident at the lower photon energies, whereas the nickel d band is more strongly evident at the higher photon energies.

9. The energy level of a nonbonding surface orbital relative to that of the parent orbital in the free atom is shifted by image-force interaction in the final state of the orbital-ionization process. The energy by which an orbital lies below the vacuum level is the orbital ionization energy, i.e., the energy difference between the occupied orbital and the ionized orbital with one electron removed. It can be seen in Fig. 5 that the nonbonding mercury 5d orbitals are each shifted by about 2 eV toward the vacuum level from the free-atom orbital energy. The fact that this reduction in orbital ionization energy accompanying adsorption is essentially the same in INS as it is in UPS means that the two holes remaining after the two-electron, ion-neutralization process has occurred cannot be situated on the same surface atom.

10. In cases where bonding character is not evident a priori, orbital peak shape and energy position relative to the free orbital energy can yield information on bonding character as well as bond energy. Thus, in the $N_P(E)$ spectrum for Ni(100) c(2x2)Te in Fig. 2, the shape and image-energy shift of the −2.1 eV orbital resemble those of the nonbonding mercury d orbitals in Fig. 5. It is thus not unreasonable to identify this orbital as non-

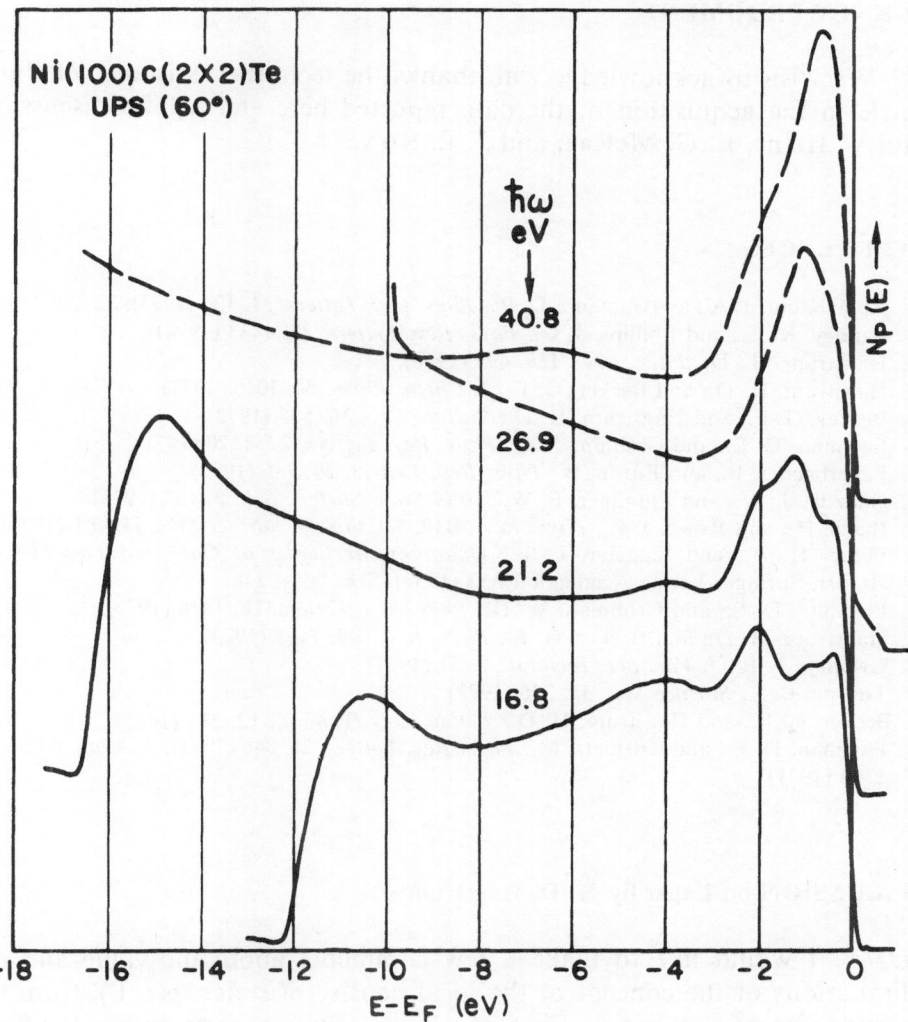

Fig. 8. UPS spectra for the NI(100) c(2x2)Te surface taken at 60 degrees light incidence for four values of photon energy, $\hbar\omega$. The curves at $\hbar\omega$ = 16.8 and 21.2 eV are photographic reproductions of X-Y recorder plots as are all such N_P curves in the other figures of this paper. The dashed curves for $\hbar\omega$ = 26.9 and 40.8 eV are tracings through recorder plots of relatively noisier data.

bonding. The orbital at −4.2 eV appears then to be a bonding orbital. Assuming it to have the same image-energy shift as the nonbonding orbital, one gets a rough estimate of bond energy for the bonding orbital as the energy separation of the two orbital peaks.

ACKNOWLEDGMENT

We wish to acknowledge with thanks the technical assistance of Philip Petrie in the acquisition of the data reported here and helpful discussions with V. Heine, E. G. McRae, and J. E. Rowe.

REFERENCES

1. Appelbaum, J. A., and Hamann, D. R., *Phys. Rev. Letters*, **31**, 106 (1973); **32**, 225 (1974).
2. Pandey, K. C., and Phillips, J. C., *Phys. Rev. Letters*, **32**, 1433 (1974).
3. Hagstrum, H. D., *Phys. Rev.*, **150**, 495 (1966).
4. Hagstrum, H. D., and Becker, G. E., *J. Chem. Phys.*, **54**, 1015 (1971).
5. Becker, G. E., and Hagstrum, H. D., *Surface Sci.*, **30**, 505 (1972).
6. Eastman, D. E., and Cashion, J. K., *Phys. Rev. Letters*, **27**, 1520 (1971).
7. Feuerbacher, B., and Fitton, B., *Phys. Rev. Letters*, **29**, 786 (1972).
8. Gadzuk, J. W., and Plummer, E. W., *Solid State Surface Sci.*, **3**, 165 (1973).
9. Ibach, H., and Rowe, J. E., *Phys. Rev.*, **B10**, 710 (1974); *Surface Sci.*, **43**, 481 (1974).
10. Bethe, H. A., and Salpeter, E. E., *Quantum Mechanics of One- and Two-Electron Atoms*, Springer Verlag-Academic Press (1957), Sec. IV, p 248.
11. Eastman, D. E., and Grobman, W. D., *Phys. Rev. Letters*, **28**, 1378 (1972).
12. Hagstrum, H. D., and Becker, G. E., *Phys. Rev.*, **B8**, 1592 (1973).
13. Grimley, T. B., *J. Vac. Sci. Technol.*, **8**, 31 (1971).
14. Thorpe, B. J., *Surface Sci.*, **33**, 306 (1972).
15. Becker, G. E., and Hagstrum, H. D., *J. Vac. Sci. Technol.*, **12**, 234 (1975).
16. Eastman, D. E., and Kuznietz, M., *Phys. Rev. Letters*, **26**, 846 (1971); *J. Appl. Phys.*, **42**, 1396 (1971).

DISCUSSION on Paper by H. D. Hagstrum

KOHN: I would like to make a few comments about the value and the limitations of the concept of the local density of states, $\rho(\mathbf{r}, E)$, from the standpoint of a theorist. First of all $\rho(\mathbf{r}, E)$ is a perfectly well-defined quantity, derivable from the one-particle Green's function. Second, in discussing surface phenomena where, of course, the properties of the system change radically in the direction perpendicular to the surface, it is of very much greater interest than the space-averaged density of states, $\rho(E)$. Finally, it is a sufficently simple quantity which, nevertheless, contains some very important information, so that it provides a very useful guide for experimental and theoretical work and a good meeting point between experiment and theory.

It would, of course, be nice if $\rho(\mathbf{r}, E)$ could be unambiguously extracted from experiments and if all interesting questions concerning electronic structure could be answered in terms of it. As we all know, neither is the case.

For example, electron emission experiments give certain information about a response function (or, equivalently, two-particle Green's function) of the system, but without further assumption or knowledge about the more detailed electronic structure (including many-body effects), one cannot obtain $\rho(\mathbf{r}, E)$, strictly speaking. Of course, especially where $\rho(\mathbf{r}, E)$ shows strong structure, measurements such as yours can provide very useful semiquantitative information about it.

The converse of this situation is also true. If, as a theorist, I were given the exact $\rho(\mathbf{r}, E)$ of a surface system and asked to calculate from it, for example, the photoemission spectrum, I could not do it without additional assumptions or knowledge of the electronic structure.

Needless to say, other interesting parameters of the electronic structure, such as total electronic energy which is of great interest in catalysis, also require information beyond $\rho(\mathbf{r}, E)$, although a knowledge of this quantity may provide some essential clues.

SCHRIEFFER: I have two comments.

In connection with Professor Kohn's comment, I would like to point out that one is generally interested in the nonlocal density of states $\rho(\mathbf{r}, \mathbf{r}', E)$ in relating theory and spectroscopic experiments, rather than in the local quantity $\rho(\mathbf{r}, E)$. The angular dependence of photoemission spectra in turn depends critically on this nonlocal function. More precisely, however, one is measuring a many-particle correlation function, which for photoemission involves a product of three current operators, suitably averaged over space. Ion-neutralization spectra involve an even more complicated quantity. Thus, one must be very cautious in comparing experiment with model density-of-states calculations because of nonlocal and many-body correlation effects.

In regard to Dr. Hagstrum's comparison of the spectra of tellurium on Ni(100) and Si(111), I would like only to clarify the theoretical implication of the sharp structure observed for Te-Ni and the broad spectra observed for Te-Si. While the sharp levels for Te-Ni suggest a strong Te-Ni bond corresponding roughly to resonant surface complex levels, the broad Te-Si levels suggest an intermediate interaction regime, but not weak interaction which leads again to a relatively sharp (adsorbate) level. Thus, weak and strong adsorption should lead to rather sharp levels, while broad spectra reflect an intermediate coupling strength.

GRIMLEY: The density of states $\rho(\mathbf{r}, E)$ is, as Professor Kohn has just said, well-defined in terms of Green functions. But when you identify particular features in your experimental curves with particular orbitals, d bonds, p orbitals of tellurium and so on, you are speaking in terms of overlapping orbitals, and now there are two densities of states, the net and the gross densities which are connected with familiar quantities in quantum chemistry, namely, with net and gross populations. These two densities of states are different, and we may, therefore, need to be more careful in our terminology.

MASON: The confident assignment of electronic levels in the photoelectron spectra of chemisorbed atoms and molecules requires binding energies to be known with respect to the vacuum level. Are these available with good accuracy when corrections have been made only for the work-function change? The widths of transitions associated with ionizations from non-bonding orbitals are very narrow for the noncoordinated molecules but may be 1 eV or so for, say, chemisorbed NO and CO. Can this be simply related to the metal ligand σ band being largely due to the coupling of the lone-pair orbital with the diffuse s band rather than the more localized d orbitals?

HAGSTRUM: The energy of the free-atom orbital with respect to the Fermi level in our spectra has been determined by subtracting the total work function of the surface with adsorbate present from the known free-space ionization energy. The change in work function from the clean surface is measured, to be sure, but it must be combined with the known work function for the clean surface to obtain the work function desired. As to the second question, it appears reasonable to believe that a so-called "nonbonding" orbital in a surface complex is broadened by interaction with the band or bands of electrons of the same energy in the underlying solid. This broadening will be proportional to the density of states in the substrate at this energy. There could also be interactions with neighboring complexes in the ordered surface array which could also lead to broadening of a nominally lone-pair orbital in the surface complex.

RHODIN: The effect of the field of the ion in INS on the local charge and wave functions during ion neutralization should be rather large. To what extent might this effect introduce perturbations in the interpretation of the chemical bonding of the adsorbate as determined by ion-neutralization spectroscopy? Is it likely to be small in some cases such as for strong, localized bonding of chalcogens on nickel but perhaps more significant in other cases such as, for example, in the observation of surface states on silicon?

HAGSTRUM: The probing ion in INS is indeed charged and produces a field at the surface. The magnitude of this field may be judged by the fact that the relaxation or image interaction with the metal reduces the effective neutralization energy by about 2 eV at the distance from the surface where the ion is neutralized (see answer to next question). We have seen no profound effect in the INS unfold function, U(E), which we would attribute to the perturbing effect of this field. For example, in Fig. 5 of the paper it is seen that the mercury 5d orbitals seen by INS appear at the same energies as do those seen by UPS. The greater breadth of the orbitals in the U(E) function could in part be due to broadening caused by the motion of the charge of the ion toward the surface. The magnitude of any perturbing effect seems to be constant from one surface to another since the change in neutralization energy from the free-space value appears always to be about 2 eV. The fact that the surface states at the Si(111)7X7 surface do not appear in the U(E) function (Fig. 6) is undoubtedly related to another effect entirely, namely, to the inability of the surface-state band to provide the two electrons needed within 10^{-14} second in order to sustain a two-electron Auger process (see Ref. 12 of paper). Thus the surface-state band apparently cannot produce a self-fold term in the kinetic-energy distribution of INS. There are electrons, presumably from cross folding between the surface-state band and the valence band, which do appear in the energy range of the surface states, however. This interesting characteristic of INS, discussed in Ref. 12 of the paper, is being studied further by means of intercomparison of UPS and INS, both of which we now have in our apparatus.

KOHN: Could you give us an idea of the electric field due to the ion at the position of the adsorbate?

HAGSTRUM: We have, of course, no direct measurement of electric field at the surface before the probing ion is neutralized. We do know the magnitude of a related parameter, however. This parameter is the change in neutralization energy of the He^+ ion from free space to the distance from the surface at which the Auger-neutralization process takes place. This change is close to 2 eV, reducing the neutralization energy from 24.5 to 22.5 eV. This number could be used to arrive at a field estimate by assuming (1) an ion-solid interaction energy, such as the image interaction, for example, from which a separation s from the solid could be obtained, and (2) the atomic geometry at the surface from which the field could be derived using the value of s obtained above.

MOLECULAR INTERACTIONS IN ADSORBED LAYERS

G. Ertl

Physikalisch-Chemisches Institut der Universität
München, West Germany

ABSTRACT

Interactions between adsorbed species cause variations of their effective binding energies with the surface. These changes are usually roughly one order of magnitude smaller than the adsorption energies and control the equilibrium arrangements of the adsorbed particles as well as other thermodynamic properties like the adsorption isotherms. Kinetic phenomena like displacement reactions or the poisoning of adsorption sites are not well understood but may be of vital importance for the mechanism of catalytic reactions. The most striking effects occur for the transition states of chemical reactions: the interactions may become very strong, and thus lead to a dramatic lowering of the activation barrier. These different aspects are illustrated with a series of experimental results.

1 INTRODUCTION

The role of a catalyst consists of varying the rate constant of a chemical reaction. This is achieved by an alteration of the reaction path and is connected with a lowering of the activation energy. Since in heterogeneous catalysis these processes take place in the adsorbed phase, some insight into the mutual interactions between adsorbed particles is of considerable importance. Such interactions may influence the geometrical arrangement, as well as the kinetics, energetics, and reactivity of adsorbed layers. Some of the different phenomena will be illustrated in this paper with a few selected examples.

2 INTERACTION BETWEEN IDENTICAL PARTICLES

Interactions between adsorbed particles may be direct (e.g., dipole-dipole or overlap of wave functions) or indirect (i.e., through coupling with the metallic electrons). The latter effect has been treated theoretically by different authors[1-4] who arrived at the general conclusion that the energies of these interactions are about one order of magnitude smaller than the binding energies between the adsorbed particles and the surface, and that the range is limited to about 2 to 3 lattice constants. A direct experimental verification of the latter prediction was recently given by Graham and Ehrlich[5] measuring the correlation between pairs of adsorbed tungsten atoms by means of the field-ion microscope.

The existence of interactions leads to alterations of the binding energies at sites in the vicinity of an already adsorbed particle and thus influences the equilibrium configuration of adsorbed atoms and the variation of the differential heat of adsorption with coverage. The surface arrangement manifests itself in the occurrence of "extra" spots in the LEED patterns, which in turn could be simulated by choosing a proper set of interaction energies. This will be demonstrated with the system O/W (211).

A series of experimental studies[6-8] revealed that, with increasing oxygen coverage, at first "streaky" half-order spots appear, which coalesce to sharp spots with maximum intensity at $\theta = \frac{1}{2}$. These additional spots are gradually weakened in intensity until at $\theta = 1$ only the diffraction pattern of a 1×1 structure is visible. In a simulation of the surface arrangements of adsorbed particles by means of the Monte Carlo technique[9], the following interaction energies were assumed: (a) in the $[1\bar{1}0]$ direction, a weak attraction with $\epsilon_1 = -kT$ between nearest neighbors; (b) in the $[11\bar{1}]$ direction, repulsion between nearest neighbors ($\epsilon_2 = 2\,kT$) and attraction between next-nearest neighbors ($\epsilon_3 = -1.8\,kT$). Although, of course, the numerical values are somewhat arbitrary, the signs of these interaction energies are a clear consequence of the LEED observations. Using this set of parameters, the equilibrium configuration was evaluated for a series of coverages between 0 $\leqslant \theta \leqslant 1$.[10] As an example, Fig. 1 shows a typical result for $\theta = 0.3$. Using the kinematic approximation for each coverage, the average value of the relative intensity of the ($\frac{1}{2}$,0)-LEED spot and its angular width was calculated. A more rigorous treatment on the basis of a dynamical theory is not expected to influence these results fundamentally.[11] Figure 2a shows the thus derived variation with coverage of the relative intensity of the ($\frac{1}{2}$,0) spot together with the experimental data.[6,7] The theoretical[10] and experimental[7] results for the angular spread of these diffraction spots are shown in Fig. 2b. Although in both cases there is some scattering of the data, it is evident that the LEED results may be properly explained just by assuming anisotropic interactions between neighboring adsorbed oxygen atoms—without the need of further speculations on the kinetics, etc., as in a recent paper which tried to simulate this adsorption system by means of an optical diffraction technique.[12]

A second example will serve more directly to illustrate the operation of interactions between adsorbed atoms: with the system H_2/Ni(110), the adsorption energy was observed at first to *increase* with coverage by about 2 kcal/mole and then to remain constant over a wide range of coverages.[13] This effect cannot be due to an a priori heterogeneity of the surface, since in this case the energetically more favorable sites would be occupied first, but must be caused by the existence of attractive interactions. The LEED pattern shows already at rather low coverages the formation of streaks,

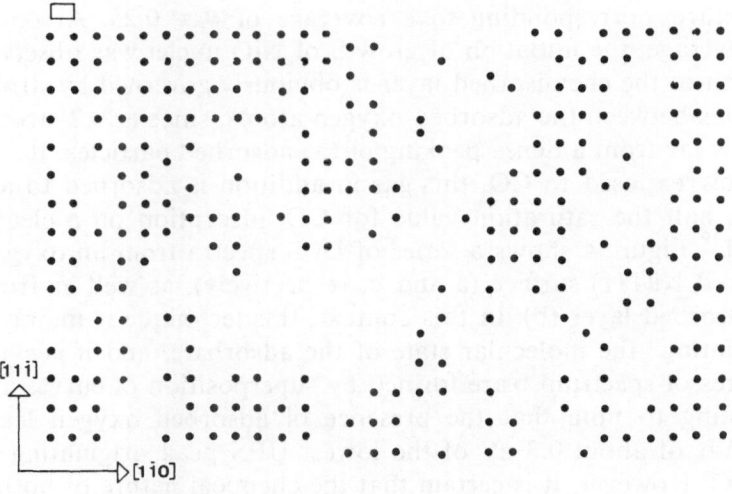

Fig. 1. Computer simulation of the equilibrium arrangement of oxygen atoms adsorbed on W(211) at oxygen = 0.3, assuming specific interaction energies.

which indicates the formation of chainlike arrangements of the adsorbed particles in this region of coverage. Figure 3 shows a set of adsorption isotherms for this system which are characteristic for a two-dimensional condensation phenomenon due to attractive interactions.

Adsorption isotherms of this type are frequently observed with physical adsorption (e.g., for xenon on graphite[14]), but have never been reported before for chemisorption systems. It is tempting to speculate that this unique effect might be correlated with the exceptional magnetic properties of the Ni(110) surface: recent scattering experiments with deuterons revealed that these nuclei become nearly completely spin polarized after capture of an electron from a Ni(110) surface.[15]

3 NONREACTIVE INTERACTIONS BETWEEN NONIDENTICAL PARTICLES

If a surface already covered by an adsorbate A is exposed to a second species B, the adsorption of the latter may be completely inhibited (competitive adsorption) or may lead to a mixed adsorbed phase (cooperative adsorption).

An example of the first type is found if a Pd(111) surface already covered with adsorbed CO ($\theta_{max} \approx 0.5$[16]) is exposed to O_2. Although the metal/oxygen bond would be energetically much more favorable, the adsorption of O_2 is completely inhibited, probably since no metallic sites for the dissociation of the oxygen molecule are available.

On a bare Ni(111) surface, oxygen forms a chemisorbed layer with a 2x2 structure, corresponding to a coverage of $\theta = 0.25$. At only slightly higher coverage, the initiation of growth of NiO nuclei was observed.[17] The completion of the chemisorbed layer is obviously governed by strong lateral interactions between the adsorbed oxygen atoms, since a 2x2 structure with $\theta = 0.25$ is far from a dense packing of the adsorbed particles. If such a surface is now exposed to CO, this gas in addition is adsorbed to about $\theta = 0.25$ [i.e., half the saturation value for CO adsorption on a clean Ni(111) surface[18]].[19] Figure 4 shows a series of UPS spectra from an oxygen- and a CO-covered Ni(111) surface (a and c, respectively), as well as from such a mixed adsorbed layer (b). In this context, this technique is mainly used for "fingerprinting" the molecular state of the adsorbate, and it is evident that the features of spectrum b are formed by superposition of curves a and c. It is interesting to note that the presence of adsorbed oxygen leads to an energy shift of about 0.3 eV of the lowest UPS peak originating from adsorbed CO. However, it is certain that the chemical nature of both kinds of adsorbed particles is essentially unaffected by their intimate contact.

The additional uptake of CO leads to some streaking of the diffraction spots from the O—2x2 LEED pattern, indicating the formation of domains

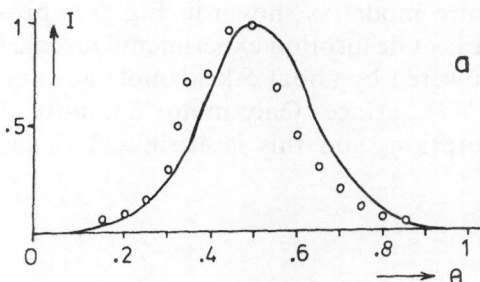

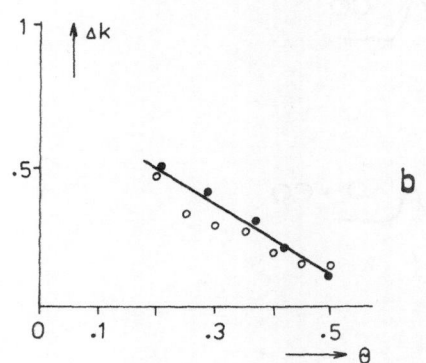

Fig. 2. Variation with coverage of the (a) relative intensity and (b) relative width of the (1/2,0) - LEED spot for the O/W(211) system. Solid curve: experimental results[6,7]; open circles: data from computer simulation[10].

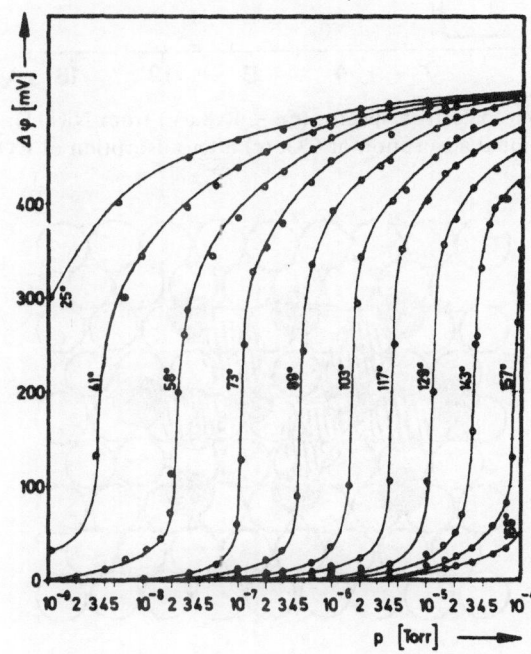

Fig. 3. Adsorption isotherms for the system H_2/Ni(110).[13]

with one-dimensional order. A structure model as shown in Fig. 5 is proposed for this mixed adsorbed layer. Flash desorption experiments revealed that the adsorption energy of CO is lowered by about 5 kcal/mole as compared with the value for a clean Ni(111) surface. Only minor amounts of CO_2 are formed during thermal desorption, and this is attributed to the relatively large Ni-O binding energy.

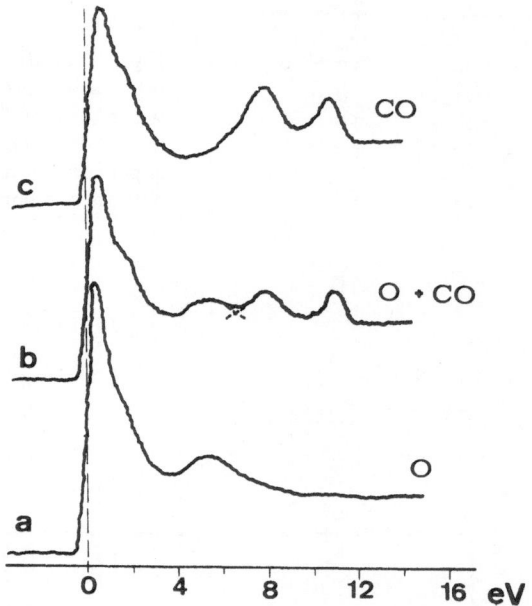

Fig. 4. Ultraviolet photoelectron spectra ($h\nu = 40.8$ eV) from Ni(111): (a) with adsorbed oxygen; (b) after additional adsorption of CO; (c) after adsorption of CO on the clean surface.

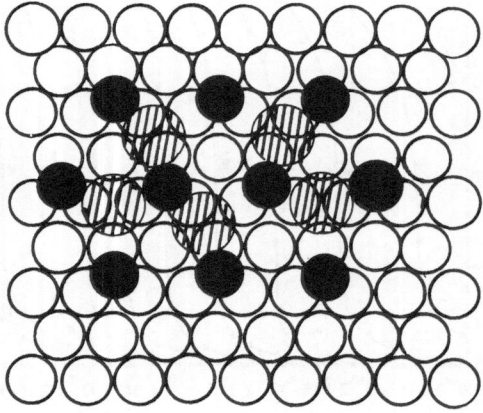

Fig. 5. Structure model for the mixed adsorbed layer of O_{ad} (dark circles) and CO_{ad} (hatched circles) on Ni(111).

Displacement reactions are a further aspect to be considered in this context. For example, exposing a NO-covered Pd(110) surface to gaseous CO leads to continuous displacement of the former species. This effect can again nicely be demonstrated by using UPS, as shown by Fig. 6.[20] The factor governing the rates (and activation energies) for such processes are still completely unknown and are felt to be rather complicated. On the other hand, such effects may play a decisive role in the overall mechanism of catalytic reactions and in any models for the action of inhibitors in catalysis.

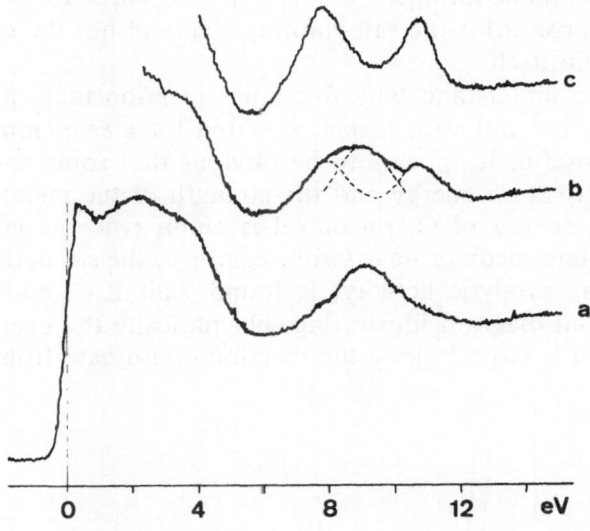

Fig. 6. Ultraviolet photoelectron spectra ($h\nu = 40.8$ eV) from Pd(110) demonstrating the displacement of adsorbed NO by CO: (a) $p_{NO} = 1 \times 10^{-8}$ Torr; (b) $p_{NO} = 1 \times 10^{-8}$ Torr; $p_{CO} = 2 \times 10^{-8}$ Torr; (c) $p_{CO} = 3.5 \times 10^{-7}$ Torr.

4 REACTIVE INTERACTION: THE CATALYTIC OXIDATION OF CO

The oxidation of CO is readily catalyzed by the platinum metals and was studied in our laboratory in some detail with palladium.[21] For example, oxygen adsorption causes on Pd(111), as on Ni(111), the formation of a 2x2 LEED pattern and a UPS peak at about 5.5 eV below the Fermi level.[20] If such a surface is exposed to CO, even at room temperature a very rapid formation of CO_2 is observed, which indicates a rather low activation energy—in contrast to nickel as described above. Besides this reaction of the Eley-Rideal type, CO_2 may also be formed above ~50°C by interaction between O_{ad} and CO_{ad}, both species being present on the surface in separate domains, as shown by a superposition of the corresponding LEED patterns.[21] The activation energy for this Langmuir-Hinshelwood reaction was determined to be 7 kcal/mole.[23]

A theoretical *ab initio* evaluation of the potential energy hypersurface (and thereby of the activation energy) for a heterogeneously catalyzed reaction appears at present to be a rather hopeless task, since this problem, even for homogeneous reactions, has so far only been solved for the simplest cases. A further complication arises from the fact that a steady-state catalytic reaction is usually composed of several elementary steps. Therefore the elucidation of the reaction mechanism is of decisive importance: for the CO oxidation on palladium, the desorption of adsorbed CO (which inhibits oxygen chemisorption) is the rate-limiting step and not the surface reaction $O_{ad} + CO \rightarrow CO_2$ itself.

In order to understand why this latter reaction takes place so readily with palladium but not with nickel, a search for a semiempirical principle appears to be useful. It appears to be obvious that some correlation exists between the activation energy and the strength of the metal-oxygen bond: the adsorption energy of O_2 on nickel is about twice as large as that on palladium. An intermediate stage (with regards to the strength of the oxygen bond as well as catalytic activity) is found with Ru[24] and Ir[25]. Figure 7 shows a potential diagram illustrating schematically the energy variation if an oxygen atom is moved along the reaction coordinate from M - O + CO

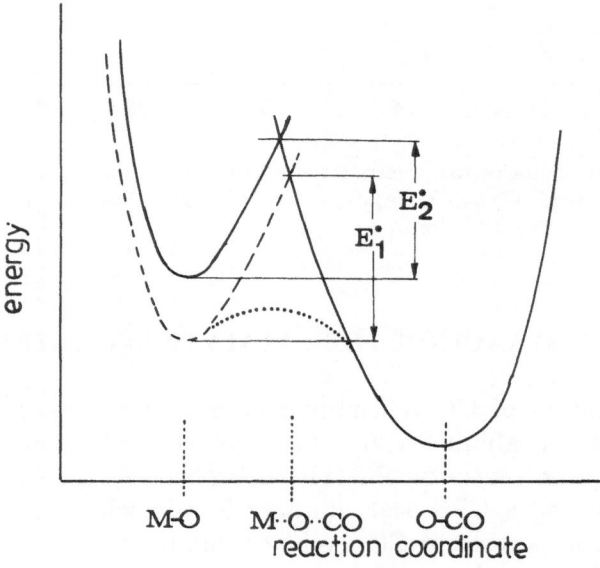

Fig. 7. Schematic potential diagram for the reaction M - O + CO → M + CO₂. The dashed curve represents a situation with higher adsorption energy and therefore also larger activation energy E$^*_1$; the case of strong stabilization of the transition state by strong electronic interactions between the reactants is indicated by a dotted line.

to M + OCO. The height of the barrier represents the activation energy, which is expected to be proportional to the depth of the left potential minimum, i.e., the M - O binding energy, provided that the shape of the potential barrier is not significantly influenced by electronic interactions between the reacting species. Relations of this type are frequently observed for the kinetics of homogeneous reactions, the well-known Brønsted law in acid-base catalysis being an example thereof.

Table I compares some data of R - O binding energies E_b with the corresponding activation energies E^* for the reaction R - O + CO → R + CO_2. With NO, NO_2, and O_2, the proportionality between E_b and E^* qualitatively holds. If the role of the catalyst would consist solely in lowering the bond strength of the oxygen atoms, then with palladium a much higher activation energy would be expected, since the binding energy of Pd_{met} - O is still rather large. The real situation is obviously similar to that found with OH, where the activation energy is negligibly small although the bond is very strong. A similar behavior was found with CrO molecules. In the case of OH, the oxygen atom contains a single unpaired electron which makes this species rather reactive and leads to a marked lowering of the activation barrier due to strong electronic interactions, as indicated by the dotted line in Fig. 7. Probably an oxygen atom adsorbed on palladium has some radical-like character causing its high activity. A detailed knowledge of the nature of the chemisorption bond (e.g., not only of the binding energy but also of the wave functions and their numbers of occupancy) is therefore of fundamental importance for an understanding of heterogeneous catalysis.

Table I. Reaction R - O + CO → R + CO_2

Reactant	Strength of the R - O Bond E_b, kcal/mole	Activation Energy E^*, kcal/mole	Refs.
O	—	3	(a)
N_2O	58	23	(b)
NO_2	65.5	28	(b)
O_2	118.9	60	(c)
HO	106.7	1	(d)
CrO	101	small	(e)
Pd(metal) . . . O_{ad}	90	3	(f)

(a) Simonaitis, R., and Heicklein, J., *J. Chem. Phys.*, **56**, 2004 (1972).
(b) Lim, M. C., and Bauer, S. H., *ibid.*, **50**, 3377 (1969).
(c) Dean, A. M., and Kistiakowsky, G. E., *ibid.*, **54**, 1718 (1971).
(d) Westenberg, A. A., and De Haas, N., *ibid.*, **58**, 4061 (1973).
(e) Izod, T.P.J., Kistiakowsky, G. B., and Matsuda, S., *ibid.*, **56**, 1083 (1972).
(f) Ertl, G., and Koch, J., unpublished.

REFERENCES

1. Grimley, T. B., *Proc. Phys. Soc.,* **90,** 751 (1967).
2. Grimley, T. B., and Walker, S., *Surface Sci.,* **14,** 395 (1969).
3. Grimley, T. B., and Torrini, M., *J. Phys.,* **C6,** 868 (1973).
4. Einstein, T. L., and Schrieffer, J. R., *Phys. Rev.,* **B7,** 3629 (1973).
5. Graham, W. R., and Ehrlich, G., *Phys. Rev. Letters,* **31,** 1403 (1973); **32,** 1309 (1974).
6. Chang, C. C., and Germer, L. H., *Surface Sci.,* **8,** 115 (1967).
7. Tracy, J. C., and Blakely, J. M., in *The Structure and Chemistry of Solid Surfaces* (G. A. Somorjai, Ed.), Wiley, New York (1969), p. 65-1.
8. Hopkins, B. J., and Watts, G. D., *Surface Sci.,* **44,** 237 (1974).
9. Ertl, G., and Küppers, J., *Surface Sci.,* **21,** 61 (1970).
10. Ertl, G., and Plancher, M., *Surface Sci.,* **48,** 364 (1975).
11. Duke, C. B., and Liebsch, A., *Phys. Rev.,* **B9,** 1126,1150 (1974).
12. McKee, C. S., Perry, D. L., and Roberts, M. W., *Surface Sci.,* **39,** 176 (1973).
13. Christmann, K., Schober, O., Ertl, G., and Neumann, M., *J. Chem. Phys.,* **60,** 4528 (1974).
14. Suzanne, J., Coulomb, J. P., and Bienfait, M., *Surface Sci.,* **44,** 141 (1974).
15. Rau, C., and Sizmann, R., to be published.
16. Ertl, G., and Koch, J., in *Adsorption-Desorption Phenomena* (F. Ricca, Ed.), Academic Press, New York (1972), p. 345.
17. Holloway, P. H., and Hudson, J. B., *Surface Sci.,* **43,** 141 (1974).
18. Christmann, K., Schober, O., and Ertl, G., *J. Chem. Phys.,* **60,** 4719 (1974).
19. Conrad, H., Ertl, G., Küppers, J., and Latta, E. E., in preparation.
20. Conrad, H., Ertl, G., Küppers, J., and Latta, E. E., *Chem. Soc. Faraday Disc.,* in press.
21. Ertl, G., and Rau, P., *Surface Sci.,* **15,** 443 (1969).
22. Ertl, G., Koch, J., in *Proc. Vth Int. Congr. on Catalysis, Palm Beach, 1972* (J. Hightower, Ed.), North Holland, Amsterdam (1973), p. 969.
23. Ertl, G., and Neumann, M., *Z. Phys. Chem. N.F.* (Frankfurt), **60,** 127 (1974).
24. Madey, T. E., Engelhardt, H. A., and Menzel, D., *Surface Sci.,* **48,** 304 (1975).
25. Küppers, J., Plagge, H., and Ertl, G., in preparation.

DISCUSSION on Paper by G. Ertl

BOUDART: I consider the mechanism you have evolved for the CO oxidation with determination of activation energies as an exciting beginning. Indeed, although we understand a lot about catalysis, we are powerless to make predictions because of the lack of quantitative information about the rates of elementary processes on well-defined surfaces. It is hoped that in a few years this situation will have improved drastically.

I am also impressed by the very low activation energy of adsorbed oxygen with gas-phase carbon monoxide. Does it mean that it is possible to run the oxidation of CO on palladium catalytically at low temperatures?

ERTL: Under steady-state conditions at low temperatures the palladium surface will become covered completely by adsorbed CO which in turn inhibits the dissociative adsorption of oxygen (a necessary prerequisite for the catalytic reaction). Thus desorption of CO (with an activation energy of about 30 kcal/mole) will be the rate-determining step.

MENZEL: Professor Ertl has mentioned the existence of repulsive interactions as well as attractive interactions between oxygen atoms adsorbed on W(112). Such anisotropic interactions seem to occur quite often for adsorbed oxygen: on W(100) a (4x1, 1x4) structure is already formed at low coverages; (nx1) structures, with n = 7. . . . 2 for increasing coverages are found on Ag(110); (2x1) structures are formed on Ru(001) (this conference). In all these cases, chains of oxygen atoms form which repel each other or lead to periodic variations of the interaction energy in the other direction. This may have to do with the oxygen p-orbitals. It would appear worthwhile to investigate this tendency theoretically.

Professor Ertl reported an increase in adsorption energy with coverage for hydrogen on nickel and explained it by attractive interactions. We have also found this behavior for oxygen on Ag(110), and explained it in the same way (this conference).

The cases of interactions between adsorbed oxygen and CO mentioned by Professor Ertl all lead to a weakening of the bonds to the surface, leading to desorption of CO_2. We have also observed the opposite behavior: on clean Ag(110), no CO is adsorbed at room temperature; if oxygen is offered at the same time, a composite layer is formed, so that CO seems to be bound more strongly to the surface by the presence of oxygen [Engelhardt, H. A., Bradshaw, A. M., and Menzel, D., *Surface Sci.,* **40,** 410 (1973)].

Professor Ertl interpreted the "2x2" pattern of oxygen on Ni(111) as a true (2x2) structure, without considering the possibility of threefold degenerate (2x1) structures. Why?

ERTL: Holloway and Hudson [*Surface Sci.,* **43,** 141 (1974)] followed the rate of oxygen uptake at Ni(111) by Auger spectroscopy and LEED and found that the intensity of the "extra" spots of the 2x2-LEED pattern becomes maximum at $\theta = 0.25$, which can only be reconciled with a true 2x2 structure, since with a 2x1 structure, the coverage should be 0.50.

EMMETT: Did you observe any interaction of the NO with the palladium to form nitrogen and a surface layer of oxygen on the palladium?

ERTL: This reaction was observed to take place at about 300°C, whereas at room temperature NO remains stable on palladium. It decomposes readily on iron and on nickel [*Surface Sci.,* in press].

EHRLICH: Have you been able to deduce a potential for interactions between hydrogen atoms from your isotherms for adsorption on Ni(110)?

ERTL: Unfortunately not, particularly since the atomic arrangement of hydrogen on Ni(110) is still unknown. The observed 1x2-LEED pattern may even be caused by a reconstruction of the surface [Estrup, P. J., and Taylor, T. N., *J. Vac. Sci. Technol.*, **11**, 244 (1974)].

EISCHENS: Infrared experiments with chemisorbed carbon monoxide show that oxygen causes the carbon-oxygen stretching frequency to shift upward by 100 to 200 cm^{-1}. For similar experimental procedures, the smallest UPS shifts you observe are about 0.3 eV. Is there any way that you can correlate the high-energy electronic shifts with the changes observed in the infrared region?

ERTL: Shifts in the IR stretching frequency of CO are ascribed to partial electron transfer from the metal to the antibonding $2\pi^*$ orbital of CO and therefore depend on the number of occupancy of this orbital. UPS probes the energies of electronic states and therefore there exists no direct correspondence with the infrared shifts. If the assignment of the shifting UPS peak with the $CO-1\pi$ level is correct, then it indicates a strengthening of the π bond in CO by coadsorbed oxygen as it is also inferred from your infrared results.

YATES: Your paper illustrates very nicely the opportunities that "clean"-surface kinetic and spectroscopic studies offer for the understanding of adsorption and reactions on actual catalytic surfaces. In this connection, one should always be alert to links between the two kinds of work, and I am wondering whether any of your UPS studies of the coadsorption of CO and NO by Pd(111) can be interpreted in terms of the production of an isocyanate complex Pd-NCO? Such complexes have been postulated by M. L. Unland [*Science*, **179**, 567 (1973)] on the basis of observation of a new infrared band at about 2260 cm^{-1} when the gases were adsorbed together on supported Ru, Rh, Pd, Ir, and Pt surfaces. Of course the pressures in his experiments were far higher than in your UPS studies.

ERTL: In our work, photoelectron spectra of the interaction of CO and NO with palladium surfaces could always be analyzed in terms of a superposition of contributions of each species alone, without any indication of significant electronic interactions or even molecular transformation. In fact Unland reported evidence for isocyanate formation after dosing his samples at 400°C with pressures of several Torr of the respective gases, which is far beyond the conditions employed in our studies.

SOMORJAI: We have studied the hydrogen-deuterium exchange on stepped and low Miller Index platinum surfaces by molecular beam scattering and have found the reaction mechanism to be identical to that reported here by Professor Ertl for the CO + O_2 surface reaction on palladium. Our technique measures the HD product concentration that forms upon a single scattering of H_2 or D_2 molecules and HD must leave the surface in 10^{-2} second or less to be detectable. From the temperature and pressure dependence of the reaction rate we have concluded that HD forms as a result of an atom-molecule reaction. At low temperatures ($<600°$K), the dominant reaction occurs between the adsorbed H atoms and the adsorbed D_2 molecules (Langmuir-Hinshelwood mechanism) with a rate constant of k = 1 x 10^5 exp(-4.5 kcal/RT). At high temperatures ($>600°$K), the concentration of adsorbed D_2 molecules is reduced to the point that the dominant mechanism is one between adsorbed H atoms and D_2 molecules incident from the gas phase (Eley-Rideal mechanism). This high-temperature reaction has a negligible or zero activation energy and its preexponential factor is 10^3. It appears that there may be a more general mechanism applicable to describe surface reactions between diatomic molecules in these metal surfaces. The catalyst action of the active metal surface is due to its ability to adsorb and dissociate the diatomic molecules and store the atoms readily, thereby converting the bimolecular reaction to an adsorbed atom-molecule reaction. Both Langmuir-Hinshelwood and Eley-Rideal mechanisms are operative and the reaction conditions (surface temperature, reactant pressures) determine which one will dominate.

WABER: Recently we have been doing a series of calculational studies on the adsorption of CO, O, S, and Se on small nickel clusters. It started with perturbed molecular calculations (PMC) such as those to be discussed during this conference. At this writing, self-consistent charge calculations have nearly been completed for CO and sulfur, but to a lesser extent for oxygen. Since the method is based on a modification of the LCAO technique which is being labelled single site orbital and the necessary integrals are being solved by the discrete variational method, some questions of reliability, use of sufficient points, etc., cause me to offer the following idea with a little reservation.

On the basis of this result, it appears that some charge transfer occurs in CO, which indicates that a free radical may form—certainly charge redistribution occurs near the internuclear line which indicates a reduction in band strength. Since there is much more convincing evidence of where sulfur is located in relation to the nickel atoms, we have done much more careful work for this case. It appears that, contrary to our expectation, excess charge does not accumulate on the sulfur atom, but some charge transfer may be toward the metal cluster. So far though, transfer appears to flow toward the cluster and does appear to occur.

These incomplete studies lead to two conclusions: (a) that substantial charge redistribution occurs in chemisorption with essentially a covalent bond being formed and (b) the adsorbate then resembles a free radical. This idea offers some comfort to me in trying to answer the question in what specific way does adsorption increase the effective rate constants manyfold? If the idea of free-radical formation does bear up under more scrutiny, one answer would be that activation energy barriers go down and rates go up by orders of magnitude (over those of reactions which occur by molecular collisions in the gas phase) when free radicals are involved. I suggest that chemisorption may promote free-radical formation.

THE APPLICABILITY OF ELECTRON-EMISSION SPECTROSCOPY TO ELUCIDATE CHEMISORPTION*

E. W. Plummer

Department of Physics,
University of Pennsylvania
Philadelphia, Pa. 19174, U.S.A.

ABSTRACT

The development of a quantum mechanical understanding of the surface-adsorbate bond and the reactions among adsorbates would surely accelerate the development of a mechanistic understanding of heterogeneous catalysis. The achievement of such a goal will depend upon a close coupling between theory and experiment. Experimentally, the geometrical location, the electronic energy levels of the surface atoms, and the adsorbate complex must be determined, i.e., the identification of the bonding configuration. In addition, the kinetics involved in formation of each surface complex must be measured. Electron-emission spectroscopy is used to measure the energy-level spectrum of the surface or the "surface density of states", in order to elucidate the nature of the chemical bond at the surface. All of the electron

*This work was supported in part by the National Science Foundation and the Advanced Research Projects Agency.

spectroscopic techniques measure the dynamic response of a surface to an external probe, be it photons, electrons, or an applied field, while the information that is desired is the ground state of the system. A discussion of the response function of the measurement technique for field emission and photoemission is presented, which is focused upon the ability to measure the "surface density of states" and the ground-state energy spectrum. Several specific systems are discussed to illustrate the general properties of chemisorption systems, and the way photoemission spectroscopy can be utilized to determine the nature of the adsorbate-substrate interaction.

1 INTRODUCTION

The premise for relating basic surface studies to heterogeneous catalysis is that a better understanding of the fundamentals of catalysis would significantly accelerate the development of successful catalytic processes. One aspect of such a fundamental understanding of catalysis is the nature of the adsorbate-surface bond. This involves the development of a quantum mechanical model of chemisorption, which by its very nature is more complicated than either the equivalent problem in a 3-D solid or in a simple molecule. Therefore, it is unlikely that a first-principles calculation will be able to predict the behavior of a real system in the foreseeable future. This necessitates an empirical approach where theory and experiment proceed hand in hand. Ideally, the experimentalist would furnish the theorist with reliable data on the geometrical position of the surface atoms and the energy spectrum before and after a given species had been adsorbed upon the surface, as well as the thermodynamic information on the formation of the surface species. The theorist would use these data to test various theoretical models to ascertain which parameters are important, hopefully then explaining the observed bonding configuration and predicting what may occur on a different system. The ultimate success of such an iterative procedure would be a theoretical model which had predictive powers. Recent developments in the theory of chemisorption[1,2] and in experimental techniques indicate that the achievement of this idealistic goal may be possible in the next few years, at least for simple prototype systems. Along with the growing maturity of this field comes an added responsibility for both the theorist and the experimentalist. The rug has to be lifted to see what was swept under. For the theorist it means that approximations and their justifications must be spelled out clearly. Internal checks must be performed for each theoretical model with the failures as well as the successes being made public. To the experimentalist it means that he must investigate what his technique really measures instead of concentrating on what he would like it to measure.

The primary objective of any electron-emission spectroscopy is to measure a well-defined average of what has been called the energy-position electron distribution function ρ (**r**, E) of the ground state of the system. If the system can be characterized by N one-electron wave functions $\psi_i(\mathbf{r})$, then

$$\rho \; (\mathbf{r}, \; E) = \sum_{i=1}^{N} \; \delta \; (E\text{-}E_i) \; |\psi_i \; (\mathbf{r})|^2 \qquad (1)$$

the "surface density of states" is usually the average of ρ (**r**, E) over the volume occupied by the surface atoms:

$$\rho_s \; (E) = \int_{\substack{\text{surface} \\ \text{atoms}}} \rho \; (\mathbf{r}, \; E) \; dV \qquad (2)$$

The advancement of the theory of surfaces and of chemisorption would appear to be intimately related to our ability to calculate and measure the properties of ρ (**r**, E). Section 2 is devoted to a critical discussion of what we now know about field emission and photoemission energy distributions. Can we justify extracting the "surface density of states" from the measured energy distribution?

There is a more phenomenological approach to electron spectroscopy which I refer to as the "fingerprint" technique. The philosophy is that you do not worry about the measurement process, just catalog the spectra for as many surface configurations as possible. These cataloged spectra are combined with gas-phase photoelectron spectra to identify the nature of the chemisorbed species in a specific surface reaction. Several examples are given in Section 3 to show how this technique can be very useful in identifying adsorption states.

Also in Section 3, data are presented to illustrate the general features of chemisorption and the resultant change in the surface density of states which has been observed in an extensive study of H_2, N_2, CO, O_2, and C adsorption on W(110) and W(100).[3].

2 SURFACE DENSITY OF STATES

There are several electron-emission processes which are now being used to study surfaces. In all of these techniques, the emitted electrons are energy analyzed so that the final-state energy of the ejected electron can be determined. The objective is then to infer the initial-state energy from the measured final-state energy and the characteristics of the probe. The three most investigated techniques are: (1) ion-neutralization spectroscopy, where a low energy ion beam is the probe; (2) field-emission spectroscopy, where an applied field is the external probe[4,5]; and (3) photoelectron spectroscopy, where an incident beam of photons is the probe.[5,6] The energy of the incident photon can range from ultraviolet to X-ray. Hagstrum* has described ion neutralization and its comparison with photoemission, so this paper focuses on field emission and photoemission.

Field emission is the process of applying a large electrostatic field, approximately 30 million V/cm, to a cold cathode so that electrons can tunnel from the solid, through the classically forbidden barrier, into the vacuum. In order to achieve these high fields at reasonable voltages, the cathode or emitter is usually etched to a sharp point of approximately 1000 A radius. The hemispherical end of the emitter exposes all crystallographic orientation so that the individual crystal phases can be located and identified in the field-emission pattern observed on the fluorescent screen. The emission characteristics of any crystal plane can be studied by placing a small "probe ole" in the screen and deflecting the field-emission pattern with electrostatic deflection plates[4] until the plane of interest is over the probe hole. The current passing through the probe hole is then energy analyzed. A schematic drawing of the potential at a metal surface with an applied electric field is shown on the left in Fig. 1a. On the bottom right of Fig. 1a is a typical field-emission energy distribution, decreasing exponentially as the energy decreases and the tunneling barrier becomes wider and higher. Field emission is to the first order an elastic process so that the total energy of the emitted electron is the same outside as it was inside; therefore, the distribution is cut off on the high-energy side by the Fermi edge. Our objective is to be able to remove the exponential tunneling probability (through the vacuum barrier) from the measured energy distribution and relate the structure in the remaining curve to the electronic properties of the surface.

Figure 1b depicts the adsorption of a foreign atom onto the schematic surface shown in Fig. 1a. The bell-shaped curve in the region of the adsorbate, centered at an energy ϵ below the Fermi edge, is intended to represent the local density of states or "virtual level" on the adsorbate. Duke and Alferieff[7] showed that tunneling through the adsorbate is equivalent to tunneling resonance, so that there is a large increase in the tunneling probability for an electron with an energy near the bound-state energy in the adsorbate. This increase or enhancement is shown schematically in the energy distribution on the right in Fig. 1b.

While field emission is basically an elastic process, photoemission [shown schematically in Figs. 1a and 1b] is an excitation process. An incoming photon excites an electron to an excited state where the difference in energy between the initial and final states is the photon energy $\hbar\omega$.** If the excited electron has sufficient energy, is headed in the right direction, and does not lose too much energy, it may escape from the solid to be subse-

---

*See paper in these proceedings.

**In a real system the energy difference between initial and final state of the total N partice system is equal to the photon energy.

quently energy analyzed. A schematic of energy distribution showing both the elastic and inelastically scattered electrons is presented on the top right of Fig. 1a. One of the fundamental problems in interpreting photoemission data is the inability to separate the primary electrons (unscattered) from the inelastically scattered electrons. The incoming light has an extinction distance in the solid of ~100 Å in the ultraviolet region, which varies with material, photon energy, and the angle of incidence of the light.[5] The surface sensitivity of photoemission is a consequence of the strong, inelastic electron scattering of the excited electron. The escape depth for an electron excited 50 eV above the vacuum level is ~5 Å in most materials[8] and increases to ~15 Å for an electron excited 1500 eV above the vacuum level. This strong attenuation of the excited electrons, which creates the surface sensitivity, also complicates the interpretation of the data.[9] In the energy range of the bulk band, there is no justifiable technique at present capable of separating the surface density of states from the bulk density of states in a photoemission spectrum.

When a foreign atom or molecule is adsorbed on the surface [Fig. 1b], an increase in signal would be expected at a kinetic energy nearly equal to the photon energy minus the ionization potential of the surface molecule. This is shown in Fig. 1b for a simple adsorbed atom. What is not shown in Fig. 1b are the effects on the energy distribution due to the redistribution of energy levels of the substrate atoms when they form a chemical bond with the adsorbate. There will always (with the possible exception of hydrogen) be energy levels of the surface complex below the bottom of the band. For these levels there will be only weak interference effects with the bulk, and interpretation of peaks in the photoelectron spectra becomes easier. Yet, relating the measured binding energies of these peaks to a calculated or measured orbital energy of a molecule is complicated by the screening of the hole state by the metallic electrons. This relaxation of the final state around the hole causes a decrease in the total final-state energy of the system and consequently an increase in the kinetic energy of the ejected electron of the order of 1 to 6 eV.

The measurement of the binding energies of core levels using X-rays as the incident radiation is usually referred to as ESCA (electron spectroscopy for chemical analysis). The shift in the observed binding energy of a core level is an indirect effect of the distribution of electrons in the valence orbitals of the molecule, both in the initial and final states. The initial-state effects are usually referred to as chemical shifts, while the final-state effects are known as relaxation. In both the UV and X-ray spectra, the measured binding energies should not agree with ground-state orbital energy calculations or with gas-phase photoionization spectra until the relaxation energy of the final state has been taken into account. For an adsorbate positioned on the vacuum side of the image plane, these relaxation shifts are just the upward shifts due to the image potential.[10]

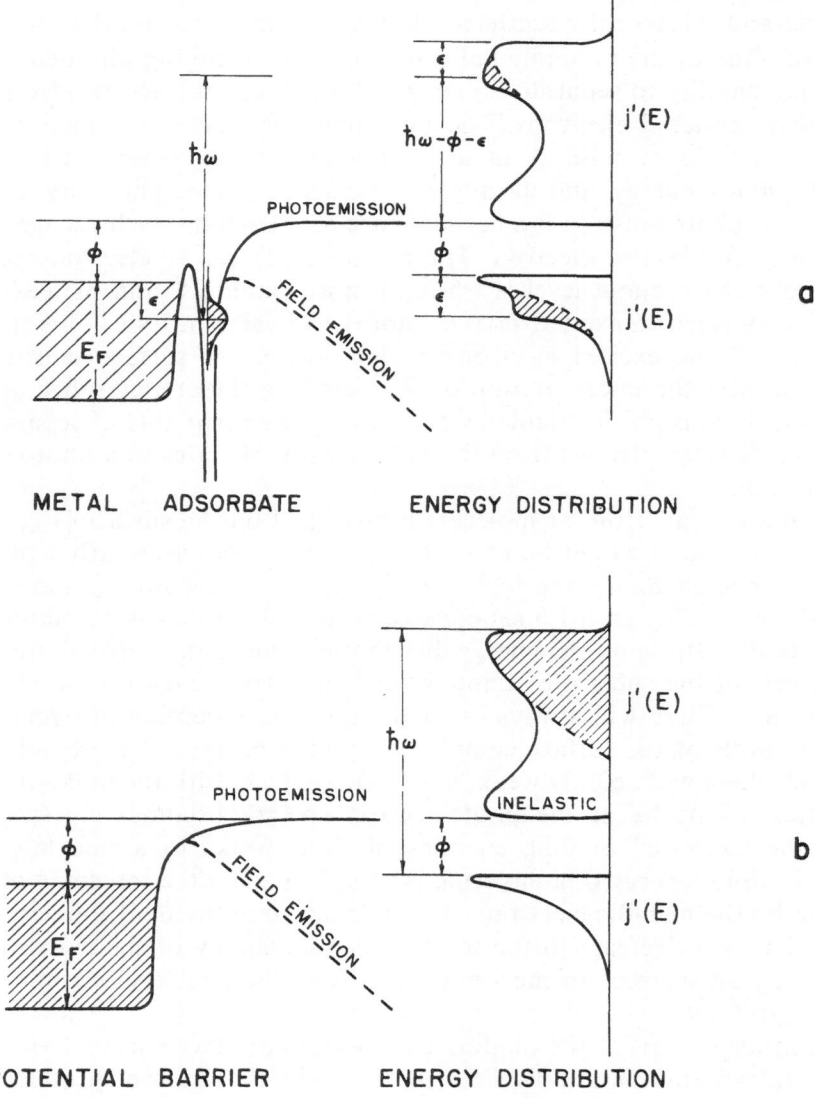

Fig. 1. (a) Schematic representation of the surface potential of a free electron metal, with and without the application of an external field, and a typical energy distribution for: (1) electrons which have tunneled elastically from the metal under the application of the applied field and (2) electrons which have been photoexcited by radiation of energy $\hbar\omega$. (b) Schematic representation of the surface potential with an idealized adsorbed atom present, with and without an applied field present. On the left are the resultant energy distributions for field emission (bottom) and photoemission (top). The shaded areas of the energy distribution depict the increase in current coming from the "virtual level" of the adsorbate, shown at an energy ϵ below the Fermi energy on the left.

2.1 Theory of Field Emission

The theory of field emission from clean and adsorbate-covered metallic surfaces seems to be in fairly decent shape at the present. In this section, the conclusions from many different works are stated and an attempt is made to list what problems still exist. For a clean, low-index crystal face where the surface potential can be approximated by a one-dimensional potential and $k_{||}$ is conserved across the barrier, many theoretical approaches[5] have produced the following result:

$$\frac{j'(E)}{j_o'(E)} = R \ (E) \propto \rho_\perp \ (z_o, E) \tag{3}$$

where $j'(E)$ is the measured energy distribution at a field F and $j'_o(E)$ is the calculated free-electron energy distribution for a field F and work function φ^4; $\rho_\perp \ (z_o,E)$ is the one-dimensional density of states $[k_{||}= 0]$ evaluated at the classical turning point z_o.[11] $\rho_\perp$ should be a weighed average of the density of states at z_o and the exponentially decreasing tunneling probability D as a function of $k_{||}$ where typically $\frac{D(k_{||})}{D(k_{||}=0)} = \frac{1}{e}$ for a transverse energy $\frac{\hbar^2 k_{||}^2}{2m}$ of 0.15 eV. In addition, $\rho_\perp$ should be averaged over the surface Brillouin zone.[12]

In the second column are plotted in Fig. 2 the measured R (E) curves for field emission from four different crystal planes of tungsten.[5] The first column shows the bulk band structure for tungsten in the specific directions, as calculated by Christensen[13], while the third column is a hand-smoothed representation of the one-dimensional density of states taken from Christensen's calculation[13]. The fourth column shows photoemission energy-distribution measurements by Feuerbacher[14], which are discussed later. At first glance, there is remarkable similarity between the bulk calculation and the field-emission measurements. All of the field-emission curves exhibit peaks near -0.7 eV and -1.3 eV which are consistently lower in energy than the peaks in the bulk one-dimensional density of states originating from the gap at Γ. There are four specific features in the field-emission curves which cannot be explained by the bulk band structure: (1) the large peak at -0.35 eV on W(100) which is a surface resonance* in the spin-orbit split gap in that direction**; (2) a peak and/or shoulder on W(110) near -0.4 eV; (3) an additional peak at -0.4 eV in the (112) direction; and (4) the current which appears in the total gap in the (111) and (112) directions.

---

*This surface resonance has been seen both in field emission[15] and in photoemission.[16,17]
The surface density of states has been calculated from this state by K. Sturm and R. Feder [*Solid State Comm.*,14**, 1317 (1974)]. Their calculation is shown by the dashed line in the third column of Fig. 2.

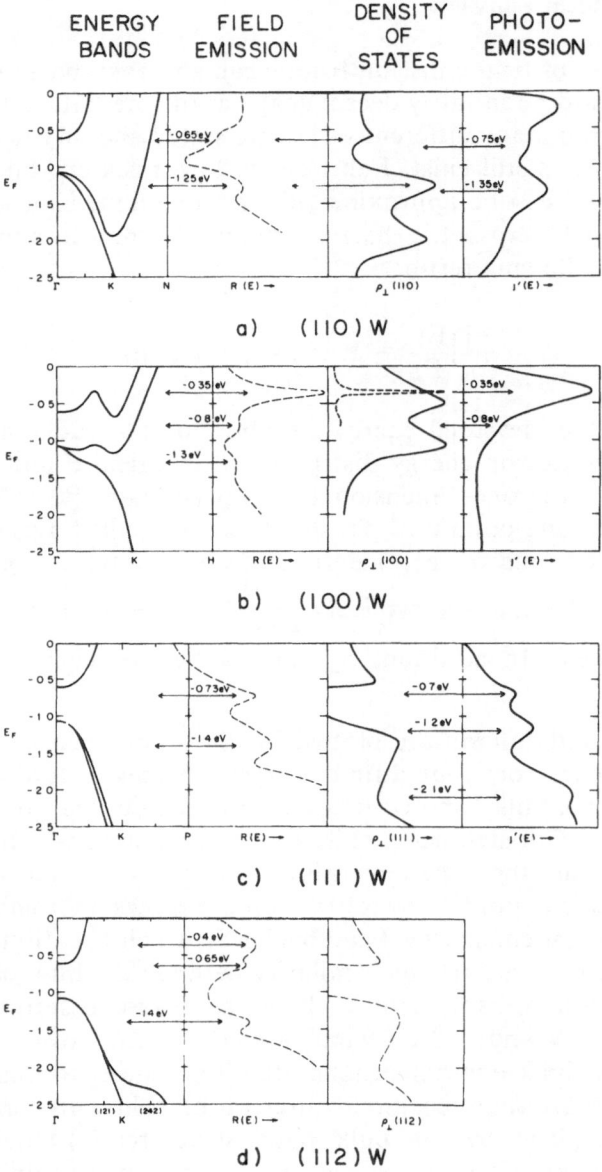

Fig. 2. Comparison of energy bands[13], field-emission R curves, one-dimensional bulk density of states[13], and normal emission–photoemission energy distributions[14] for four low-index faces of tungsten. The photon energy was 9.5 eV for W(110), 10.2 eV for W(100), and 10.2 eV for W(111). The density of states shown in the third column are the one-dimensional density of states calculated by Christensen[13] from his energy bands (shown in the first column), which have been smoothed by hand to remove the singularities. The dashed curve in the third column for W(100) is from a surface resonance calculation by Sturm and Feder (to be published).

The first three items are indications of changes in the density of states near the surface. The fourth point of disagreement above is a consequence of the measurement process. Consider the simple case where the surface is just a window through which we view the bulk. When we say that $k_{||}$ is conserved across the barrier we mean

$$\mathbf{k}_{||} \text{ (outside)} = \mathbf{k}_{||} \text{ (inside)} + \mathbf{g}_{||}$$

where $\mathbf{g}_{||}$ is a surface reciprocal lattice vector. If $k_{||}$ (outside) $= 0$ then

$$\mathbf{k}_{||} \text{ (inside)} = \mathbf{g}_{||}$$

which has one obvious solution that $k_{||}$ (inside) $= 0$. For a high-index face, the projection of the bulk Brillouin zone may be larger than the surface zone. This means that there are additional lines in k space which may contribute to the emission with $k_{||}$ (outside) $= 0$, besides the $k_{||} = 0$ bulk direction. For emission from W(111) this line is along the zone face between the points H and P. There is no gap in this cut in the energy range -0.6 to -1.1 eV.[13] How much these other directions contribute to the emission would have to be calculated, but for all crystal planes with higher index than (100), there will be new cuts in k space which must be considered. For the (112) direction, the new cut in k space enters the Brillouin zone at N and leaves the zone at a point where another [112] vector intersects. Therefore, the -0.4 eV peak on W(112) and W(110) might originate from the same bulk structure near the N point in the Brillouin zone.

Field emission is a very limited tool. The current falls off exponentially so that only a limited energy range below the Fermi energy can be observed (2 to 3 eV) before many-body effects dominate the spectrum.[5]

Specimen preparation in the past has limited field-emission energy-distribution measurements to the most refractory metals, and to compound the problem, the demands upon the energy analyzer are quite stringent. It is quite safe to say that field emission will never become a widely used tool for the study of heterogeneous catalysis, yet it may prove to be one of the most useful tools for measuring "surface density of states" as a check on theoretical calculations. Experimental measurements need to be made on systems where the surface density of states can be calculated, and the dominant features are within a few volts of the Fermi energy. We are at present preparing nickel tips by field evaporation and imaging with a channel plate. These measurements will be used as a check on the calculations of Davenport et al.* and as a comparison to the ion-neutralization data of Hagstrum*. As Hagstrum has pointed out, field emission and ion neutralization should be very similar in that the electron must tunnel from the solid. There may be a difference due to the three-dimensional potential around the incident ion, and the two-hole final state in ion neutralization.

---

*See papers by this author included in these proceedings.

Field-emission energy-distribution measurements may furnish information about the effect of particle size on the surface density of states. When an emitter is field evaporated, the size of the plane being observed can be controlled very accurately. Measurements of the surface density of states can be made as a function of the plane size. We made preliminary measurements on W(100) to determine the minimal plane size necessary to maintain the surface resonance. These measurements indicated that a cluster with the next-nearest neighbors exhibited the surface resonance. Shepherd and Peria[18] have reported the same behavior for the surface state on Ge(100).

2.2 "Theory" of Photoemission

We could begin and end this section by quoting the paper by Caroli, et al.[9], that the surface density of states cannot in general be extracted from a measured photoemission spectra. Instead, we start from a very unrealistic but simple model of photoemission and then add into this model the complications inherent in the process. We begin by assuming Koopmans' theorem[19], which says that in a one-electron picture the N-1 orbitals of the final state, after ejection of an electron, are "frozen", that is they are the same as the identical N-1 orbitals in the neutral initial state of the system. This means that the measured binding energy is identical to the Hartree Fock orbital energy of the ground state. We discuss this approximation later in terms of relaxation of the final state.

A simple view of photoemission is an independent particle picture with plane waves for the final states. Einstein[20] has shown, within this simple model, using an $\mathbf{A} \cdot \mathbf{P}$ operator for the excitation, that if the emitted electrons are angularly averaged, the energy distribution of primary electrons $j'(E)$ is

$$j'(E) \propto \int_0^\infty \rho\,(z,E)\,\exp[-z/\lambda]\,dz \qquad (4)$$

where $\rho\,(z,E)$ is the local density of states in a sheet a depth z below the surface at $z=0$; λ is the mean free path of the excited electron, which should be energy dependent. This simple model predicts a well-defined average of ρ, which is not the "surface density of states", unless the escape depth is extremely short.

The photoemission process is complicated by the details of the matrix element for excitation. We can illustrate this effect by treating the problem in perturbation theory as a transition from an initial state ψ_i to a final state ψ_f of the whole system. The matrix element M_{if} is

$$M_{if} \propto <\psi_f\,|\;\mathbf{A} \cdot \mathbf{P} + \mathbf{P} \cdot \mathbf{A}\;|\;\psi_i> \qquad (5)$$

where $\mathbf{A}$ is the vector potential of the electromagnetic field and $\mathbf{P}$ is the momentum operator. The final state is a state of the system with N-1 electrons in the solid and one electron at infinity with energy E_e. Conservation of energy gives the following relationship

$$E_e = E_{initial}^N - E_{final}^{N-1} + \hbar\omega , \qquad (6)$$

where $E_{initial}^N$ and E_{final}^{N-1} are the total energies of the N electron initial state and the N-1 hole final state.

It is usually assumed that $\nabla \mathbf{A} = 0$, but this is not true at a surface where there may be large gradients in the component of $\mathbf{A}$ perpendicular to the surface.[21,22] For the moment let us ignore the $\nabla \cdot \mathbf{A}$ term and concentrate on the $\mathbf{A} \cdot \mathbf{P}$ term. If ψ_i and ψ_f are both eigenstates of the same Hamiltonian, then by the use of simple commutation relations[23] we can convert Eq. (5) into

$$M_{if} \propto \frac{\mathbf{A}}{\omega} \cdot < \psi_f | \nabla V | \psi_i > \qquad (7)$$

where $\mathbf{A}$ has been removed from the brackets since the wavelength of the light is large compared with atomic dimensions, or that the momentum of the light is small compared with that of an electron. The potentials in the solid will be assumed to be separable into three terms:

$$V = V_B + V_S + V_{img} \qquad (8)$$

where V_B is the sum over all the bulk potentials, V_S is the potential of the last layer of atoms, and V_{img} is the surface potential which is the image potential far from the surface. This is an artificial separation but will illustrate the problem of interference effects in a photoelectron spectra. Using the potential of Eq. (8) in Eq. (7) produces

$$M_{if} = \frac{\mathbf{A}}{\omega} \cdot < \psi_f | \nabla V_B + \nabla V_S | \psi_i > + \frac{A_z}{\omega} < \psi_f | \nabla V_{img} | \psi_i > \qquad (9)$$

where ∇V_{img} has a component only in the z direction (perpendicular to the surface). The last term in Eq. (9) is usually referred to as the surface photoelectric effect.[22] It is commonly argued that the surface signal can be accentuated in UV photoemission by going to grazing angle of incidence with the light to maximize the A_z term in Eq. (9). This is probably not true in the far UV for most refractory metals like tungsten and nickel, since the index of refraction has a large imaginary term and the phase change upon reflection produces a cancellation in the standing-wave field in the z direction.[5] If we bring back the maximum in A_z occurs near 50 degrees angle of incidence [W.R.T. the normal].[5] If we bring back the $\nabla \cdot \mathbf{A}$ term and square the resultant matrix element we have

$$\omega^2 M_{if}{}^2 = | < \psi_f | -i\hbar \left[\mathbf{A} \cdot (\nabla V_B + \nabla V_S) + A_z \frac{\partial V_{imag}}{\partial z} \right]$$

$$-i\hbar \, \nabla \cdot \mathbf{A} \, | \, \psi_i > |^2 \tag{10}$$

This equation shows the possibility for interference effects between the bulk and surface or between the $\nabla \cdot \mathbf{A}$ term and the $A_z \frac{\partial V_{imag}}{\partial z}$ term. Schaich and Ashcroft have shown that the $A_z \frac{\partial V_{imag}}{\partial z}$ term will interfere constructively and destructively with the ∇V_B term by utilizing a simple Kronig-Penney model for V_B.[24] Endriz[22] has shown that if $V_B = V_S = 0$, there can still be interference between $A_z \frac{\partial V}{\partial z}$ and $\nabla \cdot \mathbf{A}$. There is no theoretically justifiable technique for separating bulk from surface emission in photoelectron spectra.

Recent papers by Feuerbacher and Christensen[13,14] suggest a possible technique for extracting something proportional to the surface density of states from a photoelectron spectra. The essential feature of their proposed technique is measurement of the electrons emitted normal to the surface at an energy where there is a gap in the final state bands in the specific direction of detection. The one-electron final state for the excitation will be the evanescent wave which decays exponentially into the bulk. If you knew the penetration depth of the evanescent wave as a function of energy, then the energy distribution would measure a function like that given in Eq. (4). We have replotted Feuerbacher's data[14] in the fourth column of Fig. 2. In general there is quite remarkable agreement between the photoemission and field-emission data. Yet it must be stressed that we do not know enough at present to draw any conclusions from this agreement.* For example, is the difference of 0.2 eV between the -1.4 eV peak seen in field emission and the -1.2 eV peak seen in photoemission significant? Could it be a consequence of the larger depth perception of photoemission compared with that of field emission?

What you measure in a photoelectron spectra also depends upon where you look. Feuerbacher and Fitton[14] showed that the energy distributions collected normal to three different low-index faces of tungsten were quite different (see Fig. 2). Our data[3,25] show that the energy distributions are different for different collection angles from the same crystal face. Figure 3 shows the work of Waclawski[25] on the angular dependence of the emission from clean W(100) and oxygen adsorbed on W(100), using a photon energy

---

*For emission normal to the surface of the (111) face of tungsten, the possible allowed states in the bulk must include the additional cut in k-space discussed for the field emission case.[5]

of $\hbar\omega = 21.2$ eV. The top set of energy-distribution curves are for clean W(100) and an exposure of 5×10^{-6} Torr sec of oxygen at a crystal temperature of 1500° K.[3] All emission within a polar angle of 45 degrees was collected. The second set of curves are for the same situation except only those electrons which were emitted at a polar angle of 33 ± 2.5 degrees perpendicular to the plane of incidence of the light were collected. The bottom set of curves are for electrons emitted normal to the surface ± 3.8 degrees. The top set of curves shows the double-peaked spectra derived from the 2p orbitals of the oxygen. The high energy peak is accentuated at normal angle of collection, while the low energy peak is much more pronounced at the 33-degree angle of collection. Also notice that the -2.0 eV

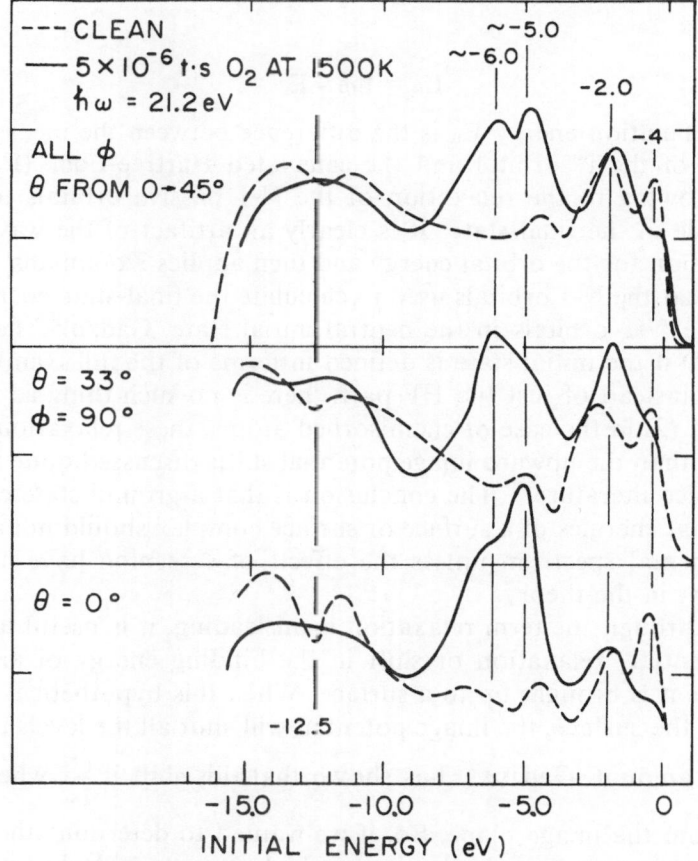

Fig. 3. Angular dependence of the photoemission spectra from 5×10^{-6} Torr-sec O_2 at 1500° K on W(100).[25] The top set of curves are for collection of all the photoemitted electrons within a polar angle of ± 45 degrees. The middle set of curves is for a collection angle of 33 ± 2.5 degrees normal to the plane of incidence of the light. The final set of curves are for normal emission ± 3.8 degrees. The incident light was at 45 degrees and the photon energy was 21.2 eV.

region in the oxygen curve is really accentuated in the 33-degree curve. There is undoubtedly a lot of information contained in angular resolved photoemission from adsorbates, but if the objective is to measure the orbital energy spectra of the surface complex, a large angle of collection or appropriate angle averaging is essential.

2.3 Relaxation Effects

The energy of the photoemitted electron is always given by Eq. (6). The binding energy E_B of an electron is given by

$$E_B = E_{final}^{N-1} - E_{initial}^N$$

or

$$E_B = \hbar\omega - E_e . \tag{11}$$

The relaxation energy E_R is the difference between the measured binding energy of the ith orbital and the calculated Hartree-Fock (HF) orbital energy, ϵ_i, owing to the relaxation of the N-1 passive orbitals toward the positive hole in the final state. It is clearly an artifact of the way one does the calculation for the orbital energy and then applies Koopmans' theorem[19] assuming that the N-1 orbitals used to calculate the final-state energy are the same as the N-1 orbitals in the neutral initial state. Gadzuk[26] has pointed out that: (1) if the initial state is defined in terms of the full Hamiltonian of the system instead of just the HF part, there is no such thing as relaxation energy; and (2) in the case of chemisorbed atoms, these relaxation shifts are none other than the upward image potential shifts discussed quite frequently in the surface literature.[10] The conclusion is that a ground-state calculation of the orbital energies of a surface or surface complex should not agree with an experimental spectrum unless the effects of screening have been taken into account in the theory.

Even through the term relaxation is misleading, it is useful to consider the "differential" relaxation or shift in the binding energy of an atom or molecule as it is brought up to a surface. When this hypothetical adsorbate is far from the surface, the image potential will shift all the levels upward by a constant amount. Gadzuk[26] has shown that this shift is $\frac{e^2}{4s}$, where s is the distance from the image plane. So, if we wanted to determine the chemical identity of a weakly bound adsorbate, which was moderately far from the image plane, we would add to the measured binding energies of the adsorbate an energy $\frac{e^2}{4s}$ and then compare the result with the gas-phase photoelectron data. For lack of a better title, we refer to this upward shift in adsorbate levels relative to the same gas-phase levels as the "surface shift". The relevant questions are the following: (1) What is the general magnitude of this "surface shift" for a chemisorbed atom or molecule? (2) How does the "surface shift" depend upon the size, shape, and position of the orbital from which an electron is to be photoejected?

The observed "surface shifts" in the valence orbitals for CO and C_2H_2 on single crystals of tungsten range from 2.8 to 3.2 eV[5], while the same shifts for chemisorbed C_2H_2 and C_2H_4 on Ni(111) range from 2.1 to 3.2 eV[27]. These shifts were obtained by adding the saturation work function for the specific states being studied to the measured binding energies relative to the Fermi energy. It is not clear at present what value of the work function one should use, but this procedure seemed to give relatively consistent shifts for different crystal faces.[5] The data of Yates et al.* indicate that the surface shift of the O_{1s} and C_{1s} levels of adsorbed CO on W(100) is larger than the corresponding shifts in the valence orbitals of CO. (See Table I.) In general, it appears that for chemisorption on tungsten and nickel one would expect to have approximately a 2 to 3 eV upward shift in the valence orbitals and approximately 5 to 6 eV upward shift in the core levels. Obviously this shift depends upon the spacing of the adsorbate from the surface[26], the size and type of orbit, and the ability of the substrate to screen the hole.

The effect of the size of the hole state on the relaxation energy can be illustrated by considering the classical self-energy of a hole state in the bulk interacting with its induced screening charge.[26,30] The classical polarization energy due to the interaction of the induced screening charge with the hole is given by $\nabla \epsilon_R$[26,30]

$$\nabla \epsilon_R = \frac{e^2}{(2\pi)^2} \int d^3q \left[\frac{1}{\epsilon(q)} - 1 \right] \rho(q) \rho(-q) \qquad (12)$$

where $\epsilon(q)$ is the static dielectric function of the conduction band and $\rho(q)$ is the Fourier transform of the hole charge density. We will approximate $\epsilon(q)$ by

$$\epsilon(q) = 1 + \frac{k_s 2}{q^2} U(k_s, q)$$

where k_s is the screening wave number. When $U = 1$ we have the Thomas-Fermi dielectric function; $U(k_s, g)$ is the RPA correction to the Thomas-Fermi dielectric function.[31] The hole state will be assumed to be a hydrogenic-type hole:

$$\rho(r) = \frac{\alpha^3}{\pi} e^{-2\alpha r}$$

---

*Paper included in these proceedings.

Table I. Binding Energies and Surface Shifts for Adsorbed CO

Configuration	Level	Binding Energy Relative to E_f, eV	Work Function, eV	Binding Energy, eV	Surface Shift, eV
Gas phase	O_{1s}			$542.6^{(b)}$	
	C_{1s}			$296.2^{(b)}$	
	$4\sigma^{(a)}$			$19.68^{(c)}$	
	$1\pi^{(a)}$			$16.54^{(c)}$	
α state W(100)	O_{1s}	533.0	4.98	$538.0^{(b)}$	4.6
	C_{1s}	286.2	4.98	$291.2^{(b)}$	5.0
	$1\pi^{(a)}$	8.7	4.98	$13.7^{(d)}$	2.8
Virgin state Mo	O_{1s}	$532.1^{(e)}$			ϵ_0
	C_{1s}	$284.6^{(e)}$			$\epsilon_0 + 1.1$
	$1\pi^{(a)}$	$6.7^{(e)}$			$\epsilon_0 - 0.7$
	$4\sigma^{(a)}$	$10.2^{(e)}$			$\epsilon_0 - 1.0$

(a) The two peaks seen in the photoelectron spectra of adsorbed molecular CO have usually been identified as the 1π and 5σ levels of the gas phase CO molecule. The lowest binding energy state, 5σ, is primarily a lone-pair orbital on the carbon atom and therefore should be perturbed upon bonding. Our assignment of the adsorbed CO levels is that the lowest binding energy level is the 1π level, while the higher binding energy level is either the 4σ or the orbital derived from the 5σ level. This is based on the calculations of K. Johnson (see paper by R. P. Messmer, these proceedings) for CO and $W(CO)_6$, the correct relaxation energies obtained using this assignment, and the change in the strength of these two levels as the photon energy is changed.
(b) See paper by Yates, et al., these proceedings.
(c) See ref. 28.
(d) See ref. 9.
(e) See ref. 27.

Within these approximations, the relaxation energy is

$$\nabla \epsilon_R = \frac{256}{\pi} \beta^8 k_s e^2 \int_0^\infty \frac{U dy}{(y^2 + U)[4\beta^2 + y^2]^4} \tag{13}$$

with $\beta = \dfrac{\alpha}{k_s}$.

Figure 4 is a plot of the relaxation energy $\nabla \epsilon_R$ as a function of $\dfrac{k_s}{\alpha}$. The solid curve is for the Thomas-Fermi screening and the dashed curve is for a static dielectric constant calculated in the RPA.[30] The large relaxation energy for small $\dfrac{k_s}{\alpha}$ is a consequence of the Thomas-Fermi theory over estimating ϵ for large q. The curve for the RPA dielectric constant was calculated for $k_s = 2$ A^{-1}. The RPA curve decreases as k_s decreases. Consider two orbits, one representing the O_{1s} level with a radius $r = \dfrac{1}{\alpha} = .03$ Å and the other representing a valence orbit of CO with r ~1.0 Å, then the relaxation energy shift of the valence orbit would be 50 to 60 percent of the relaxation energy shift of the core level. This is in qualitative agreement with what is seen experimentally. On the other hand, if the orbit being considered is a bonding orbital extending over a surface cluster of approximately four surface atoms, the relaxation shift is only 20 to 30 percent of the core-level shift. Qualitatively, we could have the following situation occurring upon chemisorption of an atom or molecule. The core levels would be shifted upward by 6 eV while the valence orbits not involved in the bonding would be shifted by 4 eV. The bonding orbitals could be shifted by 1 → 4 eV depending upon the localization of the bonding orbit.

The above model is for a hole imbedded in a free electron gas. At the surface, the screening will be less but will be dependent upon the spacing of the molecule from the surface as well as on the size of the orbits.[26] The expectations of ESCA measurements is that the surface shift of a core level will be a monotonic function of the binding energy of the adsorbate.* The larger the shift of, say, the O_{1s} level is, the larger the binding energy of the oxygen atom will be. Surely to first order, this must be correct, in that the more intimate the contact of oxygen atom is with the substrate the more the hole will be screened. Yet a very strong chemical bond could reduce the

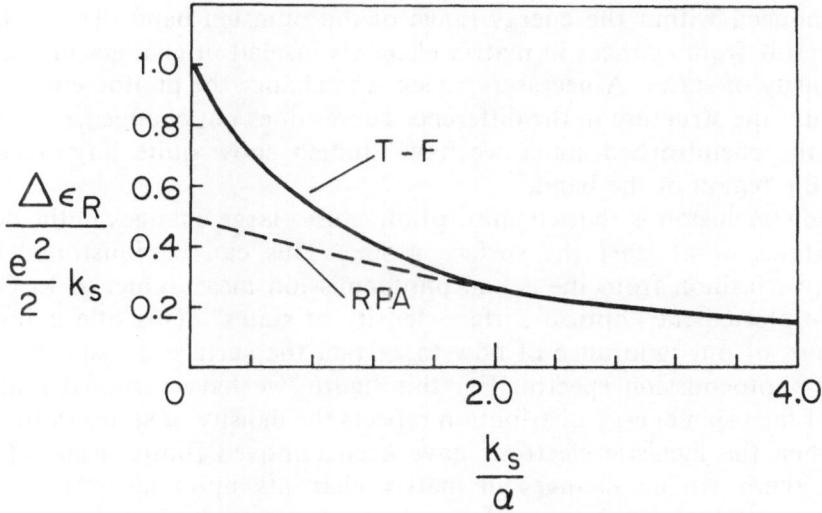

Fig. 4. Plot of the relaxation energy $\Delta\epsilon_R$ calculated from the polarization energy of an electron gas about a positive hole.[26,28] The hole state was represented by a hydrogenic orbit with a radius = 1. k_s is the Thomas-Fermi screening length. The solid curve is for a Thomas-Fermi dielectric function and the dashed curve for a RPA dielectric function.[31]

---

*See paper by Yates et al. included in these proceedings.

number of electrons in the conduction band, and thus decrease the relaxation energy. We surely need much more theoretical work in this area.

3 EXPERIMENTAL OBSERVATIONS

3.1 Changes in the Surface Density of States

Field-emission energy distributions always exhibit a dramatic change in the structure shown in Fig. 2 upon chemisorption of anything. Since the field-emission technique has no depth perception, it is not possible to draw any conclusions from these data about the changes in the local density of states averaged over some finite volume near the crystal surface. Photoemission has a finite depth perception and still shows large changes in the photoemission energy distribution, in the region of the bulk band, upon adsorption. Figure 5 shows the photoemission energy distribution from clean W(100) and a W(100) surface after exposure to 5 x 10^{-6} Torr sec O_2 at 300° K and subsequent heating to 1500° K.[3] The top curve is the difference between the two bottom curves, showing the changes induced in the spectra by adsorption. The peaks at -4.8 and -6.2 eV are levels primarily derived from the oxygen 2p levels. The changes in the spectra for higher energies are those induced within the energy range of the tungsten band. These effects could result from changes in matrix elements instead of changes in the surface density of states. A necessary check is to change the photon energy and make sure the structure in the difference curves does not change [see Fig. 8]. All of the chemisorbed states we have studied show quite large changes within the region of the band.[3]

The conclusion is that chemisorption causes large changes in the density of states of at least the surface atoms. This can be illustrated in a qualitative fashion from the actual photoemission measurements. In Fig. 6 we have plotted the "optical surface density of states". This title is just an admission of our ignorance of how to extract the surface density of states from a photoemission spectra. For this figure, we have assumed that the shape of the clean energy distribution reflects the density of states of the surface, when the inelastic electrons have been removed [bottom curve]. We assume there are no changes in matrix elements upon adsorption, that 40 percent of the signal comes from the surface atoms, and that only the surface atoms have changed their local density of states when an atom is adsorbed.* The measured difference curves for adsorption of oxygen and carbon on W(110) were used in this model to calculate the new surface density of states after adsorption. Thus two "optical surface density of states" curves

---

*The 40 percent signal from the surface is probably too large but a smaller number accentuates the changes in the surface density of states in this model.

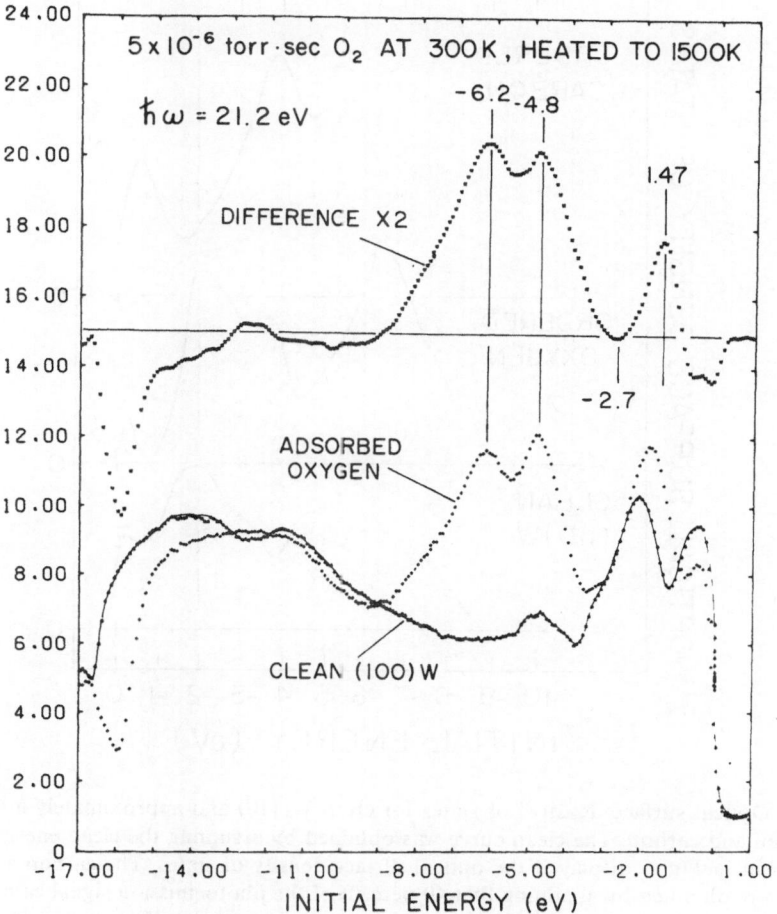

Fig. 5. Photoemission energy distributions for clean W(100) and a W(100) surface which had been exposed to 5 x 10^{-6} Torr sec O_2 at 300° K, then heated to 1500° K. The top curve shows the difference curve for these two energy distributions.

are shown in the top curves. These curves should never be construed to be the surface density of states, but they do show quite dramatically that there are large changes occurring in the "local density of states" of surface metal atoms. Any theory purporting to explain chemisorption must take this effect into account.

There is one other general observation which has emerged from our work on tungsten.[3] When the difference curves for adsorption of hydrogen, nitrogen, carbon, and oxygen into a c (2x2) configuration on W(100) are compared, it is apparent that there are marked similarities between carbon and nitrogen and oxygen and hydrogen. This is shown in Fig. 7. The peaks

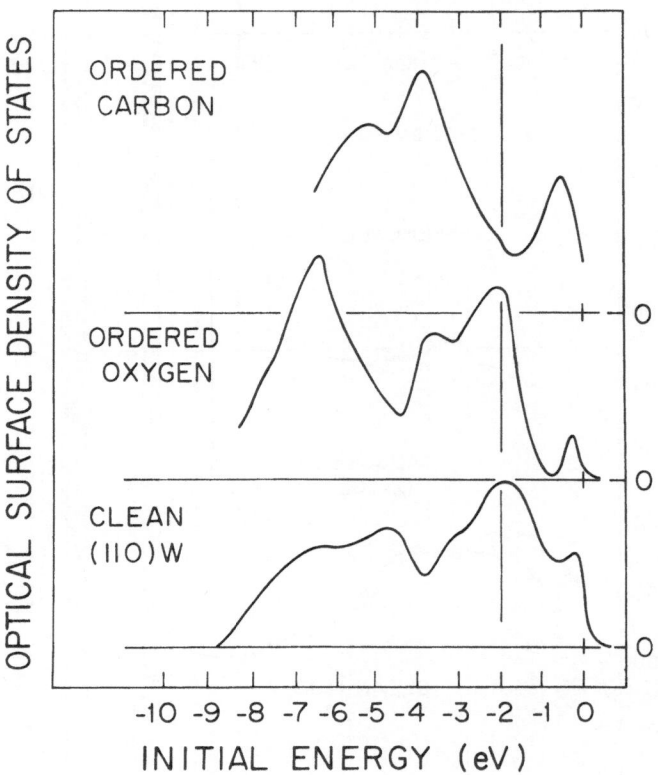

Fig. 6. "Optical surface density" of states for clean W(110) and approximately a monolayer of oxygen and carbon. The clean curve was obtained by assuming the clean energy distribution shown in Fig. 5 displayed the optical surface density of states. The carbon and oxygen curves were obtained by assuming that 40 percent of the photoemission signal came from the "surface layer" where changes were occurring upon adsorption. The measured difference curves were then used to calculate the optical surface density of states after adsorption. These curves should not be interpreted as "surface density of states".

below -5 eV are different for the different adsorbates, as one would expect from the fact that they have different electronic configurations; but in the region of the band [from 0 to -5 eV], adsorbed carbon and nitrogen look very much alike, and hydrogen and oxygen produce a peak near -1.4 eV. It is possible that these similarities reflect the fact that nitrogen and carbon are bound in the same site, which is different from the adsorption site of hydrogen and oxygen. Where an atom sits will determine which group orbitals will be involved in the bonding. It could be that in this strong bonding limit that the changes in the density of states of the surface atom reflect the geometry of the bonding.

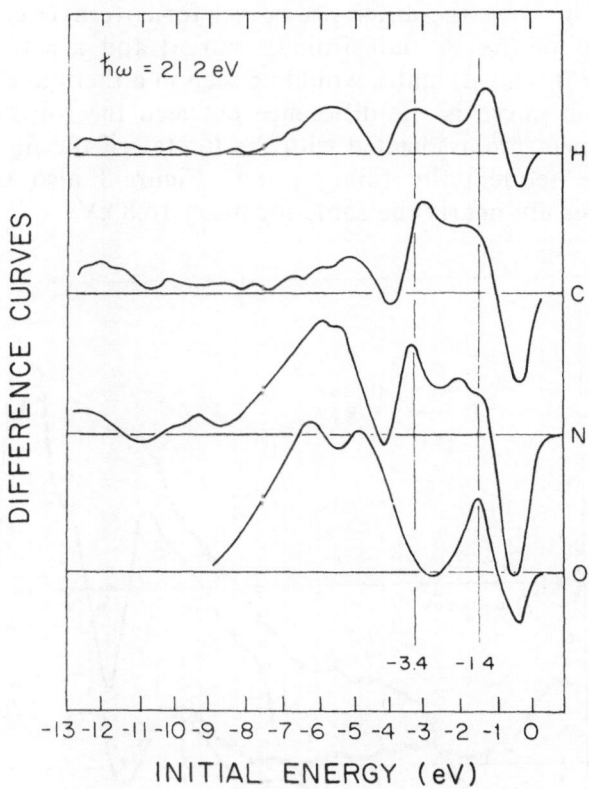

Fig. 7. Difference curves for adsorption of hydrogen, carbon, nitrogen, and oxygen into a c(2x2) structure on W(100). Notice that carbon and nitrogen produce the same structure in the first 4 to 5 eV from the Fermi energy. Oxygen and hydrogen produce a peak near -1.4 eV

3.2 "Fingerprint" Technique

In this section we give three examples of how electron-emission spectroscopy can be used to elucidate the properties of chemisorption by the "Fingerprint" technique. The first example is the adsorption of hydrogen on W(100). Thermal desorption spectra reveal two peaks which have been labeled the β_1 and the β_2 states of hydrogen. The area under the β_1 peak is twice as large as that under the β_2 peak. The β_2 desorption state desorbs at ~550° K and is the only state observed below 0.2 of a monolayer, while the β_1 desorption state sequentially fills on top of the β_2 state. We have used the changes in the photoelectron spectra as a function of coverage to check the previous field-emission data[32] and to test two models of this adsorption system. Figure 8 shows the difference curves for coverages corresponding to near saturation of the β_2 state [middle curve] and a saturated exposure where both the β_1 and β_2 states would be seen in a thermal desorption spectra. The bottom curve is the difference between the top two curves, and would be the spectrum associated with the β_1 state if the β_1 and β_2 states of hydrogen were sequentially filling states. Figure 8 also shows that the difference curves are nearly the same for $\hbar\omega = 16.8$ eV.

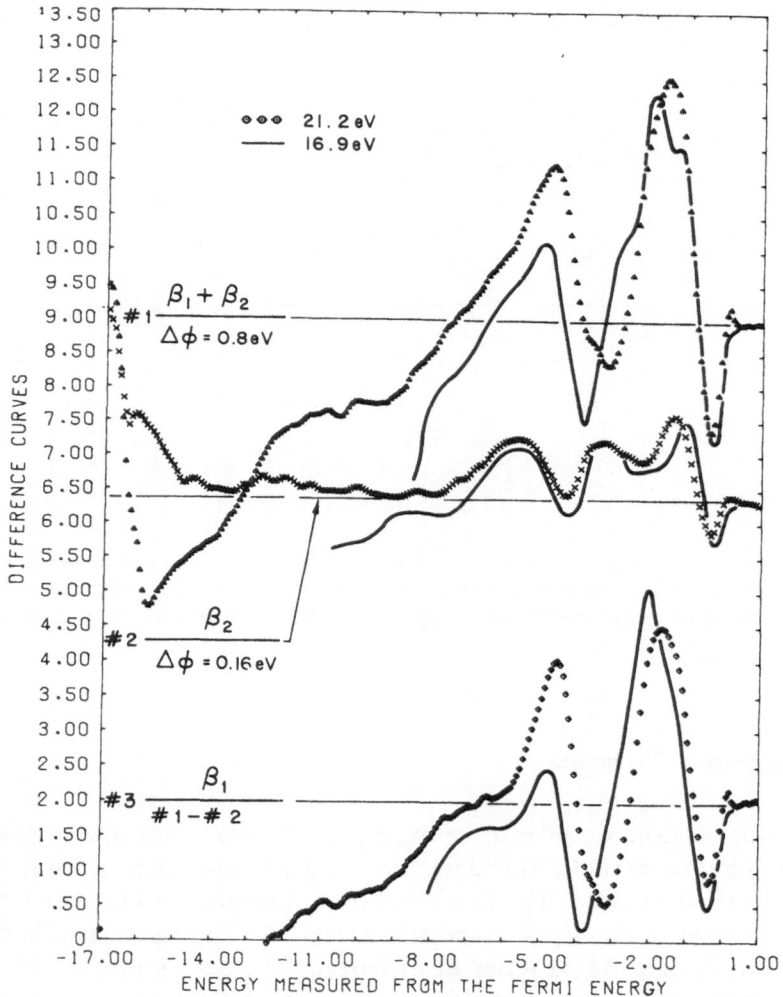

Fig. 8. Difference curves[3] for β_1, and β_1 and β_2 states of adsorbed D_2 on W(100) at 300° K: Curve 1 is saturation coverage of D_2 where both the β_1 and β_2 states will desorb; curve 2 is low coverage ($\theta \sim 0.2$) where only the β_2 state will desorb; the bottom curves show the difference between curve 1 and curve 2, or the β_1 state. The solid curves are for $h\omega = 16.8$ eV and the broken [data] curve is for $h\omega = 21.2$ eV.

Figure 9 shows the plot of the peak positions in the difference curves as a function of the work-function change, which is linearly related to coverage.[33] The data shown in Fig. 8 were used in two models of adsorption, in an attempt to fit the peak position versus coverage dependence shown in Fig. 9. The first model shown by the dashed line in Fig. 9 was a sequential filling model. The middle curve of Fig. 8 was taken as the spectrum for β_2 hydrogen and the bottom curve as the spectrum for β_1 hydrogen. We assumed the β_2 state saturated at a coverage of $\theta = 0.2$ or $[\triangle\varphi = 0.16$ eV], and from that coverage to saturation, the β_1 spectrum was linearly added to the β_2 spectrum. The second model, shown by the dashed-dot line, was a two-state conversion model, where again it was assumed that the β_2 state saturated at $\triangle\varphi = 0.16$ eV. Beyond this coverage, every new atom which is adsorbed converts one adsorbed atom to the final state which is characterized by the spectrum at the top of Fig. 8. The sequentially filling state model fits the lower and upper peaks fairly well but not the -3.5 eV peak. The conversion model fits only the upper peak. We can conclude that there is a density-dependent conversion occurring for adsorbed hydrogen beyond a coverage of approximately 0.2, and that this conversion is not simply a two-state conversion. The fact that the adsorbed layer seems to disorder or at least change its order beyond $\theta = 0.2$[34] would indicate that this

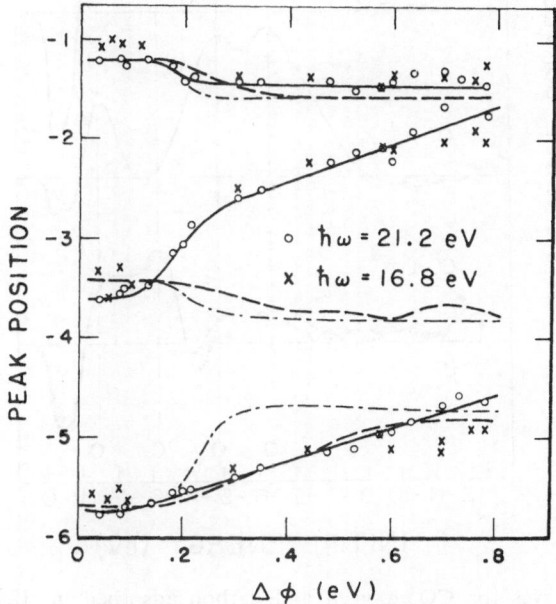

Fig. 9. Peak position as a function of work-function change for the peaks in the difference curves for D_2 adsorption on W(100).[3] Circles for $h\omega = 21.2$ eV and X for $h\omega = 16.8$ eV. The dashed lines are the peak-position variation predicted from a sequentially filling state model, while the dot-dashed lines are the predictions of a two-state conversion model.

density-dependent conversion is accompanied by something like an order-disorder transition.[32]

The second example is the dissociation of CO on W(100). Adsorption of CO on W(100) and W(110) has revealed binding states which are clearly molecular in nature.[3] When this adsorbed layer is heated to 1100° K on W(100), part of the CO is desorbed, but what remains is in a c(2x2) configuration. A comparison of the CO spectrum with the spectra of adsorbed oxygen and adsorbed carbon in the same ordered pattern should be a critical test of dissociation. Figure 10 shows the difference curves for CO, oxygen, and carbon in the c(2x2) configuration, while the dotted curve is the best fit obtained by subtracting the oxygen curve from the CO curve. The coverage for each curve was not known accurately, so the carbon curve was fit using two scaling parameters (A_{CO}, A_O) to obtain a best fit to the carbon difference curve D (E). The function

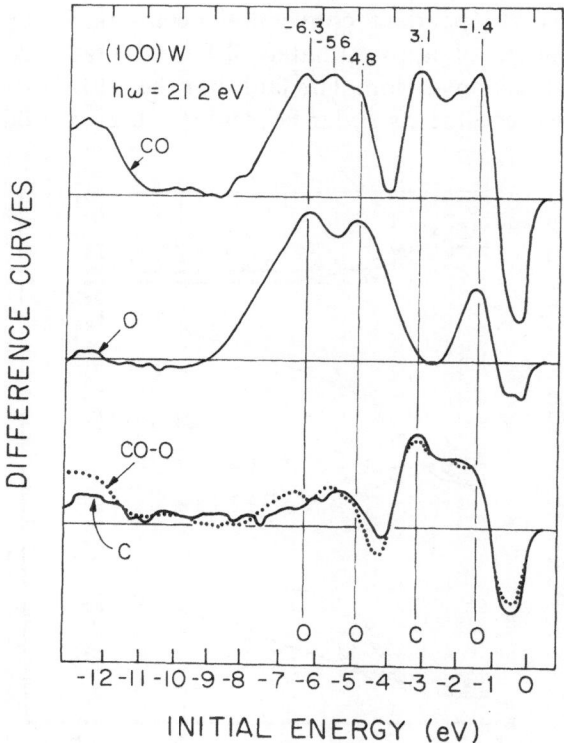

Fig. 10. Difference curves for CO, oxygen and carbon adsorbed in c(2x2) structure on W(100); the CO curve is for 5 x 10^{-6} Torr sec CO exposure at 300° K with subsequent heating to 1100° K; the oxygen curve is for 5 x 10^{-6} Torr sec oxygen exposure at 300° K, with subsequent heating to 1500° K, the carbon curve is obtained from a 5 x 10^{-6} Torr sec exposure of C_2H_4 at 300° K with subsequent heating to 1500° K.

$$\text{R.M.S.} = \frac{\sum\limits_{0}^{E \ = \ -9} \left| A_{CO}D_{CO}(E) - A_O D_O(E) - D_C(E) \right|}{\sum\limits_{E \ = \ 0}^{E \ = \ -9} \left| D_C(E) \right|} \tag{14}$$

was minimized with respect to A_{CO} and A_O. The curve in Fig. 10 was obtained for $A_{CO} = 0.6$ and $A_O = 0.2$ with R.M.S. $= 0.25$. The structure in the CO curve seems to result entirely from structure in the carbon and oxygen curves. In contrast, an attempt to do this fit with a CO layer which had been heated to $700°$K produced a worse fit, with R.M.S. $= 0.83$. It would appear that CO is dissociated in this high-temperature state. Two cautionary remarks are needed. First, there is a long-range interaction or these adsorbates would not order; therefore, it is possible that this indirect bonding effect is present in all three curves. For example, the region between -1 and -2 eV is high in all the curves. The second comment concerns the similarity between carbon and nitrogen which was pointed out in Fig. 7. We can fit the nitrogen curve with the oxygen and CO monoxide with an R.M.S. $= 0.28$.

The final example is the decomposition of C_2H_4 on W(110).[35] Figure 11 displays the difference curves for adsorbed C_2H_4 on W(110) for several heat treatments. Curve 1 is the difference curve relative to the clean W(110) spectra after a saturated exposure of C_2H_2 at $300°$K. The molecule is partially dehydrogenated since hydrogen is released upon adsorption. Our contention is that the adsorbed molecule is a C_2H_2-type molecule. The vertical lines at the top are the gas-phase ionization potentials[29] for C_2H_2 shifted by 3.1 eV to account for relaxation[27]. The two levels at -9.2 eV and -11.2 eV are the σ levels associated with the C-H bonds and the -4.3 eV level is the π orbital. When this adsorbed layer is heated to ~$500°$K, all of the hydrogen is desorbed, and curve 2 shows the difference between these two cases. The -9.2 eV and -11.1 eV levels have been removed and a small peak at -5.5 eV appears. The -9.2 eV and -11.2 eV levels are indicative of the C-H bond which has been broken, while the -5.5 eV level could be either adsorbed hydrogen or a rearrangement of the C-C bond. Upon heating to $> 1200°$K, an ordered LEED pattern forms but nothing is desorbed. Curve 3 shows the difference between heating to $500°$K and $1500°$K. The levels at -6.2 eV and -2.4 eV indicate breaking of the C-C bond on the surface. Finally, curve 4 is the difference curve relative to clean W(110) for adsorbed carbon where all the hydrogen has been removed and the C-C bond has been broken.

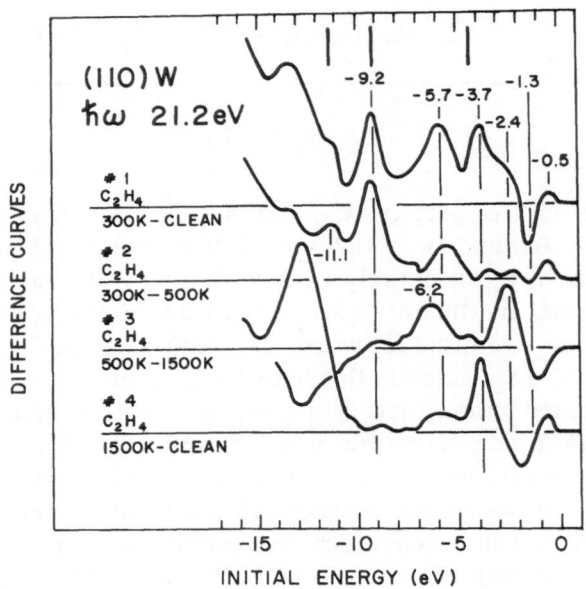

Fig. 11. Difference curves[3] for C_2H_4 adsorption on W(110) at $h\omega$ = 21.2 eV: curve 1 is the difference between spectra after 5 x 10^{-6} Torr sec exposure to C_2H_4 and clean W(110); curve 2 is the difference resulting from heating to 500° K the layer adsorbed at 300° k; curve 3 is the difference between an adsorbed layer of C_2H_4 heated to 500° K and then 1500° K; and curve 4 is the difference between the 1500° K heated layer and clean W(110). The vertical lines at the top are the gas-phase photoionization potentials for C_2H_2; they have been shifted upwards by 3.1 eV to account for relaxation.

REFERENCES

1. Grimley, T. B., *Proc. Internat. School of Physics, Enrico Fermi Course LVIII*, Academic Press, New York (June 1973); *J. Vacuum Sci. Technol*, 9, 12 (1972).
2. Newns, D. M., *Phys. Rev.*, 178, 1123 (1969).
3. Plummer, E. W., Waclawski, B. J., and Vorgurger, T. V., submitted to *Surface Sci.*
4. Gadzuk, J. W., and Plummer, E. W., *Rev. Mod. Phys.*, 45, 487 (1973).
5. Plummer, E. W., to be published in *Topics in Applied Physics*, R. Gomer (Ed.), Verlag Springer.
6. Eastman, D. E., *Electron Spectroscopy*, D. A. Shirley (Ed.), North-Holland, Amsterdam (1972), p. 487.
7. Duke, C. B., and Alferieff, M., *J. Chem. Phys.*, 46, 923 (1967).
8. Powell, C. J., *Surface Sci.*, 44 29 (1974).
9. Caroli, C., Lederer, D., Rozenblatt, Roulet, B., and Saint-James, D., *Phys. Rev.*, B8, 4552 (1973).
10. See Hagstrum, H. D., and Becker, G. E., *Phys. Rev. Letters*, 26, 1104 (1971), for the effects of the image potential on ion neutralization, and Gadzuk, J. W., *Phys. Rev.*, B1, 2110 (1970), for a discussion of image potential shifts related to field emission.
11. Penn, D. R., and Plummer, E. W., *Phys. Rev.*, B9, 1216 (1974).
12. Penn, D. R., private communication.

13. Christensen, N. E., and Feuerbacher, B., *Phys. Rev.*, **B10**, 2349 (1974).
14. Feuerbacher, B., and Christensen, N. E., *Phys. Rev.*, **B10**, 2373 (1974); Feuerbacher, B., and Fitton, B., *Phys. Rev. Letters*, **30**, 923 (1973).
15. Plummer, E. W., and Gadzuk, J. W., *Phys. Rev. Letters*, **25**, 1493 (1970.
16. Waclawski, B. J., and Plummer, E. W., *Phys. Rev. Letters*, **29**, 783 (1972).
17. Feuerbacher, B., and Fitton, B., *Phys. Rev. Letters*, **29**, 786 (1972).
18. Shepherd, W. B., and Peria, W. T., *Surface Sci.*, **38**, 461 (1973).
19. Koopmans, T., *Physica*, I, 104 (1933).
20. Einstein, T. L., *Surface Sci.*, to be published.
21. Makinson, R.E.B., *Proc. Roy. Soc.*, **A162**, 367 (1937).
22. Endriz, J. G., *Phys. Rev.*, **B7**, 3463 (1973).
23. Bethe, H. A., and Salpeter, E. E., *Handbuch der Physik*, **XXXV**, 88 (1967).
24. Schaich, W. L., and Ashcroft, N. W., *Phys. Rev.*, **B3**, 2452 (1971).
25. Waclawski, B. J., Vorburger, T. V., and Stein, R. J., *J. Vac. Sci. Technol.*, **12**, 301 (1975).
26. Gadzuk, J. W., *J. Vac. Sci. Technol.*, **11**, 275 (1974).
27. Demuth, J., and Eastman, D. E., *Phys. Rev. Letters*, **32**, 1123 (1974).
28. Atkinson, S. J., Brundle, C. R., and Roberts, M. W., *Disc. Faraday Soc.*, London (1974).
29. Turner, D. W., Baker, C., Baker, A. D., and Brundle, C. R., *Molecular Photoelectron Spectroscopy*, Interscience, New York (1970).
30. Hedin, L., and Lundqvist, S., *Solid State Physics*, **23**, 1 (1969).
31. Schrieffer, J. R., *Theory of Superconductivity*, Benjamin, Inc. (1964), p. 140.
32. Plummer, E. W., and Bell, A. E., *J. Vac. Sci. Technol.*, **9**, 583 (1972).
33. Madey, T. E., and Yates, J. T., Jr., *Structure et Proprietes des Surfaces des Solides*, Editions du Centre National de la Recherche Scientifique, Paris (1970), No. 187, p. 155.
34. Yonehara, K., and Schmidt, L. D., *Surface Sci*, **25**, 238 (1971).
35. Plummer, E. W., Waclawski, B. J., and Vorburger, T., *Chem. Phys. Letters*, **28**, 510 (1974).

DISCUSSION on Paper by E. W. Plummer

KOHN: You raised the question how a charge is screened in the surface region. A calculation of this screening has in fact recently been carried out by Smith, Ying, and Kohn.

   Now, a question to you: as you know, in photoemission the metal from which the electron is emitted is left in one of many possible final states. Does this cause a significant problem in the interpretation of the data?

PLUMMER: There is a complete spectrum of allowed values for $E_{final}^{N=1}$, the lowest energy being the completely relaxed system. The spectrum of emitted electrons will depend upon how the initial state couples to the final states; therefore, as Prof. Kohn points out, there is a possibility for the peak in a photoelectron spectra from an adsorbate (or from a molecule) to shift with photon energy. To my knowledge there is no experimental evidence for a shift in the threshold peak with photon energy.

EISCHENS: Can you explain why Demuth and Eastman did not observe chemisorption of ethane on nickel? Is this a shortcoming of UPS or is there a reason for the unexpected lack of chemisorption?

PLUMMER: I think it is clearly a problem of low sticking coefficient. At high exposures, 20 Torr sec, we thought we were seeing adsorbed C_2H_6 but analysis of the spectra showed the presence of CO.

EHRLICH: How were you able to eliminate effects from plane edges and the plane below in studying the effect of plane size on surface states? Richard Polizzotti, while still at Illinois, showed that in work-function measurements, these contribute very significantly, even for planes of 100 Å diameter.

PLUMMER: You cannot eliminate these effects, but you can check their importance. The current through the probe hole from the edges depends upon the geometry; it is an electron trajectory problem. This means the edge current depends upon the angle subtended by the plane, not its physical size. You can do experiments with fixed plane size r_0 and a variable emitter radius R, or with a variable plane size and emitter radius with $\frac{r_0}{R} = $ const. The latter experiment fixes the edge emission but changes r_0, while the first experiment fixes r_0 (surface density of states) but varies the edge emission.

FIELD ION MICROSCOPE STUDIES OF INTERACTIONS BETWEEN ATOMS ADSORBED ON TUNGSTEN (110) SURFACES

D. W. Bassett and D. R. Tice*

Imperial College of Science and Technology
London, England

ABSTRACT

Field ion microscope studies of lateral interactions between adsorbed atoms using transition metals as model adsorbates are described. Interaction energies determined for adatoms on nearest-neighbor adsorption sites on tungsten (110) surfaces are generally small but depend on adsorbate electron configuration. Dissociation energies for such adatom dimers decrease from >60 kJ mole^{-1} for Ta$_2$ to approximately zero for Re$_2$, and rise to 15 kJ mole^{-1} for Pt$_2$. Low dimer stability is considered to reflect the nonadditivity of bonding involving the substrate conduction electrons. Long-range interaction between rhenium adatoms due to indirect coupling via the substrate was explored by determining the pair distribution, but it is not considered that the expected oscillatory character of the interaction has been demonstrated conclusively.

---

*Now at English Electric Valve Co., Chelmsford, England.

1 INTRODUCTION

In catalysis by metals, the metal surface is arguably the most important reactant and its nature merits the extensive study it receives. Among the many approaches to surface studies, the atomistic approach that is possible using the field ion microscope (FIM) is unique because one can investigate processes on metal surfaces using just a few atoms. For studies relating to catalysis it would obviously be desirable if the experiments were done with nonmetals as adsorbates. Unfortunately, the possibilities are restricted severely by the limitations of the FIM, and the atomistic approach is readily applicable only to metal adatoms. These can be viewed, however, as both surface probe and model adsorbate from which information of wider application can be gained.

FIM studies of single-metal-adatom behavior initiated by Ehrlich[1] have advanced knowledge of the nature of metal surfaces significantly, particularly in relation to the effect of surface site geometry on the energy barriers to surface atom migration. Such studies have shown the behavior of metals to differ substantially from many commonly held notions, as is well demonstrated by recent observations of adatom migration at cryogenic temperatures on the (111) surface of rhodium.[2] Recent studies[3-5] have shown that similar techniques can be used to probe the nature of the lateral interaction between adatoms which is of general importance for the structure of adsorption layers. In this paper we consider FIM studies of some aspects of lateral interactions between transition-metal adatoms that show these adsorbates to behave more like the nonmetals of interest in catalysis than might be expected.

Metals of the third transition series were used as adsorbates primarily for their experimental convenience. This series of elements has the added advantage that their close-packing radii are all approximately equal, so that differences in behavior should be dominated not by differences in atomic size, but by electronic configuration which determines the position of the adatom d-levels relative to the substrate Fermi level. The tungsten (110) surface is also experimentally convenient since large (110) planes can be prepared reproducibly by field evaporation, and adatom migration occurs at sufficiently low temperatures (250 to 350K) that the surface can be readily kept clean. There is the disadvantage that the adsorption site is not well defined, although FIM evidence suggests that the two nonlattice positions in the surface unit mesh where an adatom can contact three tungsten atoms are the preferred location.

For transition-metal adatoms at separations comparable with close-packing distances, one might expect direct overlap of valence orbitals to dominate their lateral interaction. Such overlap produces strong bonding in isolated transition-metal diatomic molecules because with incomplete atomic d-shells, more valence electrons go into bonding than into antibonding molecular orbitals. One would expect a similar effect upon addition of an extra near-neighbor metal atom to an adatom on a W(110) surface if bonding is to be pairwise additive. However, the situation is more complicated because there may be Coulomb repulsion between adatom dipoles, indirect coupling of adatom orbitals through the substrate conduction electrons, and changes in the interaction with the substrate if the adatoms are displaced from their normal adsorption site when interacting with a neighboring adatom. Since the adatom d-states lie close to the Fermi energy, indirect coupling of the adatoms via the substrate may be strong and should be associated with a long-range interaction of oscillatory character.[6,7] This possibility makes the study of transition metals as model adsorbates particularly interesting.

2 EXPERIMENTAL TECHNIQUE

The technique used was based on that introduced by Ehrlich and Hudda[1] for studies of single-adatom migration. The surface was prepared by field evaporation and adsorbate was deposited on the tungsten FIM specimen so that just two adatoms were present on the (110) plane. The substrate was then heated at a temperature such that the adatoms migrated randomly about the (110) plane until a molecular cluster formed. The adatom separation and the thermal stability of such clusters provide information about the form of the interaction potential. Micrographs yield the interatomic distance for the clusters only if the image magnification is known. Calibration of the image magnification for clusters on W(110) is particularly difficult because the substrate is not resolved and cluster images may be severely distorted. It is sometimes difficult to distinguish between spacings of one or two atomic diameters except when some other factor, such as the cluster orientation, makes this obvious.

The thermal stability of clusters was determined by measuring the cluster lifetime at temperatures such that dissociation occurred in 1 to 10 minutes. Lifetimes were determined by many repetitions of a simple experiment. The cluster was first recorded in a micrograph taken while the surface was imaged at 77K, and the substrate was then heated at the chosen temperature T for a fixed period t_1 seconds, with the imaging voltage reduced to zero. The final state of the adatom deposit on the (110) plane was then determined by reimaging the surface after it had cooled to 77K. A cluster was judged to have dissociated in the heating period if both adatoms remained on the (110) plane, but at a changed separation. With certain adsorbates the chance that the adatoms released by dimer dissociation would migrate off the (110) plane over the step at its edge was high and it was necessary to include in the data ill-defined events in which one or both adatoms were missing from the (110) plane after heating.

Rate constants k_1 for first-order dissociation were deduced, as described previously[3], from the fraction f_d of heating cycles leaving a cluster intact at the end. Neglecting recombination events,

$$\ell n \, f_d = -k_1 t_1 \tag{1}$$

Recombination of dissociated clusters before the end of a heating period can be corrected for when the equilibrium constant for the reversible dissociation reaction is known or can be estimated.[3]

It has proved difficult to accumulate sufficient data for a wide enough temperature range to get the activation energy for dimer dissociation ΔE_d directly from the temperature dependence of k_1. ΔE_d was therefore deduced by assuming

$$k_1 = \nu_o \exp - \frac{\Delta E_d}{RT} \text{ with } \nu_o = \frac{kT}{h} \tag{2}$$

Since in dimer dissociation an adatom must gain sufficient energy to both break the dimer bond and move away over the energy barrier to migration between surface sites:

$$\Delta E_d = E_d + \Delta E_s \tag{3}$$

where E_d is the dimer dissociation energy and ΔE_s is the activation energy for surface diffusion.

The assumption that $\nu_o = kT/h$ in Eq. (2) should be tested, but the values of ΔE_d obtained are consistent with the limited data available for the temperature dependence of k_1 and other evidence of the dissociation reaction energetics provided by the extent of reaction at equilibrium. The equilibrium position can be determined for dissociation of W_2 and WRe, as described previously[3], and the equilibrium constant yields the dissociation free energy ΔA_d°. Assuming the dissociation entropies to be small, the values of ΔA_d° are consistent with values of E_d deduced using Eqs. (2) and (3).

The dissociation energy E_d measures the lateral interaction energy $U(r)$ for just one value of the separation r [$E_d = -U(r)$]. However, observations of the equilibrium state for deposits of two adatoms on a crystal face, such as W(110), can yield more information about $U(r)$ if the locations of adatoms relative to each other and the substrate lattice are deduced from micrographs. The pair distribution $P(r)dr$, which is related to $U(r)$, can then be obtained. For just one pair of adatoms on a (110) plane of diameter d,

$$P(r)dr = \frac{8r}{d^2} F\left(\frac{r}{d}\right) \exp\left[-\frac{U(r)}{RT}\right] dr \qquad (4)$$

$$F\left(\frac{r}{d}\right) = \frac{1}{\pi}\left[\theta - \sin\theta\right] \text{ with } \cos\frac{\theta}{2} = \frac{r}{d} \qquad (5)$$

where $F\left(\frac{r}{d}\right)$ corrects for the limited size of the area sampled. More than 10^3 heating cycles and micrographs are probably needed to obtain a meaningful distribution and this has not been done so far. Like Tsong[8], we have used larger deposits with five or more adatoms on the (110) plane simultaneously, so that each micrograph provides 20 or more pair spacings. $U(r)$ in Eq. (3) must then be replaced by the potential of mean force $\phi(r)$, and the pair distribution can be interpreted directly in terms of $U(r)$ only if the low density limit is reached, when $\phi(r)$ reduces to $U(r)$.

3 RESULTS

3.1 The Nature of Clusters

Homonuclear and heteronuclear adatom dimers thought to be "close packed" could be formed readily from two adatoms on a W(110) plane for most pairs of transition metals. Exceptions were Ir_2, owing to its short lifetime, and Re_2 which was rarely observed. Tsong[4,8] interprets the rarity of Re_2 in terms of an activation energy for its formation associated with a repulsive section in $U(r)$. During diffusion studies, Parsley[9] observed two Re_2 molecules only, but both dissociated in the next 120-second heating period, one at 343K and the other at 329K. At these temperatures, even isolated rhenium adatoms make few jumps between sites, which suggests that Re_2 could be thermodynamically unstable. The value of $U(r)$ for two rhenium adatoms at a separation of 0.27 nm given by Tsong[8] thus seems questionable, and the stability of Re_2 needs to be examined further.

No accurate interatomic distances or locations for adatoms relative to the substrate lattice can be given for the dimers studied. FIM micrographs suggest significant differences among the various dimers. Individual atoms of Pt_2 and Ir_2 were easily resolved in helium images at 77K (Fig. 1), but Ta_2, W_2, and Re_2 were normally unresolved. These differences appear to reflect differences in atomic properties, such as polarizability, rather than spacing differences. However, while Pt_2 and Ir_2 were accurately aligned parallel to the close-packed rows of tungsten atoms in the substrate, the axes of Ta_2 and W_2 images were at angles of ≈10 and 25 degrees, respectively, to this direction. This suggests an interatomic distance close to 0.274 nm for Pt_2 and Ir_2, but a different and unknown bond length for Ta_2 and W_2.

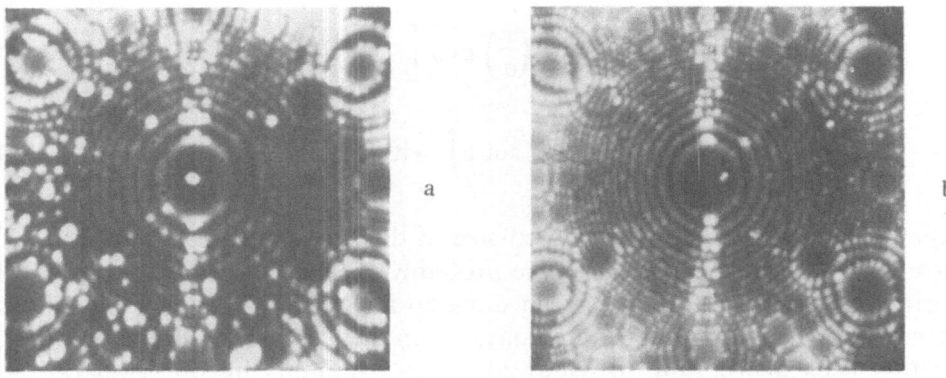

Fig. 1. Adatom dimers on (110) tungsten imaged in the FIM using helium at 77°K: (a) Ta$_2$, (b) Ir$_2$.

3.2 Dimer Dissociation Energies

The thermal stability of several different "close-packed" adatom dimers on W (110) was studied and the energy parameters relating to dissociation, including those reported previously for W$_2$ and WRe[3], are summarized in Table I. In deducing values of E$_d$ from Eqs. (2) and (3), values of $\triangle$E$_s$ from Bassett and Parsley[10] were used, and for heteronuclear molecules, $\triangle$E$_s$ for the more mobile adatom was used. Table I also includes dissociation energies E$_d'$ calculated from Eqs. (2) and (3) using $\nu_o = \dfrac{4D_o}{\ell^2}$, the frequency factor for adatom migration, instead of $\dfrac{kT}{h}$. In Table I, E$_d$ for Ta$_2$ represents only a lower limit to the dissociation energy. Ta$_2$ was so stable that almost all dissociation events recorded may have resulted from interaction of the dimer with the step bounding the (110) plane, since one or both adatoms were usually lost from the plane. The temperature dependence of the apparent dissociation rate indicated an activation energy approximately equal to that for single-adatom migration on (110), which is consistent with migration of the dimer to the plane edge being the rate-determining process in the "dissociations" observed.

Table I also includes some values for the free energy of dimer dissociation deduced from the position of equilibrium in the dimer dissociation process. Where the position of equilibrium could not be determined directly because of high rates of adatom loss from the (110) plane, it was estimated by extrapolating measurements of the dimer dissociation and formation rates to obtain the conditions for equal formation and dissociation rates.

Table I. Energy Parameters for Adatom Dimer
Dissociation (kJ mole^{-1})

	$\Delta E_,$	E_d	E_d'	ΔA_d°
Ta$_2$	75	60	70	–
W$_2$	83	31	32	30±5
Re$_2$	100	≈0	–	–
Ir$_2$	75	17	10	12±2
Pt$_2$	59	16	13	15[a]
WRe	83	19	18	14±2
TaIr	75	40	–	–

(a) Estimated value.

3.3 Long-Range Interaction Energies

The dimer dissociation energies measure U(r) for an adatom separation of approximately 0.27 nm, the close-packing distance, and for tantalum, tungsten, and platinum, this is probably the most that can be learned conveniently about U(r) from studies of adatom behavior on W(110). Studies of adatoms on other planes, such as (211) with channelled structures that prevent the adatoms from moving easily into contact should be more informative. Thus Graham and Ehrlich[5] were able to show recently that although the lateral interaction between tungsten adatoms is substantial at 0.45 nm, its effects are not detectable at distances greater than 0.7 nm.

Information about U(r) for distances other than 0.27 nm should be obtainable for both rhenium and iridium on W(110) using the pair distribution approach, but no distribution for deposits of two adatoms has been reported so far. Within the temperature range 300 to 350K, the behavior of rhenium adatom deposits on W(110) is significantly affected by the long-range part of U(r) and "long-bond" Re$_2$ clusters have appreciable stability.[4] Tsong[8] found peaks in pair distributions for five rhenium adatoms on a W (110) plane that correspond to such clusters, and deduced that U(r) ≈ −5 kJ mole^{-1} for r ≈ 0.7 nm from their amplitude. We have also obtained distributions showing similar peak structure (Fig. 2a), provided one disregards the difference between the r scales that presumably arises from some unexplained difference in calibration procedure. Both the distributions in Fig. 2 and those given by Tsong[8] show significant rejection of adatoms from short-range positions, which indicates that the interaction is repulsive.

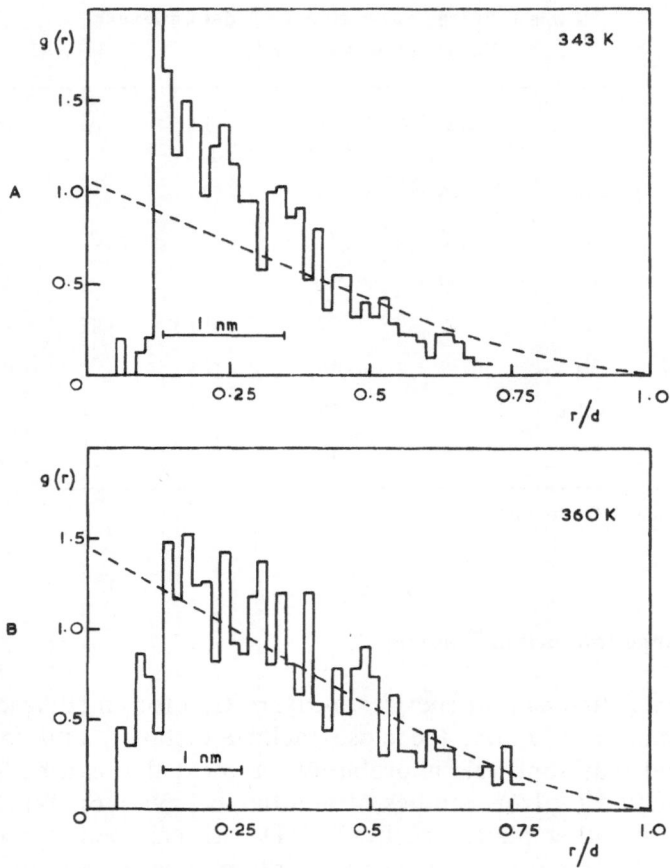

Fig. 2. Pair distributions, g(r) = P(r)dr/2πrdr in arbitrary units, for rhenium adatoms in a circular sample area on a W (110) plane: (a) deposits equilibrated at 343°K, (b) deposits equilibrated at 360°K; the dashed lines show the expected distribution for random arrangement on a structureless surface.

The data used to obtain the pair distributions have unsatisfactory features that require comment. The micrographs used to construct P(r) should each represent an independent sampling of the states accessible to the rhenium adatom ensemble at equilibrium. The amplitude of peaks in U(r) associated with an interaction energy of −5 kJ mole^{-1} should then decrease relatively little if the equilibrium temperature is increased by 20K, but P(r) for deposits heated at 360K (Fig. 2b) shows no significant peak structure. This suggests that the number of adatom jumps during the lower temperature equilibration periods was too small to destroy the correlation of one micrograph with the next, with the result that peak structure in P(r) is exaggerated. Exaggerated structure in P(r) is also produced by the incipient or partial ordering sometimes observed in rhenium adatom deposits, and it is then incorrect to assume that $\Phi(r) = U(r)$, despite the low coverage.

With doubts about both the validity of the available pair distributions and the stability of close-packed Re_2, the evidence that $U(r)$ for rhenium adatoms on W(110) has the oscillatory character associated with indirect coupling[6,7] is certainly not conclusive. However, the results are consistent with repulsion for separations less than 0.5 nm and weak attraction at separations of 0.7 to 1.0 nm with $U(r) < 5$ kJ mole^{-1}.

3.4 Impurity Effects

In relation to the practicability of atomistic experiments with nonmetal adsorbates, we note two effects observed during cluster stability studies. The Ta_2 dimers occasionally converted during heating at above 450K to an abnormal form aligned parallel with the substrate $<100>$ direction. This molecule presumably included an unseen impurity atom and was less mobile and more stable to dissociation or field desorption than Ta_2. Likely impurities might be oxygen or carbon reaching the (110) plane by diffusion from the uncleaned tip shank. An impurity, thought to be hydrogen, also caused Ni_3 clusters to change from a linear configuration like that of Ir_3, Pd_3, and Pt_3 to a triangular form. Possibly, bridging hydrogen atoms stabilize the close-packed configuration. These observations offer hope that the effect of carbon, hydrogen, and oxygen on metal surface stability could be studied using the atomistic approach.

4 DISCUSSION

While we do not yet understand their significant dependence on adsorbate electron configuration, the dimer dissociation energies in Table I confirm and extend the evidence from previous work[3,11] that adatom dimers on W(110) are rather unstable. For isolated diatomic molecules, the dissociation energies could be treated as bond energies but that approach is inappropriate here. The measured dissociation energy is the change in the adsorbate/substrate system energy and may include changes in the energy of several bonds. Thus, if the energies associated with bonding between adatom and substrate before and after dissociation are E_a' and E_a respectively and the adatom-adatom bond energy is E_{aa}, the dissociation energy is:

$$E_d = E_{aa} + 2(E_a' - E_a)$$

Any of the several factors that might produce stronger bonding of isolated adatoms to the substrate would reduce E_d below E_{aa}, so that E_d does not directly reflect the character of the dimer bond. Since there is significant valence-orbital overlap between close-packed adatoms, this bond should be comparable with a metal-metal bond, with adsorbates in the middle of the transition series having the largest E_{aa}. Repulsion between dipoles associated with charges on the adatoms will reduce E_{aa}, but this does not seem likely to be the factor determining the trend in dissociation energies with electron configuration. The electronegativities of adsorbates and substrate suggest that adatom charges may be larger for tantalum and iridium or platinum than for tungsten or rhenium, yet the trend is for the dissociation energy to reach a minimum in the middle of the series. The stability of heteroatomic dimers also suggests that charge-transfer effects are not dominant since charge transfer between the adatoms ought to stabilize these dimers. The dissociation energies, however, are nearly the arithmetic mean of those for homonuclear dimers of the constituents

$$E_d(AB) \approx \frac{1}{2} \left[E_d(A_2) + E_d(B_2) \right]$$

It therefore seems most probable that dimer stability is dominated by the change in adatom bonding to the substrate that accompanies dissociation.

Changes in adatom bonding to the substrate during dimer dissociation could arise from changes in either the location of adatoms relative to substrate atoms or the distribution of valence electrons. Displacement of adatoms in dimers off the adsorption site for isolated adatoms probably contributes to low dissociation energies, but again the trend in E_d with adsorbate electron configuration suggests that it is not the major effect. The exact alignment of the FIM images of Ir_2 and Pt_2 with the substrate close-packing direction suggests that the adatom displacements from isolated adatom adsorption sites are probably small. In contrast, the alignments of Ta_2 and W_2 images suggest that the adatoms may be displaced from the isolated adatom sites, which is not consistent with Ta_2 and W_2 being the more stable dimers since the amplitude of potential energy changes with lateral displacement should be similar for all these adsorbates. If surface molecule geometry is not the major factor in dimer stability, this leaves redistribution of electrons from the adatom-adatom bond into adatom-substrate bonds on dissociation as the important factor.

Since the adatom valence electrons couple directly to the conduction-band states in the substrate, significant nonpairwise-additive contributions to transition-metal adatom dimer bonding are to be expected. Indirect coupling of adatom orbitals via substrate conduction electrons has been discussed theoretically by Grimley[6] and by Einstein and Schrieffer[7] and can result in repulsion between adatoms on nearest-neighbor sites. However, direct orbital overlap between the adatoms was not included in the calculations and it is not clear to us how it would change the results. Einstein and Schrieffer showed that the interaction energy depends sensitively on the position of the adatom energy level relative to the substrate Fermi level E_f. This also seems to be a feature of our results, since the trend is for dimer dissociation energies to fall rapidly as the adatom level drops towards the Fermi level, perhaps to change sign for Re_2 when the level is centered just below E_f, and then to rise again but rather slowly as the level moves down into the filled part of the band. It remains to be seen whether this result would be predicted by theoretical calculations including a direct overlap contribution.

If low dimer stability relative to isolated adatoms reflects significant electron redistribution into adatom-substrate bonds when a dimer dissociates, the adsorption energies for single adatoms should be larger than expected on the basis of pairwise-interaction calculations. Since no reliable adsorption energies for the single adatoms studied are available, this cannot be checked, but it is interesting that the dimer dissociation energies are sufficiently small relative to estimated adsorption energies that the total binding energy of an adatom in a dimer shows the same trend with electron configuration as the single adatom adsorption energies. The total binding energy thus rises to a peak at rhenium (Fig. 3), consistent with all the bonding electrons available contributing to the adatom binding, irrespective of the number of nearest neighbors. FIM studies of the thermal stability of adatom trimers and larger clusters should show how the situation changes with the addition of extra neighbors. The situation is certainly complex since although the adsorbates all form trimers more stable than the corresponding dimer, only Ta_3 and W_3 always adopt a triangular "close-packed" configuration. Re_3 adopts both triangular and linear geometries, while Ir_3 and Pt_3 are always linear. A possible explanation is that as the adsorbate level moves down across the Fermi level, the balance between the relative strength of the first-neighbor interaction $U(r_0)$ and the longer range interaction $U(2r_0)$ changes to favor the extended configuration over the close-packed arrangement. Evidence that this is an electronic effect is provided by the fact that both Ni_3 and Pd_3 on W(110) are also linear, despite mismatch in atomic size between adsorbate and substrate.

In view of the importance of the long-range behavior of U(r) for the stability of adatom clusters larger than dimers and its more general relevance for the structure of chemisorbed layers, it is disappointing that FIM studies of this aspect of lateral interactions remain so inconclusive. Nevertheless, it is certain that the effects of even very weak interactions on adatom behavior can be observed. With a judicious balance between studies of molecular dynamics and quasi-equilibrium situations, present FIM techniques should yield sufficient information to establish U(r) for rhenium adatoms on W(110) and other planes in some detail, and this system merits further intensive study.

We do not think it would be justifiable at this stage to try to establish

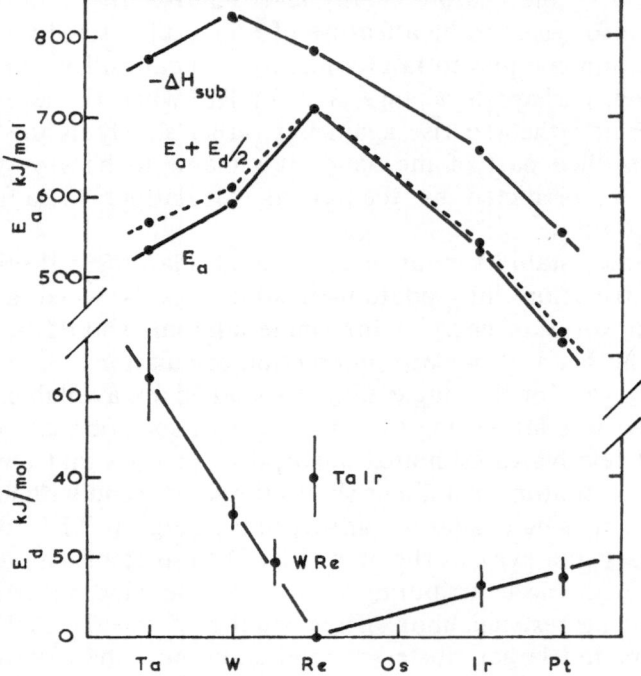

Fig. 3. Dissociation energies E_d for transition-metal adatom dimers on W (110), with desorption energies for isolated adatoms E_a and adatoms in dimers $E_a = \frac{1}{2}E_d$ estimated from activation energies for migration using $E_a = 7.1\Delta E_s$. Sublimation energies for the bulk metals $-\Delta H_{sub}$ are also shown.

some rather tenuous connection between our observations and the role of metal surfaces in catalytic reactions. However, direct experimental information on the bonding and mobility of surface metal atoms under defined conditions should provide a better basis for interpreting surface phenomena than the simple, but clearly incorrect, assumptions of the pair-bond approach. Much further study of adsorbates on surfaces of other orientation than (110) and on other substrates is required to establish what features of the behavior of metal adsorbates on W (110) are generally applicable. For the future, there remains the possibility of extending the present FIM techniques to nonmetal adatoms, such as carbon or oxygen, which are more directly relevant for catalysis.

ACKNOWLEDGMENT

We wish to thank the Royal Society and the Science Research Council for supporting this work.

REFERENCES

1. Ehrlich, G., and Hudda, F. G., *J. Chem. Phys.*, **44**, 1039 (1966).
2. Ayrault, G., and Ehrlich, G., *J. Chem. Phys.*, **60**, 281 (1974).
3. Tice, D. R., and Bassett, D. W., *Surface Sci.*, **40**, 499 (1973).
4. Tsong, T. T., *Phys. Rev.*, **B6**, 417 (1972).
5. Graham, W. R., and Ehrlich, G., *Phys. Rev. Letters*, **32**, 1309 (1974).
6. Grimley, T. B., and Walker, S. M., *Surface Sci.*, **14**, 395 (1969).
7. Einstein, T. L., and Schrieffer, J. R., *Phys. Rev.*, **B7**, 3629 (1973).
8. Tsong, T. T., *Phys. Rev. Letters*, **31**, 207 (1973).
9. Parsley, M. J., Ph.D. Thesis, University of London (1969).
10. Bassett, D. W., and Parsley, M. J., *J. Phys., D (Appl. Phys.)*, **3**, 707 (1970).
11. Bassett, D. W., and Parsley, M. J., *Nature*, **221**, 1046 (1969).

DISCUSSION on Paper by D. W. Bassett

GRIMLEY: These binding energies which you have determined for diatomic molecules on W(110) are far too small to be calculated theoretically by any method which does not include correlation effects. I remind you only that the diatomic molecule F_2 with binding energy some ten times larger than these determined here is unstable unless electron correlation is included in the calculation.

BASSETT: I would hope however that it might be possible for theory to predict trends resulting from changes in the chemical identity of the adsorbed species.

BLOCK: The formation of bright spots of adsorbed metal atoms during imaging a tungsten surface in FIM and the absence of these spots with adsorbed oxygen is combined with rather unknown intermolecular interactions. Other cases are also observed, for instance, oxygen on iron, where this adsorption is visible if imaged with krypton or xenon.

Could the orientation of adatoms still be influenced by high electric fields even at low temperatures?

BASSETT: In relation to your second point, we began by expecting that the orientation of the dimer molecular axis seen in the FIM image might be field dependent. We have looked for such effects over the field range available for helium and neon imaging but no field dependence has been observed.

ERHLICH: Two comments, about different subjects, are in order. In relation to the binding energies of atom pairs on the (110) plane, it should be noted that significant work function changes have been noted on tungsten (110) upon adsorption of tungsten atoms. Besocke and Wagner [*Phys.*

Rev., **B8,** 4597 (1973)] deduce a dipole moment on the order of ≈1 Debye per adatom. If this is correct, then two tungsten adatoms separated by one lattice spacing would be subject to considerable Coulomb repulsion. This could be a factor contributing to the low binding energies observed.

My second comment is directed toward the methods used for discerning interactions at long distances from measurements of the pair-distribution function. In studies of the pair-distribution function by W. R. Graham and D. A. Reed at Urbana, we have found it necessary to first carry out Monte Carlo simulations of the distribution expected for noninteracting atoms under the conditions of the projected experiment. Only by such simulation can one be certain that the experimental conditions are adequate for equilibration. Furthermore, comparison of the experimental distribution with the Monte Carlo distribution for noninteracting atoms is really the only way of establishing the existence of interactions, or of understanding the limits of the technique. The work of Dave Reed indicates, for example, that interaction energies less than kT will not be detected by such measurements.

Now the distribution found by Dr. Bassett at a low temperature looks as though significant interactions were occurring. However, only a very small increase in equilibration temperature gives an apparently random distribution. Since the thermal energy, and therefore the interaction that can be detected, change very little, this suggests to me that the distributions at the lower temperature may well be indicative of insufficient equilibration, rather than of interactions. We have noted such effects in simulations done for low temperatures, at which the number of atom jumps during an equilibration interval is small in comparison with the number of sites available along a coordinate. In our experiments on tungsten, we certainly have not found any indication of interactions at distances beyond 7 Å.

BASSETT: Coulomb repulsion may contribute significantly to the weak binding of dimers we observed. More experimental information on the dipoles of both isolated adatoms and adatoms in clusters of varying size for all the adsorbates is needed to decide whether this Coulomb repulsion is the major factor, although at present we think this is unlikely.

I agree that experiments to determine pair distributions must be carried out in such a way that the micrographs used represent truly independent samplings of accessible configurations. This was not the case with our lower temperature equilibration, as I tried to point out, and the use of Monte Carlo simulations to establish the appropriate conditions appears very advantageous. Much further work will be needed to establish the nature of the pair interaction potential as the interaction energies for separations beyond 0.7 nm are certainly very small.

SMITH: If my memory serves me correctly, the plot of surface energy versus element is much closer to the corresponding cohesive energy plot than is your curve for dissociation energies.

BASSETT: This is correct and it should be noted that the total binding energy of the adatoms follows a similar trend. In relation to surface energy and the interfacial energy for metal overlayers on W(110) the way the total binding energy for an adatom changes as the number of adatoms in a cluster increases is of particular interest. This is not known at present, but the structures and thermal stability of trimers and larger clusters indicate a complicated situation, and much further work is required.

WYNBLATT: You said that you cannot always make good estimates of the recombination probability for adatom pairs. Could you please elaborate?

BASSETT: The effect of recombination following dissociation before a dimer is next observed can be corrected for satisfactorily if the position of equilibrium can be established. One then uses the standard kinetic equations for a reaction proceeding towards equilibrium to extract the dissociation rate constant, as in our study of W_2 and WRe. In other cases, when equilibrium cannot be established because adatoms are too easily lost from the plane at the required temperature, we deduce a rate constant for dimer recombination from measured rates of dimer formation. However, as we do not know the position of equilibrium, the correction for recombination can be made only approximately.

WABER: I suggest that one possible reason for the minimum in the diatomic binding energy of the adsorbed atoms of 5d transition metals might be traced to spin-orbit splitting of the d-bands. That is, in j-j coupling. There are two filled subshells 5d 3/2 with 4 electrons and the upper 5d 5/2 which contains $2j + 1 - 6$ electrons. The suggestion is that the electronic structure of the rhenium atom would be $(5d\ 3/2)^4\ (6s\ 1/2)^2$, and since these are both closed shells, bond formation between two rhenium atoms would require some promotion energy to place an electron in the upper j=5/2 subshell. Assuming this to be true, the decline of the binding energy from tantalum to rhenium would be due to reduction of holes in the lower subshell; following completion, there would be an increase in binding energy for three more atoms until the j=5/2 subshell is half filled. It would be some justification for this speculative model if there was a second decline after that element.

Since I incorrectly remembered that tantalum has the electronic structure $(d\ 3/2)^2\ (s\ 1/2)^2$, the argument for rhenium could not be valid unless one d electron were transferred to the metal or in some other way were unavailable for forming a bond. I now regard this idea as purely speculative.

AGENDA DISCUSSION: EXPERIMENTS ON CLEAN METAL SURFACES WITH ADSORBATES

*T. N. Rhodin**

School of Applied and Engineering Physics
Cornell University
Ithaca, New York

*J. F. Antonini***

Battelle
Geneva Research Centre
Geneva, Switzerland

1 INTRODUCTION

Major discussion centered around problem areas posed by the Chairman as points for critical consideration. They were as follows: determination of surface atomic structure with particular consideration of surface crystallography, comparative usefulness of different approaches to the electron spectroscopy of surfaces, the nature and role of the study of simple catalytic reactions on metals, and the connection between simple catalytic reactions on metals and industrial catalytic processes. Considerations of additional related subjects such as analysis of molecular processes on metal surfaces, relation of information on the surface physics of metals to problems in practical catalysis, and suggestions of new types of experimental information on metal surface reactions needed to further general understanding of catalytic materials were also covered.

2 STRUCTURAL CONFIGURATION OF CLEAN METAL SURFACES AND CHEMICAL OVERLAYERS

Discussion centered mainly on the present and future status of measurement and calculational methods for the application of low-energy electron diffraction (LEED) to surface crystallography. Also considered were the desirability of using alternative electron or atomic scattering methods for surface structure analysis and the applicability of the presently available scattering methods to highly dispersed materials.

There seemed to be two general assessments of the problem of surface-structure measurement. One group felt rather pessimistic as to the apparently formidable computer effort now involved in dynamical analytical methods, the relative slow progress in developing streamlined methods of analysis, and the inherent difficulties of extending present methods to more complex structures. Andersson, for example, considered that high-energy electron diffraction (HEED) might be preferred even though the main difficulty here is in the preparation of sufficiently good specimens. The atomic scattering factors are properly known and the scattering problem can be effectively handled even for scattering from rather rough surfaces. Such techniques are indeed well defined from contrast analysis in electron-microscope technology. In this vein, it was pointed out by Rhodin that although structural analysis using LEED has been quite successful for relatively simple adsorbate structures on flat single-crystal surfaces[1], the problem of defining proper scattering potentials and trial geometries for complex surface structures could be considerably more difficult.

The following pertinent question raised by Ehrlich directed attention to two important factors: the apparent high computer cost typical of a complete KKR or matrix inversion calculations and the related complications introduced by the need to distinguish between adsorbed layers and oxide layers on a metal. In regard to the computer effort, Rhodin pointed out that this cost will be substantially reduced as more streamlined and effective computational programs are developed and evaluated. These relate not only to purely dynamical schemes but those based on much more efficient multiple-scattering approaches as proposed and applied by Pendry[2] and by Tong et al.[3] In addition, the Fourier inversion method described by Landman and Adams[4] has the promising additional feature that it does not require the use of a trial geometry.

In reply to the second part of the question, Webb discussed the experience of his group with reference to the applicability of the method of averaging data at constant momentum transfer as evaluated by them for oxygen chemisorption on Ni(100). He concluded that the electron-diffraction measurements themselves can be done with adequate precision and accuracy but some cases still exist which prevent achieving satisfactory agreement between the data and results of the averaging method for this particular system. It was pointed out by Rhodin that it is quite likely that the nickel-oxygen chemisorption system happens to be one of those cases alluded to by Ehrlich which is characterized by the possible occurrence of surface lattice modification associated with distortion and penetration by the adsorbate, depending on the substrate crystallography and exposure. In this regard, Andersson pointed out that the occurrence of three-dimensional oxides on the surface is not likely to pose a serious analytical problem since their Fourier coefficients can be readily characterized. There are of course observed cases where oxides do form as indicated by the work reported by Ertl and his co-workers. On the other hand, the conditions for lattice penetration and surface reconstruction do occur for specific chemisorption systems such as nickel-oxygen. Their occurrence depends very sensitively on the crystal plane, the intensity of oxygen exposure, and the adsorption time and temperature.[5] It is of course important in achieving reliable LEED intensity analysis to carefully relate the trial geometries used to the particular conditions of the chemisorption process. With reference to structure determination of hydrocarbon overlayers on metal substrates such as metal-ethylene or metal-carbonyl complexes, Mason pointed out that the problem of predicting a trial geometry in dynamical structure analysis is also complicated. The Fourier transform methods, although incomplete in themselves, have the advantage of limiting the number of structures which have to be assessed by more complete intensity calculations.

The more optimistic features of recent accomplishments in the use of LEED for structure characterization were also emphasized. Somorjai stressed the fact that substantial contributions to the understanding of the kinetics of surface reactions by the application of LEED theory to structure analysis of low-index single-crystal faces of many fcc metals and of correspondingly simple chemical overlayer systems have been made in the last 10 years. Similar successful extensions to the routine structure analysis of overlayers of organic molecules which form ordered structures on simple metal surfaces such as ethylene and benzene is suggested as a real possibility in the next 10 years. The experimental simplicity of LEED compared with reflection HEED is a distinct advantage. It was concluded that LEED has indeed been extraordinarily useful in applications to simple systems and will continue to be so. Webb pointed out that the reflection techniques themselves also have serious drawbacks associated with sensitivity to surface morphology and scattering-effect corrections in the surface region. It is possible that reflection methods using high energy electrons will be more applicable to studying the surface structure of complex organic molecules. In this regard, Grimley pointed out that the surface atomic and electronic structure of a working catalyst is basic to understanding a catalytic reaction. Since it is usually characterized by a complex surface structure, the important question as to whether LEED methods can eventually be made effective for such systems is unanswered. It was concluded that additional techniques for the atomic (and electronic) structural analysis of surfaces of practical catalysts need to be developed.

3 ELECTRON SPECTROSCOPY OF SOLID SURFACES

Spectroscopic methods in which the emission of relatively low energy electrons from solid surfaces excited by photons in the X-ray (ESCA-XPS) or ultraviolet (UPS) wavelength range, by higher energy electrons (Auger or ILEED), by ion neutralization (INS), or simply by tunneling in the presence of a high local field (FEED) promise to give useful information on the electron structure of the surfaces as well as on the excitation processes which occur.[6] Some important questions were posed as to the extent to which these methods are mutually supportive, the importance of the response function of the system under study, the degree to which the observed spectra can be related to a surface density of states, and the effects of surface polarization and charge relaxation on peak positions observed in the spectra. The first question is of particular importance with reference to the quantitative indexing of peaks observed in UPS spectra in terms of energy levels of molecular orbitals.

In terms of obtaining new information on the surface electron structure, Plummer raised a critical question as to the possibility of separating the volume from the surface contributions to photoemission in such an interpretation. Schrieffer pointed out that the UPS approach can be applied to disordered adsorption systems and also shows promise for providing useful information on the geometry of the adsorbed atoms as well as on the symmetry of the adsorption site. (Note: The use of the angular behavior of photoemission for this purpose has been discussed elsewhere by Gustaffson[7], by Liebsch[8], and by Liebsch and Plummer[9]). In response to Plummer's critical question referring to the separation of bulk from surface emission, Grimley expressed the view that theoreticians now working on the problem will succeed in developing the theory of photoemission to the point where the response function can be expressed with sufficient accuracy in terms of the parameters of the system so that the density of states can be unfolded from the photoemission data for the clean surfaces and the adsorbate-covered surfaces. In this regard, angular-resolved, energy-dependent, photoemission data will be essential.

With reference to the applicability of UPS to chemisorption at this time, Menzel pointed out that two very valuable applications are now available:

1. The so-called "fingerprinting" technique where specific features of the UPS or the XPS spectra are associated with specific binding states of surface molecules, which allows the differentiation of states, especially if combined with other methods.*,** This lends itself to following the kinetics of molecular processes on or at surfaces as discussed by Ertl in this Colloquium.

2. Comparison with gas phase UPS or XPS spectra by which useful information can be gained on the orbitals involved at least in those cases where the perturbation of the adparticle by the surface bond is small.*

An example of the development of means of extracting surface information from UPS is contained in the work of Rowe and Ibach[10], cited by Hagstrum. Here a function related to the surface density of states for Si(111) was successfully obtained from the photoemission spectra and showed good agreement with the self-consistent field calculations of Appelbaum and Hamann.[11] Rowe and Ibach took the mean of the photoelectric spectra from several surfaces to be an adequate representation of the response of the bulk. The difference between this mean spectrum and a given surface spectrum was then taken to be the photoelectric difference spectrum for this surface. It was possible, by comparison with the theory, to identify the contributions of the dangling bond as well as the back bond orbitals present in the surface region. This example illustrates how photoemission results can be interpreted in detail by theoretical calculations of the electronic character of the surface bond. Hagstrum[12] also pointed out that it is effective to use two differing types of spectroscopy, such as INS and UPS, in the study of adsorption systems. Here one makes use of the differing matrix elements and sensitivity volumes of the two types of spectroscopy in drawing conclusions. Hagstrum pointed out that such use of INS and UPS in the study of oxygen chemisorption on nickel had confirmed that oxygen can indeed be transported into the bulk on heating.

The theoretical interpretations of UPS spectra are limited by certain basic features, according to Newns and to Schrieffer. The former suggested that systems be chosen for study where the energy levels associated with ad-

---

*Bradshaw, A. M., Mengel, D., and Steinkilberg, M., *Jap. J. Appl. Phys.*, Suppl. 2, Pt. 2, 841 (1974); *Chem. Phys. Letters,* **28,** 515 (1974).
Menzel, D., *J. Vacuum Sci. Technol.,* **12, 313 (1975).

sorption lay well below the d-bands and, ideally, also the sp-bands of the substrate. The levels are less commonly found below the d-band for tungsten than for some other transition metals of interest as adsorbents. The latter raised serious questions as to the intrinsic suitability of using UPS and Auger spectroscopy to pursue understanding of the catalytic process itself. These spectroscopic techniques involve inelastic loss processes which are quite different from those in catalytic reactions involving a neutral molecule in terms of an appropriate potential energy surface.

Schrieffer suggested that looking at UPS directional effects associated with elastic scattering for more ideal systems, such as inert gas adsorption, might be more fruitful at this stage.

In terms of identification of the orbitals important in chemisorption on metals, Messmer pointed out that the σ-levels are significantly changed upon chemisorption on nickel and tungsten. For example, the 4σ-level associated with the lone pair on carbon is very important in forming a surface bond with the metal, whereas the energy position of the π-level of the oxygen is relatively insensitive to chemisorption. There was general agreement that a good deal of useful information on molecular orbitals involved in surface bonding can be obtained by comparing molecular-UPS spectra with adsorption-UPS spectra.

The significance of spectroscopic data on surface band structure, density of states and molecular orbitals and their implications with reference to the corresponding concepts of the physical and organic chemist needs to be carefully defined. In this sense, Grimley quoted Mulliken, "the quantum mechanical notion of the covalent bond is not as easy to understand as one might think", e.g., the solution of the problem of bonding is difficult from either viewpoint.

The following explanatory statement by Schrieffer clarified the coupling between the analogous concepts of the "band" and "bond" schemes to describe chemisorption. The band concept is often cast in the frame of the Bloch theorem for a periodic lattice. Physicists had thought that way for some time until they considered alloy and defect problems. One defect however can be a surface. Surfaces spoil momentum conservation yet one still talks about energy bands in the presence of a surface.

One has molecular orbitals for the enormously large solid just as one does in any molecule. The energy spectrum of these molecular orbitals has both continuum regions corresponding to bands and discrete states corresponding to surface states. We note that for a flat surface one can have sharp surface states lying in the continuous portion of the spectrum. The changes in the spectrum of the entire solid due to the surface are minuscule, however, since the surface constitutes a small perturbation on the solid as a whole. Nevertheless, this change in spectrum when suitably weighted by the amplitude of the wave functions at the surface constitutes a large shift in this projected density of states. This generalized band point of view, as practiced by physicists these days, is not a band point of view in the Bloch sense at all, but is, instead, a molecular orbital description of the entire molecule composed of solid plus adsorbate. The connection with the bond point of view is essentially through the usual Mulliken bond population analysis as in molecular orbital theory. In the band scheme, the Mulliken ideas are extended to a continuous spectrum through the projected or local density of states. It is this quantity which plays a central role in modern band theory as applied to surface problems and unifies the band and bond approach so that one can reason interchangeably in these languages. In this colloquium there has been a good deal of discussion about the projected or local density of states from both the experimental and theoretical point of view. One can extend this concept to include interelectronic correlation effects which are handled in a natural way in a valence bond scheme.

The discussion on spectroscopic methods (specifically, UPS, ESCA-XPS, FEED, and INS) raised important considerations of applicability and interpretation with reference to understanding the process of chemisorption. It is quite clear that clarification of the chemisorption process for simplified systems in terms of the electronic nature of the chemical bond at the metal surface is an important first step in understanding the nature of chemical reactions at surfaces. It does not in itself adequately answer the more difficult problem posed by the next question, namely the development of measurements and models for understanding the mechanisms of simple catalytic reactions on metals.

4 SIMPLE CATALYTIC REACTIONS ON METALS

Study of the mechanisms of chemical reactions on well-defined surfaces must follow clarification of the critical atomistic and microscopic parameters which define both the clean simple surface and the associated chemical overlayers. The problem of determining electronic and atomic structure is rendered much more difficult when a catalytic reaction takes place because the molecules are both larger and more complicated, coadsorption or sequential chemisorption occur, and order-disorder phenomena are often involved. This poses the very central question: Is it likely and if so under what circumstances can the study of critical parameters for catalytic reactions on well-defined surfaces provide significant understanding of simple catalytic reactons typical of more realistic circumstances? The work of Ertl and his coworkers on the interaction of simple molecules such as O_2, CO, and H_2 with metal surfaces presents the following suggestive answers to this question. Reference was made to the "pressure gap" of approximately ten orders of magnitude between experimental sutdies and practical reactions. This may not be so critical for certain simple reactions such as the oxidation of CO, since the rate-determining step depends primarily on the CO/O_2 ratio and not on the absolute pressure. With reference to differences in crystallography, reaction rates on polycrystalline material such as finely divided metallic particles do not appear to differ greatly from those on flat, single-crystal faces. The role of the clean surface in experimental work compared with the nature of the surface of a working catalyst also may not introduce such a major unreality between real and ideal systems as generally envisioned. It has been established (by Ertl, et al.), with reference to the oxidation of CO on palladium, that initial contamination with sulfur and carbon is essentially nullified by the reaction prior to the initiation of the oxidation process. Ertl concluded, at least for this example, that contaminated metal surfaces tend to be self-cleaning during the course of catalytic reactions. It may therefore be concluded that the results of study of simple chemical reactions on well-defined metal surfaces may be more related to practical processes than generally considered. In this vein, the value of systematic measurements of reaction-rate constants of important gas-surface reactions in terms of activation energies and preexponential factors in order to better predict the course of more complicated catalytic processes was strongly endorsed by Boudart.

Another approach to the problem of correlating chemisorption parameters with catalysis is exemplified by the approach of Tamaru and his coworkers[13] (Kawai, Kunimori, Kondow, and Onishi) who have studied specific surface states of a catalytic surface with reference to the reactants as well as the intermediates and products directly connected with the catalytic material in its working state. This rational approach is briefly illustrated by Tamaru as follows.

The study of chemisorption of various gases on solid surface has been successfully carried out by means of various new physical techniques in recent years, but it may not, in many cases, be directly connected with catalysis. One must study the reactivity of these chemisorbed species as a function of coverage and coadsorbed species to correlate chemisorption with catalysis.

The properties of the catalyst surface which is directly connected with catalysis are those of the surface in its working state, rather than those of the clean surface. Under the reaction conditions, some of the reactants, products, and surface complexes are generally adsorbed on the catalyst surface, thus considerably influencing the properties of the surface.

The adsorption on a catalyst surface during the course of the reaction can not be estimated from the adsorption separately measured for each of the reactants and products. Chemisorption is not only dependent upon the interaction among the surface species, such as the formation of surface complexes, but also upon which step is rate determining.

Consequently, it is necessary to study not only the properties of the catalyst surface in its working conditions, but also to study what materials are being adsorbed and how much is being adsorbed, their structures, and also the reactivity of the adsorbed species under the reaction conditions and, simultaneously, the rate of the overall reaction. Even if one finds a certain species on the catalyst surface, it does not imply that it is actually a reaction intermediate involved in the reaction sequence. The identification of a reaction intermediate in the adsorbed state, or the elucidation of the reaction path of heterogeneous catalysis, can be carried out only by studying its kinetic behavior in the working state. This is the way to know not only *how* heterogeneous catalysis takes place, but also *why* it takes place on a specific catalyst surface.

When oxygen is adsorbed on chromia, five peaks are observed by infrared absorption technique and each of them behaves differently in desorption experiments as well as heavy oxygen adsorption. When CO is added to the system, two of five species react to form CO_2, as demonstrated in Fig. 1. In such a manner one can tell which of the adsorbed species is the real reaction intermediate of the catalytic reaction.

Another effective technique to identify the adsorbed state of molecule is

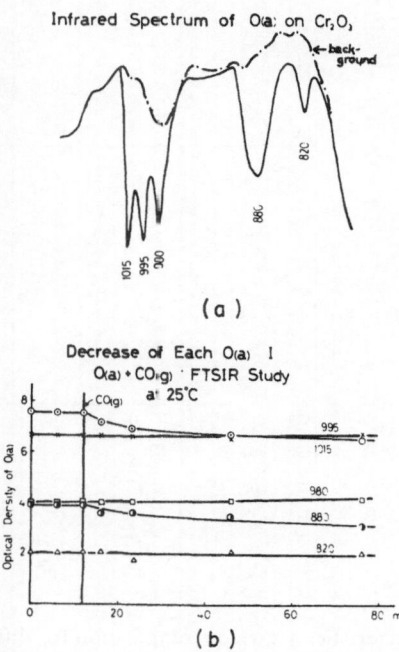

Fig. 1. (a) Infrared spectrum of oxygen adsorbed on chromia at room temperature; (b) decrease in adsorbed oxygen resulting from oxidation to carbon dioxide corresponding to each adsorption mode as a function of time.

high-resolution Auger electron spectroscopy. When ammonia is admitted to a molybdenum foil at 450°C, the nitrogen-KLL Auger electron spectra show the formation of nitride. The spectra in Fig. 2 are very similar in shape to the oxygen- and carbon-KLL spectra of molybdenum oxide and carbide, or to the KLL spectrum for atomic nitrogen calculated on the basis of an ionic model and a LS coupling scheme.

NH_3 chemisorbed on a molybdenum surface at room temperature gives a spectrum which is similar to that of gaseous NH_3 except for two points: one is the broadening of each peak and another is that the intensity of the 375-eV peak which is associated with the KLL transition decreases to an appreciable extent. Since the one-electron state corresponds to the NH σ-type bonding orbital, the decrease in this peak demonstrates that the NH bond is dissociated on chemisorption.

In this manner, high-resolution Auger electron spectroscopy can be a sensitive tool for the study of the behavior of surface species, in particular, molecular structure and valence electronic states.

The final important question to be considered is: In what areas and in what ways can experimental research on catalytic reactions make significant contributions to the development of new or more effective industrial catalytic processes?

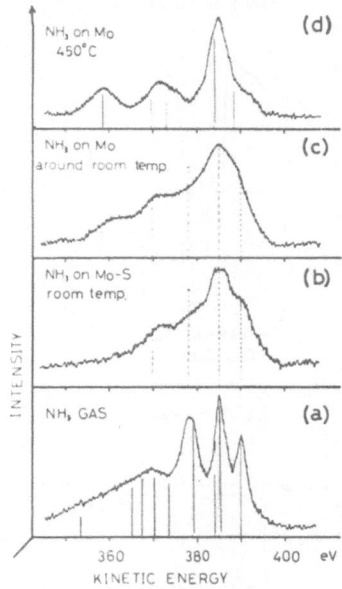

Fig. 2. Auger spectra characteristic of surface composition for different adsorbate conditions: (a) ammonia gas at room temperature; (b) ammonia adsorbed on molybdenum at room temperature; (c) ammonia adsorbed on molybdenum at somewhat higher temperature; (d) ammonia adsorbed on molybdenum at 450°C.

5 CONNECTION BETWEEN SIMPLE CATALYTIC REACTIONS ON METALS AND INDUSTRIAL CATALYTIC PROCESSES

A great deal of study is required to achieve a definitive reply to this question but the following comments tend to summarize in a useful fashion some typical initial approaches to this query. In addressing himself to the so-called "knowledge gap" between experimental studies and real catalytic processes, Kemball cited the following. Traces of promoters and intermediates can play an important role, a fact which controlled experiments might very well miss. Well-defined clean surfaces are too removed from those of practical catalysts. It is therefore recommended that experimental studies be carried out at least in part on surfaces of actual catalysts. The important point here is that the measurements be done in a systematic and careful manner using the best of modern atomistic and microscopic techniques. Study at working pressures of about ten orders of magnitude higher than pursued in normal experimental studies are required to make meaningful comparisons. That this viewpoint does not necessarily always prevail is evident from the conclusion previously stated by Ertl in a somewhat different context. A fourth area of value for study would be systematic identification of the specificity of hydrocarbon adsorption as a function of the nature and crystallography of the adsorbent.

The task of the physicist in industrial catalysis can be considered from the point of view of capitalizing on the characteristics of present physical methods. In this light, Fischer proposed that the physicist provide detailed and reliable information on surface phenomena under well-controlled and well-defined conditions. These can serve as scientific "building blocks" in the development of novel catalysts and could well lead occasionally to novel concepts. The synthesis of active sites on platinum through stepped single-crystal surfaces[14], the observation of dehydrogenation of hydrocarbons by photoemission[15], the quantitative interpretation in terms of surface segregation[16] of the catalytic activity of Cu-Ni alloys[17], and the computer modelling of the likely structure of microclusters[18] are a few examples of such contributions.

Yates neatly summarized the viewpoint by considering common concepts and techniques as the bridging mechanisms between experiment and practice. This view was further stressed by Bond who pointed out the important use of alloys to study the role of electron structure of the adsorbent in simple catalytic reactions. In addition to the projected need to systematize concepts relating to basic processes, Menzel pointed out that the systematic characterization of practical catalysts by modern experimental techniques is also largely an unexplored area; in many cases no correlations between changes in catalytic activity and physical parameters of the catalyst are known. Eischens added that the importance of investigating the active role of intermediates under dynamic conditions should not be overlooked in such systematic studies. Kokes' use of infrared spectroscopy is a good example of how fruitful this approach can be. Somehow, eventually all of these basic studies should focus on providing improved understanding of the atomic and electronic properties of complex catalysts characteristic of real operating conditions.

It appears that much more is unknown than is known about the physical and chemical principles underlying the basic experimental features of real catalytic surfaces. It is clear that the direction and scope of the responses to the questions posed in this discussion are the prelude to an expanding series of productive inquiries and responses.

ACKNOWLEDGMENT

Support by the Advanced Research Projects Agency through the Report Facility of the Cornell University Materials Science Center in the preparation of this discussion is gratefully acknowledged.

REFERENCES

1. Demuth, J. E., Jepsen, D. W., and Marcus, P. M., *Phys. Rev. Letters,* **31,** 540 (173); *ibid.,* **32,** 1182 (1974). Rhodin, T. N., and Demuth, J. E., *Proceedings of the 2nd International Conference on Solid Surfaces,* Kyoto, Japan (March, 1974). *Japan Jour. App. Phys.,* Suppl. 2, Pt. 2 (1974).
2. Pendry, J. B., *Low Energy Electron Diffraction—Theory and Applications to Determination of Surface Structure,* Academic Press, New York (1974).
3. Tong, S. Y., *Progress in Surface Science,* S. G. Davison (Ed.), Pergamon Press, Oxford (1975).
4. Landman, U., and Adams, D., *Phys. Rev. Letters,* **33,** 585 (1974).
5. Demuth, J. E., and Rhodin, T. N., *Surface Science,* **45,** 249 (1974).
6. Shirley, D. A., *Electron Spectroscopy,* Elsevier Publishing Co., New York (1972); Duke, C. B., and Park, R. L., *Physics Today,* **25,** 23 (1972); Rhodin, T. N., *Reactivity of Solids,* Anderson and Roberts (Eds.), Chapman and Hall, London (1973).
7. Gustaffson, T., Ph.D. Thesis, Chalmers University of Technology, 1973.
8. Liebsch, A., *Phys. Rev. Letters,* **32,** 1203 (1974).
9. Liebsch, A., and Plummer, E., to be published.
10. Rowe, J., and Ibach, H., *Phys. Rev. Letters,* **32,** 421 (1974).
11. Applebaum, J. A., and Hamann, D. R., *Phys. Rev. Letters,* **31,** 106 (1973).
12. Hagstrum, H. D., *Science,* **178,** 275 (1972).
13. Tamaru, K., *Advan. in Catalysis,* **15,** 65 (1964).
14. Somorjai, G. A., *Catalysis Rev.,* **7,** 87 (1972).
15. Demuth, J. E., and Eastman, D. E., *Phys. Rev. Letters,* **32,** 1123 (1974).
16. Burton, J. J., and Hyman, E. A., to be published.
17. Sinfelt, J. H., Carter, J. L., and Yates, D.J.C., *J. Catalysis,* **24,** 283 (1972).
18. Burton, J. J., *Catalysis Rev.—Sci. Eng.,* **9,** 209 (1974).

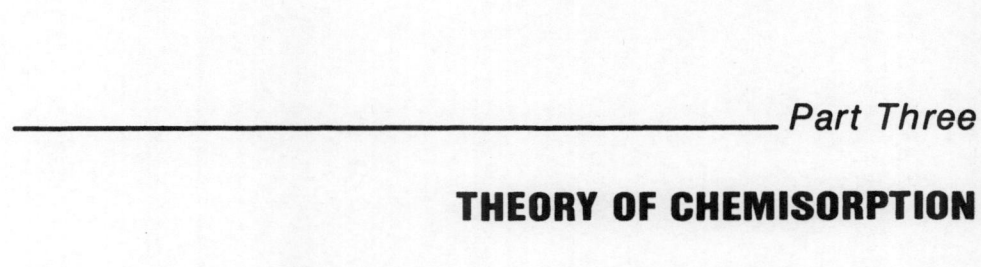

Part Three

THEORY OF CHEMISORPTION

THEORETICAL STUDIES OF METAL AGGREGATES AND ORGANOMETALLIC COMPLEXES RELEVANT TO CATALYSIS

Richard P. Messmer

General Electric Corporate Research and Development
Schenectady, New York

ABSTRACT

Theoretical studies of small transition-metal aggregates and transition-metal complexes having organic (C_2H_4) and inorganic (CO) ligands are discussed. The calculations employ the self-consistent-field $X\alpha$ scattered-wave (SCF-$X\alpha$-SW) approach to molecular orbital theory. As a result of these model calculations, a preliminary discussion is given of: (a) a possible interpretation of the effect of metal particle size in heterogeneous catalysis and (b) an interpretation of UV photoemission studies for C_2H_4 and CO chemisorbed on nickel.

1 INTRODUCTION

The subject matter of catalysis is nearly as diverse as chemistry itself. It includes, for example, the study of enzyme reactions, reactions carried out homogeneously by transition-metal complexes in solution, acid or base catalyzed polymerization of organic molecules, certain reactions in the upper atmosphere, and photocatalytic reactions, as well as reactions carried out heterogeneously on transition metals.

Nature has designed biological catalysts or enzymes, which are essential to the existence of plants and animals. These enzymes can perform tasks such as nitrogen fixation easily and with little expenditure of energy. Man, on the other hand, has effected only relatively crude methods for carrying out his desired reactions; such a case is the Haber-Bosch ammonia synthesis reviewed by Professor Emmett. Enzymes are usually extremely complex molecules and rather poorly understood. However, in certain favorable cases where the detailed structure of the enzyme and substrate are known, as in the case of trypsin, such systems represent perhaps some of the best-understood examples of catalysis.

This brings up an essential point regarding the understanding of catalysis—the lack of structural information regarding the catalyst and/or the catalyst-reactants system. This lack of information has plagued the study of catalysis in the past and will most likely continue to do so unless modern experimental techniques can be brought to bear on this problem. This situation is particularly acute for the theorist who wishes to understand catalysis from an electronic point of view. He is accustomed to having as an *Ansatz* the arrangement and positions of atoms in a solid or molecule. This provides, at least, some basic ground rules regarding the nature and extent of approximations to be used in the theory. Without such information, the possible choices of an *Ansatz* are, unfortunately, almost limitless.

Even if one concentrates on heterogeneous catalysis and specifically on catalysis by metals or metallic alloys dispersed on a support, the problems for a theorist are enormous. Should one consider the metal particles from a band-theory point of view or a molecular point of view? Going a step further, since chemisorption is an important precursor in catalytic reactions, what assumed structure should one consider for the metal particles in order to make progress on the chemisorption problem? What configurations of the adsorbate—substrate chemisorption interaction are most important? Obviously, structural information is crucial in order to address these questions meaningfully.

Assuming one had sufficient information to begin to answer some of the above questions, a new set of questions immediately comes to mind. What is the mobility or lability of the adsorbate on the metal particles? What intermediates or decomposition products of the adsorbate, if any, are important after the initial chemisorption interaction? How do these intermediates convert or interact with other adsorbed species to produce the products of the reaction? These questions relate to the kinetics of the reaction, and from a theoretical point of view are at least several orders of magnitude more complex than those regarding simple chemisorption. No quantitative first-principles calculation for the "potential surface" of even a simple catalytic reaction is likely for many years to come. The magnitude of the problem may be readily appreciated by considering the status[1] of the quantitative first-principles determination of the kinetics of the simplest possible chemical reaction, namely, $H_2 + D \rightarrow HD + H$. This situation hardly inspires confidence in the ability of theory to "predict" the path or kinetics of a catalytic reaction. However, this is not to say that theory will not be useful and even essential in elucidating the kinetics of such reactions; it serves only to point out that with proper guidance from experiment, calculations on a few configurations and not the whole "potential surface" may suffice to provide valuable information.

One further point should be raised—the stability of the metal-adsorbate or metal-intermediate* complex. It is necessary to distinguish between stability in the sense of the metal-adsorbate complex having a high bond energy on the one hand and kinetic stability on the other, in which the adsorbate is not labile and the complex exists for times long compared with times of experimental measurements. Thus, the fact that a particular metal-adsorbate complex cannot be easily detected experimentally does not necessarily mean that it has a negligible bond energy; it may mean only that it is kinetically unstable. A particularly valuable discussion of this problem for transition metal-alkyl molecules has been given by Wilkinson[2] and this knowledge should prove very valuable in certain heterogeneous catalytic systems. Theoretical calculations of bond energies could help to answer stability questions in individual cases although such calculations are by no means an easy task.

Having pointed out the limitations of theory and some of the problems presented to it by the complexity of heterogeneous catalysis, it is now important to take a more positive point of view and consider what theory might do in the immediate future to help elucidate catalysis. As it seems clear that chemisorption yields the basic precursors in a heterogeneous catalytic reaction on metals or supported metals, it is important to try to understand the possible structures and bonding of adsorbates on clusters of metal atoms so as to further our understanding of catalysis. From a theoretical point of view it also seems desirable, in order to obtain proper perspective and possible generalizations, to relate such studies of bonding of adsorbates on metals to studies of bonding in transition-metal complexes in which the adsorbate molecule is contained as a ligand. For such theoretical studies, the self-consistent-field-$X\alpha$-scattered wave (SCF-$X\alpha$-SW) method

---

*"Intermediate" is used here to designate a species which may be derived from the original adsorbate by its interaction with the metal; it is considered to be adsorbed or bonded to the metal. Hence, in the following, the term "adsorbate" is used in the more general sense to include both the original adsorbate and derived intermediates.

of Slater and Johnson[3] is rather ideally suited. The $X\alpha$-SW method is certainly the most useful and accurate realization of an abstract density functional theory as yet applied to realistic models of physical systems. The density functional approach is described by Professor Kohn in the succeeding paper.

The $X\alpha$-SW method, besides being able to provide information about adsorbate bonding to aggregates of metal atoms, can also contribute to the understanding of the effect on the electronic structure of metal aggregates arising from the particle size, the effect of the support on aggregates of metal atoms, and the relationship of electronic structure to the photoemission spectra (UPS and XPS) of chemisorbed species on metals.

One often hears the statement that the catalytic behavior of very small clusters of metal atoms is intrinsically "different" from that of large particles. Theoretical calculations on clusters of various sizes can help determine whether there is an electronic basis for such a statement due to marked differences in the electronic structure of such small clusters or whether the effect is more appropriately attributable to a more pronounced interaction of small clusters with the supporting material. As for the determination of the effect of the support on the electronic structure of small metal particles, the results of $X\alpha$-SW calculations can be used as input to a number of theoretical treatments which could provide such information. Some of these theoretical treatments are discussed by Professor Schrieffer in his paper in this volume.

A comprehensive theoretical program has been initiated both at General Electric Corporate Research and Development and at Massachusetts Institute of Technology, the latter under the supervision of Professor K. H. Johnson. This closely coordinated and cooperative effort has as its basic purposes the investigation of the electronic structure and bonding in transition-metal clusters, the nature of bonding of small molecules (e.g., C_2H_4, CO, etc.) to clusters of metal atoms, and the study of transition-metal complexes which may serve as analogues of chemisorbed species in heterogeneous catalytic reactions. The calculations employ the $X\alpha$-SW approach to molecular orbital theory, and some of the recent theoretical work in this program is summarized below.

2 RESULTS FOR METAL AGGREGATES

In order to assess the differences between the electronic structure of small metal aggregates and the bulk metal, we have undertaken a theoretical study of clusters of metal atoms of copper, nickel, palladium, and platinum. At the time of this writing, results are available for Cu_8, Cu_{13}, Ni_8, and Pd_{13} aggregates, and work is under way for Cu_{19}, Ni_{13} and Pt_{13}. The SCF-$X\alpha$-SW electronic energy levels for simple cubic Cu_8 are shown in Fig. 1(a). All of the levels shown arise from the ten 3d electrons and one 4s electron of each

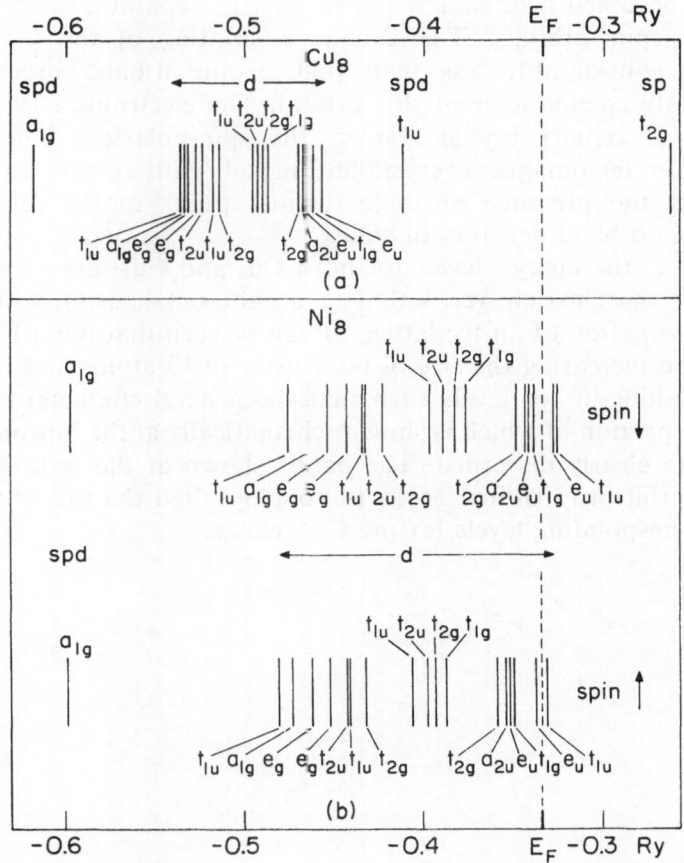

Fig. 1. SCF-Xα-SW electronic energy levels of simple cubic metal aggregates: (a) Cu₈ and (b) Ni₈.

copper atom, plus some contribution from the copper virtual 4p orbitals. The highest occupied level is the t_{1u} level at -0.411 Ry which is fully occupied, resulting in a zero net spin polarization for the Cu₈ cluster. The complete set of Cu₈ energy levels shown in Fig. 1(a) can be characterized as a dense band of d levels bounded above and below by the t_{1u} level at -0.411 Ry and the a_{1g} level at -0.617 Ry, both of which are predominantly s- and p-like in character with some d mixture. If the interval between the a_{1g} and t_{1u} levels is regarded as the precursor of the sp band in bulk crystalline copper, then already in the Cu₈ cluster the d band is totally overlapped by the sp band as in the solid.

The energy levels from a similar calculation for Ni₈ are shown in Fig. 1(b). In comparison with the Cu₈ results, the Ni₈ d band is shifted to higher energy, significantly widened and split in energy by the net paramagnetic spin polarization, which arises from the partial occupancy of the highest occupied orbital. The Fermi level E_F separates the occupied and unoccupied spin orbitals. The exchange splitting of the a_{1g} level near -0.6 Ry is considerably less than that of the d band because of the predominantly sp character of this orbital. The electronic structure of the Ni_8 cluster is remarkably similar to the spin-polarized band structure calculated for ferromagnetic crystalline nickel, with respect to sp-d band overlap and the presence of three distinct peaks in the majority- and minority-spin d-band densities of states.[4]

In Fig. 2, the energy levels for both Cu_8 and Cu_{13} are presented. The geometry of the Cu_{13} cluster is that of a cubo-octahedron, which has the proper coordination of an fcc lattice. It can be seen that the d band widens somewhat on increasing the size of the cluster to 13 atoms and that there is a general "filling-in" of levels such that the general character of the band structure, a portion of which is shown schematically at the bottom of Fig. 2, can be more clearly discerned. The bands shown at the bottom of Fig. 2 have been arbitrarily shifted so as to roughly align the top of the d band with the corresponding levels for the Cu_{13} cluster.

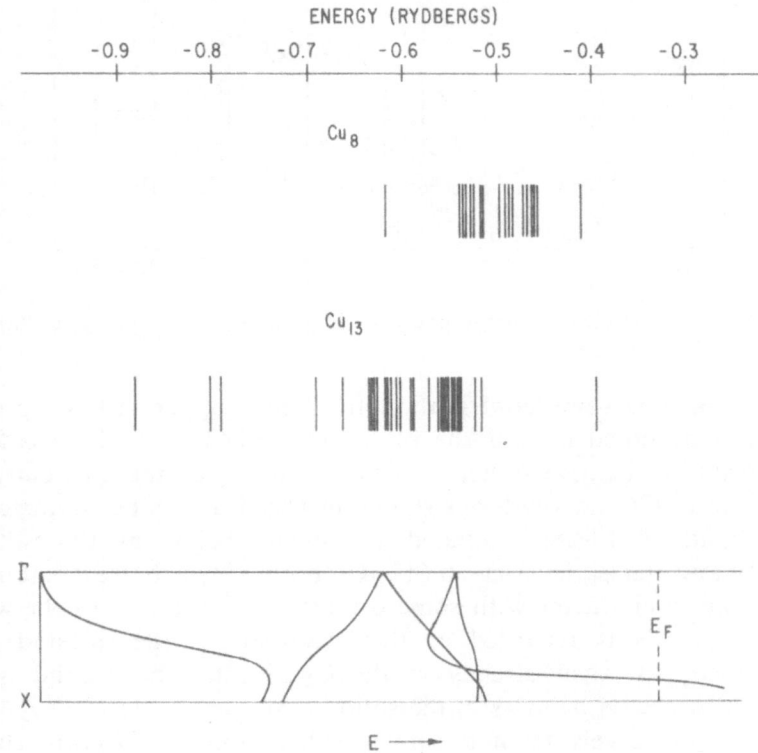

Fig. 2. SCF-Xα-SW electronic energy levels of simple cubic Cu_8 and cubo-octahedral Cu_{13} compared to schematic band structure of bulk copper.

The overall shift in the calculated d-like energy levels from Cu_8 to Cu_{13} may at first seem strange, but when one recalls that these are orbital energies and not transition-state energies[5] this shift can be easily understood. It is the transition-state energies that represent ionization potentials and thus can be compared directly to photoemission experiments for the determination of the positions and widths of bands. Experience has shown that with such systems as those under consideration, the effect of carrying out transition-state calculations is to give a set of energy levels that differ from the orbital energies by an almost constant shift of all the orbitals to lower energy. From rather general theoretical arguments, one expects that this shift between orbital and transition-state energies will decrease for larger systems, becoming zero in the limit of an infinite system. Thus the shift seen in Fig. 2 for the d band should be considerably less if transition-state energies rather than orbital energies are plotted. Such transition-state calculations are now in progress and are discussed in more detail, along with a number of other aspects of metal clusters, in work that will appear elsewhere.[6-7] Suffice it to say here that on the basis of these preliminary calculations for metal aggregates, one is struck by the similarities rather than the differences between the electronic structure of such clusters and that of the bulk metal.

3 RESULTS OF MODEL CALCULATIONS FOR METAL-OLEFIN BONDING

Olefins are involved in a number of important catalytic reactions which take place on supported transition-metal catalysts. One would therefore like to understand the interaction and bonding between the olefin and resultant intermediates with aggregates of transition-metal atoms.

As a first step toward this end, a study of the bonding of ethylene to clusters of various transition metals has been initiated. The intent of this study is to investigate the bonding of ethylene to clusters of nickel, palladium, and platinum such as those briefly described in the preceding section and to understand how the bonding changes as a function of the size of the metal cluster. However, the simplest case to start with, and the one for which a comparison to some transition-metal complexes can be made, is ethylene bonded to a single metal atom. Hence we will consider here the results of molecular orbital calculations using the $X\alpha$-SW method for ethylene interacting with a single atom of nickel, palladium, and platinum.

The calculations have been carried out with the same geometry and metal-ethylene bond distance as used in the overlapping-sphere calculation of Zeise's anion.[8] This bond distance was chosen because it is known that many ethylene complexes of nickel, palladium, and platinum have bond distances within 0.1 Å of this value. The resulting orbital energies are shown in

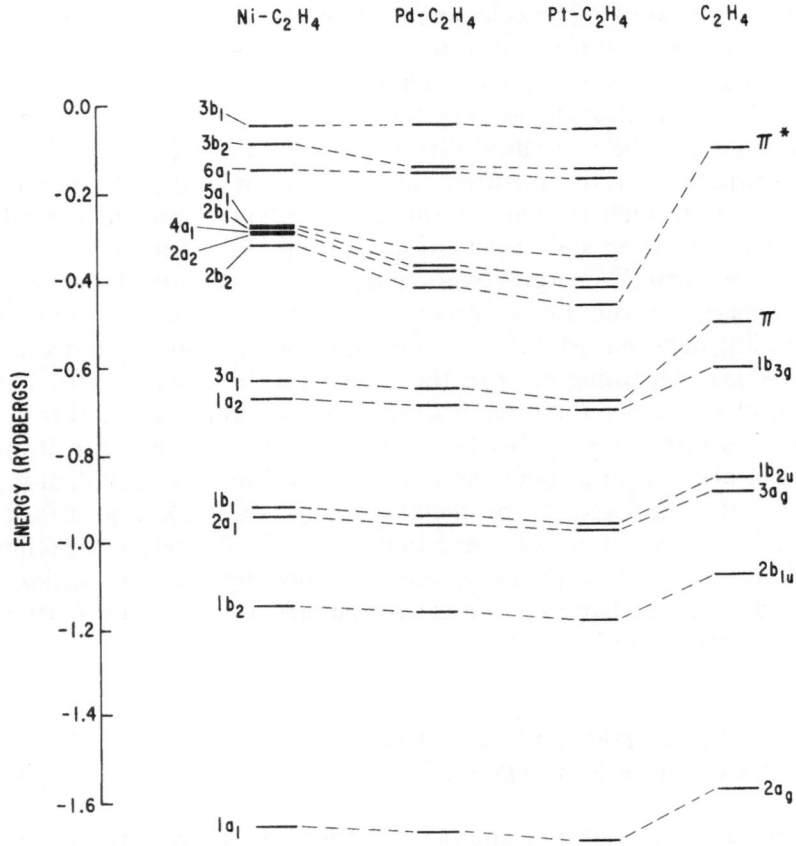

Fig. 3. Comparison of calculated electronic energy levels for Ni-C_2H_4, Pd-C_2H_4, Pt-C_2H_4 and C_2H_4; the same geometry was employed for all calculations.

Fig. 3; the same partitioning of space is used in each case. For C_2H_4, an extra "empty sphere" is employed at the location of the metal atom; the potential is spherically averaged in empty sphere regions.

In Fig. 3, the related orbitals are connected by broken lines. The ethylene orbitals from $2a_g$ up to $1b_{3g}$ undergo an almost uniform shift when ethylene interacts with a platinum atom. The positions of these shifted ethylene orbitals remain relatively constant with only a slight upward shift on going from platinum to palladium to nickel. The π and π^* levels of ethylene on the other hand experience a rather profound change on interacting with an atom of platinum, palladium, or nickel. Thus the $2b_2$ and $3a_1$ orbitals are in each case primarily responsible for the bonding. The $2b_1$, $4a_1$, and $2a_2$ orbitals are in each case nonbonding orbitals which are almost exclusively atomic d orbitals. The $3a_1$, $4a_1$, $5a_1$, and $2b_2$ orbitals for the cases of Ni-C_2H_4 and Pt-C_2H_4 are shown in Figs. 4 and 5, respectively. The $5a_1$ in

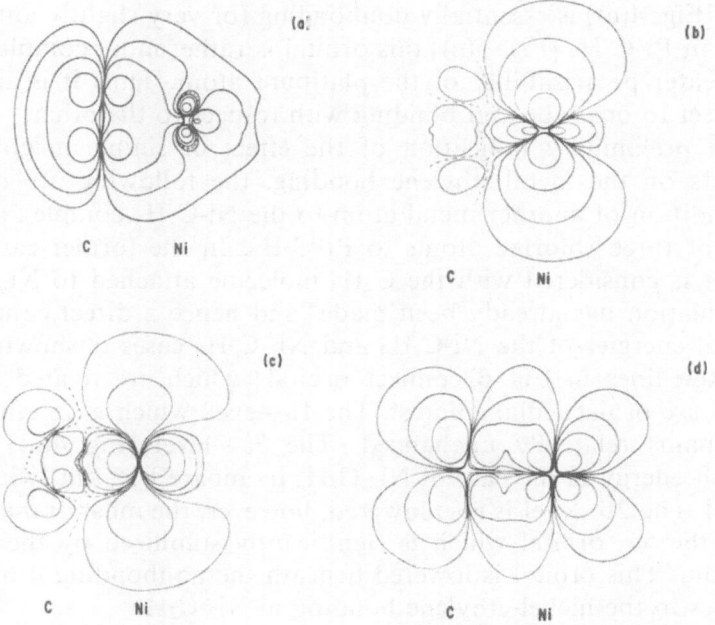

Fig. 4. Contour maps of the (a) $3a_1$, (b) $4a_1$, (c) $5a_1$, and (d) $2b_2$ orbitals of Ni-C_2H_4, nodes are indicated by dashed curves.

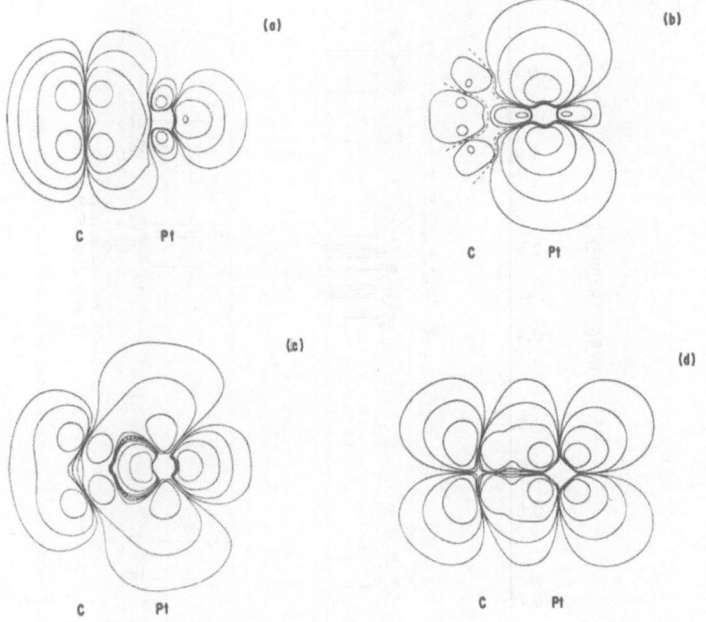

Fig. 5. Contour maps of the (a) $3a_1$, (b) $4a_1$, (c) $5a_1$, and (d) $2b_2$ orbitals of Pt-C_2H_4; nodes are indicated by dashed curves.

Ni-C_2H_4 [Fig. 4(c)] is essentially nonbonding (or very slightly antibonding); however, in Pt-C_2H_4 [Fig. 5(c)] this orbital is rather more complex, because of the greater polarizability of the platinum atom. Thus it is antibonding with respect to one lobe and bonding with respect to the other.

As a preliminary indication of the effect of simple neighboring environments on the metal-ethylene bonding, the following are considered: (1) the addition of another metal atom to the Ni-C_2H_4 complex and (2) the addition of three chlorine atoms to Pt-C_2H_4. In the former case, the Ni_2-C_2H_4 unit is considered with the C_2H_4 molecule attached to Ni_2 "end-on". This calculation has already been made[9] and hence a direct comparison of the orbital energies of the Ni-C_2H_4 and Ni_2-C_2H_4 cases is shown in Fig. 6. The broken lines in Fig. 6 connect orbitals which are related in the two cases and are of particular interest. The $1a_2$-level, which is an ethylene σ orbital, remains relatively unchanged. The $3a_1$ level [Fig. 4(a)] is slightly lowered in energy in the case of Ni_2-C_2H_4 [compare Fig. 4(a) with Fig. 2(a) of Ref. 9]. The $2b_2$ level is also lowered; however, the most dramatic change arises in the $5a_1$ orbital which is significantly stabilized by the additional nickel atom. This orbital is lowered beneath the nonbonding d orbitals and contributes to the nickel-ethylene bonding in Ni_2-C_2H_4.

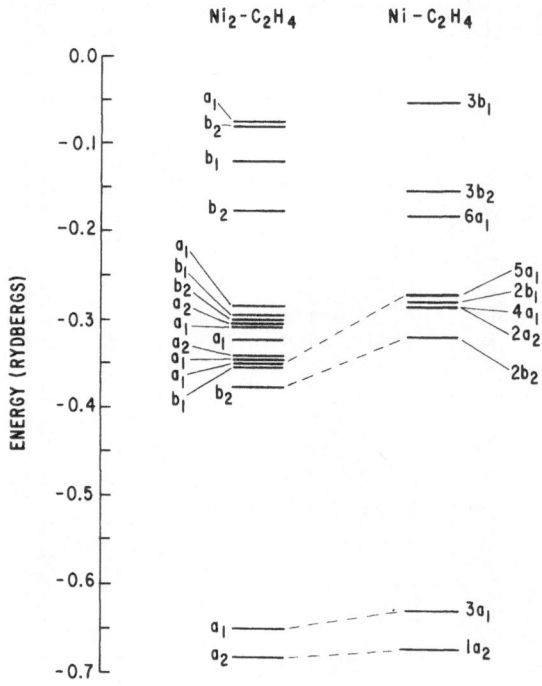

Fig. 6 Comparison of calculated electronic energy levels for Ni-C_2H_4 and Ni_2-C_2H_4; the latter one is from Rosch and Rhodin[10].

The second case to be considered is $PtC_2H_4Cl_3$. This is just the situation in Zeise's anion, except in the latter case an extra electron is added and the corresponding Madelung stabilizing potential is added. The Zeise's anion calculation has already been made[8], and hence we may compare the orbital energies of $Pt-C_2H_4$ with those of $PtCl_3C_2H_4^-$ as shown in Fig. 7.

The σ orbitals (due to ethylene) of $Pt-C_2H_4$, i.e., the orbitals $1a_1$ up to $1a_2$, remain essentially unchanged when the chlorine atoms are added to form $PtC_2H_4Cl_3^-$. The Cl 3s orbitals are found at ~ -1.5 Ry in $PtC_2H_4Cl_3^-$ as indicated in Fig. 7. A very striking effect occurs, however, for the orbitals which are derived from the π and $\pi*$ levels of ethylene; these orbitals are significantly lowered in energy in $PtC_2H_4Cl_3^-$ as compared with $Pt-C_2H_4$. The orbitals in Zeise's anion which are derived from the π orbitals of C_2H_4 are labelled "†" and those derived from the $\pi*$ orbital are labelled "*". The broken-line rectangle shows the region in which the predominantly non-bonding orbitals are found. The antibonding orbitals occur above -0.3 Ry.

Thus the chlorine atoms serve to withdraw electrons from the platinum atom and promote the bonding between ethylene and platinum by allowing the $4a_1$ and $5a_1$ orbitals to become bonding rather than nonbonding orbitals.

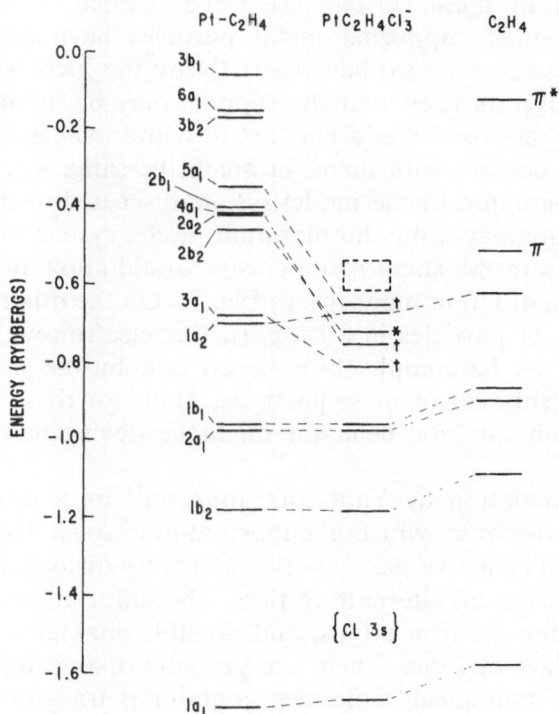

Fig. 7. Comparison of calculated electronic energy levels for $Pt-C_2H_4$, $PtC_2H_4Cl_3^-$, and C_2H_4; the levels for $PtC_2H_4Cl_3^-$ are from Rosch, Messmer, and Johnson[9].

The orbitals in Fig. 5 should be compared to those of Figs. 4 and 5 in Ref. 8; the contours chosen are the same in each case. These calculations provide a rather simple but nonetheless instructive example of the possible effect of the surrounding environment on metal-adsorbate bonding. Further details of the calculations and a more extensive discussion are to be published elsewhere.[10]

4 PRELIMINARY INTERPRETATION OF RESULTS AND SIGNIFICANCE FOR CHEMISORPTION AND CATALYSIS

Perhaps with so few and such simple calculations, it may be considered premature to venture making suggestions as to their import for catalysis. However, such tentative conclusions must be arrived at in order to provide guidance on the course of further calculations. I would therefore like to share some thoughts as regards the possible implications of these results for chemisorption and catalysis.

Thus far, the results for isolated metal clusters suggest that the electronic structures of these clusters are rather similar to the bulk metal. Hence, if very small supported metal particles have different catalytic behavior, which cannot be explained strictly by the increased surface area, the possibility suggests itself that the support may be involved. Most supporting materials are oxides and may act to withdraw electrons from small metal clusters in contact with them, in much the same way as the chlorine atoms affected platinum in the model system discussed in the previous section. A further analogy with the platinum model system suggests that the electron depletion in the small metal cluster would allow it to interact with the adsorbate in a different manner (see Fig. 7). On the other hand, for large metal aggregates or particles in a support, this electron-withdrawing effect of the support may be completely screened out, by the intervening metal atoms, from the surfaces of these particles. Thus, for these larger particles, the main effect on catalytic behavior might be determined by the surface area.

A recent calculation by Yang and Johnson[11] on a model of an iron-sulfur protein, ferredoxin, which is important in nitrogen fixation should be mentioned here. The active part involves a slightly distorted simple cube of iron and sulfur atoms on alternate vertices. The sulfur atoms can be thought of as a support for the iron atoms, and possible analogies with the effects described above are obvious. There are probably many useful analogies to be made among biological molecules containing transition-metal atoms, transition-metal complexes, and heterogeneous catalytic intermediates. Such analogies are probably *not* fortuitous. They should be sought after and the basis of the commonality elucidated.

On the basis of the calculations presented for ethylene interacting with nickel, some remarks may be ventured concerning the interpretation of photoemission experiments for ethylene chemisorbed on nickel. Some recent photoemission results of Demuth and Eastman[12] for ethylene chemisorbed on a Ni(111) surface are shown in Fig. 8. The three peaks between 6 and 10 eV arise from the σ orbitals of ethylene and have the same relative positions in the gas-phase photoemission spectrum of the ethylene molecule. The peak at ~5 eV has been shifted to higher binding energy as compared with the gas-phase spectrum. In the latter spectrum it is known that the corresponding peak is due to the occupied π orbital of ethylene. Hence, in the chemisorbed situation, the orbital at ~5 eV is derived from the π orbital of ethylene and contributes to the Ni-C_2H_4 bonding. This orbital is thus analogous to the $3a_1$ orbital of Fig. 6. A comparison of the orbitals of Fig. 6 with the spectrum of Fig. 8, suggests that the peak at ~3 eV might be attributed, at least in part, to the $2b_2$ orbital [Fig. 4(d)] which also contributes to bonding by back donation into the originally empty π^* orbital of ethylene.

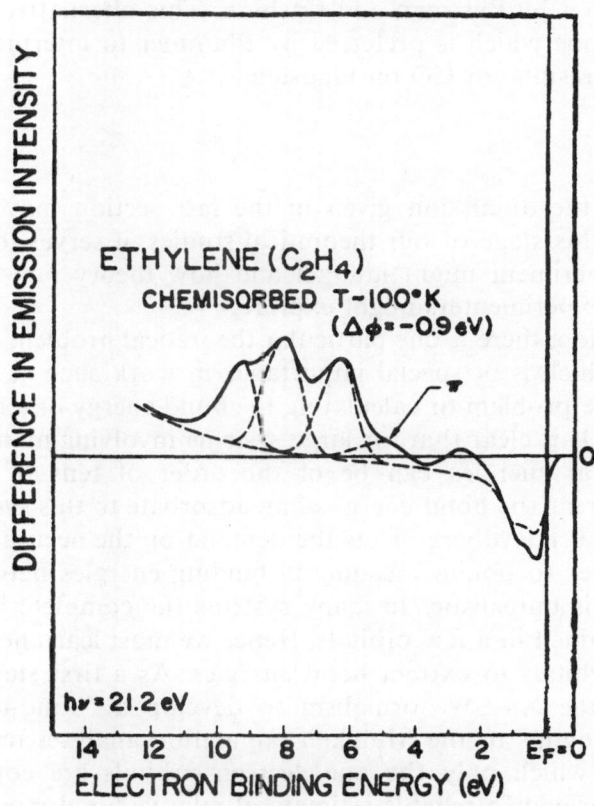

Fig. 8 Difference in emission intensity for ethylene chemisorbed on Ni(111); exposure of 1.2×10^{-6} Torr sec.[13]

Although space does not permit a detailed discussion of the recent calculations by Klemperer and Johnson[13] on transition-metal carbonyls, a few words may be said about the possible implications of these calculations with regard to chemisorption of CO. Eastman and Cashion[14] have studied the chemisorption of CO on Ni(100) films by photoemission. The photoemission spectrum shows two peaks: one at -7.5 eV, the other at -10.7 eV below the Fermi level. These peaks are interpreted by Eastman and Cashion, in analogy with the gas-phase spectrum of CO, as arising from the carbon lone pair on CO and the π_u 2p orbital of CO, respectively. The calculations of Klemperer and Johnson for $Ni(CO)_4$ and other carbonyls suggest that there is a strong stabilizing interaction between the carbon lone pair of CO and the metal, resulting in the orbital energy of an orbital derived from the carbon lone pair of CO being shifted to significantly higher binding energy, and thus becoming more bonding than the orbitals derived from the π_u 2p orbital of CO.

This suggests an alternative assignment for the peaks observed in the photoemission experiment. The alternative assignment would change the designation given by Eastman and Cashion. This alternative assignment, I believe, is the one which is preferred by Plummer to interpret some of his photoemission results for CO on tungsten.

5 SUMMARY

Although the discussion given in the last section may be somewhat speculative at this stage of our theoretical studies, it serves to indicate how theory and experiment might interact and how theory may indicate what directions the experimenters might explore.

In conclusion, there is one particular theoretical problem that should be pointed out, which is of special importance in work such as that discussed here. This is the problem of calculating the bond energy or a measure of the bond strength. It is clear that for large systems involving many heavy-metal atoms, the total energies can be of the order of tens of thousands of Rydbergs, whereas the bond energy of an adsorbate to this system will be of the order of ~ 0.1 Rydberg. Thus the demand on the accuracy of the total energies in order to obtain meaningful binding energies between substrate and adsorbate is unrealistic. In many systems the complete bonding information is contained in a few orbitals. Hence we must learn how to use these one-electron orbitals to extract bond energies. As a first step, it would be useful within the Xα-SW formalism to develop a "bond-order" scheme somewhat analogous to the Mulliken population analyses used for LCAO treatments, in which only the one-electron orbitals are considered. This could at least provide a reliable estimate of relative bond strengths between similar systems. Work is now in progress at G.E. and M.I.T. to develop such a scheme.

ACKNOWLEDGMENTS

The author would like to express his gratitude to Professor K. H. Johnson of the Massachusetts Institute of Technology for his continuing close collaboration on these problems. The author would also like to thank Professors J. R. Schrieffer, P. Soven, and E. W. Plummer for many stimulating conversations on surface physics during a brief stay at the University of Pennsylvania this Spring. This work was supported in part by the U.S. Air Force Office of Scientific Research under Contract No. F 44620-72-C-0008, for which the author is grateful.

REFERENCES

1. Koeppl, G. W., *J. Chem. Phys.*, **59**, 3425 (1973), and references therein.
2. Wilkinson, G., *Science*, **185**, 109 (1974).
3. Slater, J. C., and Johnson, K. H., *Phys. Rev.*, **B5**, 844 (1972).
4. Callaway, J., and Wang, C. S., *Phys. Rev.*, **B7**, 1096 (1973).
5. Slater, J. C., *Advan. Quantum Chemistry*, **6**, 1 (1972).
6. Messmer, R. P., Knudson, S. K., Johnson, K. H., Diamond, J. B., and Yang, C. Y., to be published.
7. Yang, C. Y., Johnson, K. H., and Messmer, R. P., to be published.
8. Rösch, N., Messmer, R. P., and Johnson, K. H., *J. Am. Chem. Soc.*, **96**, 3855 (1974).
9. Rösch, N., and Rhodin, T. N., *Phys. Rev. Letters*, **32**, 1189 (1974).
10. Messmer, R. P., to be published.
11. Yang, C. Y., and Johnson, K. H., to be published.
12. Demuth, J. E., and Eastman, D. E., *Phys. Rev. Letters*, **32**, 1123 (1974).
13. Klemperer, W. G., and Johnson, K. H., to be published.
14. Eastman, D. E., and Cashion, J. K., *Phys. Rev. Letters*, **27**, 1520 (1971).

DISCUSSION on Paper by Richard P. Messmer

DEROUANE: We have recently been involved in a parallel ESR and kinetic study of the hydrogenation of ethylene on Cu-MgO catalysts [E.G. Derouane, et al., *Proc. Symp. on Relations between Homogeneous and Heterogeneous Catalytic Processes, Brussels, 1974*, B. Delmon, and L. Jannes (Eds.), North-Holland (1974), in press]. Evidence for ethylene coordination on Cu^{2+} ions was gained from ESR data, and the catalytic sites have been identified to be Cu^{2+} ions dispersed at the surface, and not Cu^{+} ions or Cu° atoms or clusters. In view of your calculations on the Ni-C_2H_4 interaction, would you support our conclusions?

MESSMER: On the basis of our present calculations, I believe that we cannot answer your question. However, it would be perhaps very interesting to carry out calculations for model systems such as $Cu-C_2H_4$, $Cu^+-C_2H_4$ and $Cu^{++}-C_2H_4$.

BOND: Dr. Messmer's calculations seem to show that the stabilities of M-C_2H_4 complexes decrease in the sequence Pt > Pd > Ni. Can he confirm that my interpretation is correct? If I am right, calculations are encouraging because there is considerable evidence (e.g., from the stability of organometallic complexes and the kinetics of the heterogeneous $C_2H_4 + D_2$ reaction) that this is indeed the case.

MESSMER: On the basis of the trends in the one-electron energy levels, this is certainly what one would expect. However, the total energy calculations, although they indicate binding for the three systems, do not give reliable enough binding energies as yet to make a quantitative statement about relative binding.

BOND: Ugo has suggested that the configuration of coordinated and associatively chemisorbed molecules often resemble the first excited states of the free molecules. Do the energy levels calculated by Dr. Messmer for M-C_2H_4 and Zeise's anion show any correspondence with first excited state levels? Can he perform calculations employing excited states?

MESSMER: Yes, the calculated energy levels for complexes can be used to discuss the excited states of the complexes. For the case of Zeise's anion, for example, such a discussion has already been published [Rösch, Messmer, and Johnson, *J. Am. Chem. Soc.*, **96**, 3855 (1974)].

BOND: It would be of great interest to use the Xα method to explain the origin of the well-known, generally observed decrease of adsorption energy with surface coverage.

What is the point of performing calculations on M_8 clusters, which is a quite unreal and unstable configuration in comparison with say an M_6 octahedron? What hybridization of dsp orbitals is implied by the fivefold symmetry in an icosahedron?

MESSMER: As far as I'm aware, it is not clear that the assumed geometries of the M_8 clusters are unreal or unstable configurations, as it has not yet been determined experimentally what the appropriate geometry really is. As regards the appropriate hybridization for the icosahedron: this is a matter which I have not really thought about.

MASON: Many of us will be interested in the theoretical results on clusters of copper and nickel atoms and particularly the way in which individual atomic levels evolve towards the band arrangement representative of the bulk metal; the data will help us talk more precisely about the question. How large does a cluster have to be to approximate surface electronic states?

I am skeptical of the results on such complexes as $Ni(C_2H_4)$ and $Ni_2(C_2H_4)$. These molecules are nonexistent and we do not have to define any molecular dimensions. The detailed conclusions on the differences in bonding in, say, $Ni(C_2H_4)$, $Pd(C_2H_4)$, and $Pt(C_2H_4)$, must depend on the assumed geometries. In developing the $X\alpha$ method, one ought to concentrate on energy levels for the simple organometallic molecules for which spectroscopic data are at hand. In this context, I remain convinced that the $X\alpha$ method is a promising one, although one which is ordinarily not straightforward.

MESSMER: As regards the calculations on the model systems discussed in my talk, I agree that there is no experimental information about the geometries and bond distances for these systems. This is precisely the problem that I mentioned in my talk about having to make reasonable estimates about structures for which we have no experimental information. If experimenters, like yourself, would care to supply us with this information however, we would be happy to incorporate the proper geometry in our calculations.

To justify our assumption about geometry, I refer again to the fact that there are many compounds of nickel, palladium, and platinum with C_2H_4 for which the geometries are very much the same, differing by perhaps ~0.1 Å for the metal-C_2H_4 bond lengths. I should also mention that published studies using matrix isolation techniques [D. W. Green, J. Thomas, and D.M. Gruen, *J. Chem. Phys.,* **58**, 5453 (1973)] have identified such molecules as $Pt-N_2$ and that the bonding and structure are very similar to known complexes of platinum having a N_2 ligand. Therefore, it is possible that by such experimental techniques, species such as $Pd-C_2H_4$ and $Pt-C_2H_4$ could be isolated and the structure determined. If this were to be done, I would be very surprised to find that the structures differed very much from the Zeise's anion structure.

If the $X\alpha$ method were applied to only those systems where good spectroscopic data are in hand, this would essentially preclude any application to systems of catalytic chemistry or surface-science interest.

With regard to your question about cluster size, I assume you are asking how large it has to be to approximate the surface electronic states of a semi-infinite surface. At present, this is not clear, although my guess would be that it would be less than ~50 atoms.

PLUMMER: How large does a metallic cluster have to be before the ionization potential of a core level approaches the bulk value?

MESSMER: We have not really investigated this point as yet, but intend to do so in the near future. If I have to venture a guess, however, I would expect 3 to 5 layers (19 to ~40 atoms) to be quite sufficient.

PLUMMER: How large does a cluster have to before the s-band screening of the hole is like the solid?

MESSMER: Again, we have no definite information as yet; however, a cluster of approximately the size mentioned above would probably be sufficient.

GRIMLEY: In your calculations on metal clusters, Cu_8 for example, you show the Fermi level. For a given number of electrons, the Fermi level separates occupied from vacant one-electron levels. Yours appears to be a value obtained some other way.

MESSMER: In Fig. 1, the dashed line indicating the Fermi level does in fact separate the occupied and vacant one-electron levels for both Cu_8 and Ni_8.

GRIMLEY: Are your ethylene-metal complexes stable against dissociation into metal atom + ethylene? Do you have a total energy calculation?

MESSMER: Yes, a total energy calculation has been made and the ethylene-metal complexes are stable against dissociation. However, as you know, these binding-energy calculations are very difficult and the accuracy not very great, as pointed out in my talk. Therefore, I have not quoted any quantitative values for binding energies. Much further work is needed on this problem.

SMITH: This is a very promising method indeed, especially for transition metals which we know are important in catalysis. You have shown that the electron-energy-level distribution of your small particles is similar in many respects to bulk band structure. There are, as you pointed out, reactions which depend on particle size, however. An explanation that you proposed involved interaction with the support.

    I would like to propose a possible alternative explanation. Because the atomic roughness varies with particle size, so will the electron-wave-function distribution in the surface region. Since catalysis occurs in the surface, this should be important.

MESSMER: I would certainly agree with you that a different electron-wave-function distribution is an alternative explanation. As a matter of fact, Keith Johnson and I have invoked this alternative explanation in previous work on clusters. The real situation is most certainly a combination of these effects.

DENSITY FUNCTIONAL THEORY OF METAL SURFACES AND OF CHEMISORPTION ON METALS

W. Kohn

University of California, San Diego
La Jolla, California

ABSTRACT

Applications of the density functional theory to the electronic structure of bare metal surfaces and metal surfaces with adsorbed atoms are reviewed and their usefulness and limitations are discussed. Also included are brief descriptions of some ongoing work on improvement of the underlying approximations and of an adaptation of the Green's function (KKR) method to the geometry of a film.

1 INTRODUCTION

The interactions between chemisorbed molecules and metal surfaces and those between two or more chemisorbed molecules on such surfaces are obviously of fundamental interest for a sound understanding of metal catalysts. As these interactions are largely of electronic origin, there is a strong need to be able to calculate the electronic structure of such systems for given positions of all the nuclei. This problem is considerably more difficult than two problems to which it is related: the electronic structures of bulk metals and those of isolated molecules. The problem at hand has neither the simplifying feature of complete periodicity nor that of involving only a finite number of electrons; it is a full-fledged many-body problem with very low symmetry. Therefore, it is not surprising that, despite considerable theoretical activity in this area during the last few years, only the simplest systems have so far been treated. Even for these, some quite serious approximations have had to be made. At the present time the problem is being approached along several different lines, which emphasize different aspects of these systems. A number of these are being presented at this meeting.

In this paper a number of approaches are described which can be grouped under the heading of "density functional theory". In this theory, the "active" electrons are considered as an electron gas, characterized by its density $n(\mathbf{r})$; electrostatic self-consistency is strictly observed, but exchange and correlation effects are treated rather crudely. The theory is without adjustable parameters and seems capable of accounting for properties of simple metal surfaces and simple chemisorption systems with an accuracy on the order of ±20 percent.

2 SUMMARY OF THE DENSITY FUNCTIONAL THEORY

This theory (of which the most simplified version is the Thomas-Fermi method) formulates the general problem of interacting electrons in an external potential $v(\mathbf{r})$ in terms of a minimal principle for the energy, which is expressed as a functional of the electron density $n(\mathbf{r})$. It has been shown[1] that there exists a universal functional of $n(\mathbf{r})$, $F[n(\mathbf{r})]$, in terms of which one may define the energy functional

$$E_v[n(\mathbf{r})] \equiv \int d\mathbf{r}\, v(\mathbf{r})\, n(\mathbf{r}) + F[n(\mathbf{r})] \tag{1}$$

with the following properties: its minimum value is the correct ground-state energy and is attained when $n(\mathbf{r})$ is the correct ground-state density. The quantity $F[n(\mathbf{r})]$, which does not explicitly depend on $v(\mathbf{r})$, comprises the kinetic and mutual interaction energy of the electrons, including exchange and correlation effects. For the latter two, especially, various rather crude approximations have been used. However, once an adequate expression for $F[n]$ is available, the minimization of $E_v[n]$ with respect to the density distribution $n(\mathbf{r})$ is relatively simple. Several calculations, to be described below, have followed this route.

From the stationary property of $E_v[n]$ one can also derive a system of self-consistent one-particle-like equations[2],

$$\left\{ -\nabla^2 + v_{eff}\,[n(\mathbf{r}),\mathbf{r}] \right\}\, \psi_i(\mathbf{r}) = \epsilon_i \psi_i(\mathbf{r}) \tag{2a}$$

$$n(\mathbf{r}) = \Sigma\, |\psi_i(\mathbf{r})|^2 \tag{2b}$$

where

$$v_{eff}\,[n(\mathbf{r}),\mathbf{r}] = v(\mathbf{r}) + \int \frac{n(\mathbf{r}')}{|\mathbf{r}-\mathbf{r}'|}\, d\mathbf{r}' + \frac{\delta E_{xc}[n]}{\delta n(\mathbf{r})} \tag{3}$$

and E_{xc} is the exchange-correlation portion of $F[n]$. Although Eqs. (2a) and (2b) have the form of single-particle equations, they include through Eq. (3) all many-body effects which have been incorporated in $E_{xc}[n]$. Several of the approaches, which are described, have employed this method rather than the direct minimization of $E_v[n]$. In practice, the second approach offers greater accuracy.

3 THE BARE METAL SURFACE

A necessary prerequisite for understanding chemisorption on metal surfaces is an adequate theory of the bare surfaces. At the present time, it appears that in spite of rather simple approximations for many-body effects, one can obtain a reliable and quite detailed description of the surfaces of **simple** metals (such as the alkalis or aluminum). However, good theories of the surfaces of noble, transition, and other metals of more complex electronic structure are still in the future.

Reported calculations have proceded along three lines, as follows.

a. Theories Starting from the Jellium Model

The simplest model which highlights the effect of a surface on electronic structure is the so-called jellium model in which the positive ions are replaced by a uniform positive charge which terminates abruptly (Fig. 1). Charge densities and surface energies for this model were calculated by J. Smith[3], using direct minimization of $E_v[n]$, Eq. (1); and by N. Lang and W. Kohn[4], using the more accurate self-consistent Eqs. (2a) and (2b). In Smith's calculation, both kinetic and exchange-correlation energies were treated approximately, whereas in those of Lang and Kohn, the kinetic energy is treated exactly. These calculations appear to give a good description of the electron density distribution at the surfaces of real metals, as evidenced by good agreement between calculated and measured work functions. They also give good estimates of the surface energies of the alkali metals, which have a low interior electron density n_o, but give very bad, even negative, surface energies for high n_o metals such as aluminum (Fig. 2).

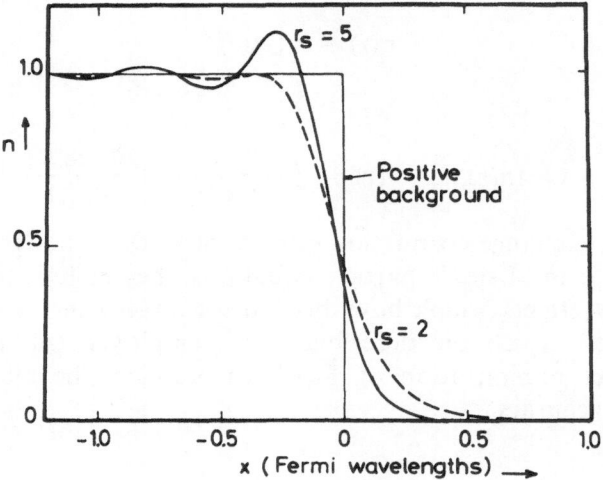

Fig. 1. Electron density near a metal surface for two values of the Wigner Seitz radius r_s.[4] One Fermi wave-length $= 2\pi/k_F$. For $r_s = 2$, it is 3.5 A; for $r_s = 5$, it is 8.7 A.

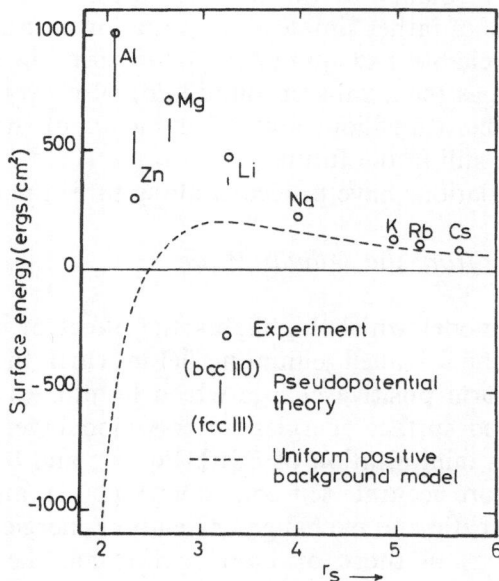

Fig. 2. Comparison between theoretical and experimental values of the surface energy.[4] The vertical lines indicate theoretical uncertainties due to the poorly known geometric arrangement at liquid metal surfaces; the experimental values are for liquid metals.

This deficiency is overcome when, in a next approximation, the ions are reintroduced in place of the positive charge background and treated by perturbation theory. The resulting surface energy then comes into semiquantitative agreement with experiment for both high- and low-electron-density metals (Fig. 2).

Recently, the simple local approximations for exchange and correlation energies used in these calculations have been criticized by Harris and Jones[5] and Jonson and Srinivasan[6]. However, while the local approximation, especially for the correlation energy, appears to be seriously in error near a surface, the local approximation for the **sum** of exchange and correlation energy (which is what is used) seems to be quite accurate.

At the present time we are working with an improved form of the exchange-correlation energy, which includes a nonlocal-gradient term,

$$E_{XC}[n(\mathbf{r})] = \int n(\mathbf{r})\left[\epsilon_{XC}(n(\mathbf{r})) + f[n(\mathbf{r})](\nabla n)^2\right]d\mathbf{r} \qquad (4)$$

The function $f(n)$ could be chosen so that the dielectric-response function $\epsilon(q;n)$ agrees with the best independently calculated and measured values for all q and n to within about 2 percent. In this form, the density functional theory is accurate both for systems of slowly varying density and for systems with small density fluctuations of arbitrarily short wavelength. Preliminary results indicate an increase in surface energies in the neighborhood of 15 percent or more.

b. Direct Numerical Solutions of the Self-Consistent Equations

Two groups have reported detailed numerical solutions of the self-consistent Eqs. (2a) and (2b) for alkali metals, in which the actual ionic potential was replaced by a weak pseudopotential.

Appelbaum and Hamman[7] take advantage of the remaining translational symmetry in the y-z plane parallel to the surface and transform the original partial differential Schroedinger equation [Eq. (2a)] into a set of ordinary coupled differential equations in the x-variable. These equations are then solved numerically in the surface region and joined to bulk solutions in the interior. The method was applied to the (100) face of metallic sodium. Their reported work function (2.70 eV) is very similar to that obtained by Lang and Kohn (2.75 eV).

Alldredge and Kleinmann[8] have obtained a solution of the self-consistent equations for a 13-layer (100) film of lithium. They write the wave functions as Fourier series in all three dimensions, imposing periodicity conditions parallel to the surface and making the wave functions vanish on two planes, x = ±b, chosen three atomic layers beyond the last planes of atoms. The calculated work function of 3.7 eV is about 1 eV higher than the measured one, possibly because of a choice of an unsatisfactory pseudopotential.

Although these two calculations were performed on two of the simplest metals, they required very considerable computing effort, and it is not clear how far such direct methods can in practice be extended. The results obtained for work functions did not represent progress over the simplest calculations described under (a) above. But, from the standpoint of catalysis, direct calculations of this kind do offer interesting new kinds of information: (1) they give a detailed map of the electron density distribution $n(\mathbf{r})$ near the surface, including its variation in the x-y plane (Fig. 3); and (2) they give information about surface states, in cases where these exist.

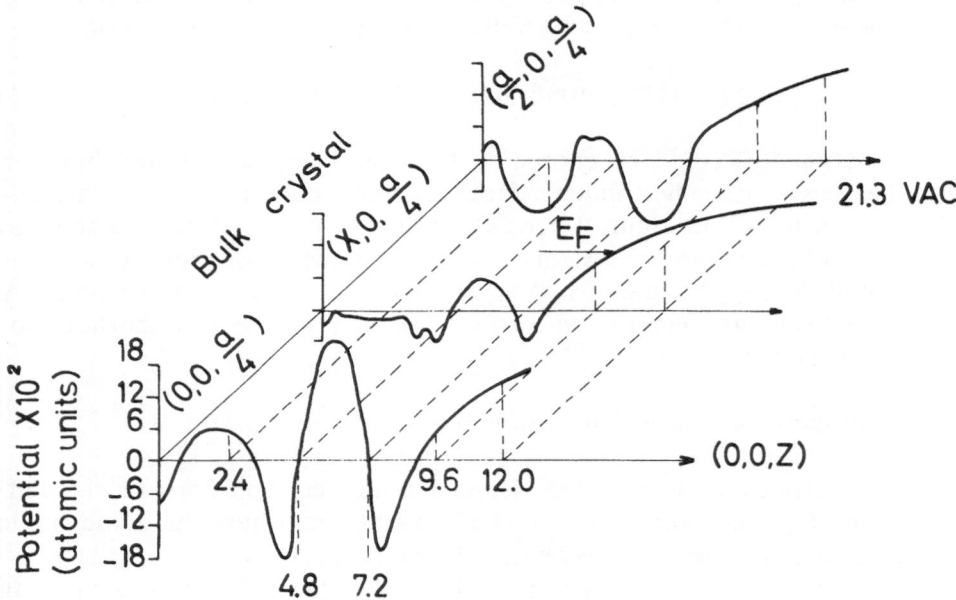

Fig. 3. Plots of the potential variation near the surface of metallic sodium.' Similar plots can also be obtained for the electron density distribution.

c. *Calculations of Small Clusters of Atoms*

In recent years, Slater[9] and Johnson[10] have developed a new and successful method, the $X - \alpha$ scattered-wave method, for the calculation of the electronic properties of molecules. This method is described in greater detail in this volume by R. P. Messmer. As far as ground-state properties are concerned, the method can be put in the context of the density functional theory by observing that, in this method, one chooses a simple form for the effective exchange-correlation energy (depending on an empirical constant α) and solves the differential equation [Eq. (2a)] by means of a multiple-scattering Green's function formalism, adapted to a collection of atoms. In this way, clusters of a small number of copper and nickel atoms have been studied, which may be considered to have small surfaces.

d. Surface Green's Function Method

The scattered-wave or Green's function method was first used in solid state physics for solving the Schroedinger equation in completely periodic lattices.[11,12] Subsequently, it was adapted[9,10] to molecules or atom clusters. We have now developed the appropriate modification for a metal film which we think will be useful for surface problems.

Details will be published elsewhere. Here we merely outline the structure of this theory. We consider a film of several parallel infinite layers of atoms, symmetrically situated near the y-z plane (Fig. 4). On two planes x = ±b sufficiently far from the surface layers, we impose the boundary conditions $\psi = 0$. We must solve a Schroedinger equation of the form

$$(-\nabla^2 + v(\mathbf{r}) - E)\,\psi = 0 \quad . \tag{5}$$

Since the film has two-dimensional translation vectors τ_2 and τ_3, the eigenfunctions can be labelled by a two-dimensional wave vector $\mathbf{g}$. The differential equation [Eq. (5)] for ψ_g can be replaced by the integral equation

$$\psi_g(\mathbf{r}) = \int_\tau G_g(\mathbf{r},\mathbf{r}')\,v(\mathbf{r}')\,\psi_g(\mathbf{r}')\,d\mathbf{r}' \quad , \tag{6}$$

where τ is the unit cell which, in the x direction, extends from x = −b to x =

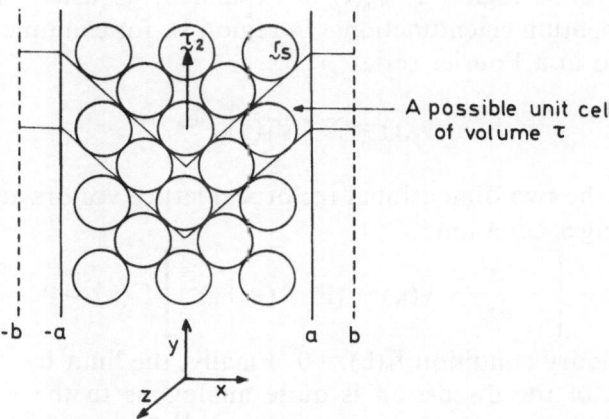

Fig. 4. A five-layer metal film; within the spheres, V is taken as spherical and for $|x| > a$, as dependent only on x. A unit cell containing five atoms is indicated.

+b (Fig. 3); G_g is the Green's function satisfying the appropriate differential equation, periodicity in the y-z plane, and boundary conditions on the surfaces x = ±b. In addition to τ_2 and τ_3, let us introduce $\tau_1 \equiv (4b, 0, 0)$. Let $G_{g.0}$ be the conventional infinite-space Green's function of the three-dimensional lattice defined by τ_1, τ_2, and τ_3, corresponding to the wave vector (0,g). Then the required Green's function G_g is given by

$$G_g(\mathbf{r},\mathbf{r}') = G_{0.g}(x{-}x', y{-}y', z{-}z')$$

$$- G_{0.g}(2b{-}x{-}x', y{-}y', z{-}z') \quad . \tag{7}$$

The space enclosed by the unit cell τ is now divided into three regions, and the following assumptions are made about the behavior of the potential in each of these:

(I) Nonoverlapping spheres surrounding each nuclear site $\mathbf{r}_s$, in which $v(\mathbf{r})$ depends only on $|\mathbf{r}{-}\mathbf{r}_s|$.

(II) Two surface strips (not overlapping these spheres), given by a < x < b and −b < x < −a in which $v(\mathbf{r})$ depends only on x. This is a reasonable approximation in the surface region.

(III) The remaining space, in which $v(\mathbf{r})$ is taken to be a constant, which may be equated to zero.

Equation (6) now becomes

$$\psi_g(\mathbf{r}) = \int_I G_g(\mathbf{r},\mathbf{r}') \, v(\mathbf{r}') \, \psi_g(\mathbf{r}') \, d\mathbf{r}'$$

$$+ \int_{II} G_g(\mathbf{r},\mathbf{r}') \, v(\mathbf{r}') \, \psi_g(\mathbf{r}') \, d\mathbf{r}' \tag{8}$$

In the spheres of region I, $\psi_g(\mathbf{r})$ is expanded, as usual, in terms of the angular momentum eigenfunctions. In region II, for example, for a < x < b, it is expanded in a Fourier series

$$\psi_g(\mathbf{r}) = \Sigma \, C_h^+ f_h^+(x) \, e^{i(g_h+g)\cdot\tau} \tag{9}$$

where $\mathbf{g}_h$ are the two-dimensional reciprocal lattice vectors and $f_h^+(x)$ satisfies the Schroedinger equation

$$\left[-\frac{d^2}{dx^2} + v(x) - (E - (\mathbf{g}_h+\mathbf{g})^2) \right] f_h^+(x) = 0 \tag{10}$$

and the boundary condition $f_h^+(b) = 0$. Finally, the limit $b \rightarrow \infty$ is taken.

The rest of the discussion is quite analogous to the KKR theory for bulk solids with several atoms per unit cell.[13] For a film of a monatomic metal with N layers, there are N atoms per unit cell, characterized by the logarithmic derivatives $L_l(E)$ of the radial solutions; in addition, the strips are similarly characterized by the logarithmic derivatives $L_h^+(E)$ of the functions $\vec{\iota}_h^+(x)$, at $x = \pm a$. There are, correspondingly, new geometric structure constants which are easily evaluated.

The result is that, for example, a film of a monatomic metal with reflection symmetry containing five atomic layers plus the two surface strips presents a mathematical problem analogous in structure to that of a bulk solid with four atoms per unit cell (three of the film atoms and one surface strip are independent). Thus it would appear that this approach could handle films of realistic thickness with a computing effort not very much larger than that needed for bulk solids. No actual computation has yet been carried out in this manner.

3 CHEMISORPTION OF HYDROGEN AND ALKALIS ON METAL SURFACES

The problem of chemisorption on metal surfaces is evidently even more complex than the problem of the bare metal surface. The symmetry is further reduced and the bonding of an adatom with the metal electron sea raises some difficult questions, especially concerning the role of correlation effects.[14] In the present formalism, all electrons (except those in the cores) form part of the total electron gas. In the first version of the theory one minimizes the total energy [Eq. (1)] directly, treating the metal ions and the core of the adatom as external potentials; in the second version [Eqs. (2) and (3)] one determines the self-consistent molecular orbitals of the combined system. In principle, either of these approaches would be exact if the exact functional F[n] were known and used. In practice one has used, so far, local exchange and correlation expressions taken from the theory of an electron gas of slowly varying density. Such a crude approximation seems impossible to justify a priori. In particular, for such systems as hydrogen adsorbed on tungsten, one expects a good deal of antiparallel correlation between the spin of the hydrogenic electron and the metal electrons[15], rather different from what one encounters in a uniform electron gas. Nevertheless, since the simple form of the energy functional has given rather good results for other systems where this would not have been expected, such as for the bare metal surfaces discussed in the previous section, and even for an isolated hydrogen atom[16], it has been deemed to be of interest to calculate the properties of a number of simple chemisorption systems. Such calculations are described below.

a. Chemisorption by Linear Response Theory

A canonical system of relevance to chemisorption is a jellium model for the surface plus a small point charge q, located at the point (x', 0,0) relative to the jellium edge. The resulting charge distribution may be written, to first order in q, as

$$n(\mathbf{r}) = n_0(x) + q \ n_1(\mathbf{r};x') \qquad (11)$$

The charge distributions $n_1(\mathbf{r};x')$ have been calculated by linear response theory, using a simple form of the functional F[n], Eq. (1).[17] The corresponding energy of the entire system as function of x' was also obtained (Fig. 5). A plot of screening charge n_1 is shown in Fig. 6 for x' equal to the equilibrium position of q at a metal surface with $r_s = 1.5$.

Taking a "great leap forward", q was next set equal to 1, r_s of the metal was taken as 1.5, and the resulting system was identified with hydrogen adsorbed on tungsten. A number of physical properties emerged in surprisingly reasonable agreement with experiment.

1. *Desorption energy.* An infinitesimal point charge desorbs naturally as an ion, with the screening charge left behind on the surface. Consequently, in this procedure one calculates, of necessity, the ionic, rather than atomic, desorption energy. The result, for q = 1, is 9 eV compared with the experimental value of 11 eV.

2. *Vibration frequency.* From the curvature of the total energy of the system as function of x, W(x), at the minimum one obtains directly the longitudinal vibration frequency

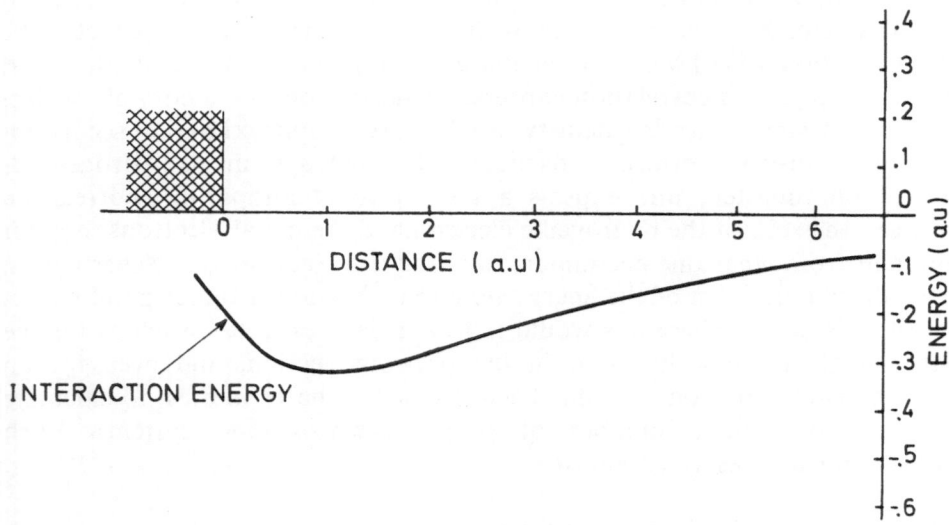

Fig. 5. Energy of a unit external point charge in the vicinity of a metal surface[16]; r_s has been taken as 1.5, corresponding to tungsten.

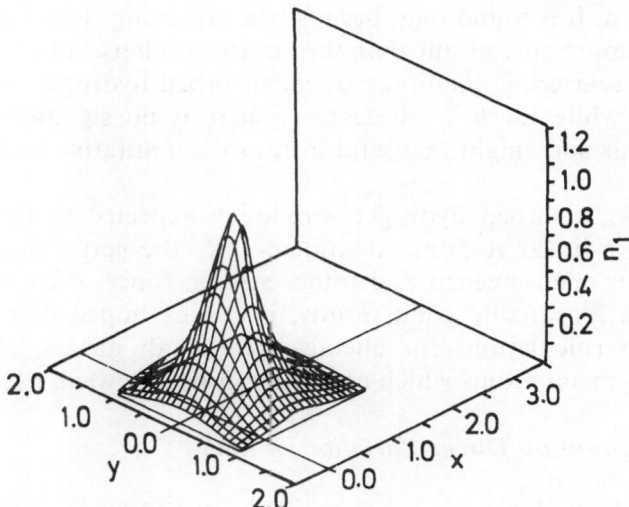

Fig. 6. Screening charge density around a point charge at its equilibrium position near a metal surface with $r_s = 1.5^{16}$; distances in 10^{-8} cm.

$$\omega = \left(\frac{1}{m} \frac{d_2 W}{dx^2} \right)^{1/2} \qquad (12)$$

which in the present calculation has the value of 200 meV. The experimental value is 140 meV.[18,19]

3. *Dipole moment.* In its natural units of e a_0, the dipole moment D of this model is very small (~ -0.01), reflecting the fact that, in equilibrium, the point charge is nearly spherically screened. The absolute value of D is so small as not to be reliable, but experiment too yields small values in the range of 0.03 a.u.

4. *Electronic resonance level.* In the present calculation, the electronic density $n(\mathbf{r})$ was obtained directly, without solving the single-particle equations [Eqs. (2a) and (2b)]. However, from this $n(\mathbf{r})$ an effective potential, v_{eff}, can be obtained from which one can determine a single-particle spectrum. An approximate calculation of this kind shows a resonance level localized on the adsorbed hydrogen at a position 5.6 eV below the Fermi level. Experiment[20,21] gives values of 5.7 and 6.3. The crudely estimated width is 3 eV, much larger than the experimental width of 0.75 eV. Part or most of this discrepancy may be due to the roughness of the calculation of this quantity.

5. *Scattering amplitude for electrons.* LEED experiments on a hydrogen-covered W(100) surface show surprisingly strong additional half-order beams.[22] This raises the question as to what are the scattering amplitudes for electrons in the 50 to 100 eV range from chemisorbed hydrogen. This question can be answered once the screening electron density

$n_1(\mathbf{r})$ is known. It is found that, because the screening charge of chemisorbed hydrogen is more spread out than the electronic charge of atomic hydrogen, the forward scattering amplitude of chemisorbed hydrogen is about 50 percent larger, while for $\theta > 90$ degrees, there is no significant difference.[23] Results of this sort might be useful in future quantitative analyses of LEED data.

Actually, adsorbed hydrogen would be expected to be rather poorly amenable to a linear response treatment since the perturbing potential due to the proton is unscreened and rather strong. Since, even in this case, the method gave reasonably good results, it can be hoped that it will also be successful in calculations for chemisorbed alkali metals, aluminum, and other simple metal atoms which can be described by weak pseudopotentials.

b. Chemisorption by Direct Variation of $n_1(\mathbf{r})$

Although the linear response theory of chemisorption of hydrogen has yielded surprisingly satisfactory results, there certainly is a need for other calculation which do not assume that the perturbing potential due to the external potential is weak. One such calculation has recently been reported by Huntington, Turk and White.[24] These authors take as their zeroth order density distribution

$$n(\mathbf{r}) = n_o(x) + n_{at}(\mathbf{r} - \mathbf{r}_a) \tag{13}$$

where n_{at} is the electron density of the isolated atom. On the basis of heuristic considerations they then introduce an additional density, $\triangle n(\mathbf{r}; \gamma_1, \gamma_2 \ldots \gamma_5)$, depending on five variational parameters and minimize the energy, Eq. (1), with respect to these.

In the actual application to a sodium atom on sodium metal, the addition of $\triangle n$ had only a 10 percent effect on the binding energy, which was calculated to be 0.374 eV. This will be seen to be approximately twice the sodium simple energy per atom, $\sigma = 0.188$ eV, this factor corresponding to a simple bond-breaking picture. The equilibrium distance from the jellium edge is 1.66 Å, rather larger than half the interplanar spacing for the close-packed (110) planes of sodium, which is 1.52 Å.

A calculation of this sort, with $n(\mathbf{r})$ of the simple form [Eq. (12)], is extremely easy to carry out (compared, for example, with the linear response theory discussed in the previous subsection) and hence this method opens up the possibility for application to many other systems.

c. Submonolayers of Adsorbed Alkalis

The first chemisorption systems to be studied by the density functional theory were submonolayers of alkali ions on transition metals. The work function Φ of such systems first decreases as function of the coverge θ, but then increases again. This system was modeled by Lang[25] as follows: the substrate is represented by a positive jellium and the adsorbed alkali ions by a charged slab of suitably chosen fixed thickness d and variable charge density $\rho(\theta)$, such that for each θ the product $\rho(\theta) \times d$ represents the average charge of the adsorbed ions per unit area. The simplest satisfactory choice for d is the spacing between the most densely packed planes of the bulk alkali–metal. With this simple model, Lang could account well for the following features of the experimental results:

1. The existence of a work-function minimum (Fig. 7)
2. The minimum work function, and its variation along the alkali series
3. The initial decrease of work function with alkali coverage
4. The fact that at one monolayer ($\theta = 1$) the work function very nearly equals that of the bulk alkali–metal.

The fact that this model works well even at infinitesimal coverage can be regarded as support for the linear response theory discussed earlier. For in such a theory it does not matter for the work function if the alkali-ion charges are localized in the y-z plane or uniformly spread out.

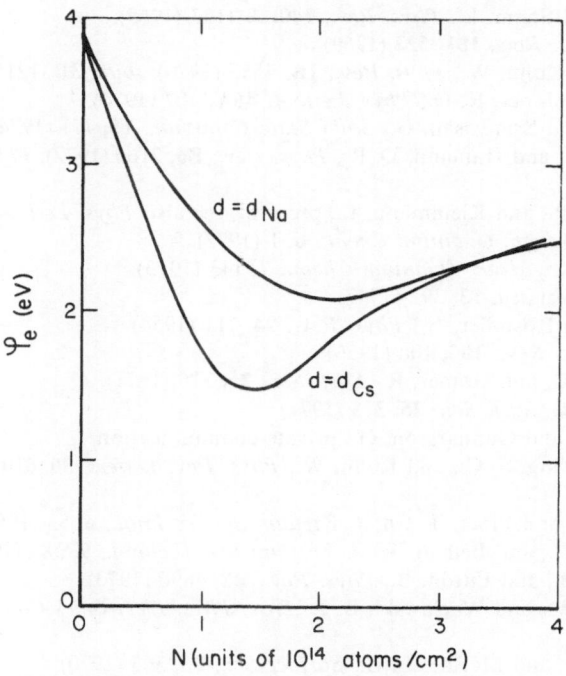

Fig. 7. Variation of the calculated work function ϕ_e of a metal, with $r_s = 2$, as function of the degree of coverage by sodium and cesium atoms.[24]

5 CONCLUDING REMARKS

The density functional formalism (which is formally exact) has been applied with rather crude approximations, especially for many-body effects, to simple metal surfaces both bare and with chemisorbed atoms. The results have been rather better than might have been expected. Among the tasks for the future would appear to be:

1. Application to other simple systems
2. Improvements of the underlying density functional F[n] for simple systems
3. Modification of the simple forms of the density functional for systems involving noble and transition metals
4. Application to systems involving chemisorbed molecules
5. Application of the Green's function method to bare surfaces and to surfaces with chemisorbed monolayers.

REFERENCES

1. Hohenberg, P., and Kohn, W., *Phys. Rev.,* **136,** B864 (1964).
2. Kohn, W., and Sham, L., *Phys. Rev.,* **140,** A 1133 (1965).
3. Smith, J., *Phys. Rev.,* **181,** 522 (1969).
4. Lang, N., and Kohn, W., *Phys. Rev.,* **1B,** 4555 (1970); *ibid.,* **3B,** 1215 (1971).
5. Harris, J., and Jones, R. O., *Phys. Lett. A,* **46A,** 407 (1974).
6. Jonson, M., and Srinivasan, G., *Solid State Commun.,* **15,** 771 (1974).
7. Appelbaum, J., and Hamann, D. R., *Phys. Rev.,* **B6,** 2166 (1972); *Phys. Rev. Letters,* **31,** 106 (1973).
8. Alldredge, G. P., and Kleinmann, L., preprint; see also *Phys. Lett. A,* **48A,** 337 (1974).
9. Slater, J. C., *Advan. Quantum Chem.,* **6,** 1 (1972).
10. Johnson, K. H., *Advan. Quantum Chem.,* **7,** 143 (1973).
11. Korringa, J., *Physica,* **13,** 392 (1947).
12. Kohn, W., and Rostoker, N., *Phys. Rev.,* **94,** 111 (1954).
13. Segal, B., *Phys. Rev.,* **105,** 108 (1956).
14. Schrieffer, J. R., and Gomer, R., *Surf. Sci.,* **25,** 315 (1971).
15. Schrieffer, J. R., *Surf. Sci.,* **25,** 315 (1971).
16. Lundquist, S., and Gunnarsson, O., private communication.
17. Smith, J. R., Ying, S. C., and Kohn, W., *Phys. Rev. Letters,* **30,** 610 (1973); *Phys. Rev.,* in press.
18. Propst, F. M., and Piper, T. Cn, *J. Vacuum Sci. Technol.,* **4,** 53 (1967).
19. Plummer, E. W., and Bell, A. E., *J. Vacuum Sci. Technol.,* **9,** 583 (1972).
20. Feuerbacher, B., and Fitton, B., *Phys. Rev.,* **B8,** 4890 (1973).
21. Plummer, E. W., and Waclawski, B. J., *Proc. Phys. Electronics Conference,* 1973, to be published.
22. Jennings, P. J., and McRae, E. G., *Surface Sci.,* **23,** 363 (1970).
24. Ying, S. C., Smith, J. R., and Kohn, W., *Phys. Rev.,* **B11,** 1483 (1975).
23. Huntington, H. B., Turk, L. A., and White, W. W., III, to be published.
25. Lang, N. D., *Phys. Rev.,* **B4,** 4234 (1971).

DISCUSSION on Paper by W. Kohn

GRIMLEY: I do not remember exactly the r_s values, but would not your method give much the same results for hydrogen on aluminum as those you have shown us for hydrogen or tungsten?

KOHN: We used the value $r_s = 1.5$ to represent tungsten, on the basis of earlier work by J. Smith on other surface properties of tungsten. The r_s value of aluminum is about 2.0. This means the electron density is less than one-half that of tungsten. This might make a considerable difference. For example, cleavage energies of metals depend very strongly on r_s and so I would not be surprised if the same were true for the binding of hydrogen.

THEORY OF CHEMISORPTION IN RELATION TO HETEROGENEOUS CATALYSIS*

J. W. Davenport, T. L. Einstein,
J. R. Schrieffer, and P. Soven

Department of Physics, University of Pennsylvania
Philadelphia, Pa.

ABSTRACT

The theory of chemisorption is discussed within the LCAO framework in terms of the local and nonlocal density of electronic states of the coupled adsorbate-substrate system. The concept of a surface complex is developed and related to states split off from the substrate bands. Explicit calculations of the spatially dependent change of the density of states on chemisorption illustrate the quasi-localized nature of the chemisorption bond. Surface densities of states for (100) fcc surfaces are shown to differ significantly from the bulk density of states. Finally, a simple model of a dissociation reaction is considered for a gas-phase complex and for the corresponding surface complex. The variation of the chemisorption energy and the activation energy with system parameters is discussed.

*This work was supported in part by the National Science Foundation and the Advanced Research Projects Agency.

1 INTRODUCTION

While a first-principles theory of heterogeneous catalysis appears to be some years away even for simple reactions, there has been considerable advance within the past few years in developing theoretical concepts and calculational techniques which hopefully will lay.the ground work for such a theory.[1-5] However, in treating these problems, difficulties arise from a number of sources. From the solid state physics point of view; (1) a clean surface has lower symmetry than the corresponding bulk material, with the situation being worse when chemisorbed species are present; (b) the most interesting catalysts are frequently complicated materials, e.g., transition metals or transition-metal oxides with strong participation of d orbitals in the chemisorption bond; (c) the scale of energies important in distinguishing competing reaction paths in a given system or the variation of activation energies in related systems is small compared to the traditional level of accuracy achieved in calculations even for bulk systems; the situation is aggravated by the increased importance of interelectronic correlation effects near the surface resulting from the lower coordination number of surface atoms and hence the narrowing of the atomic virtual levels; and (d) the geometrical arrangement of important intermediate species and/or the dominant reaction pathways are rarely known experimentally.

From the point of view of quantum chemistry, similar difficulties arise, with the spacially extended nature of the catalyst being an added complication.

One simplifying feature which occurs is the rather localized nature in space of the chemisorption bond, in that the wave functions of the adsorbate and substrate overlap significantly only over the first coordination shell of atoms surrounding the nominal bonding region. Further, the major change in the potential acting on the electrons is also primarily localized to this region. Thus, the perturbing Hamiltonian which describes the coupling of the adsorbate and substrate is spatially localized, although the effects on the system wave function of this localized perturbation extend beyond this region. As we will discuss, Green function methods are quite useful in handling such problems.

In quantum chemistry, one is familiar with the LCAO-MO description of the electronic structure of molecules.[6] This scheme has the advantage of incorporating the correct symmetry of the one-electron wave functions and allows one to visualize these functions as being made up as linear combinations of orbitals localized on the atomic sites. Phenomenological matrix elements of an effective Hamiltonian are assumed in this atomic-like basis, as in the extended Hückel and related schemes, and reasonable results for many properties, such as optical spectra, are obtained. Results for other properties such as potential energy surfaces are frequently unreliable.

Similarly, in solid state physics, the LCAO scheme is useful in describing the energy bands of bulk solids[7] for the same reasons as in quantum chemistry. Namely, the correct wave-function symmetry is incorporated in this scheme, and by fitting a few phenomenological matrix elements in this atomic-like basis, a reasonable fit is obtained to the energy bands calculated, for example, in a self-consistent scattered-wave scheme, which is much more time consuming than the corresponding LCAO analysis.

Fortunately, much of the recent work on chemisorption can be interpreted in terms of this LCAO-MO point of view.

2 THE NEWNS-ANDERSON MODEL

The simplest model of an atom binding to a metal surface is provided by the Newns-Anderson Hamiltonian[2,8], in which electrons hop between metal orbitals, ψ_k, and the adsorbate valence orbital(s), φ_a, via a matrix element, V_{ka}. In the simplest case, Coulomb interactions between electrons are not explicitly included, except between electrons on the adsorbate, in which case a single parameter U accounts for their (screened) Coulomb interaction. The Hamiltonian for the system is (for a single adsorbate orbital)

$$H = \sum_{ks} \epsilon_k n_{ks} + \sum_{ks} \left(V_{ka} c_{ks}^+ c_{as} + V_{ka}^* c_{as}^+ c_{ks} \right) + \sum_s \epsilon_a n_{as} + U n_{a\uparrow} n_{a\downarrow} \qquad (1)$$

The c^- and c create and destroy electrons in the states k or a of spin orientation $s(=\pm 1/2)$ and n is the occupation number operator. The parameter ϵ_k is the single-particle energy of the substrate orbital ψ_k in the absence of the adsorbate, while ϵ_a is the energy for one electron in this orbital φ_a for the isolated adsorbate. Traditionally, one makes the Hartree-Fock approximation for the U term, namely,

$$U n_{a\uparrow} n_{a\downarrow} \longrightarrow U \left[n_{a\uparrow} \bar{n}_{a\downarrow} + n_{a\downarrow} \bar{n}_{a\uparrow} - \bar{n}_{a\uparrow} \bar{n}_{a\downarrow} \right] \qquad (2)$$

where the average $\bar{n}_{a\sigma}$ is calculated self-consistently. In the unrestricted Hartree-Fock approximation, $\bar{n}_{a\uparrow} \neq \bar{n}_{a\downarrow}$ in general and this lack of equality allows the system to correctly approach the energy of the isolated substrate and adsorbate as their separation $R_a \rightarrow \infty$.

As discussed above, ψ_k and φ_a overlap only in a limited region of space surrounding the adsorption site. To exploit this fact, it is useful to represent ψ_k as

$$\psi_k = \sum_n a_{kn} \varphi_n \qquad (3)$$

where φ_n is an atomic-like (or Wannier) function localized on substrate atom n. More generally, n represents both site and orbital labels, e.g., a $3d_{xy}$

function on a given site. For simplicity, we assume that φ_a is orthogonal to all the substrate orbitals φ_n and that $V_{na} = 0$ for all n except for orbitals on atoms in contact with the adsorbate. For an s-band substrate, with a hydrogen atom adsorbed directly on top of a substrate atom (labelled 1), V_{ka} is

$$V_{ka} = a_{k1} <1|V|a> \tag{4}$$

If the substrate is Ni(100), for adsorption on top of a nickel atom,

$$V_{ka} = a_{k,1,d\sigma} <1,d_\sigma|V|a> + a_{k,1,4s} <1,4s|V|a> \tag{5}$$

where $d\sigma$ is the $3d_{z^2}$ orbital (z is normal to the surface). Thus, if the k dependence of the a_k's is known, the k dependence of V_{ka} is also known.

It is straightforward to show[2,9] that the chemisorption energy is given for the matrix element [Eq. (4)] by

$$\Delta W = -\frac{1}{\pi} \sum_S \int_{-\infty}^{E_F} \mathrm{Im}\, \ln\left(1 - \frac{V^2 G_{11}(E)}{\epsilon - E_{as} - i\delta}\right) dE - U\bar{n}_{a\downarrow}\, \bar{n}_{a\uparrow} \quad (\delta = 0^+) \tag{6}$$

where $G_{11}(E)$ is the Green function for atom 1 for the clean surface, E_F is the Fermi energy, and $<1|V|a> \equiv V$. E_{as} is the shifted atomic level

$$E_{as} = \epsilon_a + U\bar{n}_{a,-s} \tag{7}$$

and

$$\bar{n}_{as} = \frac{1}{\pi} \int_{-\infty}^{E_F} \mathrm{Im}\, \frac{1}{E - E_{a\sigma} - V^2 G_{11}(E)}\, dE \tag{8}$$

Equations (7) and (8) can be solved self-consistently for the quantity $\bar{n}_{as}$ which allows ΔW to be explicitly determined from Eq. (6).

It is important to note that the only properties of the substrate which enter ΔW are the matrix element V and the surface-atom Green function $G_{11}(E)$. However, the fundamental quantity of physical importance is not $G_{11}(E)$ but rather the surface density of states $\rho_{11}(E)$, defined by

$$\rho_{11}(E) = \sum_k |a_{k1}|^2\, \delta(E - \epsilon_k) \tag{9a}$$

and more generally

$$\rho_{nn'} = \sum a_{kn}\, a^*_{kn'}\, \delta(E - \epsilon_k) \tag{9b}$$

The corresponding bulk density of states is given by replacing the $|a_{k1}|^2$ by a constant, since by Bloch's theorem the a's for the bulk are of the form $e^{ik\cdot R_n}$.

Generally, G_{nn} and $\rho_{nn'}$ are related by

$$G_{nn'}(E) = \int_{-\infty}^{\infty} \frac{\rho_{nn'}(E')}{E-E'-i\delta} \, dE' \qquad (10)$$

Hence, if one knows $\rho_{11}(E)$ for the clean surface, one can, within this model, determine the binding energy ΔW for given values of ϵ_a, U, and V.

Kalkstein and Soven[10] calculated $\rho_{11}(E)$ for the (100) face of a simple cubic crystal (cubium) with a single orbital (s band) per site. Only nearest-neighbor hopping T was included. Their results are shown in Fig. 1 for an atom on the surface plane and on the next-to-surface plane, as well as for a bulk atom. Units are chosen so that $2T = 1$, leading to a band width of 6 in these units. A narrowing of the density of states occurs at the surface, as is consistent with the second-moment sum rule

$$\int_{-\infty}^{\infty} \rho_{nn}(E) \, E^2 dE = Z_n \, T^2 \qquad (11)$$

where Z_n is the coordination number of atom n and the center of the band is taken as the origin of energy. These workers showed that surface states can split off from the band if the potential on surface atoms differs from that on subsurface atoms by a significant amount.

The binding energy arising from $\rho_{11}(E)$ shown in Fig. 1 was calculated by Einstein and Schrieffer[9] as a function of band filling (E_F) for a range of parameters V/T and E_a. Their results for the reasonable value $V/T = 3$ and with $E_a = -0.6T$ are shown in Fig. 2 as the curve marked A. A similar curve (A) for the rather large ratio $V/T = 5$ is shown in Fig. 3. Self-consistency was not explicitly included in these calculations. As one might expect, the binding is strongest when E_a falls at approximately E_F so that strong forward and back bonding can occur, i.e., electrons below E_F in the metal hop virtually to the unfilled portion of the φ_a level and vice versa. For a half-filled symmetric band with E_a at the center of the band, no net charge transfer occurs.

It is interesting to note that while the density of states for the coupled system is continuous for $V/T = 3$, a split-off state occurs for $V/T = 5$, cor-

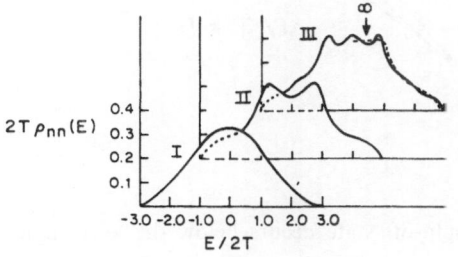

Fig. 1. Local density of states for an atom on the surface plane and succeeding planes into the bulk for the (100) face of an s-band simple cubic solid.

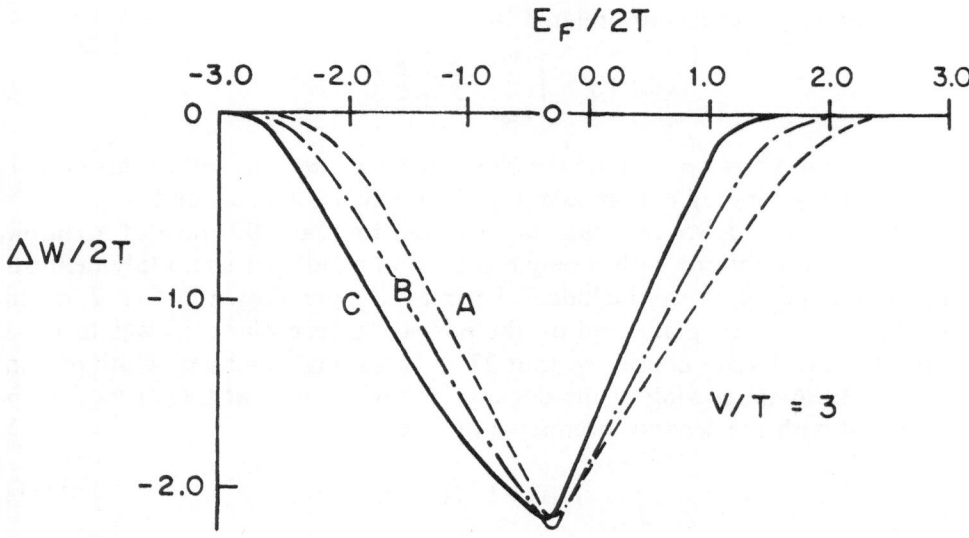

Fig. 2. The chemisorption energy for an atom adsorbed (A) atop a single substrate atom, (B) in the bridge position, and (C) at the fourfold coordinated centered site on the (100) face of s-band cubium, plotted as a function of the Fermi energy. All energies are expressed in units 2T, where the bulk band width is 12T. The interaction strength is taken $V/T = 3$ and $E_a = -0.6T$.

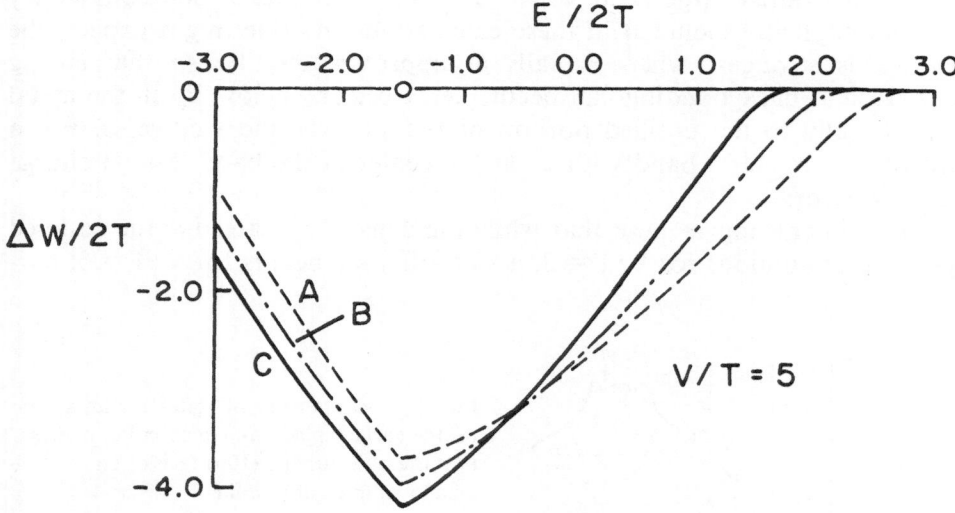

Fig. 3. Same as in Fig. 2, except $V/T = 5$. A split-off state occurs below the lower band edge for all three binding sites.

responding to the bonding molecular orbital of a surface complex formed from atoms a and 1. These local densities of states are discussed below.

What happens if one moves the adsorbate to another location on the surface? One finds that the above expressions continue to hold if $V^2 G_{11}(E)$ is replaced by

$$V^2 G \ (E) \longrightarrow \Sigma_a(E) = \sum_{nn'} <a|V|n> G_{nn'}(E) <n'|V|a> \qquad (12)$$

This general expression simplifies at sites of high symmetry. For example, if the adsorbate sits in the bridge position (B) halfway between two neighboring surface atoms labelled 1 and 2, and φ_a is an s-like orbital (even on reflection), by symmetry one has

$$<1|V|a> = <2|V|a> \equiv V_B/\sqrt{2} \qquad (13)$$

Thus, the adsorbate "self-energy" function for the bridge site is

$$\Sigma_{aB}(E) = V_B^2 \, G_B(E) \qquad (14a)$$

where

$$G_B(E) = 1/2 \left[G_{11}(E) + 2G_{12}(E) + G_{22}(E) \right] \qquad (14b)$$

The form of Eq. (14b) suggests that one introduce the concept of a "group orbital", which in this case is

$$\psi_B = \frac{\varphi_1 + \varphi_2}{\sqrt{2}} \qquad (15)$$

From (Eq. 13) we see that φ_a couples only to ψ_B rather than to $(\varphi_1 - \varphi_2)/\sqrt{2}$. Thus, the relevant quantity for bridge bonding is the density of states for the "bonding" bridge orbital [Eq. (15)], which is shown as curve B in Fig. 4. Finally, for binding in the four-fold coordinated (or centered) site C, the group orbital bonding to an s-like adsorbate orbital is constructed from the four atoms in the surface plane surrounding φ_a as

$$\psi_C = \frac{1}{\sqrt{4}} (\varphi_1 + \varphi_2 + \varphi_3 + \varphi_4) \qquad (16)$$

Denoting $<n|V|a> \equiv \dfrac{V_C}{\ }$ (n = 1, 2, 3, 4), one has

$$\Sigma_{aC}(E) = V_C^2 \, G_C(E) \qquad (17)$$

where $\rho_C(E)$ is also shown in Fig. 4. While the areas under the ρ curves are constrained to be unity by normalization of the group orbital, the peak energy shifts to lower values of E as one involves more substrate atoms in

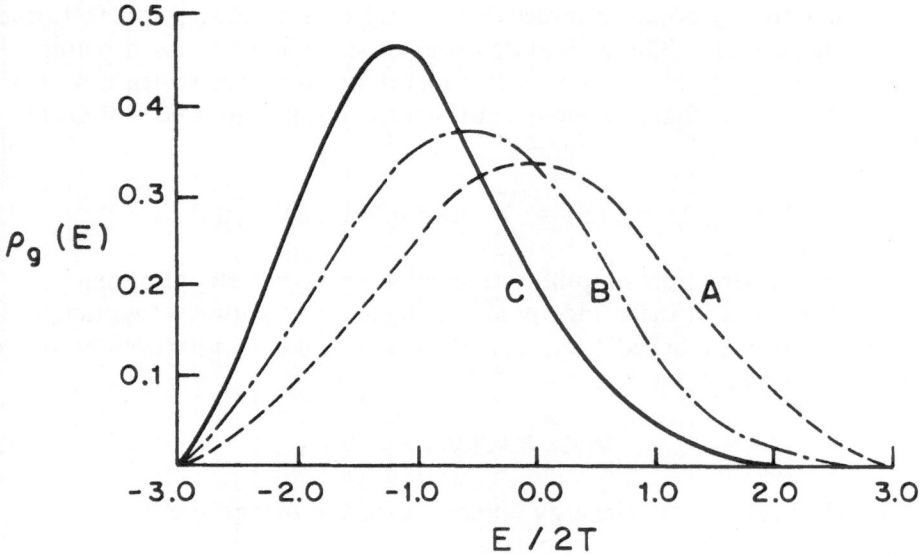

Fig. 4. The group-orbital density of states corresponding to a σ-like bond (A) atop, (B) at the bridge, and (C) at the centered sites for (100) s-band cubium.

this σ bond. This is reasonable, since only metal states ψ_k having essentially the same phase on all atoms entering the group orbital can contribute significantly to the bonding. However, these ψ_k's tend to have low energy because their phase varies slowly in space.

Alternatively, for a π-bonded adsorbate, the group orbital density of states would be skewed toward higher energy for B and C sites since φ_a would couple, for example, to $(\varphi_1 - \varphi_2)/\sqrt{2}$ in the bridge site, and rapidly oscillating ψ_k's would enter with greater weight.

The binding energy for the B and C sites is shown in Figs. 2 and 3, where V refers to V_B and V_C, respectively, in these cases.

If the matrix element V coupling the adsorbate orbital φ_a and substrate group orbital ψ_g is sufficiently strong, it is clear that a good starting point is a surface complex formed by occupying the molecular orbitals based on ψ_g and φ_a. By splitting the group orbital off from the remaining states of the solid, one strongly perturbs the electronic structure of the "indented solid", much as if a vacancy were located in the surface plane. The electronic states of the surface complex then rebond to the indented solid to form the rebonded surface complex. In practice, this rebonding energy is generally significant, as is the energy required to split the group orbital off from the solid. Mathematically, the extreme surface-complex limit corresponds to states being split off from the continuum of band states.

3 THE PROJECTED DENSITY OF STATES
AND THE SURFACE COMPLEX

To gain further insight into the chemisorption bond, it is helpful to understand the change in total density of states of the system ($\Delta\rho$) upon chemisorption, as well as how this change is distributed in space. Such calculations have been carried out by Einstein.[11,12] Within the LCAO framework discussed above, one finds for φ_a bonded to a group orbital ψ_g

$$\Delta\rho(E) = \frac{1}{\pi} \operatorname{Im} \frac{\partial}{\partial E} \ln \left(1 - \Sigma_a(E)\, G_a(E)\right) \tag{18}$$

Here, $G_a(E) = (E - E_a - i\delta)^{-1}$ is the zero-order Green function for the adsorbate and, as in Eqs. (14a) and (17), the adsorbate self-energy is

$$\Sigma_a(E) = V_{ag}^2\, G_g(E) \tag{19}$$

For the special case of bonding atop a single surface atom for the s-band cubium (100) substrate, $\Sigma_a = V^2 G_{11}(E)$ and $\Delta\rho$ is plotted in Fig. 5 for several values of V/T, with E_a fixed just below the center of the band. For plotting convenience, the free adsorbate density of states $\delta(E - E_a)$ is suppressed. For weak V (= 0.5), the delta function at E_a broadens to a Lorentzian "virtual level", whose width is proportional to V^2. For intermediate values of V (= 1.5), the E_a resonance splits into a bonding and an antibonding resonance

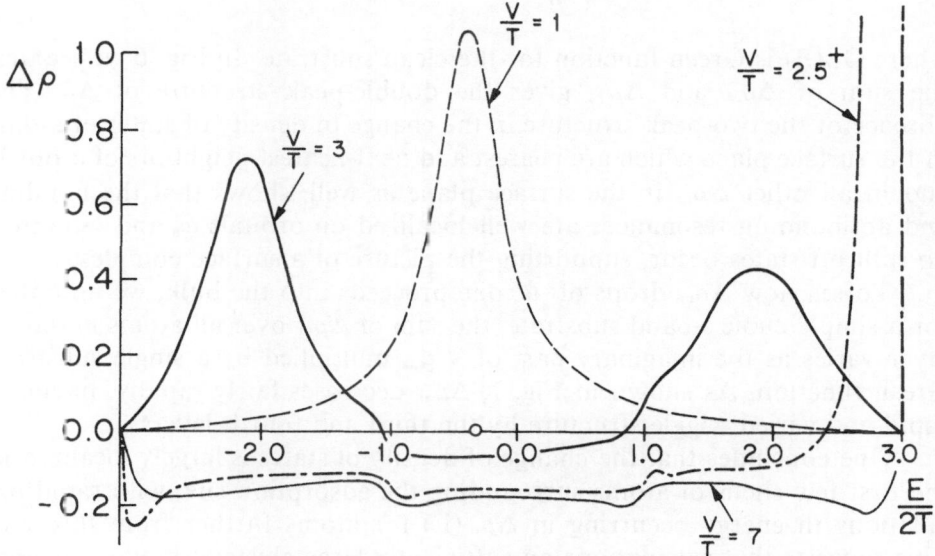

Fig. 5. Change of total density of states on adsorption of an atom on the (100) surface of s-band cubium, for several values of V. Split-off states are not shown. The small circle denotes E_a.

characteristic of the onset of surface-complex formation. For large V (= 2.5), a narrow antibonding resonance occurs at the upper band edge and a bonding state is barely split off from below the bottom of the band. For very large V, $\Delta\rho$ for E within the band ($-3 \leq E \leq 3$) is insensitive to V corresponding to a solid missing orbital φ_1 (i.e., a surface vacancy), with φ_1 and φ_a forming the bonding and antibonding states at energies $\simeq \mp V$. For systems of physical interest, V is typically between 1.5 and 2.5.

The total change in density of states $\Delta\rho$ can be decomposed into $\Delta\rho_{aa}$ and changes in the substrate local density of states $\Delta\rho_{nn}$

$$\Delta\rho = \Delta\rho_{aa} + \sum_n \Delta\rho_{nn} \tag{20}$$

Fixing attention on V = 1.5 so that all of $\Delta\rho$ lies within the band, we ask how the $\Delta\rho_{nn}$'s drop off from the adsorption site. One finds

$$\Delta\rho_{aa}(E) = \frac{1}{\pi} \text{Im } \mathcal{G}_{aa}(E) - \delta(E - E_a) \tag{21}$$

where $\mathcal{G}_{aa}$ is the adsorbate Green function for the coupled system

$$\mathcal{G}_{aa}(E) = \left[E - E_a - V^2 G_{11}(E) \right]^{-1} \tag{22}$$

Also, one has

$$\Delta\rho_{nn} = \frac{1}{\pi} \text{Im } V^2 \mathcal{G}_{aa}(E) \, G_{1n}^2(E) \tag{23}$$

where $G_{1n}(E)$ is Green function for the clean substrate. In Fig. 6 we see that the sum of $\Delta\rho_{aa}$ and $\Delta\rho_{11}$ gives the double-peak structure of $\Delta\rho$. The absence of the two-peak structure in the change of density of states of atoms in the surface plane which are nearest and next-nearest neighbors of atom 1, and in all other $\Delta\rho_{nn}$ in the surface plane as well, shows that the bonding and antibonding resonances are well localized on orbitals φ_a and φ_1 even if no split-off states occur, supporting the picture of a surface complex.

To see how $\Delta\rho_{nn}$ drops off as one proceeds into the bulk, we note that for a simple cubic s-band substrate, the sum of $\Delta\rho_{nn}$ over all atoms in the nth layer varies as the imaginary part of $V^2\mathcal{G}_{aa}$ multiplied by a single substrate Green function. As shown in Fig. 7, $\Delta\rho_{nn}$ decreases fairly rapidly, having a rapid and varied wiggle structure by the third and fourth layers.

One concludes that the change of density of states is largely localized in the first few shells of atoms surrounding the adsorption site, with rapid oscillations in energy occurring in $\Delta\rho_{nn}(E)$ for atoms further from this site. This suggests that calculations on sufficiently large clusters should converge well to the solid substrate limit. Also, it suggests that chemisorption on thin films and bulk solids should be closely related for the same single-crystal

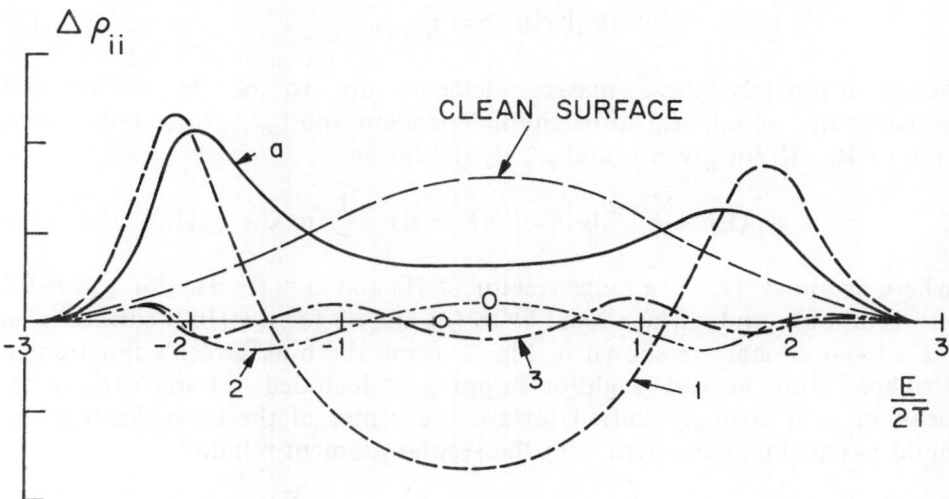

Fig. 6. The change of local density of states for the (1) adsorbate, the substrate atom 1, to which a is coupled, and for the (2) nearest neighbor and (3) next nearest neighbor of atom 1 in the surface plane. Also shown is the density of states for the clean surface.

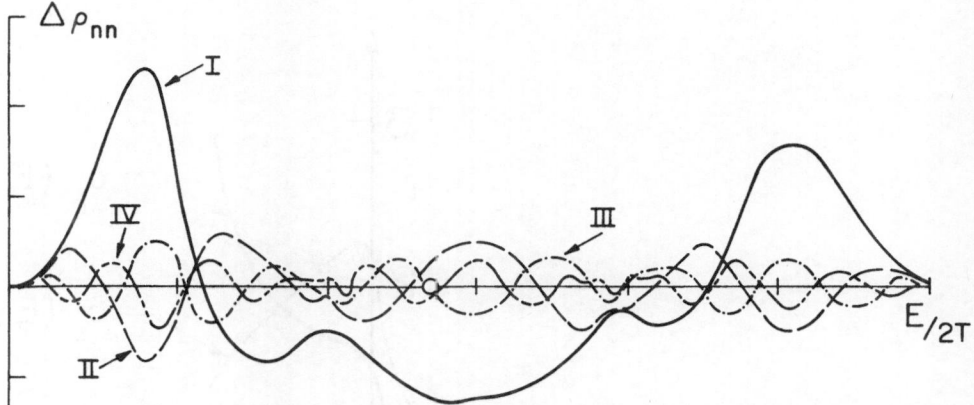

Fig. 7. Change of total density of states on the first (I), second (II), third (III), and fourth (IV) planes.

surface. It may well be, however, that surface defect concentrations are widely different for films and bulk specimens and apparent differences in chemisorption properties may occur for this reason.

Turning to more realistic substrates, Davenport[12,13] has generalized the results for s-band cubium to an fcc d-band solid. For the clean substrate, consider localized basis functions $\varphi_{i\mu}$ of symmetry type μ located on site i. The matrix elements of H_0 in this site representation are

$$<i\mu|H_0|i\mu> = \epsilon_{i\mu} \tag{24}$$

$$\langle j\mu'|H_0|i\mu\rangle = t_{j\mu', i\mu} \tag{25}$$

While ultimately these matrix elements are to be determined self-consistently, we take $\epsilon_{i\mu}$ to be site independent and $t_{j\mu', i\mu}$ to be only a function of $\mathbf{R}_i - \mathbf{R}_j$ for given μ and μ'. By definition,

$$\rho_{i\mu}(E) = \sum_k |\langle i\mu|k\rangle|^2 \delta(E - \epsilon_k) = \frac{1}{\pi} \text{Im } G_{i\mu, i\mu}(E) \tag{26}$$

where as above, $|k\rangle$ is an eigenvector of H_0 and $G = (E - H_0 - i\delta)^{-1}$, $(\delta = 0^+)$.

The bulk and surface local DOS (ρ_b and ρ_s) for the (100) surface of an fcc s-band crystal are shown in Fig. 8, using the bulk Green's functions of Frikkee. Only nearest-neighbor hopping is included. At the surface, the peak in ρ_b is strongly shifted toward the center of the band, leading to a band narrowing, consistent with the second-moment relation

$$M_{i\mu}(2) \equiv \int_{-\infty}^{\infty} (E - \epsilon_{i\mu})^2 \, \rho_{i\mu}(E) \, dE = \sum_{j\mu'} |t_{j\mu', i\mu}|^2 \tag{27}$$

since the number of nearest neighbors is reduced by a factor of 2/3 at the surface.

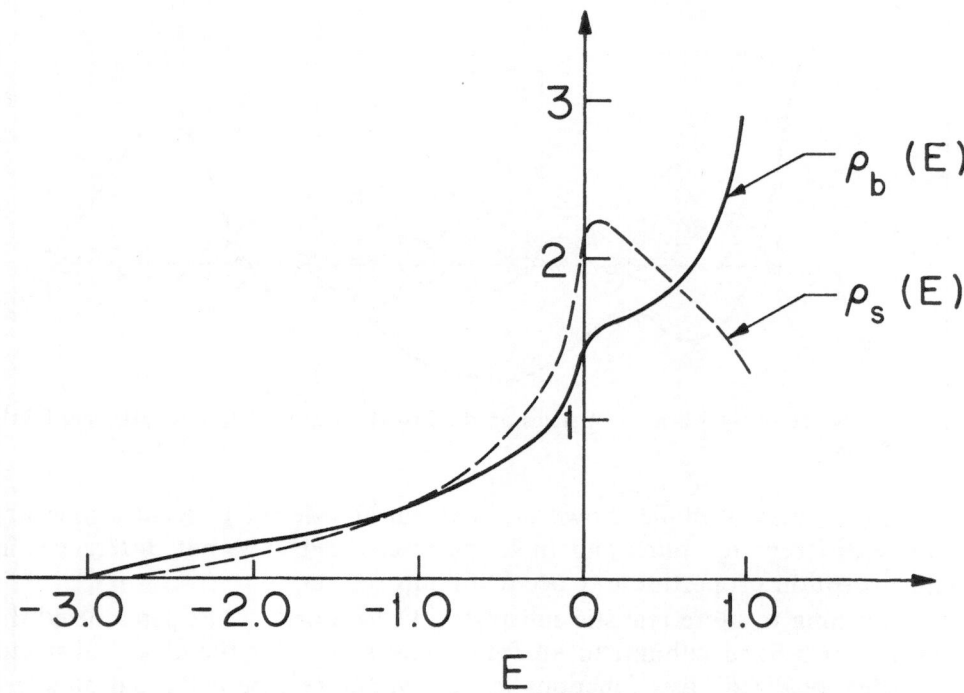

Fig. 8. The bulk and surface density, ρ_b and ρ_s, for the (100) surface of a face-centered cubic s-band solid. The bulk band width is chosen to be 4 units of energy, corresponding to a nearest-neighbor hopping $-1/4$.

For the fcc(100) d-band case, taking z as the positive normal to the surface, we consider the five d-functions having symmetry xy, yz, zx, x^2-y^2,
and $3z^2-r^2$ on each site. Considering again only nearest-neighbor hopping,
symmetry reduces $t_{j\mu',i\mu}$ to

$$t_{j\mu',i\mu} = \alpha t_\sigma + \beta t_\pi + \gamma t_\delta , \qquad (28)$$

where t_m corresponds to hopping between orbitals of magnetic quantum
number m = 0, 1, 2 along the symmetry axis joining i and j. α, β, and γ
have been tabulated by Slater and Koster. Estimates of t_m can be obtained
from an LCAO fit to the bulk band structure. For nickel one finds
$t_\sigma = -0.338$ eV, $t_\pi = +0.182$ eV, and $t_\delta = -0.026$ eV. Using these values,
one obtains the results shown in Fig. 9 for the total d-band density of states.
The large DOS near the top of the bulk band, arising mainly from the xy,
yz, and zx orbitals, is again shifted toward the band center. However, the
orbitally projected surface DOS, $\rho_{s\mu}$, shows that the main narrowing effects
occur for the "out-of-plane" functions, yz, zx, and $3z^2-r^2$, which stick out
from the surface, while the "in-plane" functions, xy and x^2-y^2, are much less
affected. These results are consistent with the second-moment relations
[Eq. (27)] since, for example, the yz function, lying mainly in the yz plane,
loses half of its important nearest neighbors at the surface, while the xy
function loses none. Note the large peak in ρ_s near the band center. No such
peak occurs for the bulk.

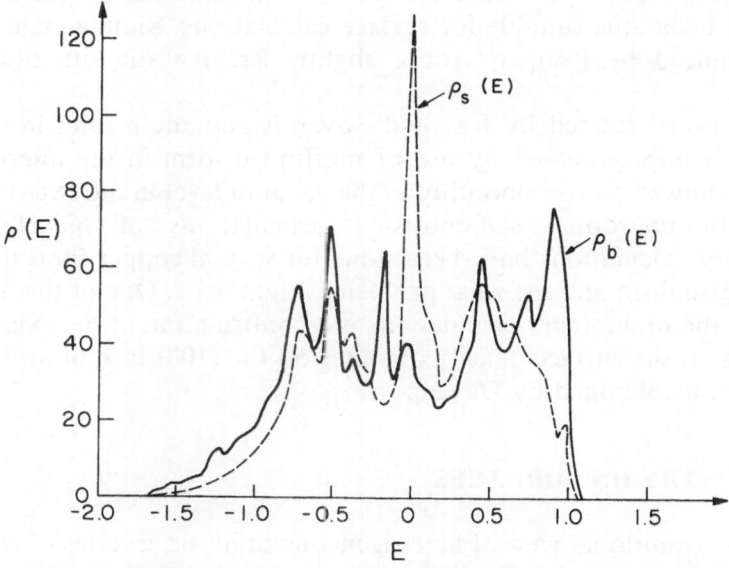

Fig. 9. The local density of states for a bulk atom and an atom of the (100) surface of a face-
centered cubic d-band solid. Parameters are chosen to fit the bulk bands of nickel. Self-
consistency affects these results, as discussed in text.

It is clear that self-consistency effects can be very important in determining $\rho_{s\mu}$. For nickel, if one uses the above $\rho_s(E)$ with the Fermi level fixed by the bulk, the bulk hole count of roughly 0.6 is reduced to approximately 0.2 at the surface since the out-of-plane orbitals yz and zx are nearly filled (rather than having 0.2 holes as in the bulk). In a self-consistent calculation, this would lead to an increase in $\epsilon_{i\mu}$ on the surface plane by several eV, thereby raising and distorting $\rho_{s\mu}(E)$ so that the total surface hole count would return to approximately its bulk value. An essential difference is that at the surface, the holes are no longer equally shared among the t_{2g} orbitals (xy, yz, and zx) but are concentrated somewhat in the xy orbital. Furthermore, the increase in $\epsilon_{s\mu}$ can lead to increased peaking of $\rho_{s,xy}(E)$ at the top of the band, and for sufficiently large $\epsilon_{s\mu}$, a surface state of xy symmetry will split off from the top of the d band. Although delta-function-like in this model, the actual split-off state would be broadened by hybridization with the s-p background. At present, it is unclear whether this split-off state occurs at self-consistency.

One of the difficulties associated with a calculation such as Davenport's is knowing exactly what values to choose for the hopping integral. While we can presumably obtain the "bulk" values of these integrals by fitting to, say, an APW calculation for the infinite crystal, there is little doubt that different values are appropriate for the surface layers of a semi-infinite crystal.

In order to circumvent this problem, as well as to aid in the solution of other surface-related problems, Kar and Soven[14] have developed a multiple-scattering technique suitable for surface calculations. Similiar schemes have been advanced by Cooper[15] for a slightly less realistic situation and by Kowalski[16].

The model treated by Kar and Soven is actually a film, in which the potential is approximated by use of muffin-tin form in the interior of the film and allowed to rise smoothly to the vacuum level in the exterior region. They plan on doing self-consistent calculations of nickel surfaces. Preliminary calculations have been done for several copper films in order to test the formalism and see what problems might arise. One of the interesting results of the preliminary calculations is a confirmation of the existence of a large peak in the surface density of states on the (100) face of an fcc transition metal, as obtained by Davenport.

4 CHEMISTRY ON SURFACES

In the traditional view of heterogeneous catalysis, a series of elementary reaction steps take place, with various complexes being chemisorbed as reaction intermediates. An understanding of the chemisorption energy of these intermediates and the activation energies for the reaction steps depends on an understanding of the interaction between adsorbates while on or near the surface. In general, each reaction step should be viewed as a concerted reaction in which adsorbate-adsorbate, adsorbate-substrate, and substrate-substrate bonds are formed and broken in a simultaneous continuous process. The bonds are not pairwise additive, i.e., "bond resonance" is important.

If the adsorbates are sufficiently widely spaced on the surface, the primary interaction between adsorbates is indirectly through the solid[9,17], by means of a modification of the substrate wave functions. In addition, long-range electric dipole forces are of importance if there is significant charge transfer in the chemisorption bonds. These long-range interactions are responsible for the complicated overlayer structures observed on single crystal surfaces. For closely spaced adsorbates, the direct and indirect interactions are both important and are not simply additive.

As in the above discussion, we consider the several adsorbate problems within an LCAO framework. The Hamiltonian [Eq. (1)] is generalized to

$$H = \sum_{ks} \epsilon_k n_{ks} + \sum_{ks\alpha} \left(V_{k\alpha} c_{ks}^+ c_{\alpha s} + V_{k\alpha}^* c_{\alpha s}^+ c_{ks} \right)$$
$$+ \sum_{\alpha\beta s} t_{\alpha\beta} c_{\alpha s}^+ c_{\beta s} + \sum_{\alpha s} E_\alpha n_{\alpha s} + E_{CT} \qquad (29)$$

where α and β label the adsorbates and, as in Eq. (7), E_α is to be determined self-consistently. E_{CT} is the usual counterterm occurring in the Hartree-Fock scheme, as in Eq. (6).

Some of the important questions we wish to address are: (a) What is the relation between gas-phase reactions involving a complex reaction versus similar reactions for the corresponding surface complex? (b) What is the relation, if any, between the chemisorption energies and activation energies for simple surface reactions? (c) What are the qualitative features (electronic and geometric) which influence the potential energy function for surface reactions?

To fix ideas, consider a symmetric diatomic molecule (a–b) like H_2, interacting with a surface at a bridge site, as in Fig. 10, where z is normal to the surface. For simplicity, consider only an s-like orbital on each atom; p, d ... orbitals can be handled in a similar manner. By symmetry, the bonding (+) and antibonding(−) molecular orbitals

$$\varphi_\pm = \frac{\varphi_a \pm \varphi_b}{\sqrt{2}}$$
$$(30)$$

couple only to the corresponding group orbitals

$$\psi_\pm = \frac{\varphi_1 \pm \varphi_2}{\sqrt{2}}$$
$$(31)$$

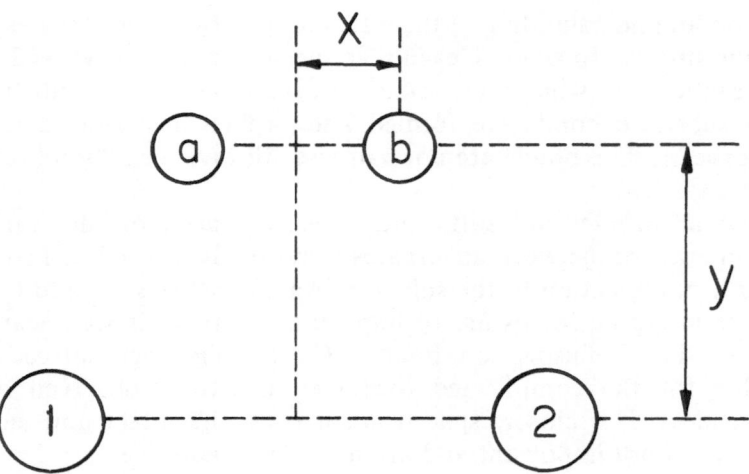

Fig. 10. Geometry for the bridge site (1–2) dissociation reaction of a diatomic molecule a–b.

Since by symmetry the + and − states are not mixed by hydrogen, the diatomic problem breaks into two separate problems, each of which appears to be chemisorption involving a single adsorbate orbital, as in Section 2. Thus, the chemisorption energy is in analogy to Eq. (6)

$$\Delta W = -\frac{1}{\pi} \sum_{g,s} \int_{-\infty}^{E_F} Im\ ln\ \left(1 - \frac{V_g^2 G_{gg}(E)}{E - E_{ag} - i\delta}\right) dE \qquad (32)$$

where $g = \pm$. The matrix elements are given by

$$V_{\pm} = <1|V|a> = <2|V|a> \qquad (33)$$

and

$$E_{a\pm} = E_a \mp t \qquad (34)$$

where t (>0) is the hopping interaction for the molecule.

It is instructive to first consider the limit of an isolated complex formed by the four atoms 1, 2, a, and b. For the group orbital DOS $\rho_{\pm}(E) = \delta(E \mp T)$, $E_a = 0$, the energy for four electrons occupying the levels is the lesser of

$$W = \begin{cases} -2\,(T + t) \\ \\ -\sqrt{(T-t)^2 + 4V_+^2} - \sqrt{(T-t)^2 + 4V_-^2} \end{cases} \qquad (35)$$

For T $=0$ and $V_\pm = V$, Eq. (35) reduces to $W = -2\sqrt{t^2 + 4V^2}$, which shows that the t and V bonds compete. If one thinks of t and V being functions of x and y, once can see how activated dissociation occurs if the a—b bond must be greatly stretched to form the 1—a bond, i.e., when t decreases significantly in magnitude before V has begun to increase significantly. Alternatively, a long-range V can lead to a small (or negative) activation energy for dissociation.

When the complex is on the surface (1 and 2 now being substrate atoms), several changes occur. Since the complex is in contact with the substrate, a nonintegral number of electrons will, in general, occupy the surface complex as a whole, and the $+$ and $-$ state in particular. Also, the sharp group-orbital levels shift and broaden into resonances, as in Fig. 4. These effects generally weaken the chemisorption bonds relative to those in the gas-phase complex. As for the gas phase, by stretching the a—b bond, the (unoccupied) antibonding level E_{a-} drops in energy and interacts more effectively with the antibonding DOS ρ_{--}. The analogous effect occurs for the bonding level E_{a+} in regard to ρ_{++}. As $t \to 0$, ΔW reduces to the chemisorption energy of a and b plus the indirect interaction between these adsorbates.

While numerical calculations must be carried out for a specific system, a general understanding follows by studying the interaction of a single adsorbate orbital φ_g of energy ϵ_g with a substrate group orbital ψ_g whose DOS is a Lorentzian of width Γ_g centered at E_g,

$$\rho_{gg}(E) = \frac{\Gamma_g/\pi}{(E - E_g)^2 + \Gamma_g^2} \tag{36}$$

Energies from different symmetry orbitals can then be added. For this Lorentzian model, the chemisorption energy is

$$\Delta W = 2\sum_\sigma \left[E_\sigma \left(\frac{1}{2} - \frac{1}{\pi}\tan^{-1}\frac{E_\sigma}{\Gamma_g} \right) + \frac{\Gamma_g}{2\pi}\ln\left(E_\sigma^2 + \Gamma_g^2 \right) \right] \tag{37a}$$

$$- 2\left[E_g\left(\frac{1}{2} - \frac{1}{\pi}\tan^{-1}\frac{E_g}{\Gamma_g} \right) + \frac{\Gamma_g}{2\pi}\ln\left(E_g^2 + \Gamma_g^2 \right) \right] - \epsilon_g\left(1 - \frac{\epsilon_g}{|\epsilon_g|} \right)$$

where $\sigma = \pm 1$ and

$$E\sigma = \epsilon_g + E_g + \sigma\mathrm{Re}\left[(\epsilon_g - E_g - i\Gamma_g)^2 + V_g^2 \right]^{1/2} \tag{37b}$$

$$\Gamma_\sigma = \Gamma_g + \sigma\mathrm{Im}\left[(\epsilon_g - E_g - i\Gamma_g)^2 + V_g^2 \right]^{1/2} \tag{37c}$$

The Fermi energy is chosen to be the origin of energy in these expressions. In Fig. 11, ΔW is plotted for $E_g = 0$ (half-filled surface DOS) as a function of V_g for several values of ϵ_g, illustrating that $\Delta W \propto V_g^2$ for small V_g (the perturbation result) and $\propto -2V_g + E_{in}$ for large V_g (surface-complex limit), where E_{in} is the energy required to split the group orbital off from the solid, leaving the indented solid behind. ΔW decreases slowly as $|\epsilon_g|$ increases from zero and finally varies as $1/|\epsilon_g|$ for $|\epsilon_g|/\Gamma_g \gg 1$, (perturbation regime).

By combining Eq. (37) with Eqs. (32) through (34) we obtain the energy of the adsorbed molecule. For simplicity we choose $E_g = 0$ (half-filled surface bands) $\Gamma_{\pm} = \Gamma$, $V_{\pm} = V$, $E_a = 0$ and plot in Fig. 12c the system energy as a function of a reaction coordinate $0 \leq Q \leq 1$. Three cases are shown, having the same limiting energies (for $Q = 0$ and $Q = 1$), corresponding to exothermic chemisorption. The parameterizations are:

$$
\begin{array}{ccc}
 & V/\Gamma & t/\Gamma \\[2mm]
\text{I} & \dfrac{1.5}{e^2}\,(e^{2Q}-1) & \dfrac{1.5}{e^2}\,(e^{2(1-Q)}-1) \\[3mm]
\text{II} & 1.5\,Q & 1.5\,(1-Q) \\[2mm]
\text{III} & 1.5\,\sqrt{Q} & 1.5\,\sqrt{1-Q'}
\end{array}
\tag{38}
$$

In case I, a large activation energy occurs, corresponding to a nearly broken a—b bond occurring before the adsorbate-substrate bonds form, because of

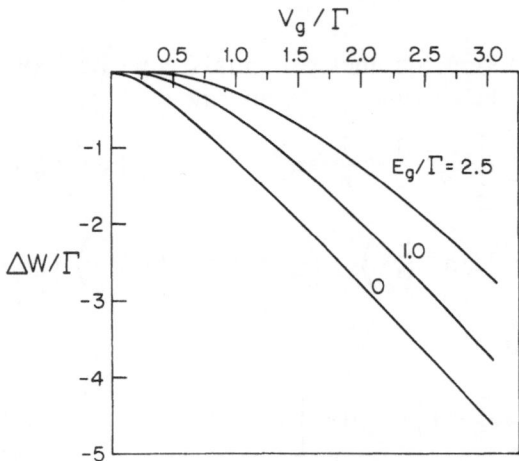

Fig. 11. The bond energy for an orbital φ_g, interacting via the matrix element V_g with a half-filled group orbital ψ_g, whose density of states is a Lorentzian centered at $E_g = 0$. Curves for several values of the unperturbed adsorbate level ϵ_g are shown.

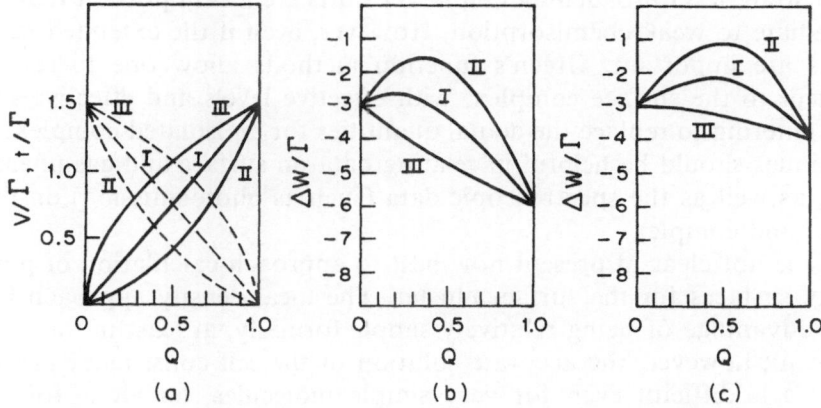

Fig. 12. (a) Three examples of the variation of t and V for the dissociation reaction coordinate (b) system energy for the gas-phase four-atom complex as a function of the reactions coordinates for the three cases shown in Fig. 12a; (c) system energy for the dissociation reaction at the bridge site of the solid surface, corresponding to the free complex reaction of Fig. 12b.

the rapid fall of t before V increases significantly. In case II, V + t = const, and a small barrier occurs. For case III, the rapid increase of V as t decreases removes the activation barrier.

The corresponding curves are plotted for the gas-phase complex in Fig. 12b. The isolated substrate atoms having lower coordination numbers are more reactive than for the surface problem. Hence the binding energy is greater for the gas-phase complex for all values of Q, thereby reducing the activation barrier for the reaction even more than for the surface reaction. Note, however, that the activation energy is reduced by a small fraction of the increase of the final-state binding energy. Similar effects should occur in more realistic models, with geometrical and electronic DOS effects both playing an important role.

5 CONCLUSION

Many of the concepts of importance in understanding gas-phase reactions can be extended to surface reactions. The influence of orbital symmetry, orbital directionality, energy-level matching, etc., all enter through the appropriately defined local density of states on the surface. Sharp levels of a metallic cluster broaden and shift as one approaches the limit of the surface of an extended solid. For strong chemisorption of the reactants in both activated and quasi-stable intermediate states, the effects of substrate-level broadening and -level shifts are less important than for intermediate to weak chemisorption. However, even if the extended substrate effects are important, Green's function methods allow one to reduce the problem to the surface complex, with effective levels and effective interactions entering to replace the actual quantities for an isolated complex. These techniques should be helpful in relating data on surface and gas-phase reactions, as well as the spectroscopic data (such as photoemission) on the free and bound complex.

It is not clear at present how best to approach calculation of potential energy surfaces for the surface cluster. The local-density approach has the great advantage of being relatively simple formally, at least in the so-called $\rho^{1/3}$ form; however, the accurate solution of the self-consistency problem it presents is difficult even for very simple molecules, let alone for surface complexes involving transition-metal atoms.[4,5] The so-called muffin-tin approximation for the self-consistent potential is generally inadequate to calculate reasonable energy surfaces in general, although recent progress on including nonmuffin-tin terms[18] looks promising for small complexes. Possibly these techniques can be combined with the Green function scheme for including the effects of the indented solid on the electronic structure of the surface complex to construct a feasible, accurate scheme.

REFERENCES

1. Grimley, T. B., *Proc. Internat. School of Physics, Enrico Fermi Course LVIII*, Academic Press, New York (June, 1973); *J. Vacuum Sci. Technol.*, **9** (2), (1972).
2. Newns, D. M., *Phys. Rev.*, **178**, 1123 (1969).
3. Schrieffer, J. R., *Proc. Internat. School of Physics, Enrico Fermi Course LVIII*, Academic Press, New York (June, 1973); *J. Vacuum Sci. Technol.*, **9** (2), 561 (1972).
4. Johnson, K. H., Norman, J. G., Jr., and Connolly, J.W.D., in *Computational Methods for Large Molecules and Localized States in Solids*, F. Herman, A. D. McLean, and R. K. Neshet (Eds.), Plenum, New York (1973), p. 161. See also Johnson, K. H., and Messmer, R. P., *J. Vacuum Sci. Technol.*, **11**, 236 (1974) and Messmer, R. P., Tucker, C. W., Jr., and Johnson, K. H., *Surface Sci.*, **42**, 341 (1974).
5. Smith, J. R., Ying, S. C., and Kohn, W., *Phys. Rev. Letters*, **30**, 610 (1973).
6. Parr, R. G., *The Quantum Theory of Molecular Electronic Structure*, W. A. Benjamin, Inc., New York (1964).
7. Mueller, F. M., *Phys. Rev.*, **153**, 659 (1967).
8. Anderson, P. W., *Phys. Rev.*, **124**, 41 (1961).
9. Einstein, T. L., and Schrieffer, J. R., *Phys. Rev.*, **B7**, 3629 (1973).
10. Kalkstein, D., and Soven, P., *Surface Sci.*, **26**, 85 (1971).
11. Einstein, T. L., *Proc. 34th Annual Conference on Physical Electronics*, February 25-27, 1974, Paper A-5; *Surface Sci.*, **45**, 713 (1974).
12. Davenport, J. W., Einstein, T. L., and Schrieffer, J. R., *Proc. Second International Congress on Surface Science*, Kyoto (1974), Paper 28aA3.
13. Davenport, J. W., *Proc. 34th Annual Conference on Physical Electronics*, February 25-27, 1974, Paper D-6.

14. Kar, N., and Soven, P., *Phys. Rev.,* **B11**, 3761 (1975).

15. Cooper, B. R., *Phys. Rev. Letters*, **30**, 1316 (1973).

16. Kasowski, R. V., *Phys. Rev. Letters*, **33**, 83 (1974).

17. Grimley, T. B., *Proc. Phys. Soc. (London)*, **90**, 751 (1967); Grimley, T. B., and Walker, S. M., *Surface Sci.*, **14** 395 (1969).

18. Connolley, J. W., and Danese, B., *J. Chem. Phys.*, **61**, 3063 (1974); Danese, J. B., *ibid.*, p. 3071.

DISCUSSION on Paper by J. R. Schrieffer

SCHWAB: Single particles do not come down on bare surfaces in a real catalytic situation. Is there any way of treating it theoretically?

In pactical catalysis a group of reactions is known to be kinetically of zero order, e.g., HCOOH decomposition over metals or NH_3 over tungsten. The interpretation is that the whole active surface is covered by the reactant, so that a coherent layer of reactant molecules is lying on top of the substrate lattice. Is there any way in the frame of existing theories to treat such a situation, and have any attempts been made to do so?

SCHRIEFFER: In principle, one can treat the fully covered surface in which the overlayer is commensurate with the surface lattice structure by the same techniques as for the electronic structure of the clean surface. The periodicity of the system can considerably simplify the problem, relative to the adsorption of a single adsorbate, although this is not always the case. I would say, however, that the weakly covered surface and the fully covered, commensurate overlayer, are the simplest limiting cases.

KOHN: Commenting further on Professor Schwab's question, I would like to say that I personally think that once we learn how to treat perfect systems efficiently (e.g., by the Green's function technique), a surface with a monolayer may be easier to calculate than a surface with a single adsorbate. The reason is, of course, that the monolayer system still has translational symmetry parallel to the surface, while the single adsorbate system has much lower symmetry.

DEROUANE: Your approach to bridge dissociation reactions is of particular interest to me. Indeed, we just completed*,** LCAO-MO *ab initio* computations on the linear system $[MgOHV_{Mg}^{II} HOMg]^{2+}$.

H_2 formation is simulated by moving the two hydrogens symmetrically towards the center of the MG^{2+} vacancy (V_{Mg}^{II}), while the

---

*Derouane, E. G., Fripiat, J. G., and André, J. M., *Chem. Phys. Letters*, in press.

**Derouane, E. G., Fripiat, J. G., and André, J. M., to be published.

dissociative chemisorption of hydrogen is studied by adding H_2 to the system $[MgOV^{II}MgOMg]^{2+}$ at the Mg^{2+} vacancy and lengthening the H-H bond.* Concerning most of the effects that your simple, and therefore very interesting, model predicts so nicely, i.e., levels crossing, electron transfer, localized perturbation effects, etc., we too found that there is some experimental evidence that supports the reality of these effects in the particular system that we investigated.

D. M. Newns*
A. Bagchi**

1 NATURE OF THE BOND IN CHEMISORPTION

The chairman opened the discussion with a comparison of the different assumptions involved in some chemisorption calculations, with special reference to dissociative hydrogen chemisorption on nickel.

1.1 Models of Chemisorption (Newns)

The density of states for the isolated nickel crystal (neglecting exchange splitting) and a hydrogen atom is shown in Fig. 1(a); on the left side is the density of states of nickel, consisting of the prominent narrow d band and less dense but broader sp band (shaded regions of bands are occupied). The

---

*Chairman.
**Secretary.

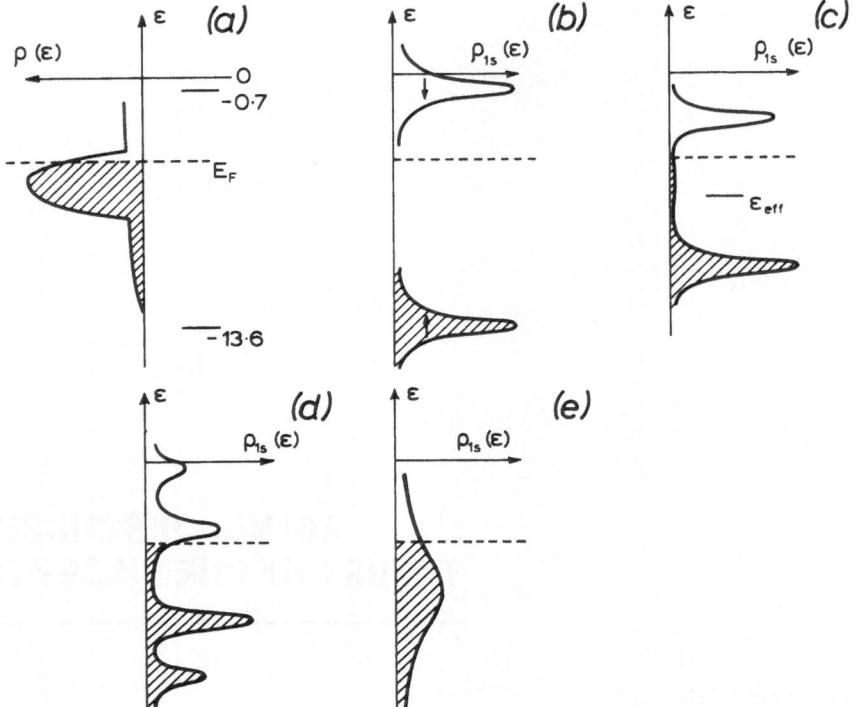

Fig. 1. Hydrogen on nickel. Schematic plots of density of states versus energy (vacuum level at $\epsilon = 0$). (a) At left, bulk density of states $\rho(\epsilon)$ of Ni; at right, ionization and affinity levels of hydrogen. (b) $\rho_{1s} (\epsilon)$ = local density of states projected into 1s orbital for weak adatom-metal coupling. (c) $\rho_{1s} (\epsilon)$ in HF theory, showing effective hydrogen level ϵ_{eff}; adatom-metal coupling chosen to fit chemisorption energy. (d) $\rho_{1s} (\epsilon)$ in the approximation of W. Brenig and K. Schonhammer (qualitative). (e) $\rho_{1s} (\epsilon)$ for fairly weak coupling to free-electron-like substrate (HF approx.).

work function ϕ varies with crystal face but a value of 4.5 eV is typical. On the right is shown the ionization level of hydrogen at -13.6 eV and the affinity level at -0.7 eV, taking vacuum as energy zero.

If the hydrogen atom interacts only weakly with the metal, then there will be some lifetime broadening of the ionization and affinity levels. This is shown in Fig. 1(b) in which the density of states $\rho_{1s}(\epsilon)$ *within the hydrogen 1s orbital* is plotted. These levels will still not be all one-electron levels, in that an up-spin electron in the ionization level allows only a down-spin electron to enter the affinity level. Although it is not convenient to draw this kind of energy-level diagram in the induced-covalent-bond theory[1] of chemisorption, the weak coupling form of this theory implies something like Fig. 1(b) and its time-reversed image tunneling into one another.

A picture which has been proposed by Newns[2], on the basis of self-consistent calculations in which the strength of metal-adatom coupling was obtained by a fit to the observed binding energy, is one which approximates to a "virtual molecule" situation. The effective hydrogen level is shifted, by electron-electron repulsion, from -13.6 eV up to a little below the Fermi level. Because of the narrowness of the nickel d band, this level may now combine with neighboring nickel d orbitals to form a molecule, the bonding and antibonding states of which are represented by the shaded and un-shaded peaks, respectively, in Fig. 1(c). Photoemission results for hydrogen on palladium[3], titanium[4], and nickel[5] all show a peak resembling the lower peak, but various interpretational difficulties occur in all three cases.

A model which includes electron correlation, but avoids the reformulation of the last-mentioned approach inherent in the induced-covalent-bond model, is that of Brenig and Schönhammer.[6] Some features of the bonding/antibonding structure of Fig. 1(c) are retained, but correlation structure reminiscent of Fig. 1(b) also appears, giving the four-peak $\rho_{1s}(\epsilon)$ shown qualitatively in Fig. 1(d). This is one interpretation of photoemission data for hydrogen on tungsten[7] which suggests two density-of-states peaks below the Fermi level. One notes that for tungsten, the d band is wider than that for nickel, and the bonding and antibonding states might appear within the d band.

Some CNDO calculations exist[8] in which not only the nickel d band but, more doubtfully, also the sp band are represented on a tight-binding model. These calculations reveal that sp hydrogen bonding predominates over d hydrogen bonding, contrary to what is assumed in models (c) and (d).

In contrast to this, the model of Smith, Ying, and Kohn[9] lumps the d and sp band together into a high-density, free-electron-type band. This is likely to be a poor approximation to the narrow d band of nickel, though perhaps slightly better for the somewhat wider d band of tungsten; only about one-quarter of the experimental binding energy has so far been obtained with this approximation. The kind of virtual level density of states found in this model is illustrated qualitatively in (e); that found in Ref. 9 is slightly narrower and further below E_F than illustrated.

Thus, quite a range of models of hydrogen chemisorption on transition metals exists, depending on the importance assigned to such factors as bonding to the sp versus d bands and to electron correlation. We might hope that experimental spectroscopy and also first-principle calculational techniques such as the XαSW cluster technique[10] will shortly narrow this range of possibilities.

Turning briefly to carbon monoxide adsorbed on nickel and some other metals, two interpretations of the photoemission spectra[11] are current. The similarities between the gas phase and adsorbed CO spectra suggest that the molecule is only weakly perturbed on adsorption. However, the observed spectra are also compatible with XαSW cluster calculations[12] in which the higher σ nonbonding and lower π bonding level of gaseous CO are inverted on chemisorption; the σ nonbonding level is essentially a carbon lone pair which is pushed down by forming a σ bond to the metal.

1.2 Discussion

It was pointed out by Emmett that the type of substrate orbital important for adsorption depends on the adsorbate; copper, for example, does not adsorb H_2 but does adsorb CO. The Chairman commented that ability to adsorb H_2 dissociatively depended critically on the H_2 dissociation energy being smaller than twice the adsorption energy for a hydrogen atom. It is slightly smaller for nickel which does adsorb H_2 but slightly larger for copper, though model (c) seems to work as well for both metals. Hence ability to adsorb H_2 may not be a direct criterion of bond type.

Concerning the relative importance of the sp and d bands, Schwab discussed a series of experiments[13] on alloys between pairs of transition metals, e.g., W-Ir or Mo-Pt. Paramagnetic susceptibility and the activation energy of formic acid dehydrogenation were measured on each alloy. A remarkable correlation of both magnitudes was found. Since the susceptibility is generally connected to the density of states near the Fermi energy, one can perhaps conclude that it is also the d states that are responsible for the catalytic action. The Chairman pointed out that modern theories of alloys, employing for example the CPA formulation[14], reject the concept of a common band in which the Fermi level shifts continuously as a function of concentration in, say, Eley's PdAu alloys[15] or Schwab's alloys. Rather, the modern picture has some features resembling a mixture of the palladium and gold components. The more sophisticated interpretation of the susceptibility variations, such as that of J. Kanamori[16], which becomes necessary makes the interrelation wilth the extremely interesting catalytic variations found by Schwab and others more difficult.

The usefulness of XαSW calculations in discussing the nature of the bonding was illustrated by Johnson, referring to a calculation[17] on a tetrahedral Pd_4 cluster with a hydrogen atom at the center. The energy-level diagram [Fig. 2 (a)] shows a strongly bonding level a_1 split off from the bottom of the band of d levels. The bonding level consists predominantly of the palladium 4d level and the 1s level of hydrogen, as shown by the contour map [Fig. 2(b)] of the orbital wave function. A weakly antibonding a_1 level can be seen to be split off from the top of the d band. It is rather remarkable that the bonding a_1 level lines up with the peak seen in photoemission by Eastman and Cashion[3] for hydrogen absorbed in palladium, even though the two systems are not identical.

The discussion on hydrogen-palladium bonding was continued by Ehrenreich giving an account of calculations[18] on hydrogen absorbed in palladium, which may possibly have a bearing on the adsorption question.

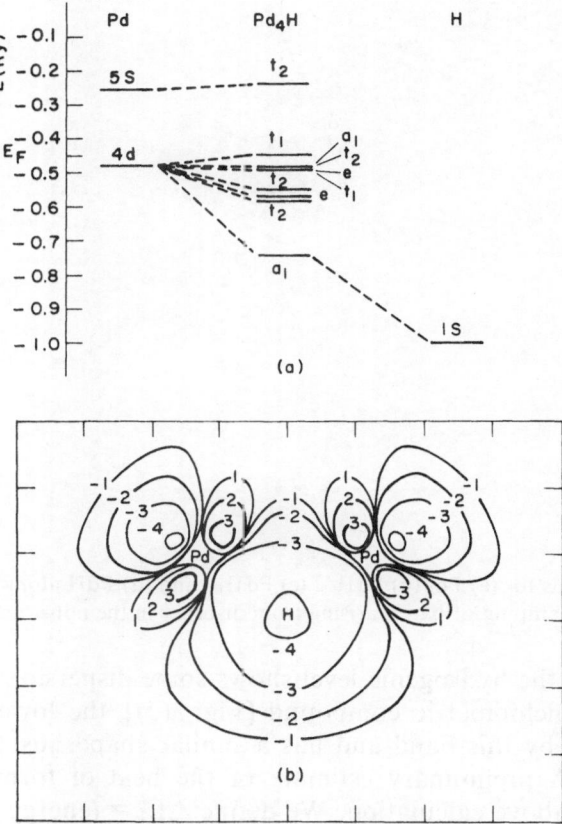

Fig. 2. Hydrogen in palladium. (a) SCF-Xα electronic energy levels for a tetrahedral cluster of four palladium atoms containing an interstitial hydrogen atom. The levels are labeled according to the irreducible representations of the tetrahedral (T_d) symmetry group, and the top of the d band is marked by the "Fermi level" E_F. Also shown are the energy levels for a palladium atom (in the $4d^9\,5s$ configuration) and a free hydrogen atom. (b) Contour map of the a_1 orbital wavefunction of Pd_4H for the energy level -0.746 Ry shown in (a) and plotted in a plane containing two palladium atoms and the hydrogen atom. The contour values decrease in absolute magnitude with decreasing absolute values of the contour labels. The sign of the contour label gives the sign of the wave function. To ensure clarity of presentation, the interior nodes of the wave function near each palladium nucleus are not shown.

Figure 3(a) illustrates the palladium band structure in the (100) direction, showing the broad d band and the conduction bands. An average t-matrix calculation on the Pd-H alloy systems is performed, an approximation which allows for the broadening of energy levels as well as the determination of their position. With 5 percent hydrogen in palladium [Fig. 3(b)], a hydrogenic level appears which a charge-density calculation indicates to be largely associated with hydrogen with a little d admixture. The lower band is broadened and the broadening increases on going to 10 percent hydrogen,

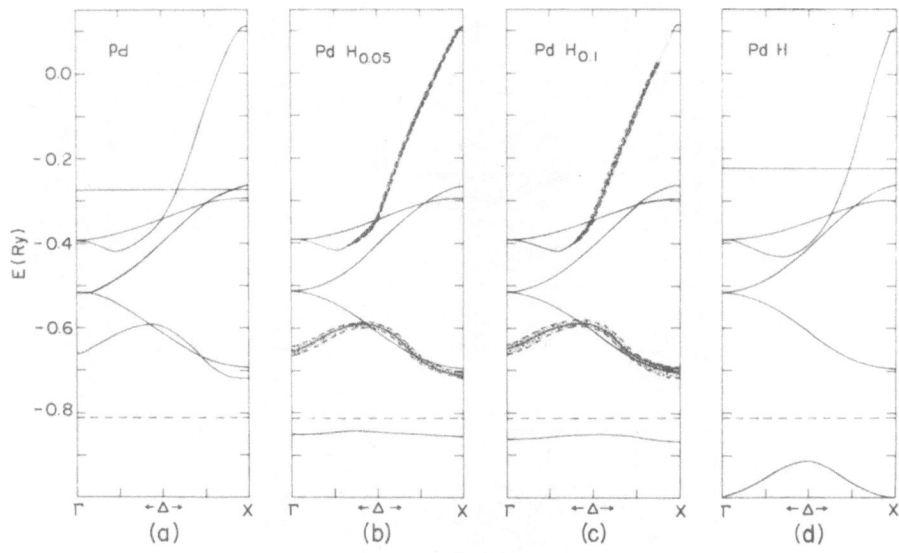

Fig. 3. Energy bands for (a) Pd, (b) PdH05, (c) PdH10, and (d) PdH along the [100] direction. Shading indicates damping of bands arising from disorder in the nonstoichiometric hydrides.

in which region the hydrogenic level shows some dispersion. [Fig. 3(c)]. On going to the stoichiometric compound [Fig. 3(d)], the lower s-like band is simply replaced by this band and has a similar shape, just as if the former were lowered. A preliminary estimate of the heat of formation has been made from the above calculation. We define $\triangle E_1$ = (energy of Pd + H^++e)-(energy of PdH). The point at issue is whether $\triangle E_1$ is large enough to break the H_2 molecule into its component pieces by supplying 1.17 Ry of energy per hydrogen atom. The position of the lowest part of the s band below the d band is lowered as we go from the metal to the hydride. The average d-band position is also lowered, and the extra electron, crudely speaking, is put back at the Fermi energy, which is also negative. We find that for nickel, it barely breaks the H_2 bond because $\triangle E_1$ is -1.19 Ry, whereas for Cu, $\triangle E_1$ is -1.12 Ry and the bond is just not broken. This is compatible with the weak solubility of H_2 in nickel, and the insolubility in copper, observed experimentally. For the hypothetical fcc titanium, $\triangle E_1$ is -1.31 Ry which is somewhat gratifying as TiH_2 does indeed have a larger heat of formation. However, $\triangle E_1$ for palladium is -1.12 Ry, which means that H_2 cannot dissolve in palladium, contrary to observation. It is amusing and gratifying that one can come up with rather large energies from this sort of calculation, and this type of picture certainly complements the model that Schrieffer discusses in his paper in the present volume.

Kohn elaborated on the picture of hydrogen chemisorption on a transition-metal surface within the calculation with Smith and Ying.[9] One starts with a small point charge in the vicinity of the surface, and it is screened by all the electrons, both d- and s-like. The only thing characterizing s and d electrons is the total charge density varying as a function of position. One then extrapolates this result to the full charge of a proton, but it still retains the same qualitative character. If one used only the sp electrons, one would end up with a totally different r_s value and a very different result. Insofar as the d electrons contribute to the total charge density, they are taken into account, but only as screening the proton. Although this picture cannot contain the whole truth of the situation, the fact that for a variety of physical questions, one obtains reasonable results without any adjustable parameter indicates that the picture has some pertinence.

An interesting experimental point was brought up by Eischens, viz., that infrared spectra can detect H-Pt vibrations, but cannot detect H-Ni vibrations even under similar coverage conditions.[19] Eischens' interpretation is that when vibrations are seen, the adatom is on a site of Schrieffer's type a, and when they are not seen, it is on a Schrieffer site c. The attention of theoreticians is drawn to this problem.

The Chairman briefly summarizes this discussion. Schwab has given experimental reasons for believing that the d band is important in chemisorption, though one sees that there may be difficulties of interpretation. The relations pointed out by Emmett between ability to chemisorb and bond type may not always be simple. Thus H_2 will adsorb on nickel but not on copper, but *atomic* hydrogen adsorbs quite exothermically[20], at temperatures where the adlayer is immobile, on copper and elements to its right in the periodic table. Evidently a hydrogen-metal bond can exist not only on copper, whose d band may well be important, but also on elements for which this is unlikely. The correlation suggested by Eischens between infrared absorption and bond type seems promising; also one must look to the electronic spectroscopy techniques for help here.

The calculations discussed by Johnson and by Ehrenreich give us some a priori theoretical insight, though this as yet bears more on the system hydrogen absorbed in palladium. Johnson's interpretation of his Pd_4H calculation seems to be rather close to the virtual molecule picture in Fig. 1(c), but Ehrenreich's calculation does not seem obviously to imply this. It is at least clear that the virtual molecule picture is orthogonal to Kohn's approach to chemisorption, which is a linear response theory neglecting the tight-binding-like properties of the d band.

2 THE SCF-Xα-SW TECHNIQUE

The discussion was opened by an invited introductory contribution from Messmer. The first essential component of the SCF-Xα-SW technique is the Slater local approximation to exchange and correlation in which a parameter α interpolates between the Kohn-Sham ($\alpha=2/3$) and the original Slater ($\alpha=1$) formulas for local exchange energies. A set of Hartree-Fock-like SCF equations is generated. Instead of solving the equations within an LCAO basis, the potential is approximated by using muffin-tin spheres around the atoms, and inscribing the entire molecule within a larger sphere in the manner of Johnson. The problem is then solved by multiple-scattering theory, in a manner reminiscent of the KKR method for band-structure calculations. An advantage of this technique is that it avoids the multicenter integrals characteristic of the SCF-LCAO theory.

An example of a system for which calculations have been done is the complex [$Cl_3PtC_2H_4$]. Energy-level diagrams for this system and for PtC_2H_4 are shown in Fig. 7 of the paper by Messmer, together with contour maps of the wave functions for PtC_2H_4 in Fig. 5 of that paper. The orbitals in [$PtC_2H_4Cl_3$]⁻ are shown in Fig. 4.

A discussion on the above ensued, involving Messmer, Mason, Johnson, and Schrieffer. Referring to the presence of nodes on the contour maps in orbitals, such as in Figs. 4(c) and 5(c) of Messmer's paper, it was noted

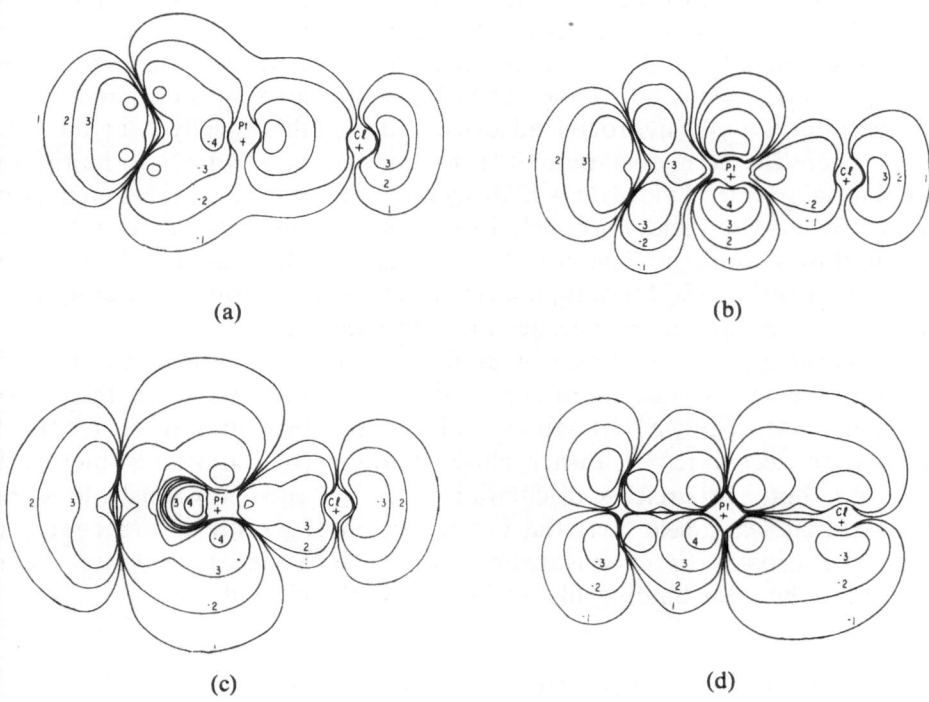

(a)   (b)

(c)   (d)

Fig. 4. Contour plots of the (a) $5a_1$, (b) $6a_1$, (c) $7a_1$, and (d) $2b_2$ orbitals of [$PtC_2H_4Cl_3$]⁻ which can be compared directly with the $3a_1$, $4a_1$, $5a_1$, and $2b_2$ orbitals of PtC_2H_4 in Fig. 5 of the paper by Messmer. The contours in each figure are chosen to be the same.

that an orbital may be bonding with respect to the Pt-C bond but antibonding with respect to the C-C bond. The relationship between portions of the MO's and platinum d_z^2 and d_{xy} orbitals which enables the scattered-wave MO's to be interpreted on an LCAO description was pointed out by Johnson.

A discussion of binding energies developed from a binding energy/distance curve for Li_2[21] presented by Messmer (Fig. 5). The SCF-$X\alpha$-SW calculation is seen to give a lower energy than a conventional SCF-HF calculation, and to be closer to the experimental result. Johnson pointed out the failure of a Hartree-Fock calculation on a molecule to give proper dissociation behavior, a long range aspect of the correlation error felt all the way down to the equilibrium nuclear distance. Corrections to HF involve the use of configuration-interaction theory or perturbation theory. In contrast to this, the $X\alpha$ technique gives the proper dissociation because of the local exchange approximation to the density functional and the equitable division of space.[22] Mason doubted that the Hartree-Fock calculation was strictly comparable with the $X\alpha$ calculation, pointing out that HF was not necessarily designed to give the lowest energy under all the conditions where $X\alpha$ was compared with it; indeed the variational principle was employed differently in the two methods. Johnson replied to this that the $X\alpha$ method is variational with respect to its own Hamiltonian. Nevertheless the orbital energies which are total energy differences between determinental wave functions in HF theory, are first derivatives of total energy with respect to occupation number in $X\alpha$ theory, in a manner analogous to Fermi liquid theory.[22] Hence orbital energies are not strictly comparable in the two theories.

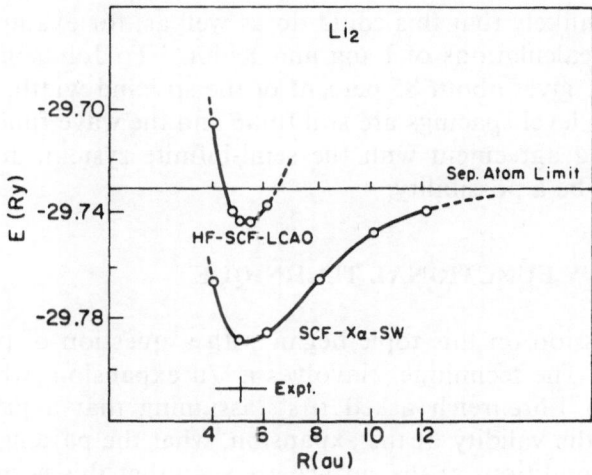

Fig. 5. Total energy of the Li_2 molecule as a funciton of internuclear distance, calculated by the SCF-HF-LCAO and SCF-$X\alpha$-SW methods, respectively.

A comment inserted by the Chairman on this discussion is that in conventional HF theory, which employs the exact Hamiltonian, a low energy is an accurate criterion of a good wave function. However, since $X\alpha SW$ theory contains muffin-tin and local exchange approximations, its total energy is indeed not strictly comparable with HF, and one should be cautious in interpreting a low energy. In a case such as H_2, HF can also be made to give proper dissociation by going over to unrestricted HF.

The discussion then turned to the importance of correcting the simple muffin-tin approximation, by allowing overlapping spheres or employing the procedure of Danese.[23]. Messmer presented a comparison of calculations on C_2 employing the simple muffin-tin model, the Danese procedure, and the conventional CI method (see Fig. 6 of Ref. 23), the simple HF method giving almost zero binding in this case. The value of correcting the muffin-tin approximation by allowing overlapping spheres was also emphasized by Herman, who took the TCNQ molecule as an example.

Kohn said he was impressed by the calculations on copper and nickel clusters, for example, but wondered whether Messmer had plotted integrated densities of states. Messmer replied that he had noted in a previous talk the correlation between regions of high density of states in the Wang and Callaway calculation[24] for nickel and groups of levels in the nickel-cluster calculation.

Kohn introduced the problem of electrostatic potential in the surface region and work function, which determines the absolute position of energy levels with respect to vacuum. He inquired to what extent the surface region is adequately treated by the cluster method, and whether this is important.

The Chairman suggested that the sp band, which is very important in screening, might not be very well treated in a cluster calculation, in which case it seems unlikely that this could do as well as, for example, the careful work-function calculations of Lang and Kohn.[25] To Johnson's point that a 13-atom cluster gives about 85 percent of the sp band width, the Chairman replied that the level spacings are still finite and the wave functions may not be in very good agreement with the semi-infinite system. Johnson agreed that this might be a possibility.

3 THE DENSITY FUNCTIONAL TECHNIQUE

The discussion on this topic began with a question of principle posed by Ehrenreich. The technique[25] involves a ∇n expansion, where $n(\mathbf{r})$ is the charge density. Ehrenreich asked first, assuming that a parameter exists characterizing the validity of the expansion, what the parameter is, and second, whether conditions at the surface are such that this is small.

Kohn replied that the exchange and correlation energy is written as the sum of two terms, one depending solely on the local density and the other being proportional to its gradient squared. It is not a gradient expansion in the sense that this is the correct expansion for small gradients; the gradient term is chosen in such a way that the wave number dependent dielectric constant comes out as correct as possible. The standard of correctness for this quantity is taken to be the best available calculation, probably found in the work of Singwi and others.[26] A fit of 2 percent over all wave numbers may be obtained with the above approximation. This approximation to the density functional is valid in two limiting situations: (1) when the density is sufficiently slowly varying in space but of arbitarily large variation in amplitude and (2) when the variation of density is of small amplitude, no matter how rapid its spatial variation. At a surface conditions are somewhere between these situations, the density change being large and also rather rapid. Therefore, one cannot be sure that the expansion will be valid in the surface region, but one has sufficiently bracketed the surface problem to give grounds for hoping it will be reasonably accurate.

Ehrenreich inquired whether calculating the next term in the expansion, viewing it as such, could give a check on its validity. Kohn replied that including the gradient term itself changes surface energies by only about 15 percent, which gives some encouragement. Going further seems hardly worthwhile since one is already within 2 percent of the dielectric constant.

The Chairman asked whether the formulation is applicable to the d-band metals so important in chemisorption and catalysis, and in particular whether it could describe something like a bond between an adatom and a transition-metal d band. Kohn replied that the question of applicability to d bands was an extremely interesting one; calculations of work function had been carried out on the noble metals[25] with results 1 to 1.5 eV too low. Kohn felt that this was a poor result indicating that the method was not yet ready for direct application to d electrons. Kohn wondered why this was so, in view of the successful application of $X\alpha$ theory to d bands and the close similarities between $X\alpha$ theory and the density functional formalism.

Johnson agreed that the $X\alpha$ method becomes virtually identical to the Kohn-Sham concept when $\alpha=2/3$. Johnson pointed out that tests conducted by others on systems as complicated as Ne_2 and rare-gas crystals do give van der Waals binding, a complex many-body effect; whether or not $X\alpha$ includes all correlation effects, it does include within the framework of the Fermi hole concept aspects of exchange and correlation combined, which additively seem to be much better than many more elaborate procedures for correcting Hartree-Fock.

The Chairman wondered whether the latter claim is not a little too complimentary to $X\alpha$ theory. The dispersive van der Waals interaction is characteristic of systems with nonoverlapping charge densities, for which any local density theory gives essentially additive energies, and thus only electrostatic interaction. However, the work of Kim and Gordon gives surprisingly good results near the energy minimum but not in the asymptotic region.

Smith suggested that the poor results for copper work function were attributable not to the inadequacy of the density functional formalism for copper but to the use of the jellium model for copper which gets less accurate as the d band fills up. The density functional formalism had been used quite successfully for band structure calculations in transition metals.

At this point Mason raised the whole question of accuracy in chemisorption calculations. In SCF *ab initio* calculations or transition-metal complexes, fairly good agreement with ESCA and UV photoemission spectra was achieved, as was also the case using the calculational methods of Johnson. However, no such calculations achieved chemical accuracy, by which we mean the ability to achieve chemical stability. As an example, one might take the difference in binding energy between fourfold and twofold adsorption sites; direct experimental information on this was unavailable, though an attempt to obtain such information through inelastic neutron scattering was under way in Mason's group. However indirect evidence from NMR C^{13} spectroscopy on large clusters of metal atoms showed the activation energy for migration of CO from onefold-bridging-site positions to threefold-site positions on the cluster surface to be about 8 to 10 kcal. The problem is that this kind of accuracy is not achievable by present *ab initio* calculations. Turning to the simpler calculations such as those discussed by Kohn and Schrieffer, similarly crude calculations on simple molecules could not achieve useful accuracy without building in a great deal of chemical intuition; indeed this may be the reason for the relative success of this genre of model. In simple perturbation theory, bond strength is of order $S^2/\triangle E$, where S = overlap between localized wave functions forming the bond and $\triangle E$ is their difference in energy. One wonders whether perturbation theory based on the theories of Kohn, Schrieffer, or Johnson could give a realistic energy profile for migration of CO over a flat platinum surface. One notes that the assumptions as to distance above the surface in Schrieffer's fourfold and onefold positions rather critically affect the accuracy of his calculations. After thus emphasizing the inadequacy of current calculations in achieving chemical accuracy, Mason pointed out that an organic chemist's choice of the most useful theoretical development in the last 10 years would be Woodward-Hoffman theory[27], an entirely symmetry-based theory. The inorganic chemist's view would also be: if a good answer is unobtainable by symmetry arguments, then forget it. This is an exaggeration, but it is a majority view in chemistry among experimentalists. The Chairman however felt that qualitative concepts about the chemisorption bond, established with the aid of quantitative experimental data, ought to help chemists sort their way through the diverse range of phenomena presented by heterogeneous catalysis.

4 THE MOLECULAR ORBITAL APPROACH

Grimley opened the discussion on this topic with the following invited contribution:

Schrieffer mentioned in his paper the possibility of treating an adsorbate/adsorbent cluster in a high approximation, but using a low approximation to treat the rest of the adsorbent and its coupling to the cluster. I began to develop this method over a year ago, and the general method, which is conveniently formulated in terms of Dyson's equation, together with the results of our first calculation for hydrogen on simple and face-centered cubium have been published.[28] We used the Hartree-Fock approximation on the cluster and the tight binding (Hückel) theory elsewhere. Similar calculations have now been made for hydrogen on Li (100) using a single Slater-type orbital for the 2s state of lithium. Some of the results are shown in Fig. 6. The gross occupancies on the hydrogen and lithium atoms in the diatomic molecule embedded in a tight-binding substrate are $n_H = 0.947$ and $n_{Li} = .030$, and the (surface) bond order is .022. As expected, the surface bond is very polar for this system. The lower δ function in the density of states $\Delta\rho$ shown in Fig. 6(c) can be regarded as the H 1s level pushed down by the electrostatic field of the Li^+ ion in the surface diatomic molecule; the upper δ function corresponds to the lithium 2s level pushed up by the electrostatic field of the H^- ion.

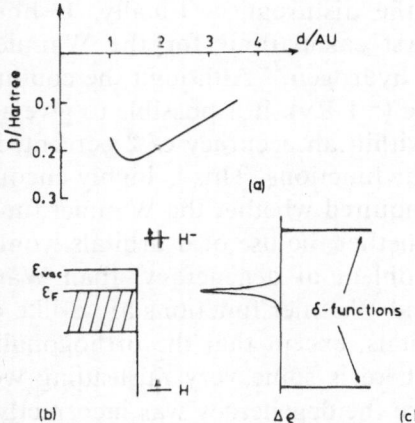

Fig. 6. On-site chemisorption of hydrogen by Li(100: (a) the potential energy curve, (b) unperturbed electron levels, (c) the density of states $\Delta\rho$ contributed to the system by the hydrogen atom calculated for the equilibrium distance 1.40 AU. The δ functions are 0.49 Hartree below and 0.15 Hartree above ϵ_{vac}, and they can be regarded as the ionization and affinity levels of the polar surface molecule Li_+H^-. $\Delta\rho$ is negative in the lithium band because

$$\int_{-\infty}^{+\infty} d\epsilon \, \Delta\rho \simeq 1$$

Kohn then described a self-consistent Wannier function approach, as an alternative way of doing molecular orbital calculations. For a periodic lattice, Wannier functions are orthogonal localized functions in terms of which the Bloch waves can be expressed. We have recently done a formal theory on the existence and properties of Wannier functions for systems that are not periodic, in particular, for example, for a metal with a surface, and a metal with a surface and an adatom.[29] The important point about these Wannier functions, as verified by a calculation of Smith, is that in the presence of a disturbance in the periodicity of the system, they differ from the corresponding bulk functions only in the immediate vicinity of the disturbance. Therefore, in trying to describe a metal with a surface and an adatom, one may use the Wannier functions of the bulk everywhere except in the close atomic neighborhood of the disturbance where a limited number of modified Wannier functions may have to be introduced. It is possible to perform a self-consistent calculation in the sense of the density functional approach entirely within the framework of these localized Wannier functions. We have done this by taking advantage of some beautiful work of Heine, Haydock, and Kelley, which makes it unnecessary to find the delocalized eigen functions. Such a calculation, if carried out exactly, will be completely equivalent to a molecular orbital calculation, with some approximation for exchange and correlation. The method appears to have both computational and conceptual advantages, and clearly shows the localized nature of the disturbance. Finally, I should add that we have recently done the first calculations for the Wannier function of a bulk system, viz., metallic hydrogen.[29] Although the conduction band of metallic hydrogen is very wide (~ 1 Ry), it is possible to give a very accurate description of the system, within an accuracy of 2 percent, in terms of simple and well-localized Wannier functions. This is highly encouraging.

The Chairman inquired whether the Wannier function would be similar to a d orbital, and whether the use of d orbitals would not be simpler for d bands, with their problem of degeneracy, than Wannier functions. Kohn replied that the d-band Wannier functions are d-like orbitals, with the same symmetries as d orbitals, except that the orthogonality relation introduces some fluctuations. There is some very misleading work on Wannier functions for copper where the degeneracy was incorrectly treated and the functions were, therefore, very delocalized.

Messmer asked whether, on bringing an atom up to the surface, there would be a change in the Wannier functions due to the interaction, for example, a reorthogonalization tending to disturb all the Wannier functions. It may be best, in some sense, to proceed with nonorthogonalized orbitals for a localized-state description of the problem, because the chances that they will not be disturbed when an adatom is brought up to the surface are much greater. There is no orthogonality requirement. Kohn agreed that when an atom is added to the surface, all Wannier functions are modified. But his group had shown in a model theory, which had been verified by Smith in an actual calculation, that this disturbance decays exponentially into the solid. It amounts to only 1 percent or so when one is one or two atoms away. So it is not a serious problem, and working with nonorthogonal functions does not seem to simplify the problem. In fact the contrary may be true. The nonorthogonal functions fall off with an exponential envelope that is the same as for the Wannier functions. They may be better localized by a factor of 2 or so, but they are not better localized in character or by an order of magnitude.

Smith commented on Messmer's question. He had expanded the orthogonal functions in terms of nonorthogonal functions so that nonorthogonal functions did indeed enter, and symmetrically orthogonalized the nonorthogonal functions by Löwdin's technique. In his calculation, the nonorthogonal functions were changed very little by cleavage that makes the surface, although the Wannier functions changed dramatically. The change in the latter, however, went away rapidly on going into the bulk.

Mason said he tended to assume that Kohn could choose his Wannier functions in a symmetry-adapted way for the adsorption problem. There seems to be a considerable overlap between his theory, if such symmetry-adapted functions are used, and Grimley's work on the group-orbital concept in chemisorption. Kohn agreed that an element of overlap existed in the description of the surface complex, the difference being that in his approach, the bulk of the solid was also described uniformly in terms of the localized functions.

5 RELAXATION EFFECTS

Since the subject of relaxation shifts seemed several times to have excited interest at the Colloquium, the Chairman thought it worthwhile to add the following preface to the remarks made on this subject in the discussion.

In an experimental spectroscopy, such as optical absorption, ultraviolet photoemission, or X-ray emission, one may measure a difference of energy levels or more generally some density of states. Problems may exist in comparing the experimental level difference or density of states with a theoretical one. In UV absorption spectroscopy of diatomic molecules, for example, one measures transitions between a vibrational state near the minimum of the lower energy/distance potential curve and a vibrational state belonging to a higher potential energy curve here assumed to possess a minimum. The Franck-Condon principle tells us that the most probable final state contains sufficient excited phonons that it lies vertically above the initial state, and thus one measures an energy greater than the "relaxed" difference between the two potential minima. On the other hand, Citrin and

Hamann[30] showed that the core ionization energies of rare-gas atoms trapped in a host metal are reduced due to screening of the final state. The "relaxed" energy is here measured because the plasma frequency, governing relaxation time, is not small with respect to the relaxation shift as is the molecular vibration frequency in the previous example.

In UPS of an adsorbed atom, according to the Franck-Condon principle the ionization energy measured from the center of a current-versus energy peak should be unrelaxed with regard to the (smeared out) phonons, representing removal of an electron without movement of the ion cores. On the other hand, a Citrin-Hamann screening relaxation shift of a localized adatom level should occur, since the surface plasmon frequency is larger than or comparable with this shift (a few volts).

So far, shifts arising from interaction with quasi-independant excitations such as phonons or plasmons have been considered. Even in an isolated atom or molecule with fixed ion cores the levels measured, for example, in photoemission will differ from differences between Fock eigenvalues because of, among other factors, relaxation of the electronic structure around the charged final state, so that experimentally the energy levels appear higher than the Fock eigenvalues. Slater's transition-state concept[22] is one approach to this problem.

A first discussion on this subject followed Schrieffer's inquiry as to whether the levels shown in the Johnson PdH₄ calculation[17] (Fig. 2), and which he compared with photoemission data, were transition-state energy levels. Johnson replied that these levels in fact are ground-state levels, but that a transition-state calculation has been performed giving relaxation shifts of the order of several electron volts. Nevertheless the levels are very similar, semiquantitatively. In response to Schrieffer's inquiry as to the sign of the relaxation shift, Johnson states that this is downward, and Schrieffer contrasted this with Hartree Fock where the shift is upward from the Fock eigenvalues. Johnson's explanation of this paradox was that the levels plotted in the diagram (Fig. 2) were close to Mulliken electronegativities, the mean of ionization and affinity levels. Photoemission measures the ionization level which is below this Mulliken mean level.

Kohn then formulated a question on the generality of the notion of transition state introduced by Slater, whose importance and success in the theory of optical transitions he emphasized. The question was whether this idea was limited by difficulty in extending it to the case of an optical transition in a metal, even one with narrow bands. Kohn inquired whether some means might be found to retain the transition-state concept in such a case. Kohn agreed with Messmer, who raised the point, that in the case of a localized excitation in a solid, one could retain it.

Johnson pointed out that the primary advantage of the transition-state method was for systems where the band description of the excitation is inappropriate, and that Slater has shown that with itinerant states, the relaxation shifts essentially vanish within the transition-state procedure. Kohn replied that, physically, the shifts do not vanish; for a system with a bandwidth of say a couple of volts or so, for which we know that the energy band description is very appropriate and also that there are large relaxation effects, one indeed finds that the effects vanish within the theory as formulated, but one wonders how to recover them.

Messmer suggested that, since in an itinerant model pulling out an electron changes the atomic environment only by order $(1/N)$, such a model may be inappropriate for this type of system. In disagreeing with this, Schrieffer made the following comment:

I disagree somewhat with Messmer's response to Kohn's comment. If one thinks of a process, spectroscopy for example, where one coherently excites an electron over many atoms so that the initial state wave packet that you create in the physical observation process is delocalized over an enormous number of atoms, then one still knows however that a correlation effect does exist, but it does not exist within a one-electron framework. It is much like the polaron problem; if one lifts an electron, or, say, puts an electron into a polar crystal, it is not that the electron is delocalized totally, but that the polaron is delocalized, i.e., the lattice polarizes around the electron in its instantaneous position in its itinerant path. It is this following cloud of correlation which is missed or which tends to be missed in a transition-state-like calculation. Now one need not say that the method is not good for very many problems, but there are significant physical effects which are missed.

In agreeing, Johnson pointed out that another limitation of the transition-state method lies in its omission of multiplet structure in certain open-shell systems.

The discussion ended with a measure of agreement that problems remain in extending the transition-state method to the band situation.

For completeness, the Chairman adds that in certain ranges of parameters, a self-energy correction to the band energy brings in the effects discussed by Kohn and Schrieffer as a level shift. Calculations from this point of view exist in the literature. In the relaxation case where surface plasmon frequency is greater than the level shift, Hewson and Newns[31] have shown that the shift for a half-filled adsorbate level is zero when level broadening greatly exceeds the surface plasmon frequency, but is equivalent to a final-state relaxation shift in the opposite case.

REFERENCES

1. Schrieffer, J. R., and Gomer, R., *Surface Sci.,* **25**, 315 (1971); Paulson, R. H., and Schrieffer, J. R., *Surface Sci.,* in press.
2. Edwards, D. M., and Newns, D. M., *Phys. Letters,* **24A**, 236 (1967); Newns, D. M., *Phys. Rev.,* **178**, 1123 (1969).

3. Eastman, D. E., Cashion, J. K., and Switendick, A. C., *Phys. Rev. Letters,* **27**, 35 (1971).
4. Eastman, D. E., *Solid State Commun.,* **10**, 933 (1972).
5. Ertl, G., private communication.
6. Brenig, W., and Schönhammer, K., *Z. Physik,* **267**, 201 (1974).
7. Plummer, E. W., and Waclawski, B. J., *Proc. Phys. Elec. Conf.* (1973); Menzel, D., private communication.
8. Fassaert, D.J.M., Verbeek, H., and Van der Avoird, A., *Surface Sci.,* **29**, 501 (1972); Blyholder, G., *Surface Sci.,* **42**, 249 (1974).
9. Smith, J. R., Ying, S. C., and Kohn, W., *Phys. Rev. Letters,* **30**, 610 (1973); *Solid State Commun.,* **15**, 1491 (1974).
10. Johnson, K. H., *J. Chem. Phys.,* **45**, 3085 (1966); Slater, J. C., and Johnson, K. H., *Physics Today,* **27** (10), 34 (1974).
11. Eastman, D. E., and Cashion, J. K., *Phys. Rev. Letters,* **27**, 1520 (1971); Ertl, G., private communication; Clarke, A., Gay, Ian D., and Mason, R., private communication.
12. Batra, I. P., and Robaux, O., private communication.
13. Brill, P., and Schwab, G.-M., *Z. Phys. Chem., Neue Folge,* **92**, 339 (1974); *ibid.,* **76**, 38 (1971).
14. Soven, P., *Phys. Rev.,* **156**, 809 (1967); Velicky, B., Kirkpatrick, S., and Ehrenreich, H., *Phys. Rev.,* **175**, 747 (1968).
15. Couper, A., and Eley, D. D., *Discussions Faraday Soc.,* **8**, 172 (1950).
16. Kanamori, J., and Terakura, K., *Proc. 7e Conf. Int. de Magnetisme* (1970); *Jour. Physique,* **32**, C1-282 (1971).
17. Slater, J. C., and Johnson, K. L., *Physics Today,* **27** (10), 34 (1974).
18. Weiss, J., Gelatt, J. C., and Ehrenreich, H., private communication.
19. Pliskin, W. A., and Eischens, R. P., *Z. Physik. Chem., Neue Folge,* **24**, 11 (1960).
20. Gasser, R.P.H., *Chem. Soc. (London) Ann. Repts.,* **LIX**, 7 (1962).
21. Rosch, N., Messmer, R. P., and Johnson, K. H., *J. Amer. Chem. Soc.,* **96**, 3855 (1974).
22. Slater, J. C., and Johnson, K. H., *Phys. Rev.,* **B5**, 844 (1972).
23. Danese, J. B., *J. Chem. Phys.,* **61**, 3071 (1974).
24. Wang, C. S., and Callaway, J., *Phys. Rev.,* **B9**, 4897 (1974).
25. Kohn, W., and Sham, L. J., *Phys. Rev.,* **140,** A1133 (1965); Lang, N. D., and Kohn, W., *Phy. Rev.,* **B1**, 4555 (1970); **B3**, 1215 (1971); **B7**, 3541 (1973).
26. Vashishta, P., and Singwi, K. S., *Phys. Rev.,* **B6**, 875 (1972) and references therein.
27. Hoffman, R., and Woodward, R. B., *Accts. Chem. Research,***1**, 17 (1968).
28. Grimley, T. B., and Pisani, C., *J. Phys. Chem.,* **7**, 2831 (1974).
29. Smith, J. R., and Gay, J. G., *Phys. Rev. Letters,* **32**, 774 (1974); Kohn, W., and Onffroy, J., *Phys. Rev.* **B8**, 2485 (1973); Andreoni, W., Kohn, W., Rao, M., and Sommers, C., (unpublished).
30. Citrin, P. H., and Hamann, D. R., *Chem. Phys. Letters,* **22**, 301 (1973).
31. Hewson, A. C., and Newns, D. M., *Proceedings of 2nd ICSS, Japanese J. Appl. Phys.,* Supp. 2, Part 2, 121 (1974). Gumhalter, B., and Newns, D. M., unpublished.

 *Part Four*

THE EFFECT OF SMALL PARTICLES AND POROUS CARRIERS

CATALYTIC AND MAGNETIC
ANISOTROPY OF IRON SURFACES

M. Boudart, H. Topsøe, and J. A. Dumesic

Stauffer Laboratories of Chemistry and Chemical Engineering
Stanford University
Stanford, California

ABSTRACT

Small (1.5 nm), medium size (4 nm), and large (30 nm) metallic iron particles were prepared on a magnesium oxide support. The turnover number, $\overline{N}$, for the ammonia synthesis on these catalysts at atmospheric pressure between 570 and 700°K. was found to increase with particle size and with nitrogen-induced surface reconstruction for the small and medium-size particles. This latter effect was also accompanied by decreases in the surface-sensitive magnetic-anisotropy energy barrier and carbon monoxide chemisorption of the corresponding iron surfaces. It thus appears that small and medium-size particles are lacking in certain sites which are produced via the nitrogen-induced surface reconstruction, and all effects taken together indicate that these sites are C_7 sites, i.e., surface atoms with seven nearest neighbors.

337

1 INTRODUCTION

The catalytic properties of small particles are of practical and theoretical interest. These small catalyst particles are of commercial interest since their high surface-to-volume ratio makes more of the catalytic material and reactor volume available for reaction. In addition, though, the surface structure of small particles is expected to be different than that of larger particles. Thus, a study of the effect of particle size on a given catalytic process may provide valuable information about the kinetically significant surface interactions, and their dependence on surface structure. However, to determine the relation between catalytic activity and surface structure in this manner, the surface properties of the small particles must be measured under reaction conditions. In this paper, the rationale for such measurements is discussed, with reference to small metallic iron particles used for the synthesis of ammonia. A more detailed discussion of the concepts contained herein is presented elsewhere.[1-5]

2 CATALYST CHARACTERIZATION

Metallic iron particles spanning the size range from 1.5 to 30 nm were prepared on a magnesium oxide support.[1] The starting material in the catalyst preparation was magnesium hydroxy-carbonate which, as evidenced by electron microscopy, had a platelike structure (shown schematically in Fig. 1a). In aqueous slurry, several percent of the Mg^{2+} ions were exchanged with an appropriate number of Fe^{3+} ions, in accord with electric-charge neutrality, to form a catalyst precursor (Fig. 1b). Metallic-iron particles were subsequently formed via dihydrogen reduction of the precursor at 700°K (Fig. 1c).

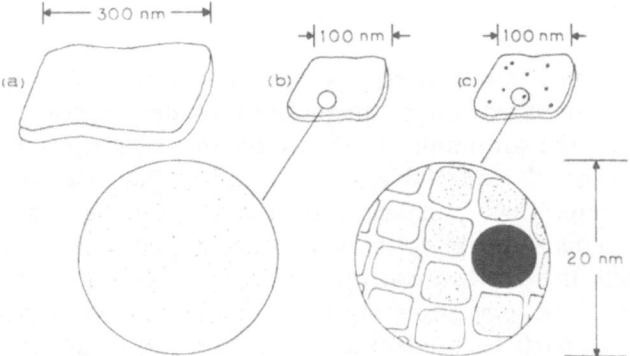

Fig. 1. Fe/MgO catalyst genesis: (a) magnesium hydroxy-carbonate; (b) catalyst precursor; (c) reduced catalyst; light shading represents Fe^{2+}- or Fe^{3+}-rich regions; dark shading represents metallic iron.

During the final reduction step, two important chemical processes take place. First, the large particles of magnesium hydroxy-carbonate decompose to form smaller particles of magnesium oxide, with the overall platelike structure of the precursor being preserved. Thus, the platelike structures, as seen in electron micrographs of the reduced catalysts, are aggregates of many smaller magnesium oxide particles with characteristic dimension of approximately 5 nm. Second, the structure of the iron phase is dramatically changed during the final reduction step. The fairly uniformly distributed iron on magnesium hydroxy-carbonate is transformed to iron-rich clusters in magnesium oxide, the chemical state of the iron in these clusters being partially Fe^{2+} and partially metallic, as will be shown presently. Evidence for this clustering is provided by Mössbauer spectroscopy, electron microscopy, and X-ray diffraction.

To use this material effectively for the ammonia synthesis, it is necessary to determine what fraction of the iron is in the metallic state, and the size of the metallic iron particles must be measured. With reference to the first point, the fraction of the iron in the metallic state was determined using Mössbauer spectroscopy, magnetic susceptibility, and volumetric oxidation studies. The results of these determinations are given in Table I.

Table I. Fe/MgO Catalyst System: Metallic Iron Characterization

Catalyst, % Fe on MgO	Fraction of Iron in Metallic State	Particle Size (d), nm	Dispersion (D), %
1	0.40	1.5	50
3	0.60	4.0	25
5	0.40	4.0	25
8	0.45	6.0	15
16	0.50	10.	8
40	0.70	30.	3

Different catalysts are denoted by the total amount of iron (in weight percent including all chemical states) on magnesium oxide referred to dry, reduced sample, e.g., 5 percent Fe/MgO. As seen in Table I, an appreciable fraction of the iron on the Fe/MgO catalysts (ca. 0.5) is indeed in the metallic state. The results of metallic-iron particle-size measurements for the Fe/MgO catalysts are also summarized in Table I; these results were obtained using Mössbauer spectroscopy, magnetic susceptibility, electron microscopy, selective chemisorption of carbon monoxide, and X-ray line broadening. The 1 percent Fe/MgO catalyst provides very small metallic-iron particles (1.5 nm), the 3, 5, and 8 percent Fe/MgO catalysts give metallic-iron particles of intermediate size (ca. 5 nm), and large metallic iron particles (>10 nm) are present on the 16 and 40 percent Fe/MgO catalysts. The metallic-iron dispersions for the different catalysts are also given in Table I, where the dispersion is the percentage of the metallic iron atoms that are surface atoms of iron particles. With the above Fe/MgO catalyst-system characterization, it is now possible to study the effect of particle size on the ammonia synthesis over metallic iron.

3 STRUCTURE SENSITIVITY OF AMMONIA SYNTHESIS

The ammonia synthesis at atmospheric pressure between 570 and 700°K was conveniently studied using a quartz integral flow reactor, the exit of which was connected to an IR spectrophotometer.[2] A stoichiometric dihydrogen:dinitrogen gas mixture (H_2:N_2 = 3.1) was passed over a known quantity of catalyst in the reactor at a given temperature, and the mole fraction of ammonia in the gas leaving the reactor, y, was measured using the IR spectrophotometer. Knowing the value of y and the gas flow rate through the reactor, the average ammonia synthesis rate per metallic-iron surface atom, or an average turnover number ($\overline{N}$), can be readily calculated. In general, the ammonia synthesis rate depends on the partial pressure of ammonia over the catalyst, and because the reaction is limited by equilibrium, values of $\overline{N}$ are best given for specified values of efficiency, η, defined as the mole fraction of ammonia leaving the reactor normalized by the equilibrium conversion.

The values of $\overline{N}$ at 678°K and $\eta = 0.15$ for the 1, 5, and 40 percent Fe/MgO catalysts are given in Table II. These values were obtained after Deoxo-H_2 reduction (dihydrogen passed through a Deoxo purifier and a molecular-sieve trap at 77°K) at 700°K, followed by the flowing of the stoichiometric H_2:N_2 mixture over the catalyst. With reference to Table II, it can be seen that $\overline{N}$ increases with increasing iron particle size from 1.5 to 30 nm, i.e., the ammonia synthesis over iron is a structure-sensitive reaction. To compare the synthesis activity of the Fe/MgO catalysts with that of conventional iron synthetic ammonia catalysts, a sample from the latter category was studied. This catalyst was doubly promoted (Al$_2$O$_3$, K$_2$O) and is part of the series developed at the Fixed Nitrogen Research Laboratory (FNRL).[6] As seen in Table II, the value of $\overline{N}$ for 40 percent Fe/MgO agrees quite well with that measured for the conventional catalyst, and as the metallic-iron particle size is decreased (1 and 5 percent Fe/MgO), $\overline{N}$ correspondingly decreases.

To serve as a guide in subsequent discussions, it seems appropriate to speculate on the origin of the observed structure sensitivity. If, after dihydrogen reduction and subsequent exposure to the reactant gas mixture,

Table II. Average Turnover Number, $\overline{N}$, at 678°K
After Deoxo-H_2 Reduction,
(Efficiency = 0.15)

Catalyst	Particle Size (d), nm	$\overline{N}$, 10^3 sec^{-1}
1% Fe/MgO	1.5	1.0
5% Fe/MgO	4.0	9.0
40% Fe/MgO	30.	35.
FNRL 441	150.	33.

the large particle surface is rich in special sites that are quite active for the synthesis of ammonia, but these special surface sites become less abundant as the particle size is decreased, then the observed ammonia synthesis structure sensitivity can be understood. This, of course, is only one possible explanation of the effect, and as a first test of this postulate, consider the following arguments. In general, the most stable metallic-iron surface structure in dihydrogen may be different than the most stable surface structure in the presence of dinitrogen.[7] Thus, the surface of a metallic-iron particle formed via dihydrogen reduction might be different than that formed in the presence of nitrogen (mono- or dinitrogen at the surface). It was postulated above that the surface of small iron particles prepared via dihydrogen reduction (and subsequent exposure to the reactant gas mixture) is lacking in special surface sites that are quite active for ammonia synthesis. However, the creation of these sites may be facilitated by formation of the small particles in the presence of nitrogen. The experimental test of these concepts follows.[2]

The 5 percent Fe/MgO precursor was first reduced (as previously described) in Deoxo-H_2, and upon exposure to the reactant gas mixture, $\overline{N}$ was measured (see Table III). The metallic-iron particles (4.0 nm) were then treated with pure ammonia at 670°K and atmospheric pressure for 1 hour so as to form iron nitride, and the nitride was then decomposed in the $H_2:N_2 = 3:1$ reactant gas mixture at the same temperature. If, indeed, the amount of nitrogen present during the metallic-iron surface formation is important in determining the ultimate structure of the surface, then this "ammonia treatment" should reconstruct the iron surface, i.e., the surface after this treatment should be different than that before the treatment. Indeed, as shown in Table III, the ammonia treatment significantly increases the value of $\overline{N}$. Subsequent treatment of the catalyst with Deoxo-H_2 for 20 hours at 670°K shows that the effect of the ammonia treatment can be completely erased (Table III) by removing the dinitrogen from the gas mixture over the

Table III. Average Turnover Number, $\overline{N}$, at 678° K
for 5 Percent Fe/MgO After Sequential
Pretreatments (Efficiency = 0.15)

Pretreatment Sequence	$\overline{N}$, 10^3 sec^{-1}
1. Deoxo-H$_2$ reduction	9.
2. Ammonia; 1 hr	13.
3. Deoxo-H$_2$; 20 hr	9.
4. Pd-diffused H$_2$; 20 hr	7.

catalyst at this temperature. Since cylinder dihydrogen usually contains traces of dinitrogen (ca. 50 ppm) that are not removed by Deoxo purification[8], the possible effect of this dinitrogen was eliminated by treating the catalyst with palladium-diffused H$_2$ (dihydrogen purified by passage through a palladium-silver thimble) for 20 hours at 675°K. As shown in Table III, this treatment slightly decreases the value of $\overline{N}$, but of greater significance is the transient response of $\overline{N}$ after this treatment. The time required to reach the stationary-state ammonia synthesis rate after exposure of the catalyst to the reactant gas mixture was approximately 6 hours after the palladium-diffused H$_2$ treatment, but only ½ hour after the Deoxo-H$_2$ treatment. Again, an effect of nitrogen present during surface formation is evident, with the metallic-iron surface structure prepared in pure dihydrogen not being stable under ammonia synthesis reaction conditions.

The preceding effects of pretreatments on catalytic activity evidence various nitrogen-induced surface reconstructions, as summarized in Fig. 2.

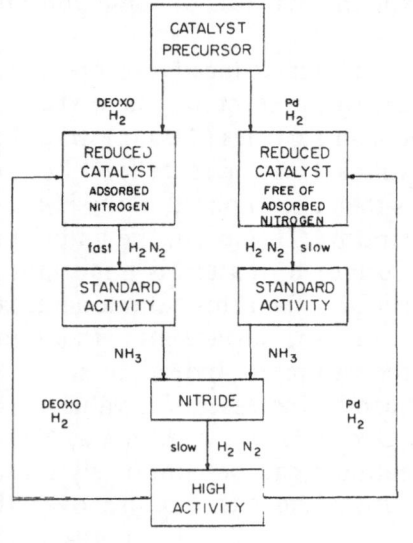

Fig. 2. Summary of catalytic-activity dependence on catalyst. Pretreatment; slow and fast refer to rate of attainment of stationary state.

Returning now to the structure-sensitivity postulate, it was said that small particle surfaces of iron (after dihydrogen reduction and exposure to the reactant gas mixture) are lacking in special surface sites that are quite active for ammonia synthesis, but now these sites can be created by the ammonia treatment. The large particle surfaces of iron were said to be rich with these special sites even after dihydrogen reduction and exposure to the reactant gas mixture, and thus only a small effect of the ammonia treatment on $\overline{N}$ is expected. This expectation is experimentally verified, with $\overline{N}$ increasing by only 10 percent upon ammonia treatment. Thus, it appears that the origin of the structure sensitivity and the nitrogen-induced surface reconstructions are related, i.e., both phenomena can be understood in terms of the surface concentration of a special site that is quite active for the synthesis of ammonia.

While a study of the ammonia-synthesis kinetics provides valuable information about the metallic-iron surface, kinetics alone does not provide a microscopic description of the actual surface-structure changes taking place with nitrogen-induced surface reconstruction and particle-size change. To obtain a more complete understanding of these phenomena, surface-structure measurements under reaction conditions must be made. Unfortunately, it is not at all clear how these measurements are to be made. This problem is addressed presently.

4 SUPERPARAMAGNETISM AND MAGNETIC ANISOTROPIES

The physical properties of iron are influenced to a great extent by magnetic effects, and it may be that interesting surface structural information will result through a systematic study of magnetic interactions. One such interaction is superparamagnetism[9], the basis for which is summarized in Fig. 3. The small iron particles supported on magnesium oxide are well below the single-domain size for metallic iron (ca. 20 nm)[10], and thus to each particle a single magnetization vector, $\mathbf{M}$, can be associated. The magnetic energy, however, is not isotropic, and for the magnetization to move in space, various magnetic-anisotropy energy barriers must be crossed. Comparing the magnitude of the average magnetic-anisotropy energy barrier E_A with that of the thermal energy kT allows a simple expression to be written for the magnetization relaxation time τ[11]

$$\tau = \tau_0 \exp\left(E_A/kT\right)$$

where τ_0 is a calculable proportionality parameter. Magnetic-anisotropic effects become detectable when the value of τ becomes comparable with or greater than the "experimental observation time" for the magnetic experiment, τ_c. For magnetic-susceptibility experiments, τ_c is the order of 10^2 seconds[12], while for Mössbauer spectroscopy τ_c is approximately equal to

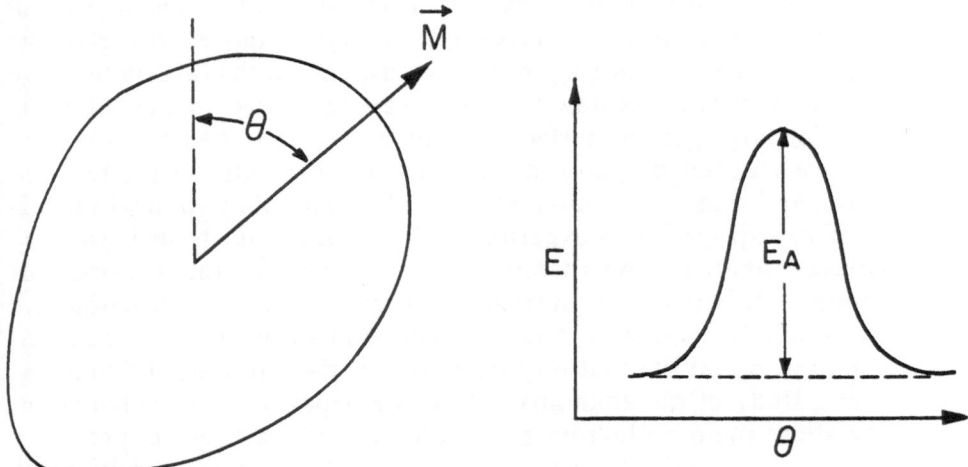

Fig. 3. Superparamagnetism and magnetic anisotropy.

the Larmor precession time of the nuclear moment about the internal magnetic field (ca. 10^{-8} second).[13,14] Thus, for short experimental observation times (Mössbauer spectroscopy) or at low temperatures (magnetic susceptibility), magnetic-anisotropic effects can be studied for the small particles, thereby allowing the magnitude of E_A to be estimated. It will be shown later that it is in E_A, its magnitude and origin, that interesting surface structural information can be found.

One origin of magnetic anisotropy is magnetocrystalline anisotropy, i.e., the magnetic energy is not isotropic with respect to the crystallographic axes. For metallic iron, the lowest energy direction of the magnetization is the [100] direction[15], and the magnetization must pass through high-energy directions in order to move from one low-energy direction to another symmetry-related direction, as shown in Fig. 4. All of the atoms in the particle contribute to this magnetic-anisotropic effect, and thus the magnitude of the energy barrier is proportional to the particle volume. The proportionality parameter is readily obtained from single-crystal measurements, and for metallic iron this parameter is of order 10^{-2} J cm^{-3}.[15]

Because of demagnetization effects, the overall shape of the particle can also produce a magnetic-anisotropic effect[16], as shown in Fig. 5. Low-energy directions for the magnetization are now dictated by the particle shape and correspond to elongated-particle directions. The magnetic-anisotropy energy barrier for magnetization relaxation is again proportional to the particle volume, the proportionality parameter calculable from the particle shape[16], and the square of the magnetization magnitude. Owing to the large magnetization of metallic iron, the shape magnetic-anisotropy proportionality parameter is quite large even for moderate shape elongations

(1 J cm^{-3} for an ellipsoidal particle with a 0.5 aspect ratio). Thus, for metallic iron, shape magnetic anisotropy may indeed overshadow magnetocrystalline-anisotropic effects.

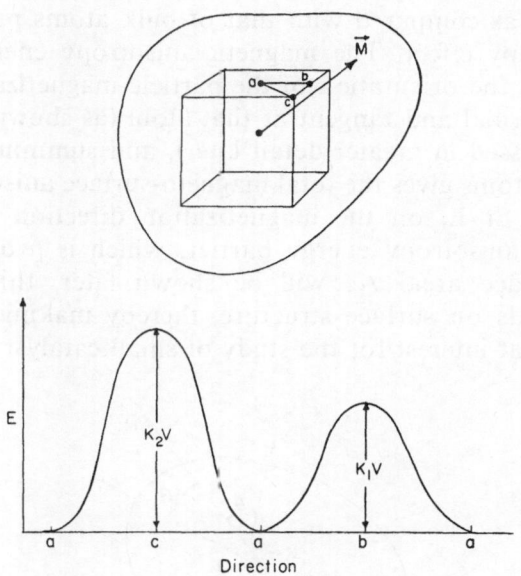

Fig. 4. Magnetocrystalline anisotropy.

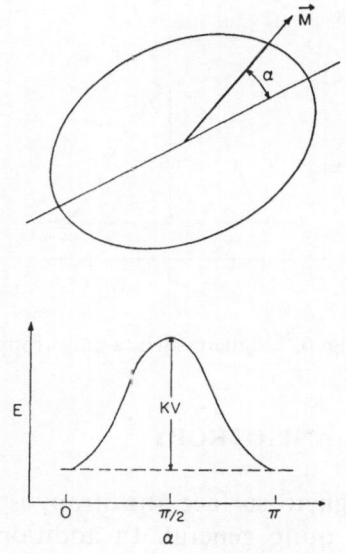

Fig. 5. Shape magnetic anisotropy.

Both of the previous magnetic anisotropies are volumetric effects, and as such do not directly relate to the problem of surface-structure measurement. However, in 1953 Néel proposed a phenomenological theory of "magneto-surface anisotropy"[17,18], according to which the lower symmetry of surface atoms as compared with that of bulk atoms provides a surface-sensitive anisotropy effect. The magnetic-anisotropy energy for a surface atom depends on the orientation of the particle magnetization with respect to the surface normal and tangent at that atom (as shown schematically in Fig. 6, and discussed in greater detail later), and summing this interaction over all surface atoms gives the total magneto-surface anisotropy energy, E_s. The dependence of E_s on the magnetization direction gives rise to the magneto-surface anisotropy energy barrier, which is proportional now to the particle surface area. As will be shown later, this proportionality parameter depends on surface structure, thereby making magneto-surface anisotropy of great interest for the study of small catalyst particles.

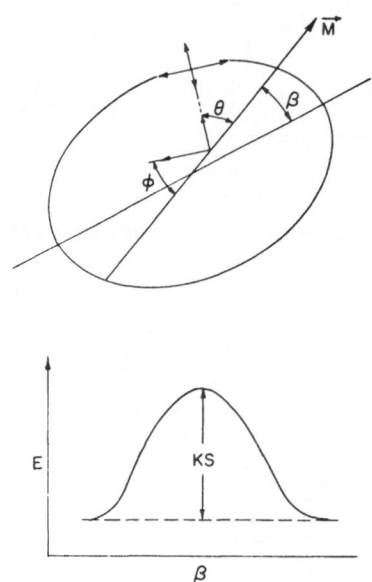

Fig. 6. Magneto-surface anisotropy.

5 MAGNETO-SURFACE ANISOTROPY

Néel's theory of magneto-surface anisotropy is based on symmetry considerations, and thus is quite general. In addition, though, Néel also estimated the magneto-surface anisotropy proportionality parameter to be 10^{-7} J cm^{-2}. Using this estimate, the particle-size dependence of the magneto-surface anisotropy energy barrier is readily obtained, and is compared with that for shape magnetic anisotropy in Fig. 7. With reference to this figure, for particle sizes less than approximately 6 nm, the surface may indeed provide the dominant magnetic-anisotropy energy barrier. As will be seen presently, Mössbauer spectroscopy is sensitive to magnetic-anisotropic effects. Using this technique, the magnitude of the magnetic-anisotropy energy barrier for the very small iron particles (ca. 2 nm) can be measured to be 10^{-20} J[4], a value that compares nicely with the magneto-surface anisotropy estimate of Néel (Fig. 7). Thus, not only are the small iron particles below the size for which magneto-surface anisotropy may be present, but the measured magnetic-anisotropy energy barrier for these particles agrees with that estimated for magneto-surface anisotropy. The presence of this anisotropic effect must, however, be experimentally verified before turning to surface-structure measurement.

To investigate the possible presence of magneto-surface anisotropy, the sensitivity of the magnetic-anisotropy energy barrier to a surface phenomenon, namely chemisorption, will be studied. This will first be done using Mössbauer spectroscopy. As mentioned earlier, particles with magnetization relaxation time, τ, greater than τ_c (ca. 10^{-8} second) show magnetic–anisotropic effects manifested by a six-peak Mössbauer spectrum.[13] Particles with τ less than τ_c behave superparamagnetically and, as

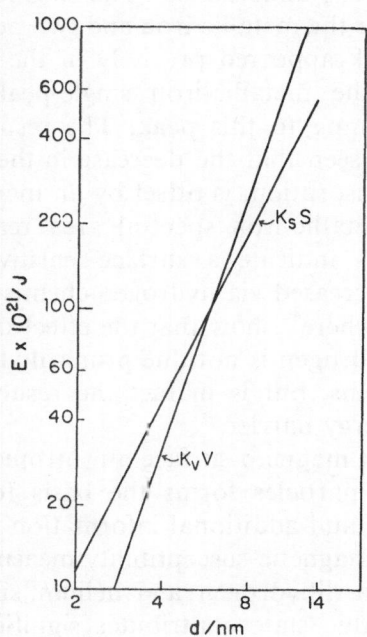

Fig. 7. Magnetic anisotropy energy barriers for small particles.

such, give rise to a single-peak Mössbauer spectrum. Thus, decreasing the magnitude of the magnetic-anisotropy energy barrier, which decreases the value of τ for each particle, causes a transition from a six-peak to a single-peak Mössbauer spectrum. This transition will take place only for those particles of the particle size distribution for which τ is sufficiently close to τ_c, as determined by the magnitude of the magnetic-anisotropy energy-barrier change.

A Mössbauer spectrum (100 mc ^{57}Co/Cu source) of the 1 percent Fe/MgO sample in dihydrogen at 683°K[4] is shown in Fig. 8a. Four of the peaks from the metallic-iron six-peak spectrum are clearly visible (the two peaks farthest to the right and the two peaks farthest to the left), with the remaining two peaks being hidden by the central spectral doublet. This central doublet results from Fe^{2+} in magnesium oxide[1], but at present it is the metallic-iron spectrum that is of greatest interest. A spectrum of 1 percent Fe/MgO at 683°K in flowing helium is shown in Fig. 8b, and by comparison with Fig. 8a it can be seen that chemisorbed hydrogen decreases the area of the six-peak spectral component. If indeed this spectral change is the result of a change in the magnetic-anisotropy energy barrier, then the decrease in six-peak area must be offset by an increase in the metallic-iron single-peak area, since hydrogen chemisorption should not change the total amount of metallic iron present in the catalyst. However, the single-peak component is hidden by the Fe^{2+} doublet, and for consistency, a computer analysis of the spectra was undertaken.[4] The spectra were each fitted with nine peaks: six peaks for the metallic iron and two peaks for the Fe^{2+}. In the final fits, the extra peak appeared precisely in the position expected if it were in fact due to the metallic-iron single-peak component, thereby providing physical meaning to this peak. The results of this analysis are given in Table IV. It is seen that the decrease in the six-peak area, accompanying hydrogen chemisorption, is offset by an increase in the single-peak area, with the total metallic-iron spectral area remaining constant. The Mössbauer spectra thus indicate a surface-sensitive magnetic-anisotropy energy barrier that is decreased via hydrogen chemisorption. Additional experiments reported elsewhere[4] show that the effect on the Mössbauer spectrum of chemisorbed hydrogen is not due primarily to cancellation (or pairing) of metallic-iron spins, but is in fact the result of a decrease in the magnetic-anisotropy energy barrier.

The conclusion that magneto-surface anisotropic effects are present on the small metallic-iron particles forms the basis for subsequent surface-structure measurements, and additional information to support this conclusion was sought. Thus, magnetic-susceptibility measurements were made on the Fe/MgO samples in dihydrogen and helium, since only that fraction of the iron in the metallic state contributes significantly to the observed susceptibility.[5]

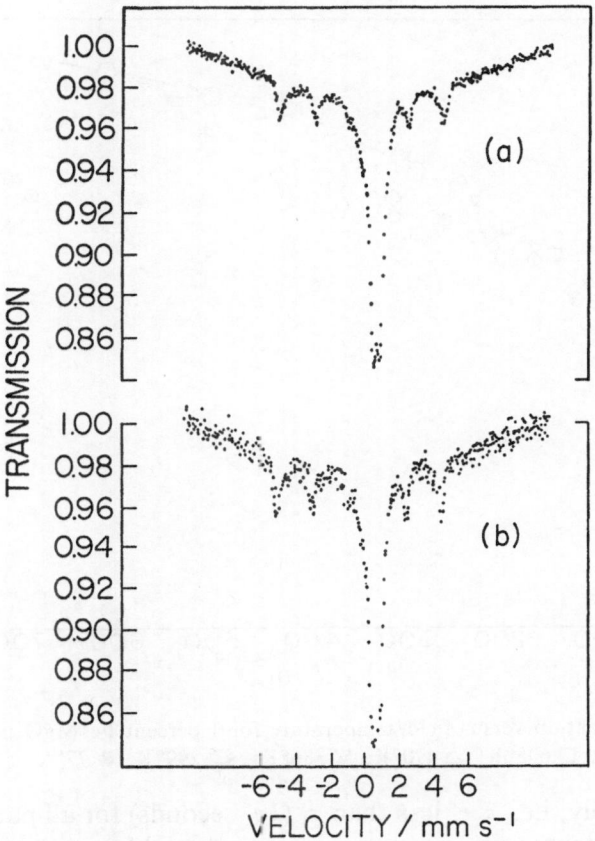

Fig. 8. Mössbauer spectra of 1 percent Fe/MgO at 683°K: (a) in H_2; (b) in helium.

Table IV. Mössbauer Spectral Areas of 1 Percent Fe/MgO at 683°K

	Metallic Iron Areas at 683°K, mm sec^{-1}	
	In Helium	In Hydrogen
Magnetically Split	0.053	0.047
Superparamagnetic	0.015	0.020
Total	0.068	0.067
Fraction Superparamagnetic	0.23	0.30

Figure 9 shows the curves of magnetization, M, versus applied field/temperature, H/T, for 1 percent Fe/MgO in dihydrogen, after dihydrogen reduction at 710°K. If the metallic-iron particles behave super-

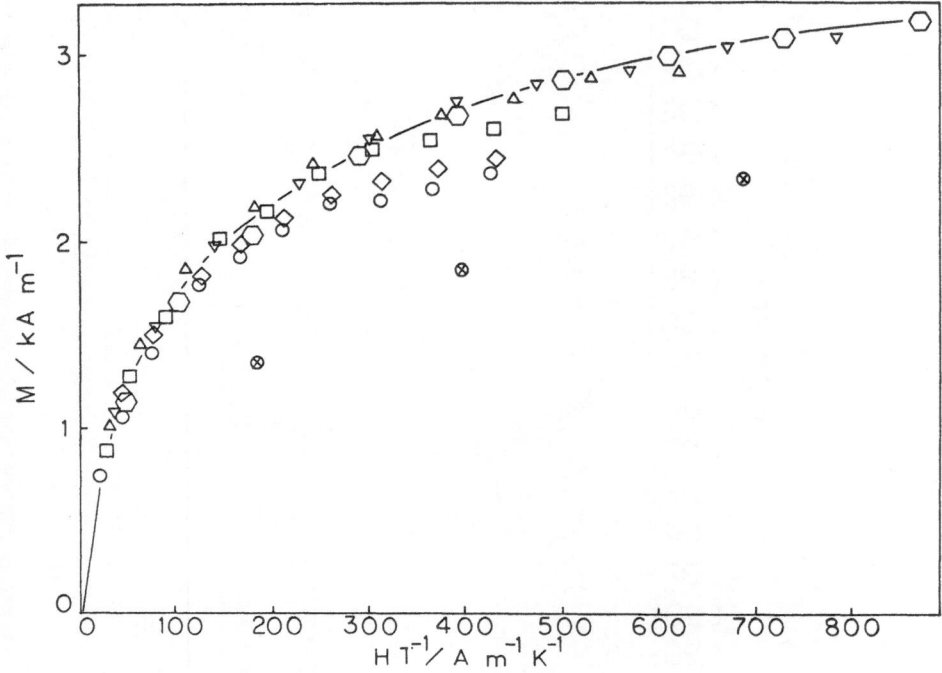

Fig. 9. Magnetization versus field/temperature for 1 percent Fe/MgO in H_2: ◯ 710°K; ◇704°K; ☐ 605°K; △ 490°K; ▽ 386°K; ◯ 299°K; ⊗ 77°K.

paramagnetically, i.e., τ is less than τ_c (10^2 seconds) for all particles, making magnetic-anisotropic effects unimportant, then M versus H/T curves at different temperatures should superimpose.[9] At and above 298°K, these curves superimpose, for the 1 percent Fe/MgO sample, the slight low M deviations observed with increasing temperature owing to the temperature dependence of the spontaneous magnetization. The M versus H/T curve at 77°K, however, is not in superposition with the others, indicating the importance of the magnetic-anisotropy energy barrier at this temperature. Purified helium was then passed over the 1 percent Fe/MgO sample for 1 hour at 710°K, after which magnetic-susceptibility data were collected at 300 and 77°K in helium. A comparison at 300°K of the effect of hydrogen chemisorption on the magnetic susceptibility can be made without reference to magnetic-anisotropic effects, while at 77°K, these latter effects must be considered.

This comparison is made at 300°K in Fig. 10. It is thereby seen that chemisorbed hydrogen (preadsorbed at 710°K) does not significantly affect the magnetization at a temperature above which magnetic-anisotropic effects can be neglected, in agreement with the conclusion reached using Mössbauer spectroscopy. As shown in Fig. 11, however, at 77°K, a

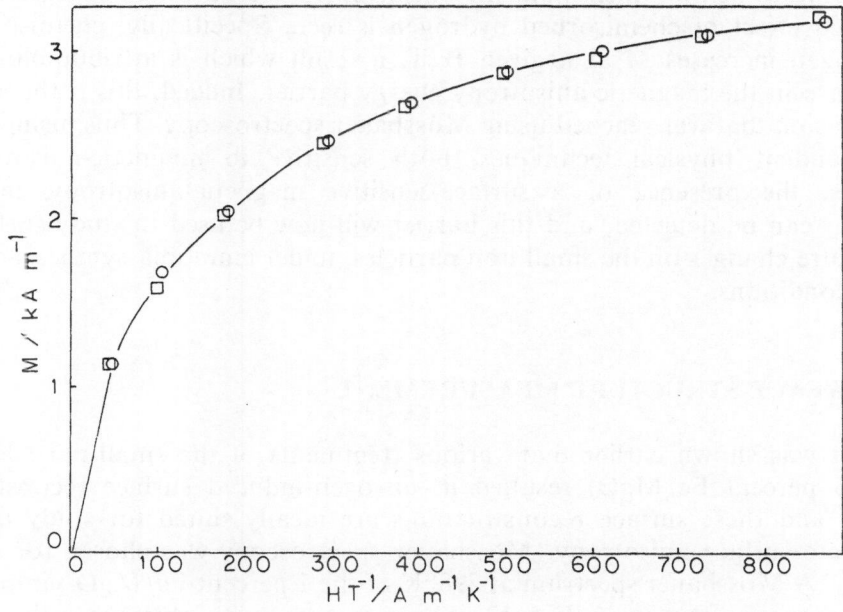

Fig. 10. Magnetization versus field/temperature for 1 percent Fe/MgO above the super-paramagnetic transition temperature: O in H_2 at 299°K; □ in helium at 303°K.

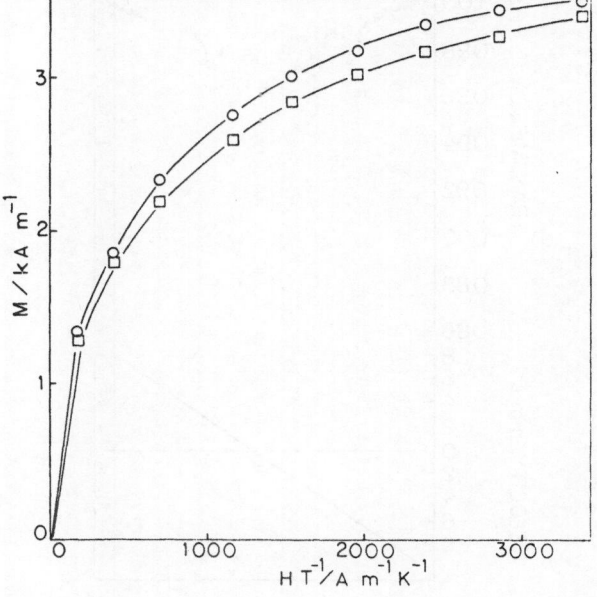

Fig. 11. Magnetization versus field/temperature for 1 percent Fe/MgO below super-paramagnetic transition temperature: O in H_2 at 77°K; □ in helium at 77°K.

temperature below which magnetic-anisotropic effects must be considered, a marked effect of chemisorbed hydrogen is seen. Specifically, chemisorbed hydrogen increases M at a given H/T, a result which is attributable to a decrease in the magnetic-anisotropy energy barrier. Indeed, this is the same conclusion that was reached using Mössbauer spectroscopy. Thus, using two independent physical techniques, both sensitive to magnetic-anisotropy effects, the presence of a surface-sensitive magnetic-anisotropy energy barrier can be detected, and this barrier will now be used to study surface-structure changes on the small iron particles, under ammonia-synthesis reaction conditions.

6 SURFACE-STRUCTURE MEASUREMENT

It was shown earlier that various treatments of the small particles (1 and 5 percent Fe/MgO) resulted in nitrogen-induced surface reconstructions, and these surface reconstructions are ideally suited for study using magneto-surface–anisotropy. Mössbauer spectroscopy was chosen for such study.[3] A Mössbauer spectrum at 298°K of the 1 percent Fe/MgO sample in dihydrogen is shown in Fig. 12. This spectrum was obtained using the constant-acceleration Doppler velocity mode, also shown in this figure.

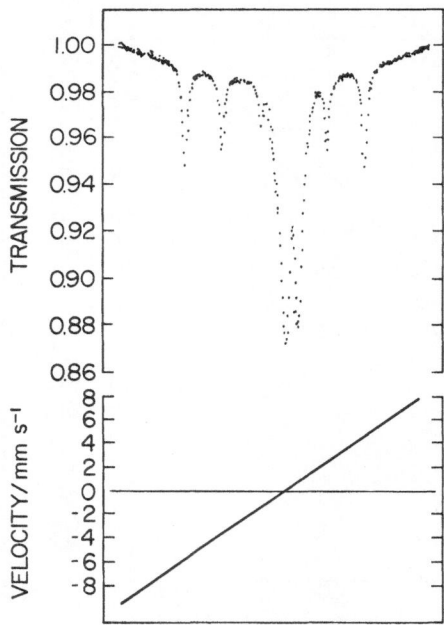

Fig. 12. Mössbauer spectrum (298°K in H₂) of reduced 1 percent Fe/MgO; constant acceleration mode.

Although this velocity mode is useful for obtaining information about all of the different chemical states of iron present in the sample[1], it is not suited for measuring small changes (ca. 10 percent) in the metallic-iron spectral parameters alone. For this reason, the velocity offset mode will be used for the metallic-iron surface-structure measurement; this mode, shown in Fig. 13, allows the two outermost metallic iron peaks to be scanned in detail.

The 3 percent Fe/MgO sample, for which the surface provides the dominant magnetic-anisotropy energy barrier[4], was reduced in palladium-diffused H_2 and then exposed to the $H_2:N_2 = 3:1$ reactant gas mixture at 683°K. Then, with the sample under the flowing reactant mixture, Mössbauer spectra were taken using the velocity offset mode. The area, A, of the metallic-iron peak, appearing at the most negative velocity, is plotted versus temperature in Fig. 14. The sample was then treated with ammonia, as previously described, and spectra were again taken with the sample under the flowing $H_2:N_2$ gas mixture. From Fig. 14, it is apparent that the ammonia treatment decreases A at a given temperature. The study of the ammonia-synthesis kinetics showed that the effect of the ammonia treatment on $\bar{N}$ could be reversed by subsequent treatment of the catalyst in dihydrogen. Carrying out this treatment, and taking Mössbauer spectra with the sample under the flowing reactant gas mixture, it is seen in Fig. 14 that

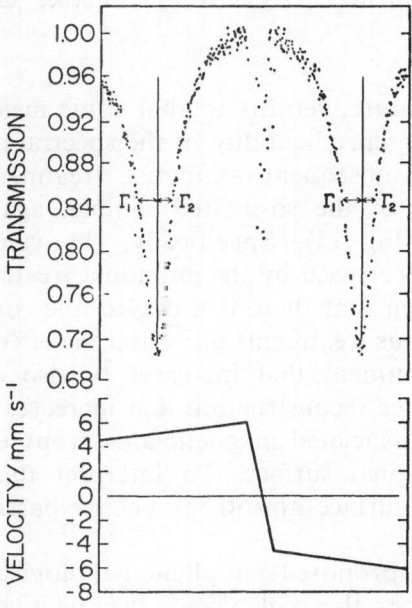

Fig. 13. Mössbauer spectrum (298°K in H_2) of reduced 3 percent Fe/MgO; velocity offset mode.

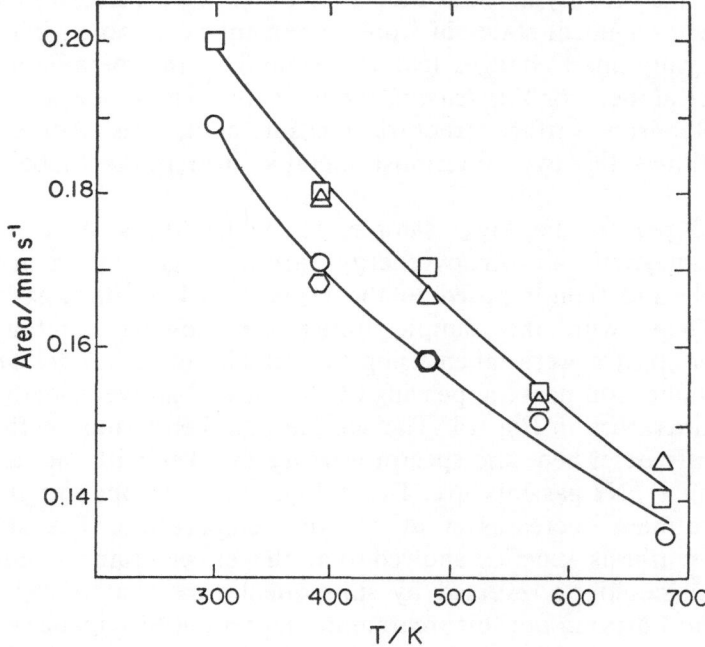

Fig. 14. Effect of sequential pretreatments on spectral area versus temperature of 3 percent Fe/MgO; Mössbauer spectra in $H_2:N_2$. Pretreatment sequence: □ H_2 reduction; O NH_3; △ H_2;   ◇ NH_3.

A, at a given temperature, returns to that value measured before the ammonia treatment. The reproducibility of the spectral-area changes is shown by the effect of a subsequent ammonia treatment on the A-versus-temperature behavior of the so-treated sample, again under the flowing $H_2:N_2$ gas mixture (Fig. 14). Specifically, the value of A at a given temperature again is decreased by the ammonia treatment.

It can now be seen that there is a one-to-one correspondence between the effects of the various treatments on $\overline{N}$ and the effects of these treatments on A. That is, a treatment that increases $\overline{N}$ also decreases A. Thus, a nitrogen-induced surface reconstruction that increases $\overline{N}$, creates a metallic-iron surface with an associated magnetic-anisotropy energy barrier smaller than that of the original surface. To interpret this effect, the relation between the magneto-surface-anisotropy energy barrier and surface structure must be discussed.

At the time Néel proposed the phenomenological theory of magneto-surface anisotropy, he also calculated the relative magnitudes of the magnetic-anisotropy energy barriers for different low-index crystal planes of various lattice structures. For bcc metallic iron, several of these low-index planes are shown in Fig. 15. The description of small particle surfaces in terms of exposed crystallographic planes, however, seems awkward in view of the many high-index planes of very small extent which are undoubtedly present. Instead, it seems more natural to specify the concentrations of the various surface sites that are present on the small particle surface, where a surface site, C_i, is defined as a surface atom with i nearest neighbors. For example, with reference to Fig. 15, the (100) plane of iron is a collection of C_4 sites, the (110) plane is composed of C_6 sites, and the (111) plane is a collection of C_7 and C_4 sites. The nature of these sites is perhaps more clearly shown in Fig. 16, in which the symmetries of the C_4, C_5, C_6, and C_7 sites can be seen to be $\overline{C}_{4v}$, $\overline{C}_s$, $\overline{C}_{2v}$, and $\overline{C}_{3v}$, respectively (Schönflies notation with the overhead—to distinguish the symmetry symbol from the name of the site C_i). It is in terms of these sites that Néel's theory of magneto-surface anisotropy must now be recast.

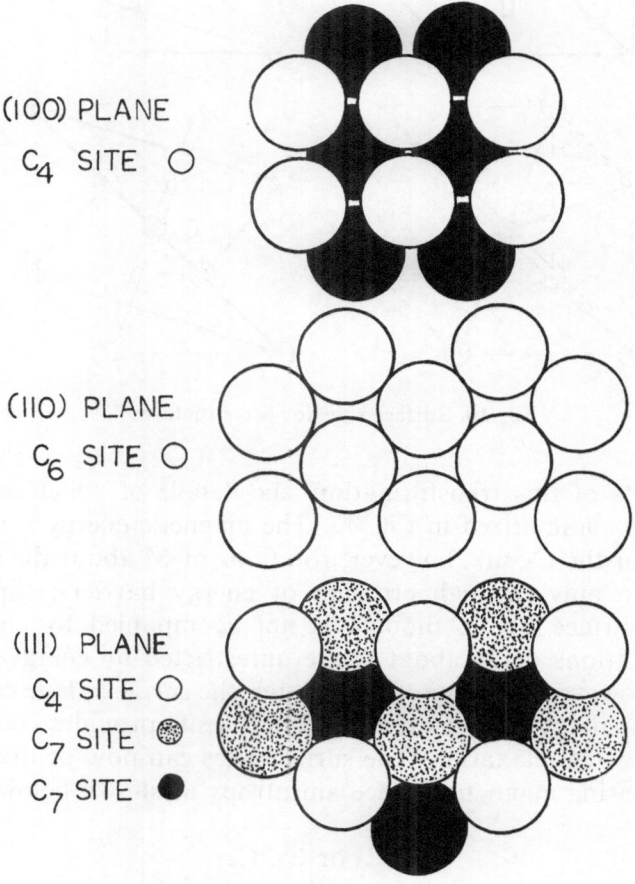

Fig. 15. Surface sites on bcc surface planes.

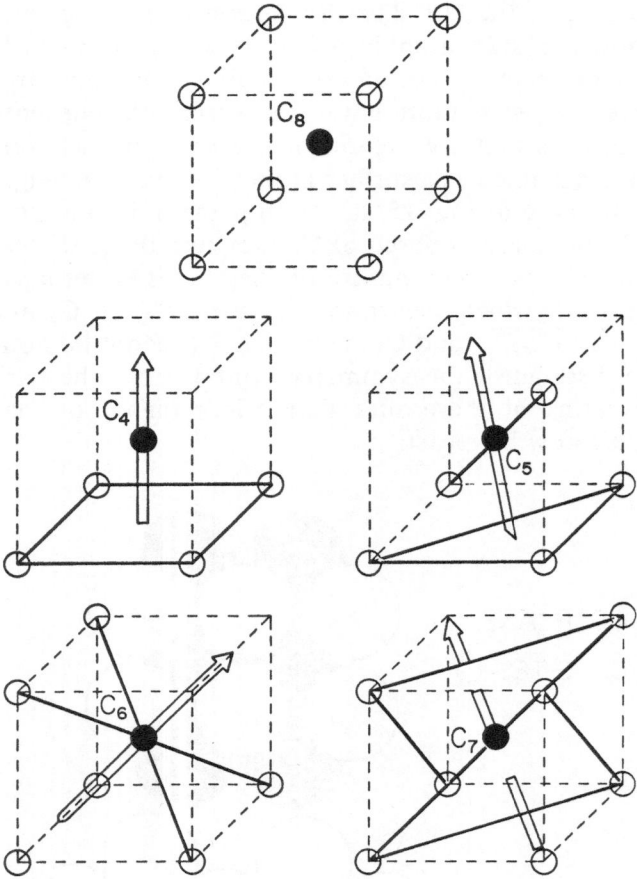

Fig. 16. Surface sites for bcc structures.

The results of this transformation, the details of which are presented elsewhere[4], are summarized in Fig. 17. The magnetic energy is isotropic for the C_4 site. For the C_5 site, however, rotations of **M** about the surface normal, **N**, occur only through crossing of energy barriers; flipping of **M** through the surface plane, though, is not acompanied by energy-barrier crossings. Rotations of **M** about **N** are unrestricted by energy barriers for the C_7 sites; however, flipping of **M** through the surface plane can now take place only via energy-barrier crossing. The C_6 site provides energy barriers for all modes of **M** relaxation. The surface sites can now be arranged in an order of decreasing magneto-surface anisotropy as shown below:

$$C_6, \ C_5 \ \text{or} \ C_7, \ C_4$$

decreasing magneto-surface anisotropy

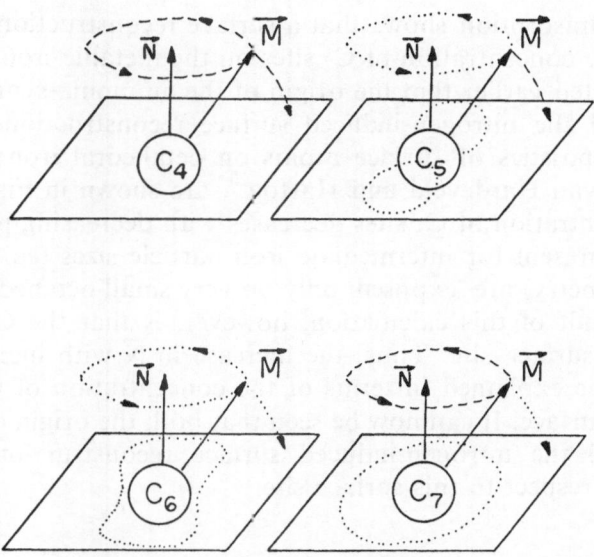

Fig. 17. Magneto-surface anisotropy energy for surface atoms: - - - - unrestricted motion; . . . . restricted motion.

Thus, the effect of a nitrogen-induced surface reconstruction that increases $\overline{N}$, is to convert sites from the left to sites toward the right on the above sequence of sites. To identify that surface site which is created via this surface reconstruction, another sequence of sites is required, and this additional sequence is provided by carbon monoxide chemisorption.

On the basis of the steric considerations in a monolayer, Brunauer and Emmett[19] postulated that different crystallographic planes of iron would chemisorb, at saturation, different amounts of carbon monoxide. Accepting the views of these workers, which appear to be quite reasonable[3], the surface sites can be arranged in an order of decreasing carbon monoxide uptake potential, as shown below:

$$C_4, \ C_5, \ C_6, \ C_7$$

decreasing CO uptake potential

Experimentally it was found that the ammonia treatment, which increases $\overline{N}$, decreases the carbon monoxide uptake of the metallic-iron surface (ca. 10 percent).[3] Using Mössbauer spectroscopy, it was also shown that the iron particles do not sinter during this treatment. Thus, the effect of the nitrogen-induced surface reconstruction that increases $\overline{N}$ is to convert sites from the left to sites toward the right on the above sequence of sites. Comparison of the sequences of sites obtained from Mössbauer spectroscopy and carbon monoxide chemisorption shows that a surface reconstruction that increases $\overline{N}$ increases the concentration of C_7 sites on the metallic-iron surface.

It was stated earlier that the origin of the ammonia-synthesis structure sensitivity and the nitrogen-induced surface reconstruction seemed to be related. The statistics of surface atoms on octahedral iron crystallites, as calculated by van Hardeveld and Hartog[20], are shown in Fig. 18. It is seen that the concentration of C_7 sites decreases with decreasing particle size, C_6 sites become present for intermediate iron particle sizes (ca. 5 nm), and C_4 sites ($\overline{C}_{4v}$ symmetry) are exposed only on very small octahedral crystallites. The major result of this calculation, however, is that the C_7 site is not a small-particle surface site. Thus, the increase in $\overline{N}$ with increasing particle size can also be explained in terms of the concentration of C_7 sites on the metallic-iron surface. It can now be seen that both the origin of the structure sensitivity and the nitrogen-induced surface reconstructions can be understood with respect to this surface site.

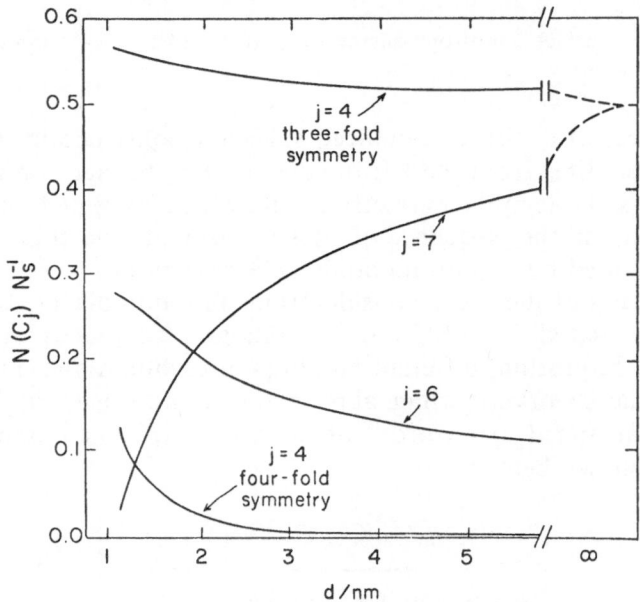

Fig. 18. Statistics of surface atoms on octahedral iron crystallites.[20] N (Cj) = number of surface atoms with j nearest neighbors; N_s = total number of surface atoms; D = particle size.

7 CONCLUSION

In this study, a correlation has been found between increasing ammonia synthesis catalytic activity over iron and increasing concentration of C_7 sites on the iron surface. Not only is this conclusion self-consistent in this study of Fe/MgO catalysts, but it is in agreement with the beliefs of others based on different catalyst systems. Brill and Kurzidim[21] observed that the catalytic activity for the ammonia synthesis on metallic iron was higher when the metal was formed via $H_2:N_2 = 3:1$ reduction of magnetite, Fe_3O_4, than when formed by magnetite reduction with dihydrogen. It was also observed with field emission microscopy that dinitrogen reconstructs an iron tip by adsorbing preferentially on (111) planes and increasing their extent.[7] Hence, it was surmised by Brill and Kurzidim that the role of dinitrogen in the reduction of magnetite, forming a more active catalytic surface, was to favor the appearance of (111) faces which, in turn, were assumed to possess a higher catalytic activity than other faces for the ammonia synthesis. More recently, McAllister and Hansen[22] found that the ammonia decomposition proceeds at a rate 10 times higher on the (111) face than on the (100) or (110) faces of tungsten single crystals. In view of the fact that the (111) face of bcc metals is the low-index plane that exposes C_7 sites, the conclusions of the present study are in line with the experimental findings of others.

Our work presents the first clear evidence of Néel's surface magnetic anisotropy and the first assignment of surface sites for a structure-sensitive reaction on a metal. This was made possible by Mössbauer spectroscopy. In particular, Mössbauer spectra of metallic ferromagnetic iron on magnesium oxide showed the same internal magnetic field and isomer shift as those of bulk iron.[1] Any sizable electronic effect between the small supported metal particles and the support is thus very unlikely, and the observed effects of particle size on reaction rate can be ascribed to structural and not electronic effects.

REFERENCES

1. Boudart, M., Delbouille, A., Dumesic, J. A., Khammouma, S., and Topsøe, H., *J. Catalysis*, in press (Part I of series: Surface, Catalytic and Magnetic Properties of Small Iron Particles).
2. Dumesic, J. A., Topsøe, H., Khammouma, S., and Boudart, M., *J. Catalysis*, in press (Part II of series).
3. Dumesic, J. A., Topsøe, H., and Boudart, M., *J. Catalysis*, in press (Part III of series).
4. Dumesic, J. A., Topsøe, H., and Boudart, M., submitted to *Surface Sci.* (Part IV of series).
5. Dumesic, J. A., Topsøe, H., Anderson, J. H., and Boudart, M., submitted to *Surface Sci* (Part V of series).
6. Bokhoven, C., van Heerden, C., Westrik, R., and Zwietering, P., in *Catalysis*, Vol. III, Emmett, P. H. (Ed.), Reinhold Pub. Co., New York (1955), p. 283.
7. Brill, R., Richter, E. L., and Ruch, E., *Angew. Chem. Intern. Ed.*, **6**, 882 (1967).
8. Kummer, J. T., and Emmett, P. H., *J. Phys. Chem.*, **55**, 337 (1951).
9. Selwood, P. W., *Adsorption and Collective Paramagnetism*, Academic Press, New York, (1952), Chapt. 3.

10. Morrish, A. H., *The Physical Principles of Magnetism,* John Wiley and Sons Inc., New York (1965), p. 344.
11. Brown, W. F., Jr., *J. Appl. Phys., 30,* 1308 (1959).
12. Bean, C. P., and Livingston, J. D., *J. Appl. Phys., 30,* 120S (1959).
13. Kündig, W., Ando, K. J., Lindquist, R. H., and Constabaris, G., *Czech. J. Phys., 17,* 467 (1967).
14. McNab, T. K., Ph.D. Dissertation, University of Western Australia, 1968.
15. Morrish, A. H., *op. cit.,* p. 312.
16. Morrish, A. H., *op. cit.,* Chapt. 7.
17. Néel, L., *Compt. Rend. Acad. Sci., 237,* 1468 (1953).
18. Néel, L., *J. Phys. Radium, 15,* 225 (1954).
19. Brunauer, S., and Emmett, P. H., *J. Am. Chem. Soc., 62,* 1732 (1940).
20. van Hardeveld, R., and Hartog, F., *Surface Sci., 15,* 189 (1969).
21. Brill, R., and Kurzidim, J., *Colloques Int., CNRS, 187,* 99 (1969).
22. McAllister, J., and Hansen, R. S., *J. Chem. Phys., 59,* 414 (1973).

DISCUSSION on Paper by M. Boudart

EISCHENS: In comparison of specific reaction rates as a function of particle size, especially for supported metals, is there a possibility that particles of different sizes may also be at different temperatures? For example, in the case of an exothermic reaction, I would expect that the metal might be at a higher temperature than the support and that smaller particles would be more prone to this excess heating. Thus, small particles would give a misleading indication of higher specific activity.

BOUDART: This is a very good point as it is very difficult to exclude the occurrence of microscopic heating of small metal particles even though assurances can be given concerning the absence of macroscopic gradients of temperature. Fortunately, in our study of ammonia synthesis on small particles of iron, it turns out that the small particles are less active per unit surface area than the larger particles. In cases where the opposite is observed, additional information is required to answer your question. Thus, in the case of our recent study (F. V. Hanson, Ph.D. Dissertation, Stanford, 1975) of the reaction between H_2 and O_2 on Pt/SiO_2, we have found that the reaction was structure insensitive in excess oxygen but structure sensitive in excess hydrogen. In the latter case, the small particles are indeed more active than the larger ones. But the study was carried out in such a way that in all cases the *measured* rate was approximately the same. Hence, the fact that, on the same samples, the reaction was found to be structure insensitive in excess oxygen eliminates the possible effect of microscopic heating in this case, and therefore also in excess hydrogen, where differences in activity with particle size were found.

BLOCK: The structure insensitivity of hydrocarbon reactions on small metal particles should be discussed in connection with field emission observations. There, most of those substances show surface selectivity in chemisorption. Could it be possible that in that early state of reaction, chemisorption will be completed and the further reaction will proceed on sites which have no memory of original crystal plane sites?

BOUDART: This is a very good possibility. To know the answer, it will be necessary to observe the catalyst surface in its working state.

WALTER: We carried out electrochemical hydrogen oxidation on platinum of diverse degree of disposition: plain sheets, wire nets, platinum black, and Raney platinum. The current-density-pertinent surface area was always the same, provided that no surface contamination was present. This finding is in good agreement with the other experimental result on catalyst activity of finely divided supported platinum catalysts.

    Is the determination of crystal particle size of iron on MgO influenced by the fact that a substantial percentage of the iron present is not reduced?

BOUDART: In reply to your question, the strong chemisorption of CO which we used to measure the dispersion of metallic iron is specific in the sense that it is not observed on the Fe^{2+} ions eventually present at the surface of the MgO support. In fact, we obtained at least for some of our Fe/MgO samples, good agreement between average metal particle size obtained by CO chemisorption, electron microscopy, X-ray line broadening, and magnetic measurements.

EMMETT: Is there a chance that the catalyst particles supported on magnesium oxide may be partially covered by magnesium oxide particles? I am referring to magnesium oxide as molecules and not as iron-magnesium oxide compounds.

BOUDART: In the case of our 1.5 nm metallic-iron particles, if a substantial fraction of the surface and therefore also of the total number of iron atoms in the particle was covered with magnesium oxide molecules, we should see changes in the Mössbauer spectra, which we do not see unless there is no chemical or electronic interaction between MgO and iron. We believe there is no such interaction and that we therefore observe phenomena pertaining to metallic iron.

YATES: We have observed a surface-structure-insensitive catalytic decomposition process on two clean single crystals of tungsten. The experiments compared W(100) and W(111) for the catalytic decomposition of formaldehyde. Thermal programmed desorption and work-function measurements were used. In addition to the major products H_2 and CO, small amounts of CH_4 an d CO_2 were produced at H_2CO coverages above $\sim \frac{1}{2}$ monolayer. The kinetics of thermal programmed liberation of Ch_4 indicated on both crystals that two stages of activated decomposition occurred—at about 300°K and about 500°K. The evolution curves for CH_4 are very similar for the two crystals. We believe that we are observing rate processes which are related to the decomposition of intermediate surface species whose properties do not strongly depend on the underlying crystal structure of the tungsten [*J. Catalysis,* **30,** 260 (1973)].

CHARACTERIZATION OF POROUS MATERIALS, IN PARTICULAR RANEY PLATINUM

G. Walter and E. Robens

Battelle-Institut e.V.
Frankfurt/Main
West Germany

ABSTRACT

A critical evaluation of the different methods of pore-structure analysis shows that a satisfactory statistical description of a porous composite is still missing. To deduce diffusion processes as encountered in heterogeneous catalysis, data from sorption measurements and microscopic observations are not sufficient. Investigations on Raney platinum by sorption and electrochemical measurement, scanning electron microscopy, and impedance measurement are discussed.

1 INTRODUCTION

Most heterogeneous catalytic processes are carried out with highly porous materials. Surface areas of some hundred m^2/cm^3 or $10^8\ m^2/m^3$ are quite usual.

Thus, on most catalysts, the surface is exhibited inside the catalyst pellets or grains. The access or flow of reactants toward the surface is thereby limited, as is expressed by the Thiele modulus, which describes the ratio of true surface reaction rate to the rate of flow toward the surface. The flow-rate expression is given by the diffusivity, which may be compared with diffusion constants of fluids not restricted by small and tortuous pores. In small pores with a diameter of, say, some 10 nm, the mean free path of gases is larger than the pore width, and hence, gas molecules collide with the wall more often than with other gas molecules. This is the Knudsen regime of flow where the components of a mixture diffuse independently, i.e., the flow rate depends on pore size, not on the mole fractions of the components. The pores neither go straight through the catalyst pellet, nor is their diameter constant. Furthermore, they are highly interconnected. As a consequence, it is usual to have recourse to the correction term tortuosity. In the papers known to the authors, the gas flow is not calculated by the route: pore-size determination → determination of tortuosity → calculation of flow properties, but rather the tortuosity is found from independent determinations of the pore size and the flow rates of noncatalyzed gases through pellets.

2 SELECTED METHODS FOR INVESTIGATING PORE STRUCTURES

2.1 The Pore-System Concept

For pores with a radius of more than, say, 20 nm, mercury porosimetry is widely in use. It is based on Pascal's equation

$$p = \frac{2\sigma}{r_m} \tag{1}$$

that connects a pressure difference p across a curved liquid surface and its mean curvature r_m (σ = surface tension). Smaller pores are studied by the application of Kelvin's equation to capillary condensation. It reads as follows:

$$RT \ln \frac{p}{p_s} = \frac{2\sigma V_M}{r_m} \tag{2}$$

and describes the influence of the mean radius of curvature r_m on the ratio of vapor pressure p to p_s, the vapor pressure of a flat surface (R = gas constant, T = temperature, V_M = mole volume). Of course, Pascal's and Kelvin's equations are closely related thermodynamically.

In cylindrical pores at complete wetting of the pore walls, r_m is identical with the pore radius minus the thickness of an adsorbed layer t. In the following discussion, t will be neglected, although it may be significant with small pores having radii of, say, less than 5 nm. Micropores, i.e., pores of radii less than 1 nm will not be considered here. Experimentally, the pore-size determination by the Kelvin equation is an extension of the surface-area determination by the BET method toward higher relative pressures p/p_s.

Soon after the widespread application of the BET technique and capillary condensation method, it was realized that pore systems are not simple assemblies of straight capillaries of different lengths and widths. This is best documented by the 1958 Bristol Conference on Porous Solids.[1] There, a large number of pore structures and models were proposed. However, it turned out that, although it is possible to deduct capillary condensation phenomena from pore structure, it is not possible, unfortunately, to derive a specific pore system from the adsorption-condensation isotherm. This is easily understood by realizing that the mean radius of curvature depends on *two* principal radii, r_1 and r_2, of a surface:

$$\frac{1}{r_m} = \frac{1}{2}\left(\frac{1}{r_1} + \frac{1}{r_2}\right) \tag{3}$$

which may have different signs.

For instance, the pendular condensate rings between two contacting spheres will assume the shape of a catenoid with $r_1 = -r_2$ and, consequently, infinite mean radius of curvature at saturation pressure (Fig. 1). Hence, in a porous material loosely composed of strings of spherical particles, capillary condensation will stop before all interior space is filled with condensate, as shown schematically in Fig. 2. With assemblies of spheres of closer packing, capillary condensation will occur in all interstices before saturation is reached because the pendular condensate rings merge. The geometry of pendular rings between spheres or cones at lower relative pressures is somewhat complicated, although not beyond the calculating power of computers.

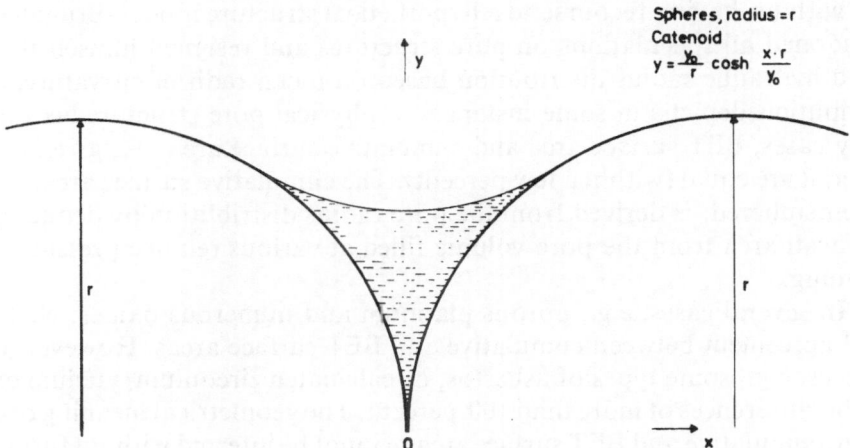

Fig. 1. Meniscus between adjacent spheres; mean curvature zero; $p = p_s$.

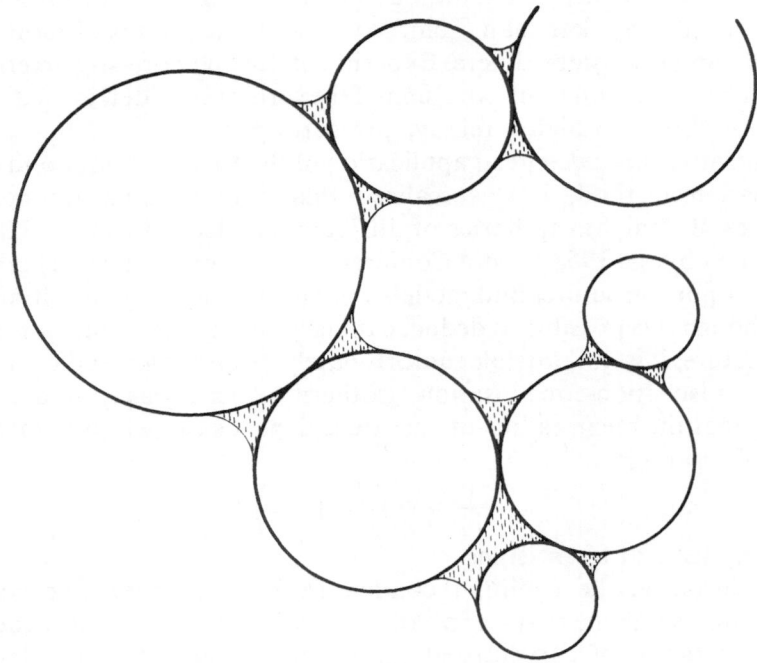

Fig. 2. Assembly of spheres; annular capillary condensation of vapor.

The above discussion illustrates that "pore radii" derived from capillary condensation between spheres do not reflect the paths of gas molecules as is suggested by the term "pore". The same argument applies to any assembly of packed particles such as crystals.

Since it is not possible to deduct pore structures from capillary condensation without having recourse to a hypothetical structure model, Brunauer has abandoned all speculations on pore structures and resigned himself to a so-called hydraulic radius distribution based on mean radii of curvature.[2] This distribution depicts, in some instances, a physical pore structure because, in many cases, BET surface area and cumulative surface area of a given porous material are equal (within a few percent). The cumulative surface area, as may be remembered, is derived from the pore radius distribution by deducing the pore wall area from the pore volume filled at various relative pressures, and summing.

In several cases, e.g., porous platinum and numerous oxides, we found good agreement between cumulative and BET surface areas. However, other samples, e.g., some types of asbestos, or calcinated zirconium/yttrium oxide, exhibit differences of more than 100 percent. The geometrical meaning of coinciding cumulative and BET surface areas cannot be inferred without further independent experimental evidence of the structure of a given porous material.

2.2 The Corpuscular System Concept

Quite another approach to a description of porous materials is given by the "dusty gas" model. It does not start from the concept of tubular pore systems but, rather, from isolated solid particles scattering the incoming gas molecules. The first calculation of a highly diluted suspension of spheres in a gas was made by Derjaguin[3] in 1946. Later on, Prager and Weissberg[4-6], and Watson and Mason[7-9], hit on the same idea. They extended the model to more concentrated suspensions, e.g., true pore systems. However, mathematical difficulties led Weissberg[4] to a model in which the position of the sphere centers is assumed to be statistically independent. Even at moderate void fractions, the model leads to considerable overlap of the spheres, as is exemplified by Weissberg's estimate that the void fraction U is

$$U = e^{-V} \tag{4}$$

where V is the volume fraction of all spheres making up the porous material. With U = 0.5, V is calculated at 0.69, about 40 percent too large because, physically, U cannot exceed 1-V. The distribution function of the sphere radii is adjusted in such a way that the experimentally determined surface area equals the total surface area of all spheres, multiplied by the void fraction. Thus, the model will not be very useful for porous systems of moderate or low porosity and, furthermore, the particle radius is essentially an adjustable parameter.

Van Eekelen[10] has recently shown that in Weissberg's model, very different sphere radius distributions can lead to very similar pore radius distributions. He stresses the fact that it is impossible to deduce particle size distributions from pore radius distributions when assuming Weissberg's model.

So we are left again with the same unsolved problem as in the case of models based on pore systems: a satisfactory statistical description of a porous composite of isometric crystals or spheres is still lacking.

3 INVESTIGATION OF THE PORE STRUCTURE OF RANEY PLATINUM

A practical porous system, with which we have been working for several years, is Raney platinum.[11-17] Raney platinum is made by treating alloys such as $PtAl_3$ or $PtAl_4$ with aqueous potassium hydroxide solution. The aluminum is dissolved and the platinum recrystallizes to form a sponge of tiny crystals. Shortly after formation of the sponge, the specific surface area S_g is determined to be about 40 m^2/g. Upon exposure to water vapor at 200°C for more than 10 hours, the surface area is found to be 8 to 10 m^2/g.

When comparing the surface area of Raney platinum with that of other porous materials one has to keep in mind the high density of metallic platinum (21.4 g/cm^2). Thus, in terms of surface area per unit solid volume S_v, a sample of Raney platinum of $S_g = 40$ m^2/g and a sample of alumina of $S_g = 250$ m^2/g are equally dispersed ($V_s = 850$ m^2/cm^3). It is, therefore, felt that Raney platinum is representative of a large group of disperse microcrystalline materials.

Earlier we measured the pore radius distribution of Raney platinum, assuming cylindrical pores and using the capillary condensation method.[16] The pore radius distribution was found to be sharply peaked at 2 nm.

From the specific surface area (40 m^2/g), assuming a very narrow particle size distribution and cubic particle shape, a mean particle size of 7 nm is estimated.

The random sphere model yields an average sphere radius of

$$r_m = \frac{3U}{S_v} \cdot \ln \frac{1}{U} \tag{5}$$

With the values of U (= 0.5) and S_v determined on Raney platinum, r_m is calculated to be 10 nm, a value in better accordance with the estimated particle size.

We have also tried to obtain information on the presence of larger pores by SEM. Figs. 3a and 3b represent the surface of a Raney Pt sponge made by treatment of grain of PtAl$_3$ alloy, without mechanically disturbing it, during and after dissolution of the aluminum. The grain boundaries of the original PtAl$_3$ are still clearly visible, but no pores of a width exceeding the resolution limit of our instrument (~2.5 nm) were found on the various sections of the exposed surface of the sponge.

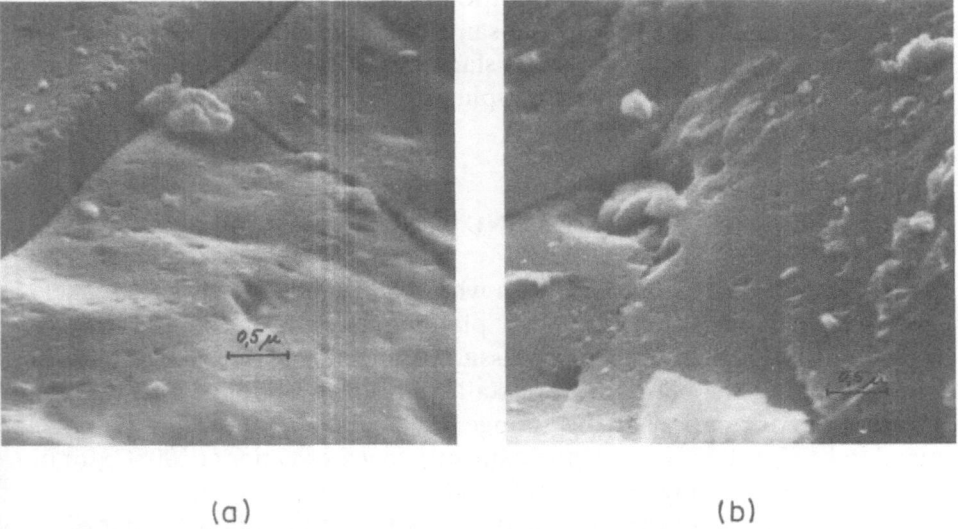

(a)   (b)

Fig. 3. Scanning electron micrographs of the surface of Raney platinum.

In addition, we conducted impedance measurements on Raney platinum immersed in an electrolyte solution in order to determine its mean pore radii. Electrodes were made in the following way. A powder of 50 percent by volume of $PtAl_3$ having a mean particle size of 30 μm was compacted with gold powder of the same particle size to form a cylindrical pellet (thickness 0.5 mm, radius 4 mm). The pellet was first immersed in 3N KOH in order to dissolve the aluminum from the alloy, and then washed with distilled water and immersed in 0.02N H_2SO_4. As the gold is not affected by this treatment, the surface area is almost exclusively generated by the finely divided Raney platinum. In an earlier investigation, close agreement was found between the BET area of such a pellet and its surface area estimated from the amount of chemisorbed hydrogen electrochemically deposited from an acid solution.*

The mean pore radius may be calculated from the impedance measurements by the following equation[18]:

$$r_0 = 2\pi \cdot d^2 \cdot K \cdot \rho \cdot f_0 \qquad (6)$$

where d = thickness of the electrode; K = capacitance per unit area of the electrolytic double layer; ρ = conductivity of the electrolyte; and f_0 = limiting frequency. Essentially, the measurement consists in the determination of the electrode capacitance as a function of frequency. (The details of the calculation will be discussed in a forthcoming paper.)

At low frequencies, the entire surface area is exposed to the electrolyte and acts as plate of the capacitor metal-electrolyte; whereas, at frequencies of f_0, screening of the inner surface takes place so that the capacitance measured is reduced.

With d = 0.07 cm, K = $2 \cdot 10^5$ $F \cdot cm^2$, ρ = 300 $\Omega \cdot cm$, and $f_0 = 3 \cdot 10^1$; r_0 is found to be 5.4 nm. The accuracy of this method has not yet been assessed, but we estimate that it will be of the order of 50 percent. The main source of error is the specific capacitance K, which is very sensitive to contaminations of the catalyst surface.

4 CONCLUSIONS

The various methods and structural models applied to Raney platinum yield data which may deviate by one order of magnitude. It is felt that this situation is representative of most of the porous materials encountered in catalysis research: it is as yet impossible to give a statistical description of porous matter such that the parameters can be experimentally determined to provide the basis for accurate calculations of flow and diffusion rates of fluids.

---

*The amount of hydrogen deposited was determined from the electric charge taken up by the pellet within a suitable (electrochemical) potential range.[16]

ACKNOWLEDGEMENT

We would like to thank our colleagues, Dr. R. Knödler and Mr. A. Köhling, for the preparation of the samples and the impedance measurements, and Dr. R. König for the SEM investigations.

REFERENCES

1. Everett, D. H., and Stone, F. S. (Eds.), *Structure and Properties of Porous Materials*, Butterworths, London (1958).
2. Brunauer, S., Mikhail, R. Sh., and Bodor, E. E., *J. Colloid Interface Sci.*, **24**, 451 (1967).
3. Derjaguin, B., *Comp. Rend. Acad. Sci. USSR*, **53**, 623 (1946).
4. Weissberg, H. L., *J. Appl. Phys.*, **34**, 2636 (1963).
5. Strieder, W. C., and Prager, S., *Physics of Fluids*, **11**, 2544 (1968).
6. Strieder, W., *J. Chem. Phys.*, **54**, 4050 (1971).
7. Evans, R. B., III, Watson, G. M., and Mason, E. A., *J. Chem. Phys.*, **35**, 2076 (1961).
8. Evans, R. B., III, Watson, G. M., and Mason, E. A., *J. Chem. Phys.*, **36**, 1894 (1962).
9. Mason, E. A., Evans, R. B., III, and Watson, G. M., *J. Chem. Phys.*, **38**, 1808 (1963).
10. Van Eekelen, H.A.M., *J. Catalysis*, **29**, 75 (1973).
11. Krupp, H., Rabenhorst, H., Sandstede, G., Walter, G., and Jones, R. Mc., *J. Electrochem. Soc.*, **109**, 553 (1962).
12. Binder, H., Köhling, A., Krupp, H., Richter, K., and Sandstede, G., *J. Electrochem. Soc.*, **112**, 355 (1965).
13. Binder, H., Köhling, A., and Sandstede, G., *Adv. Energy Conversion*, **6**, 135 (1966).
14. Sandstede, G., Walter, G., and Wurzbacher, G., *Nature*, **216**, 476 (1967).
15. Walter, G., Wurzbacher, G., and Krafczyk, B., *J. Catalysis*, **10**, 336 (1968).
16. Binder, H., Köhling, A., Metzelthin, K., Sandstede, G., and Schrecker, M.-L., *Chem.-Ing.-Technik*, **40** 586 (1968).
17. Binder, H., Kohling, A., and Sandstede, G., in *From Electrocatalysis to Fuel Cells*, G. Sandstede (Ed.) Univ. of Washington Press, Seattle and London (1972), p. 15.
18. de Levie, R., *Electrochim. Acta*, **9**, 1231, (1964); see also Euler, K. J., *Electrochim. Acta*, **17**, 619 (1972).

DISCUSSION on Paper by G. Walter

EMMETT: Why do you refrain from using the pore-size-distribution method of Barrett, Joyner, and Hallenda and use instead the Kelvin-equation calculations?

WALTER: As hinted at in the paper, the Kelvin equation has to be applied to the meniscus in mesopores after correction for the thickness t of an adsorbate layer. The correction may be made by the Barrett, Joyner, and Hallenda method (1951) or the more recent Kiselev-Brunauer, Mikhail, and Bodor method (1969). This, however, does not solve the problem of assigning pore radii to mean radii of curvature of the capillary condensate.

EMMETT: You probably understand that you must expect a difference in pore volume and pore size distribution as measured by the fluid-flow methods, and as measured by adsorption methods. Clearly the adsorption methods will include pores that dead end and that will not be measured by your flow systems.

WALTER: The steady-state fluid-flow (e.g., Kozeny-Carman) methods will certainly not reflect the presence of dead-end pores; the nonsteady-state method of Barrer, however, will. It is felt, nevertheless, that because of the complex flow pattern in interconnected pore systems of pore widths comparable to the mean free path of gas molecules, the interpretation of flow experiments cannot do without the introduction of model assumptions.

IMELIK: I certainly agree that not very much progress has been made in porous-texture theory and that improvements are needed. However, for gas-phase studies, the BJH method gives quite reliable results if the amplitude of distribution of pore radius is not too wide. The agreement between cumulative surface (S_{cum}) and the BET surface (S_{BET}) is then often in the range of about 5 percent. The BJH data agree also rather well with pore size distributions obtained by low-angle X-ray scattering measurements. However, the Kelvin equation must be used only for a pore-radius range where it can be applied. In this respect, the lower limit you indicated for this equation seems to be too low. It is known from N_2 adsorption isotherms that the hysteresis loop does not occur below p/p_0 about 0.4, which corresponds to a pore radius of 18 to 20 Å. Since the existence of the hysteresis loop is considered as proof of the capillary condensation, it seems that for pores smaller than 18 to 20 Å radius, the Kelvin equation cannot be used. If $S_{BET} \neq S_{cum}$, qualitative information may still be obtained: (a) $S_{BET} > S_{cum}$, which indicates that the sample contains micropores (R < 20 Å); and (b) $S_{BET} < S_{cum}$, which happens especially in two cases:

(1) Primary pores intercommunicating in a way that the big pore has a tiny pore as the opening pathway.

(2) Secondary-layer texture, where the capillary condensation is able to push away the layers, which close again in desorption conditions at low p/p_0 values, giving an exaggerated pore volume for small pores. However, the BJH method is quite unsufficient in these cases and only combined

data obtained by several methods (small-angle X-ray scattering, electron microscopy) permits establishment of a realistic picture of the texture.

WALTER: The paper was intended to point to open questions in the treatment of complex porous structures. Established facts of experimental techniques and their limitations had to be treated quite summarily. Prof. Imelik's comments are, therefore, very welcome. Of course, at the lower limit of application of Kelvin's equation given in my paper, the essentially continuum-mechanical equation has lost its meaning. The limit quoted by Prof. Imelik is certainly the more practical one, above which Kelvin's equation may be applied rather safely, inserting values of σ ($r_m \rightarrow \infty$) and neglecting the influence of the wall's force field on r_m.

AGENDA DISCUSSION: SMALL PARTICLES

K. H. Johnson*
G. Hochstrasser**

1 EXPERIMENTAL CHARACTERIZATION OF
SMALL METAL PARTICLES

Chairman Johnson opened the discussion with a set of questions related to the experimental characterization of small metal particles, namely: What experimental techniques can we use to characterize the structures and properties of small catalytic metal aggregates (e.g., platinum, palladium, nickel, and other Group-VIII metallic clusters supported in porous refractory materials such as silica, magnesium oxide, and zeolite)? Can we measure the actual sizes and geometrical arrangements of these small clusters? Can photoelectron spectroscopy tell us something about the electronic structures or densities of states of small metal particles? What about the use of other physical techniques such as magnetic susceptibility, electron paramagnetic resonance, and Mössbauer spectroscopy? How can

*Chairman.
**Secretary

selective chemisorption be used to investigate the dispersion of metallic clusters? What are the relative advantages of using structure-insensitive and structure-sensitive catalytic reactions in the study of the chemical properties of supported metal particles?

M. Boudart first pointed out that particles of more than 10 Å, hereafter called "small particles", can already be well synthetized and characterized, which is not the case for particles of less than 10 Å, hereafter called "clusters". For such clusters, the electronic interaction with the substrate becomes important. Such interaction has not been detected for larger particles.

Derouane then described the usefulness of ESR for gathering information on the electronic structure of clusters: mono-, di-, tri-, etc., atomic clusters of silver, monopositively ionized, can be prepared by irradiation of silver nitrate emulsions. These are paramagnetic; moreover, a silver isotope has a nuclear spin. Hence, looking at the electron-nucleus hyperfine interaction, and evaluating the "s", "p", and "d" character of the unpaired electron would give details of the electronic structure of the clusters and its evolution along Ag_n^+ with n increasing.[1]

Imelik mentioned that this does not work for platinum and palladium clusters in zeolites.

Derouane added that the NMR determination of proton Knight's shift values in "hydrides" or for hydrogen dissolved in metallic systems is useful and this as a function of temperature and hydrogen pressure. A typical example is the Pd-H system, for which such Knight's shift data have been recently reported[2] which show, through a secondary (spin-orbit coupling) effect, that d-band levels are responsible for the hydrogen interaction with palladium.

Wynblatt pointed out that the technique of transmission electron microscopy should be mentioned. It allows not only the determination of particle size and shape, but also the characterization of structure through diffraction effects. For example, the identification of so-called "non-crystallographic" or multiply twinned particles (which adopt icosahedral or pentagonal shape) is possible by means of contrast effects in dark-field electron microscopy.

Bond stressed the importance of the preparation of small particles. Further practical and theoretical advances will be very slow until one discovers how to make monodispersed systems. It will be difficult to determine genuine particle size effects without knowledge of the preparation method of very small particles of narrow size distribution. At the moment, the technique of colloid chemistry seems to be the best hope for preparation of monodispersed particles.

Emmett then called attention to a method that has been used for obtaining pure iron crystals that are uniformly about 100 Å. The work which he referred to was carried out by a graduate student, Chong, at the University of California at Santa Barbara in 1966. Chong's procedure involved the vaporizing of iron pentacarbonyl in a helium stream and joining this stream with one consisting of heated helium. By choosing the concentrations and temperature correctly, it was found possible to produce a deposit of iron of uniform size (about 100 Å).

Johnson pointed out that one of the problems that faces the theoretician, who needs a starting point for electronic-structure calculations, is the nature of the arrangement of atoms in a cluster. Are there experimental techniques that can provide direct or indirect information on the actual arrangement of atoms?

Imelik agreed that problems of preparation and characterization are different for small particles and for clusters. For small particles, preparation is easier in this range (even for industrial catalysts) and a lot of possibilities and techniques exist for the characterization (diameter and shape), such as electron microscopy. A problem in this range is the actual distribution of the size, how to measure it, and how to reduce the dispersion as much as possible. For clusters, there is a real problem of preparation as well as of characterization.

Rhodin referred to the pioneering work of Professor John Turkevich and his associates[3-5] in which homogeneous distributions of metallic particles of palladium, gold, and platinum were prepared, characterized, and evaluated for catalytic activity. It is significant that this was one of the more unusual efforts to relate particle size distribution to both rates of reaction and to mechanisms of catalysis. It was conceived and carried out almost 25 years ago. Organic reducing agents were used to convert the salts of the metallic catalysts into particles of well-defined size and particle distribution. There is some possibility that the organic reagents may have left some residual material on the particles. It is not clear what effect if any, this might have had on their catalytic activity. The particle size distributions were carefully evaluated using high-resolution electron microscopy when this technique was in its infancy. It would be very interesting to reinvestigate these same materials for the same reactions using more modern methods of both atomic and electron structural characterization.

In reply to Johnson's question regarding the type of clusters which could be of interest to a theoretical physicist, Eischens mentioned that some chemical laboratories prepare clusters by matrix isolation. Metal is evaporated into a cooled rare gas. Reasonable claims are made that the particle size can be well controlled and thus individual atoms can be obtained by this process. A great deal of exciting IR evidence has been obtained with such clusters. An open question is: Can you obtain an alloy, using such a method?

Messmer indicated that Gruen at Argonne has applied a sputtering technique to create atoms or small clusters (up to five atoms) which were allowed to react, e.g., with N_2, in a rare gas matrix and examined by IR spectroscopy. Similar studies have been done elsewhere using CO and O_2.

Johnson commented on the application of X-ray photoelectron spectroscopy (XPS, ESCA) to small particles. Theoretical calculations made on clusters (e.g., twelve atoms surrounding one atom at the center) suggest that core orbitals of central atoms of the clusters have quite different binding energies from core orbitals of the "surface" atoms. This is a type of chemical shift. These shifts are very sensitive to the nature of the coordination, as they are in molecules. ESCA could be used, in principle, to measure these shifts. Such experimental data, in combination with theoretical calculations of electronic spectra, could lead to information about the possible atomic arrangements in the clusters.

2 ELECTRONIC EFFECTS OF THE SUPPORTING ENVIRONMENT

Johnson, referring to the already-mentioned possible effect of supporting environment and to Boudart's mention of structure-insensitive and -sensitive catalytic reactions, proposed a second set of questions, related to electronic effects of supporting environment: Are there electronic interactions between the porous supporting material (e.g., silica, alumina, zeolite, or magnesium oxide) and the metal particle which can modify catalytic activity or selectivity? What is the fundamental nature of these interactions? Is there effective electronic charge transfer between the metal cluster and the surrounding matrix? Are there covalent contributions to the interaction? What types of experimental techniques and theoretical models can be used to investigate particle-support interactions?

Boudart referred to the paper presented by R. A. Dalla Betta at the International Congress on Catalysis of 1972. Clusters of five to six platinum atoms prepared in the supercage of Y zeolites showed, for two reactions, a catalytic activity higher by one to two orders of magnitude than any other platinum catalysts that have been looked at. Platinum acted like its neighbor to the left in the periodic table, as if there had been partial charge transfer from the cluster to the zeolitic framework.

Ehrlich inquired whether there have been indications from magnetization measurements on very small particles of the effect on the substrate of adsorbed layers.

Johnson, starting from the known electronic structures of SiO_2, Al_2O_3, and MgO, mentioned that an important aspect of the valence band structures of these refractory materials is the presence of "nonbonding" oxygen 2p orbitals. These orbitals are, however, available for bonding with the d orbitals of transition-metal atoms or clusters which are supported on the materials. Thus the electronic structures and possibly the catalytic properties of small transition-metal clusters (e.g., nickel, palladium, and platinum) could be strongly influenced by the supporting environment.

Imelik pointed out that support effects are normally observed for acidic or basic supports. This means that Lewis acids or charge transfer complexes are present. Therefore, it is not oxygen atoms but oxygen vacancies that are implied in electron transfers from metal atoms to the supporting material.

Schwab reported a distinct support effect in the CO oxidation by nickel oxide. NiO catalyzes CO oxidation with an activation energy of 16 kcals/mole. Silver does not catalyze this reaction under the conditions used. Now, when a silver foil is covered by a nickel oxide layer 100 to 150 Å thick, the activation energy of the CO oxidation by NiO is raised from 16 to more than 40 kcals/mole; the thinner the layer, the higher is the activation energy. Here, silver is the support and NiO is the catalyst. The explanation is that, on the outer side of the semiconductor layer, the accumulated layer of electrons emitted from the metal into the semiconductor is still perceptible and thus raises the Fermi level of the catalyst relative to its bands and hence increases its activation energy. Similarly, silicon carbide can be used as a support for a silver layer. Depending on whether the SiC is doped n-type or p-type, the activation energy of the formic acid decomposition on the silver varies. (This change of the Fermi level of a metal can be understood only on the basis that there is much more silicon carbide than silver.) In the first example, the activation energy of the NiO can also be influenced by manipulating the electron concentration and hence the work function of the silver (or gold) by alloy formation.

3 ELECTRONIC STRUCTURES OF BIMETALLIC AND MULTIMETALLIC ALLOY CLUSTERS

Johnson then suggested that the remainder of the discussion be devoted to the nature of the electronic structures of bimetallic and multimetallic alloy clusters and opened it with the following questions: Can we begin to understand the observed changes in the catalytic behavior of small metal particles as a function of "alloying" (e.g., Group-VIII elements with Group-IB elements) in terms of the electronic structures, densities of states, and chemisorptive properties of bimetallic and multimetallic clusters? Are conventional theories of alloying in bulk crystalline metals applicable to small bimetallic and multimetallic clusters? How important is surface segregation in determining the chemisorptive and catalytic properties of these clusters? Why are some catalytic reactions much more sensitive to alloying than others? Is there any hope of reliably predicting catalytic activity or selectivity as a function of alloying?

Fischer pointed out the research at Exxon on ethane hydrogenolysis by Cu-Ni alloys. Sinfelt, Carter, and Yates showed that alloying 5 percent of copper to nickel drops the activity of nickel catalyst by 2 orders of magnitude; alloying 10 percent copper drops the activity 5 orders of magnitude. The catalyst is a self-supported dispersed alloy having reasonably large grains. The properties of the grains might therefore be assumed to be the same as those of the bulk. Burton and Hyman considered the possibility of surface segregation from the bulk. It was concluded that, for the temperature at which Sinfelt observed the catalysis, the catalyst with 5 percent copper in the bulk presents a surface having 99 percent copper atoms. Burton and Hyman calculated the surface concentration of nickel atoms and found that the activity of the catalyst is proportional to the square of the concentration of nickel atoms at the surface. The conclusion was that the active sites are pairs of nickel atoms in a surface monolayer of almost pure copper. This shows that it is possible to separate segregation effects in alloys from other concentration effects. It has been concluded that, even in highly dispersed bimetallic and multimetallic clusters, where approximately all the atoms are surface atoms, one should not throw away the concept of surface segregation, because of the importance of coordination and saturation, as already mentioned by Bond. The more volatile element will go to those sites which have lower coordination, and the less volatile element will go to those sites that have higher coordination.

Ehrenreich followed with two remarks, first mentioning that copper and nickel are not miscible over an appreciable concentration range. Cu-Ni alloys have therefore to segregate when cooled down slowly from the melt. Second, each nickel atom in a sea of copper should carry its d^9s configuration with it. There are, therefore, unfilled d shells associated with the nickel atom, and this is consistent with Van Vleck's minimum polarity hypothesis. If such d shells are important in catalysis, this could be an important ingredient.

Kohn mentioned work done at the University of Washington in Seattle. Work-function measurements have been made on Ag-Au alloys, showing that there is no linear but a parabolic, change of the work function when one goes from pure gold to pure silver. A possible explanation is that there is some segregation at the surface. Auger measurements were made. These, however, showed that the surface concentration is the same as that in the bulk. Although this has nothing to do with clusters, it was the starting point of semiempirical (Hückel-type) molecular-orbital calculations made by Davidson, a quantum chemist, on gold-silver clusters models nearly 100 atoms in size. Results are not conclusive, but the methodology is interesting enough to be pursued further.

Grimley commented on how one might approach the theory of chemisorption on two nickel atoms floating in a "sea" of copper atoms at a Cu-Ni alloy surface. Grimley stressed that the method he described earlier is perfectly suited to treating this problem. He preferred to consider only one nickel atom in the surface sea of copper atoms segregated from the bulk nickel, but had no doubt that the calculation will show that hydrogen is chemisorbed almost as strongly at such a site as it is on pure nickel. Grimley emphasized the importance of actually carrying out such a calculation.

In regard to segregation in general, Wynblatt cited studies of Cu-Pt alloys. Theories such as those of Frank Williams predict that the component with lower heat of sublimation (or lower surface energy) will segregate to the surface of a binary alloy. Although platinum and copper differ widely in their sublimation energies, no surface segregation is observed in Pt-Cu alloys in vacuum. On the other hand, copper does segregate to the surface in the presence of oxygen. Thus the simple theories do not always predict the correct trend in experimental behavior. However, it is tempting to use this type of theoretical concept to rationalize the experimental observations of Sinfelt on Cu-Ru and Cu-Os "bimetallic" clusters. One can possibly make sense of the alloying that has occurred between copper and ruthenium (or copper and osmium) in small clusters, in spite of the fact that these materials are normally immiscible. Since ruthenium and osmium have heats of sublimation higher than those of copper, it is quite conceivable that some copper can adsorb on the surface of rhenium or osmium clusters, and thus decrease the surface free energy of the system.

Boudart mentioned the usefulness of Mössbauer spectroscopy for the study of clusters and the need for a greater number of isotopes suitable for such a method.

Emmett recalled the work done 30 years ago on carbon monoxide chemisorption over alloys of copper and nickel.

Finally, Somorjai pointed out that the application of surface thermodynamic concepts to simple systems leads to good agreement with the observed segregation of the component of lower surface free energy or lower heat of sublimation on the surface. Two systems were studied, Pb-In whose heat of mixing is positive (endothermic reaction) and Ag-Au whose heat of mixing is negative. According to Williams' theory as well as to simple statistical models, concentration gradients of the excess component into the bulk (in case of endothermic reaction) and enrichment of the component of lower surface free energy on the surface (in the other case) were observed. Since Auger electron spectroscopy is a technique that integrates over more than one layer, it is necessary to correct the data. Such corrections might settle many controversies appearing in the literature. As an example, silver in Ag-Au alloys should accumulate on the surface, and work-function data behave just as they should if silver would cover 90 percent of the surface when one has a 50-50 silver-gold alloy. There is little known about the surface composition of multicomponent systems with complex phase diagrams. Somorjai stressed that there is exciting and necessary fundamental work to be done, using any sensitive surface technique, such as ESCA or Auger, in order to establish correct surface phase diagrams, i.e., to determine the surface composition as a function of bulk composition and, of course, temperature.

REFERENCES

1. Forbes, C. E., and Symons, M.C.R., *Molecular Phys.,* **27**, 467 (1974).
2. Brill, P., and Voitländer, J., *Ber. Bunsenges.,* **77**, 1097 (1973).
3. Turkevich, J., Stevenson, P. C., and Hillier, J., *Faraday Soc. Disc.,* **11**, 55 (1951).
4. Turkevich, J., and Kim, G., *Science,* **169**, 873 (1970).
5. Turkevich, J., *Advan. Catalysis,* **23**, 91 (1973).

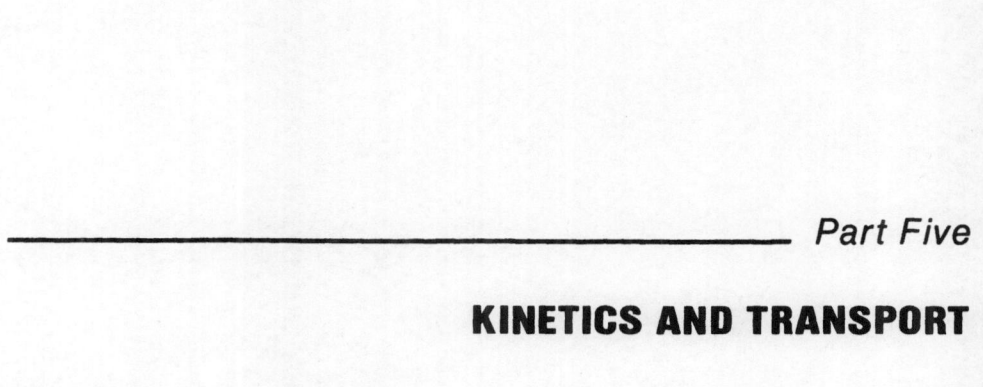

Part Five

KINETICS AND TRANSPORT

FIELD IONIZATION AT SURFACES INVESTIGATED BY MASS SPECTROMETRY

J. H. Block and W. A. Schmidt

Fritz-Haber-Institut der
Max-Planck-Gesellschaft, Berlin-Dahlem
West Germany

ABSTRACT

Experimental results are reported which elucidate details of field-ionization mechanisms at surfaces. On group 1B metals, complex ions are formed because of field interaction. Oxygen on silver is not detectable as field ion. The physical adsorption of sulfur and the molecular transformation $S_2 \rightarrow \ldots \rightarrow S_8$ is not field dependent.

1 INTRODUCTION

Single surface particles within a geometrical resolution in the atomic scale can be analyzed by field-ion mass spectrometry (FIMS) in connection with field-ion microscopy (FIM). The atom probe, developed by E. W. Müller[1], has been applied in order to identify individual structures at a surface. In heterogeneous catalysis, the identification of these surface structures is a fundamental question. In particular, the crystal-face selectivity of chemisorbed molecular systems can be analyzed by this sensitive tool. Furthermore, field pulse techniques yield information on time-dependent surface processes with a time resolution in the microsecond scale. and thus represent a direct method for kinetic measurements of surface reactions. Thus, the formation of a monolayer of a chemisorbed state can be measured and subsequent chemical reactions can be analyzed within the atomic scale of single-surface sites and a time resolution of microseconds.

In FIMS, surface molecules are converted into ions owing to the extreme external field of ≈ 1 V/Å. A straightforward correlation between the usual surface structures and mass-spectrometrically analyzed field ions is complicated by intriguing field effects. The behavior of molecules in electric fields in the V/Å range is still unexplored. For a quantitative analysis of surface structures by FIMS, these field phenomena are a major point. The understanding of these field effects, therefore, is presently one of the main intentions of our research work.

2 EXPERIMENTAL PROCEDURES

Two different types of field-ion mass spectrometers have been used in the investigations to be described.

1. Experimental results on silver, copper, and gold surfaces have been obtained in a CH_4 magnetic-sector-type instrument. The mass resolution was of the order of $M/\triangle M = 500$ (10 percent peak height), and the sensitivity for ion currents was of the order of 10^{-22} amp ($\approx$ions/minute). The new ion source allowed FIM and FIMS simultaneously. As demonstrated in Fig. 1, a channel plate C with luminescent screen S and an ion focusing lens L are part of the UHV chamber V (residual pressure in the lower 10^{-10} Torr range). The emitter tip T may be placed exactly in the position 1 for mass-spectrometric analysis or precisely in position 2 for field-ion microscopy. Easy and quick control of the surface emissivity, surface orientation, and selective chemisorption is possible at any stage of the mass-spectrometric investigations.

2. Measurements on adsorption processes of sulfur on tungsten have been performed in a quadrupole mass filter equipped with a field ion-source. This compact design easily allows extreme UHV conditions as necessary for surface studies. A rapid mass scan up to 1000 amu/sec enables an immediate comparison of ion intensities at different regions of the mass scale. Furthermore, the analysis yields the true mass number of ions and the mass scale is independent of ion energies. The quadrupole mass filter is superior for fast kinetic measurements, although the mass resolution ($M/\triangle M \approx 400$) and the sensitivity (presently only $\approx 10^{-17}$ amp) are inferior. The combina-

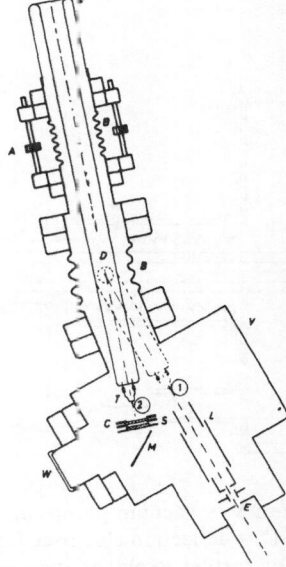

Fig. 1. Field-ion source combined with a channel-plate field-ion microscope: F = cooling finger, A = adjustment screw for tip position, B = bellows, D = turning point for the mechanical support (not shown), M = mirror, W = window, E = entrance slit of the 60-degree-magnetic-sector mass spectrometer.

tion of a field-ion source with a quadrupole mass filter is shown in Fig. 2. The emitter tip T is mounted on a cooling finger F which can be adjusted to the optical axis of the instrument by a bellows B. The emitter is heatable by current leads. The temperature can be controlled and automatically regulated through additional connections to an electronic device which controls the voltage drop of the emitter loop. The lens system L creates the necessary field, decelerates the ions to acceptable energies (<100 V), and contains an auxiliary repeller R. The quadrupole filter Q of a commercial residual-gas analyzer is used for ion detection. A Faraday cage F, an electron multiplier EM, and a channeltron are provided. The out-of-line position of sensitive detectors is necessary to avoid background noise from electrons OR x-rays. In particular case of Fig. 2 an electrochemical source S for the formation of an S_2 beam is installed.

The ion emitters were prepared from high-purity metal wires by electrochemical etching (silver, copper, and gold in KCN solution, tungsten in NaOH). Tip radii, usually less than 1000 Å at the beginning of experiments, were controlled by field emission and known onset fields for field ionization of test substances. In most cases, final cleaning of surfaces was achieved by field evaporation.

3 IONIZATION MECHANISMS

Pure field ionization by electron tunneling in a field of 10^8 V/cm and without surface interaction has been described theoretically.[2] Ionization is

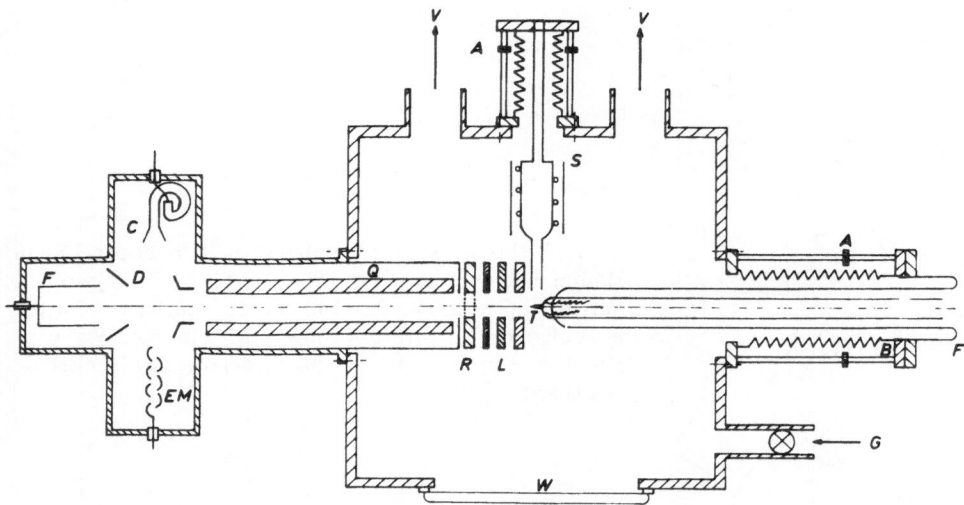

Fig. 2. A field-ion source combined with a quadrupole mass filter: V = vacuum pumps and pressure measurement, A = adjustment screw for tip orientation, D = deflection electrode for ion beams, G = gas inlet system, W = observation window. For further explanations, see text.

achieved simply by deformation of molecular Coulomb potentials. With increased field strength at the emitter surface, ions may be formed at distances of considerably more than 10 A from the solid surface, where chemical interactions are negligible. As shown by energy distribution measurements, this simple mechanism of electron transfer is however an exception in FIMS. Intermolecular interactions such as proton transfer, donor-acceptor complex formation, chemical ionization, or ion-molecule reactions, usually determine the mass spectra.[3,4]

Since molecular ions have orbital symmetries which are different from neutrals, they are frequently energetically unstable. Therefore, the fragmentation of field ions, although far less important than with electron or proton impact, has to be taken into account when the chemical composition of a surface is evaluated by field ions. This fragmentation phenomenon is well established for various kinds of substances.

Recent results[5] clearly indicate that because of field interactions new ionic compounds may also be formed which are not present at zero-field conditions. The complex chemistry of the emitter metal ions is of a major importance in regard to the kind of ions which are formed at the surface. In the general line of observations, this result corresponds with the rule that the field ions are mainly determined by the energetics of the formation process.

Since desorbing molecular ions differ in chemical composition from usual adsorbates or chemical surface structures, the details of the ionization mechanisms have to be elaborated before FIMS can be used for a quantitative surface analysis of individual molecules in the geometrical resolution of the atomic scale.

4 FIELD-INDUCED REACTIONS

4.1 Field Desorption from Copper, Silver, and Gold

If a surface with a chemisorbed layer is exposed to an external electric field of $>10^8$ V/cm, ions are desorbed after heterolytic bond cleavage of surface compounds. This bond cleavage may occur between the chemisorbed structure and the metal surface. However, many cases are known where a metal atom, together with the chemisorbed molecule, is desorbed in ionic form. This is particularly very common if the field-desorption field strength is close to the field strength of field evaporation of the metal. Obviously, the field evaporation process is facilitated if the chemical bonds of the desorbing ion are partially saturated by ligands.

These energetic considerations also have to be applied to explain the desorption of complex ions of group 1B metals. All the Cu^+, Ag^+, and Au^+ ions have a d_{10} electron configuration and are coordinatively highly unsaturated. Therefore, ionic complexes are formed with impinging gas molecules which are usually not chemisorbed at the metal surfaces.

Field desorption from copper, silver, and gold surfaces has been studied in the presence of H_2O, NH_3, C_2H_4, CH_4, CO, N_2, H_2, Kr, Xe, and some other substances. Although many of these gases are known not to be chemisorbed on these metals, metal complex ions are formed with high intensities. The data compiled in Table 1 are not complete with respect to temperature and pressure dependences of ion intensities. The value n gives the number of ligands which have been observed. The fact that maxima in n are still restricted by gas supply cannot be excluded. The coefficient Cpl^+/Me^+ indicates the ratio of complex ions to uncoordinated metal ions which are measured under the experimental conditions. A high value of this ratio demonstrates that complex desorption is highly favored over normal field evaporation. Within the experimental accuracy, the evaporation rate of uncoordinated metal ions was not altered in any case (no promotion).

A more quantitative dependence of ion intensities in the system Ag/H_2O is given in Fig. 3. The temperature dependence indicates two temperature regions. At $130°K$, a first maximum in intensities with high coordination number n is observed; at $220°K$, the maximum is related only to compounds with $n \leqslant 6$; similar dependences have been obtained for ammonia where, however, n is always $\leqslant 8$.

From complex chemistry, H_2O and NH_3 are known to form coordinative bonds with the Ag^+ ion. They are ligands of the Ag^+, which accepts

Table I. Ionic Species Desorbed From Group 1B Metals

Gas	Copper Species	Temp., °K	Cpl'/Me+	Silver Species	Temp., °K	Cpl', Me+	Gold Species	Temp., °K	Cpl' Me+
H_2	$Cu(H_2)_n^+$, $n \leqslant 3$	140	10^3	$Ag(H_2)_n^+$, $n \leqslant 2$	90	1	$Au(H_2)_n^+$, $n = 1$	320	1.5
N_2	$Cu(N_2)_n^+$, $n \leqslant 2$	140	≈ 1	$Ag(N_2)_n^+$, $n = 1$	180	0.4	$Au(N_2)_n^+$, $n = 1$	260	5
CO	$Cu(CO)_n^+$, $n \leqslant 2$	140	<1	$Ag(CO)_n^+$, $n \leqslant 2$	210	10^3	$Au(CO)_n^+$, $n \leqslant 2$	190	15
H_2O	$Cu(H_2O)_n^+$, $n = 1$	260	1	$Ag(H_2O)_n^+$, $n \leqslant 12$	130	$>10^6$	No complex		
NH_3	$Cu(NH_3)_n^+$, $n = 1$	180	1	$Ag(NH_3)_n^+$, $n \leqslant 8$	130	$>10^4$	$Au(NH_3)_n^+$, $n = 1$	260	5
CH_4				$Ag(CH_4)_n^+$, $n = 1$	210	4	–		
C_2H_4	$Cu(C_2H_4)_n^+$, $n \leqslant 3$	260	10^3	$Ag(C_2H_4)_n^+$, $n \leqslant 3$	210	≈ 2	–		
Kr, Xe	–			No Complex					
Pyridine	–			$Ag(Py)_n^+$, $n \leqslant 4$	295	$>10^4$	–		

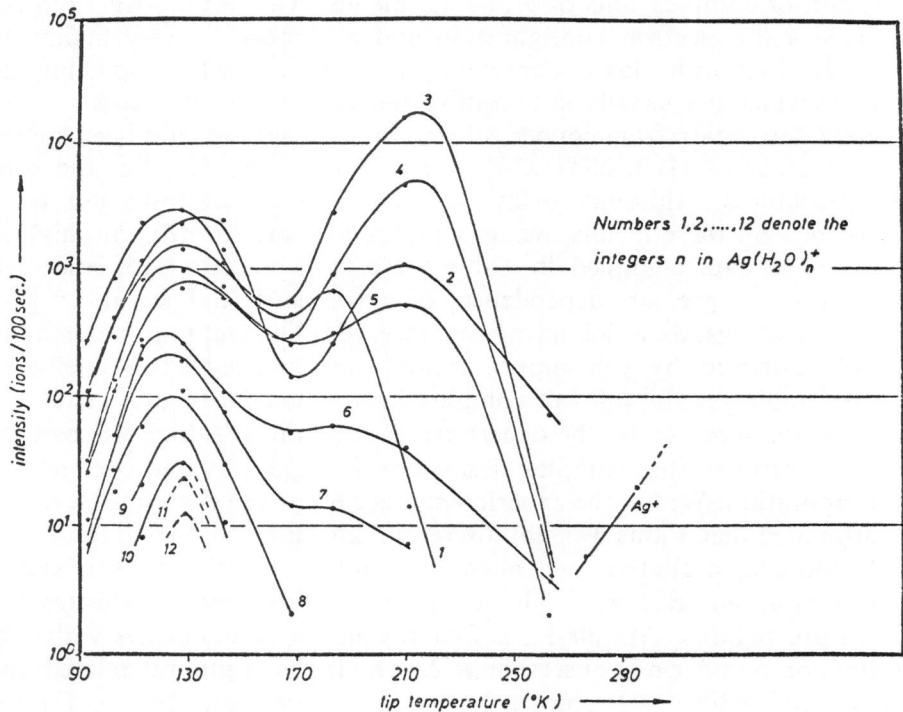

Fig. 3. The temperature dependence of the different intensitieis Ag $(H_2O)_n^+$ at a water vapor pressure of 2 x 10^{-6} Torr. The formation of ice crystals in the ionization zone was prevented at steady field conditions.

electrons from H_2O or NH_3. Since silver diamine and silver tetraaquo complexes are well known, the ions with maximum n in Table 1 can tentatively be formulated as $[Ag\,(NH_3)_2]^+ \cdot 6NH_3$ and $[Ag\,(H_2O)_4]^+ \cdot 8H_2O$, respectively. In a qualitative manner, the temperature dependence (Fig. 3), in accordance, indicates a higher thermal stability of complex ions with ligands in the first tightly bound coordination sphere. Complex ions which are formed with other gases, as compiled in Table 1, display rather unusual structures. Considering these structures, two phenomena have to be taken into account.

1. These complex ions may have a rather low energy of formation and many of them most probably will react chemically if collision partners are available. They are formed under field-ionization conditions since the field evaporation of these complexes is energetically still more favorable than the evaporation of uncoordinated Ag^+, Cu^+, or Au^+ ions.

2. A lifetime of 10^{-6} second is sufficient to substantiate these ions as stable compounds after desorption in FIMS. The absence of doubly charged ions, which are usually obtained in complex chemistry with Cu^{++} ions or in the particular case of silver-pyridine coordination compounds, indicates that fast kinetic processes prevent the formation of the usual energetically favored products.

The field desorption and field evaporation of surface structures in FIMS has certain similarities with the electrochemical dissolution of metals in the surface double layer of an electrode. In this case, the solvation energy of ions obviously facilitates the conversion of metal surface atoms into ions of the liquid.

Details of ionic structures which are observesd with different gases in group 1B metals are presently under consideration.

4.2 Interaction of Oxygen with Silver

The adsorption of oxygen on silver surfaces involves unique structures since silver is the exceptional metal for the epoxidation of ethylene.[6] Three different adsorption states are known: (a) an almost nonactivated ($E_a < 3$ kcal/mole) dissociative adsorption which covers nearly half a monolayer at room temperature, (b) an activated ($E_a \approx 8$ kcal/mole) nondissociative adsorption, and (c) an activated dissociative adsorption (details in ref. 6). The surface selectivity of chemisorption at lower partial pressures could be derived from field electron-emission studies.[7,8]

The attempt to analyze these chemisorption states by FIMS gave no indication for Ag-O-surface compounds. Between 80 and 425°K, none of the expected compounds was traceable, neither during experiments with dynamic gas supply of 10^{-8} to 10^{-5} Torr O_2 and steady electric field, nor at zero-field adsorption ($< 6 \cdot 10^4$ L) and subsequent field application. Ag^+ from field evaporation and O_2^+ from field ionization (only first case) were the only ions measured with high intensities. An exception was the field ionization of water which yielded a few Ag O^+ ions among large amounts of Ag $(H_2O)_n^+$.

The nonappearance of oxygen-silver chemisorption states in FIMS could be explained by a potential-energy diagram (Fig. 4a). There are different experimental indications that oxygen layers on silver are negatively charged.[10] This surface polarization is disturbed by the external positive field, as illustrated in Fig. 4.

We have to consider the potential-energy diagram as it was used by Gomer.[2] An oxygen molecule adsorbed at a silver surface faces different desorption modes when exposed to a positive external field. It is assumed that the O_2^- species on the surface still has energy states resembling those which are known from the gaseous molecular ion. This assumption is supported by the fact that IR-studies actually indicate the peroxide structure and that the g-values of EPR measurements of O_2^- on the surfaces of different substances with different vectors of the crystal field are only slightly altered. The O_2^- molecular ion has 15 states of vibrational excitation with O.II eV energy difference within the electronic ground level. The attachment energy EA of the electron is 0.43 eV; an excited electronic level could not be observed.

For such a structure on the silver surface, the activation energy of desorption amounts to 18 kcal/mole ($= 0.77$ V) which is slightly higher than EA (≈ 10 kcal/mole). In order to draw the relative potential curves of the states $Ag + O_2$, $Ag^+ + O_2^-$, and $Ag^- + O_2^+$, we further need the work function of Ag ($\phi = 4.3$ eV) and the ionization potential of O_2 ($I = 12.2$ eV). Interaction energies of $Ag + O_2$ and $Ag^- + O_2^+$ are unknown; however, these reactions are presumably less favored than $Ag^+ + O_2^-$.

Potential energies between the various oxygen species and silver are plotted in Fig. 4a, where no external field is applied. At medium electric fields (Fig. 4b), negatively adsorbed oxygen becomes unstable. Field ionization or the field desorption of O_2^+, however, will occur only at very high electric fields (Fig. 4c), where the potential $Ag^- + O_2^+$ intersects the neutral potential $Ag + O_2$ within the tunneling length of electrons.

Although the schematic representation of Fig. 4 cannot describe interaction potentials in a quantitative manner, there are obviously three different regions of field strength. The O_2^- species will be adsorbed at the silver surface only at a very low field strength. At increasing fields, neutralization of the surface species will occur (at $F < 0.5$ V/Å in Fig. 4). If this results in a nonbonding orbital configuration, a neutral oxygen molecule will desorb. For the desorption of positive molecular ions O_2^+,

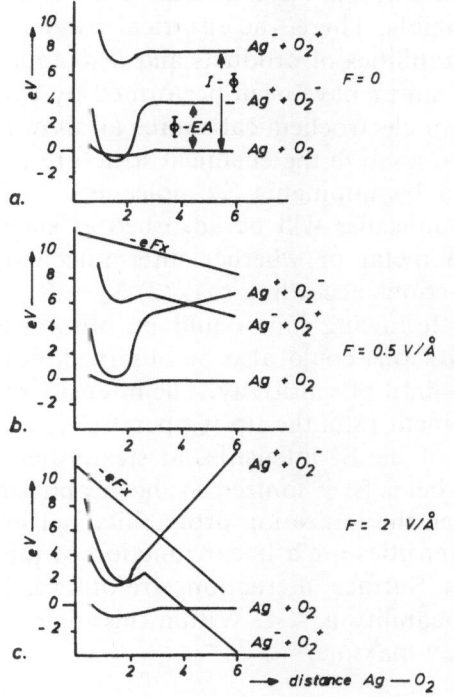

Fig. 4. The potential-energy diagram of various charged adsorption structures of molecular oxygen on silver.

much higher fields are required (F > 2 V/Å in Fig. 4). Thus, the field-induced desorption of neutral molecules seems to be possible in the present case.

For the atomic oxygen which is mainly adsorbed on silver at lower pressures, energy data are less precise. Considering the potential-energy diagram, the desorption field would be higher, since EA for atomic oxygen amounts to 1.465 eV and the ionization potential to 13.6 eV.

5 FIELD-INDEPENDENT REACTIONS

The difference between oxygen and sulfur as appearing in FIMS is due to the different electronic structures. An oxygen atom can use p-orbitals and s-p-hybrids only; a sulfur atom, in addition, has access to five empty $3d^0$ orbitals. This explains the ability of sulfur to form polyatomic molecules like S_8. The adsorption of sulfur on a metal surface, like tungsten, leads to a strongly bound chemisorption state (α-sulfur), and a loosely bound physisorbed state. The latter is easily field desorbed at rather low electric fields of $\approx 5 \cdot 10^7$ V/cm. The formation of this physisorbed layer is connected with a chemical reaction which is field independent.

Surface reactions are independent of external electric fields if the electrical momentum of the reaction is negligible. There, the electrical momentum is defined by the dipoles and polarizabilities of products and reactants.

Investigations on the interaction of sulfur have been performed by using an S_2 molecular beam, produced in an electrochemical source as shown in Fig. 2. The question of interest was to analyze the chemical structure of the physisorbed sulfur which is formed by impinging S_2 molecules. The problem to be solved was whether S_2 molecules will be adsorbed as such with still interfering interaction to the metal or whether intermolecular forces within the adlayer will lead to reactions according to $(x/2) S_2 \rightarrow S_x$.

Under steady-field conditions, the following ions could be observed (Fig. 5) S_2^+, S_5^+, S_6^+, S_7^+, and S_8^+; the S_4^+ ions could also be observed, but only under certain conditions and at the limit of sensitivity. The intensity of a particular ion depends on the impingement rate, the tip temperature, and the ionizing field. The field dependence of the S_x^+ intensities at steady field conditions is shown in Fig. 5. At higher fields S_2 is ionized in the gas phase prior to the adsorption process, since then the ionization probability is 1 in front of the surface. Accordingly, S_2^+ intensities are a linear function of the S_2 beam intensity under these conditions. Surface interactions are observed at lower fields, where the ionization probability is < 1. Within this region, the S_5^+, S_7^+, S_6^+, and S_8^+ intensities display maxima.

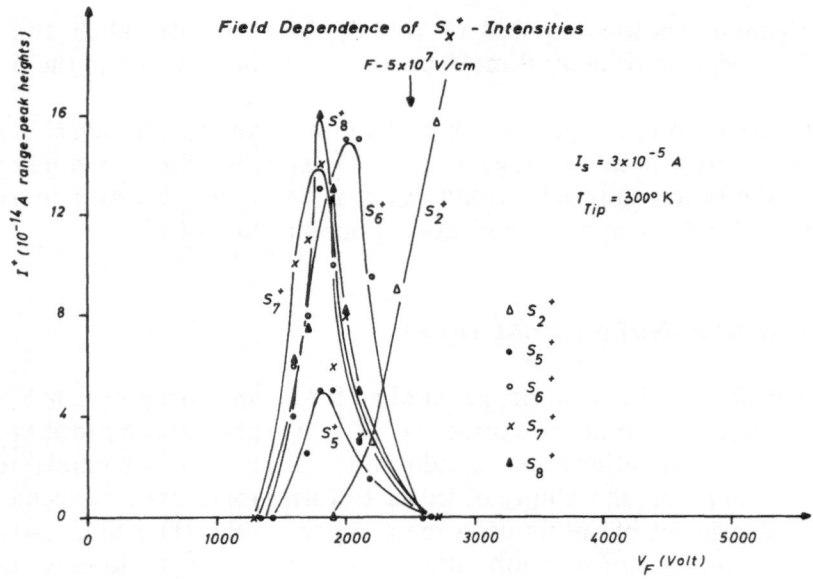

Fig. 5. Ion intensities of various sulfur molecular ions as a function of the field strength (proportional to V_F). The S_2 beam intensity (given by the source current I_S) corresponds to 2 x 10^{13} molecules cm^{-2} sec^{-1}.

The following kinetic regime can be measured:

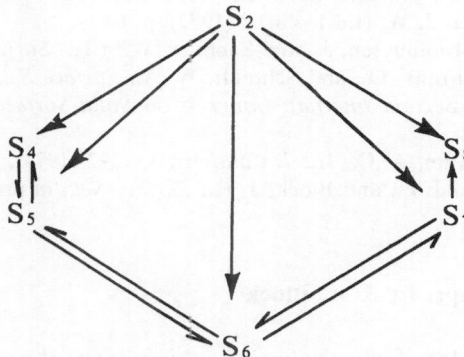

From field-pulse, temperature-jump, and diffusion measurements after source shutoff[11], it could be concluded that the rate of S_2 disappearance is fast and the formation of the thermodynamically stable S_8 modification is a relatively slow process. Field influences on these reaction rates could not be found.

6 CONCLUSIONS

The straightforward analysis of surface structures and surface reactions by FIMS is particularly difficult for those ionic structures which are field stabilized and those reactions which have an electrical moment. A profound knowledge of these field phenomena is required for the interpretation of molecular surface interactions by field ions.

ACKNOWLEDGMENT

Experimental work by G. Bozdech and O. Frank is gratefully acknowledged. The Deutsche Forschungsgemeinschaft and the Senator für Wirtschaft Berlin (West) supported this research work.

REFERENCES

1. Müller, E. W., *Naturwiss*, **57**, 222 (1970).
2. Gomer, R., "Field Emission and Field Ionization", Harvard University Press (1961).
3. Beckey, H. D., and Röllgen, F. W., *Naturwiss*, **58**, 23 (1971).
4. Block, J. H., *Proc. Fifth Internat. Congress on Catalysis*, Hightower, J. W. (Ed.), Vol. 1 (1973), p. 91.

5. Schmidt, W. A., Frank, O., and Block, J. H., *Surface Sci.*, **44**, 185 (1974).
6. Kilty, P. A., Rol, N. C., and Sachtler, W.M.H., *Proc. Fifth Internat. Congress on Catalysis*, Hightower, J. W. (Ed.), Vol. 2 (1973), p. 64.
7. Janssen, M.M.P., Moolhuysen, J., and Sachtler, W.M.H., *Surface Sci.*, **33**, 625 (1972).
8. Czanderna, A. W., Frank, O., and Schmidt, W. A., *Surface Sci*, **38**, 129 (1973).
9. Block, J. H., *Proc. Second Internat. Congress on Solid Surfaces*, Kyoto (1974); *Jap. J. Appl. Phys.*, in press.
10. Clarkson, R. B., Cirillo, A. C., Jr., *J. Catalysis*, **33**, 392 (1974).
11. Davis, P. R., Bechtold, E., and Block, J. H., *Surface Sci.*, in press.

DISCUSSION on Paper by J. A. Block

EHRLICH: I wonder if the fields at which you observe the evolution of water complexes from a silver surface are comparable to those operating at the metal-electrolyte interface? If so, this technique should allow really interesting explorations of electrode processes.

BLOCK: The evolution of Ag^+-water complexes is observed in the 10^7 V/cm field range, which is comparable to that of a double layer at an electrode surface. However, the field and reaction conditions in an electrode process are much more complicated than in our case.

BOUDART: Would you elaborate on the mechanism by which Ag^+ collects as many as 12 H_2O ligands before leaving the tip?

BLOCK: The desorption of $Ag^+ (H_2O)_n$ complexes most probably involves a two-step mechanism: (a) the transition of a lattice silver atom into a highly polarized adsorbed state, and (b) the desorption of this adsorbed state. There are indications that the second step is rate determining. This includes a sufficient residence time of $Ag\delta^+$ (partly positively charged) adsorbed particles which can form complex ions with surrounding ligands.

EMMETT: Would you explain how the presence of water vapor enables you to obtain the silver-oxygen ions, whereas oxygen in the absence of water vapor fails to give you such ions?

BLOCK: According to the model we derived, the silver-oxygen surface dipole (with the negative charge in oxygen) may be destabilized by a positive electric field, such that neutral oxygen desorbs. If a multilayer of water is present at the surface, field interactions at the surface bonds will be different. It seems possible that the water ligands will shield the Ag-O bond to a certain extent. Thus, AgO^+ ions, which are known to be stable molecular ions, can desorb in small intensities.

THE MECHANISM OF HYDROCARBON CATALYSIS ON PLATINUM CRYSTAL SURFACES

G. A. Somorjai

Inorganic Materials Research Division, Lawrence Berkeley Laboratory and Department of Chemistry; University of California Berkeley, California

ABSTRACT

In the past several years we have studied the atomic structure of platinum crystal surfaces and the structure of adsorbed hydrocarbons by low-energy electron diffraction and the surface composition by Auger electron spectroscopy. Catalytic reactions of low reaction probability (dehydrocyclization, dehydrogenation) have been studied on one face of a single crystal less than 1 cm^2 in area by mass spectrometry at low pressures ($\sim 10^{-4}$ Torr) and by gas chromatography at high pressures ($\sim 10^3$ Torr).

The atomic structure of high-Miller-Index platinum surfaces is characterized by atomic height steps arranged periodically and separated by atomic terraces of low-Miller-Index orientation [(111) or (100)]. Experiments that compare the reactivity of crystal surfaces and supported platinum particles indicate that the atomic structure of polydispersed catalyst particles can be reproduced on these stepped crystal surfaces. The structure of adsorbed hydrocarbons has been studied on both low- and high-Miller-Index platinum surfaces. Atomic steps play a controlling role in dehydrogenating the hydrocarbon molecules and in dissociating hydrogen and other diatomic molecules of large binding energy. n-Heptane may undergo isomerization, hydrogenolysis, and dehydrocyclization on the various platinum surfaces. The atomic structure of the stepped crystal surfaces appears to control the product distribution for these competing reactions.

1 INTRODUCTION

Platinum is one of the most versatile elements utilized for heterogeneous catalysis. It is employed to catalyze a large variety of hydrocarbon reactions in reducing atmospheres (for example in hydrogen) and oxidation reactions ranging from mild oxidation of alcohols, to ketones, to one of the most exothermic reactions, the ammonia oxidation. Moreover, it has been possible to prepare platinum catalysts (thin films or in dispersed particle form) in such a way that it selectively catalyzes one out of many competing hydrocarbon reactions. If the catalyst is prepared differently, it catalyzes another hydrocarbon reaction selectively and yields an entirely different product distribution from the same mixture of reactants.[1-3] For years we have been studying platinum crystal surfaces by a variety of techniques, and I would like to suggest that the key to the platinum activity and selectivity that is built in by appropriate surface preparation is the atomic surface structure that forms. The microstructure in which platinum atoms have different atomic environments and different numbers and arrangements of nearest neighbors determines the way in which the complex hydrocarbon reaction proceeds. Thus, in order to understand and control hydrocarbon catalysis of platinum, we have to uncover the correlation between the atomic surface structure and the reactivity. This paper reports on the status of our research in this field in which we utilize a variety of techniques.

In order to solve such a complex problem as the understanding of the mechanism of platinum catalysis of hydrocarbons, we have divided our research into smaller segments as follows: (a) the kinetics and mechanism of hydrogen dissociation via studies of H_2-D_2 exchange reaction, (b) the adsorption of hydrocarbons on various platinum surfaces, (c) comparison of reaction rates of platinum crystals at high pressures with rates on platinum particles dispersed on porous alumina supports, and (d) correlation of reactivity (turnover number) for a given hydrocarbon reaction with the atomic structure of platinum and correlation of product distribution with atomic structure. We discuss these various research studies briefly and separately and summarize only those observations that are important to the understanding of the mechanism of platinum catalysis that we have arrived at up to the present. It is hoped that clear statements of the key observations at the end of each of these sections will facilitate theoretical formulations of outstanding problems of metal catalysis.

The study of catalysis on the atomic scale was made possible by the availability of sensitive mass-spectrometer and gas-chromatograph detectors that can monitor the products of surface reactions coming from crystal surfaces less than 1 cm^2 in area. In this way, single crystals of well-defined orientation can be used as catalysts.[4] The surface structure is monitored by low-energy electron diffraction (LEED), the surface composition by Auger electron spectroscopy, and the kinetics of the surface reactions are monitored in a steady-state flux of reactants at high (1 atmosphere) or at low (10^{-4} Torr) pressures, or in a well-defined molecular beam. With the combination of these powerful techniques at hand, the reactivity can readily be followed as a function of surface structure and surface composition.

2 THE PLATINUM CRYSTAL SURFACES AND THE VARIOUS TECHNIQUES USED TO INVESTIGATE THEIR SURFACE CHARACTERISTICS

Platinum is a face-centered cubic metal with a melting point of 1769°C. The highest density, lowest free-energy surface in vacuum is designated by the Miller Index (111) followed by the (100) crystal face. In the presence of surface impurities or adsorbates, the relative surface free energy of the various crystal faces may change. This can result in rearrangement of the atomic surface structure.[5] The LEED patterns characteristic of the clean platinum (111) and (100) surfaces are shown in Figs. 1a and 1b, along with the schematic representation of their real lattice structure. In the (111) crystal surface, each atom has six nearest neighbors and the structure is similar to that expected from the projection of the X-ray unit cell to the (111) surface. The clean (100) crystal face along with the clean (110) crystal face is reconstructed, i.e., the surface structures are different from the one expected from the X-ray unit cell. This surface reconstruction has been studied in several laboratories and it appears to be due to a hexagonal distortion of the surface layer in the (100) crystal face. The accurate location of the atoms in the reconstructed surface awaits the surface-structure analysis from the LEED beam intensities which is in progress in this laboratory and in others.

When the crystal is cut at an angle with respect to the low-Miller-Index surface along a high-Miller-Index plane, a stepped surface results. Fig. 2 shows the LEED pattern of the (775) surface cut at 9.5 degrees to the (111) face in the direction of the (100) face, along with the schematic diagram of the real space lattice that can be deduced from the diffraction

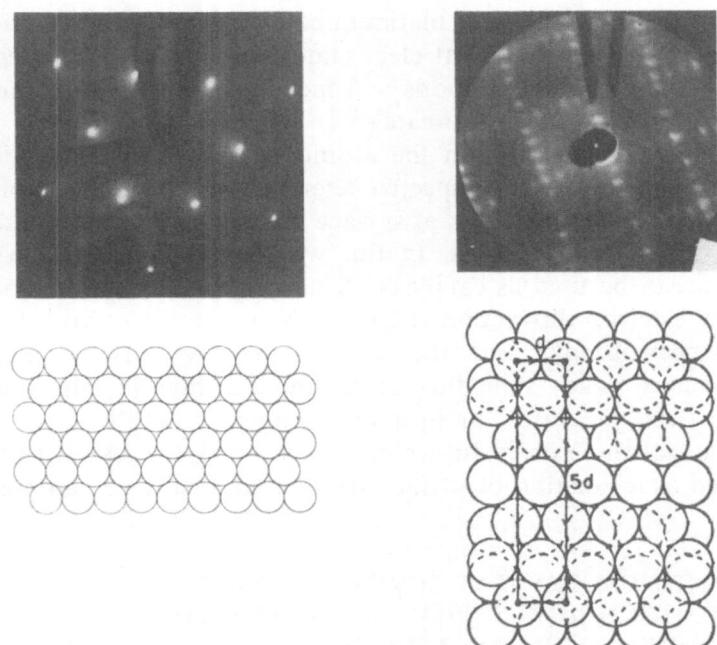

Fig. 1. (a) Low-energy electron diffraction pattern and schematic representation of the Pt(111) face. (b) Diffraction pattern of the Pt(100) surface and schematic representation of the (100) surface with a hexagonal overlayer.

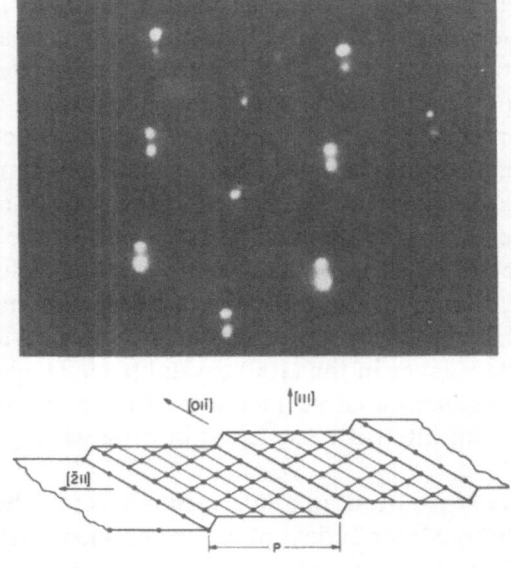

Fig. 2. Diffraction pattern from the Pt(5)-[6(111)x(100)] surface and its schematic representation.

pattern. Detailed analysis of stepped surfaces of metals and semiconductors has been made, and the methods of analysis are described elsewhere.[6,7] The atomic surface structure of the (775) face is composed of (111) orientation terraces separated by steps of one-atom height whose orientation is (100), as determined by the angle of cut. The step periodicity gives rise to the doubling of the diffraction beams at certain electron energies, as indicated by Fig. 2a. Experience in this laboratory indicates that these stepped or vicinal surfaces have a high degree of thermal stability if the terraces are, on the average, five atoms or more wide. The stepped surfaces can readily be regenerated after partial disordering by heat treatment near the melting point or ion bombardment. Stepped surfaces with three- or four-atom-wide terraces appear to undergo faceting either on heat treatment or during adsorption. In our notation, the (775) surface is designated as Pt(S) [6(111)x(100)], where S indicates a stepped surface, 6(111) indicates the widths and orientation of the terrace, and (100) indicates the orientation of the steps that are of one atom in height (the 1 to indicate the step height is deleted for brevity). This notation describes the atomic surface structure more realistically than does the Miller Index notation. Fig. 3 shows the diffraction pattern and the schematic diagram of the real lattice structure of the surface cut at 9.5 degrees from the (111) crystal face and rotated 20 degrees in the direction of the (310) face. Since the step orientation is one of high Miller Index, the steps should have a fairly high concentration of kinks in addition to those that are thermally regenerated in the step. Indeed, the chemical behavior of this Pt(S)-[7(111)x(310)] surface is different from that of the other stepped surfaces with (100) step orientation.[8]

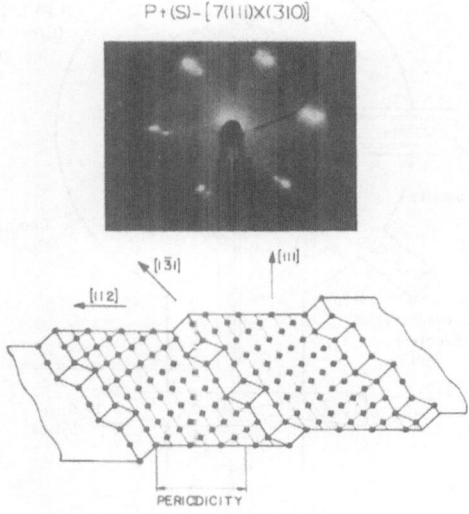

Fig. 3. Diffraction pattern and schematic representation of the Pt(S)-[7(111)x(310)] surface.

These low index and stepped surfaces are representative of the variety of surface structures used in our catalytic studies of platinum. One may vary the angle of cut, thereby changing the terrace widths and step orientation of the platinum surface. We have prepared surfaces with 5 to 18-atom-wide terraces of (111) or (100) orientation. As long as the stable stepped platinum surfaces are clean, the step height is monatomic. On adsorption of hydrocarbons, the step heights and terrace widths may change.

The surface structure of platinum and of adsorbed gases has been studied by (LEED) and the surface composition has been monitored by Auger electron spectroscopy. Transport studies, to determine the reactivity of the platinum surfaces, were carried out in three ways. By the steady-state transport method, the flux of reactants incident on the platinum surface and the reaction products were monitored by a mass spectrometer in a manner shown schematically in Fig. 4.[9] Because of the mass-spectrometer detector, the highest pressure that can be employed is of the order of 10^{-4} Torr. The schematic diagram of the apparatus that is used at reactant pressures as high as 2 atm is shown in Fig. 5. Using the gas-chromatograph detector, this "stirred batch reactor" can readily monitor the ring opening of cyclopropane on one face of a single crystal less than 1 cm^2 in area.[10] Reactions at both low and high pressures can be investigated using the apparatus shown in Fig. 5. The detailed description and working principles of these instruments are described in recent publications.[10,11]

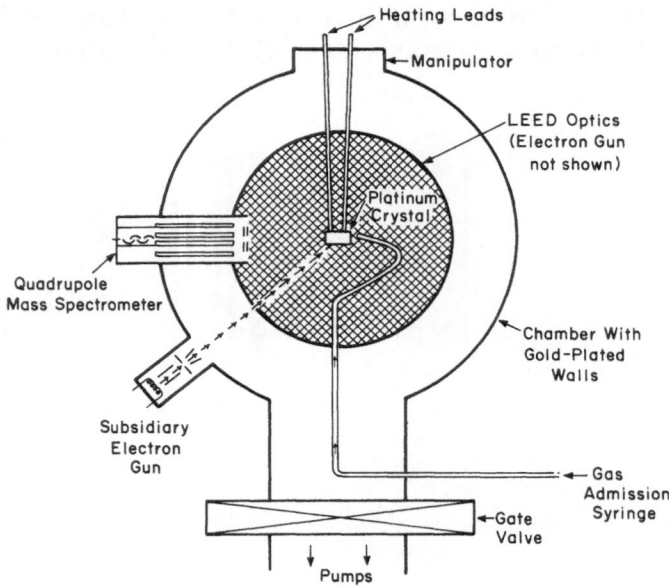

Fig. 4. Schematic diagram of the diffraction chamber that was also used to carry out the low-pressure ($\sim 10^{-4}$ Torr) surface reaction studies.

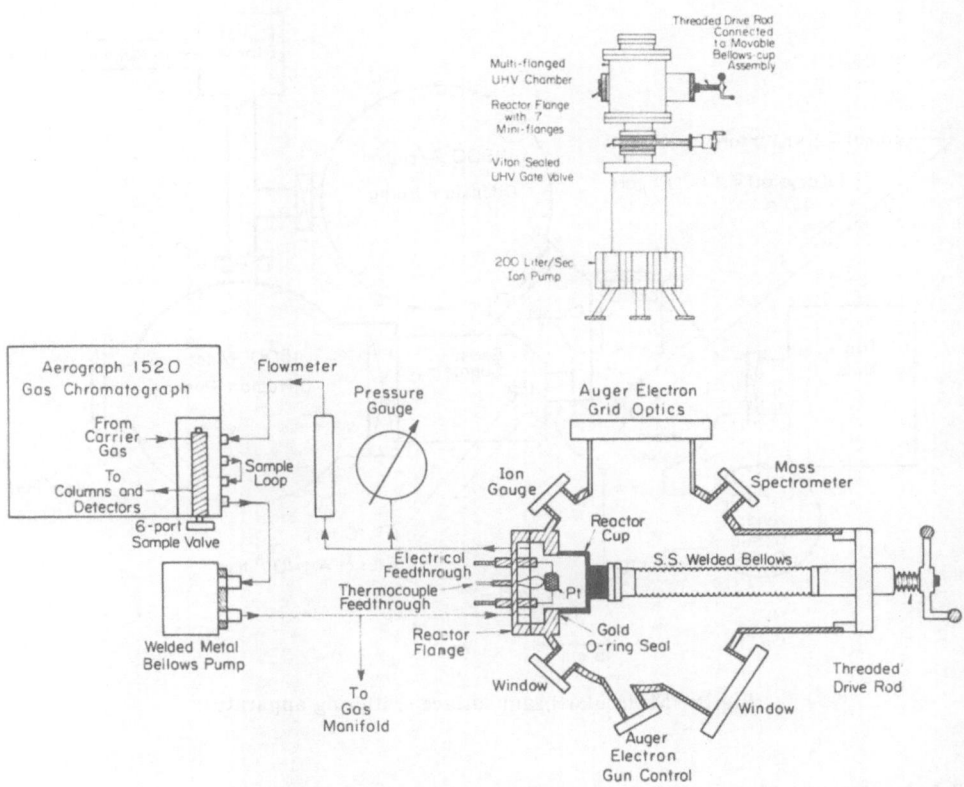

Fig. 5. Schematic of the UHV assembly for high-pressure catalysis on single-crystal platinum surfaces.

Another method of studying the kinetics of surface reactions is molecular-beam scattering. Fig. 6 shows the scheme of the molecular-beam-scattering apparatus used in studies of the H_2-D_2 exchange reaction on the various platinum surfaces.[12] A well-collimated molecular beam of the reactant gas or gas mixture is scattered from the crystal surface and the products desorbed at a given solid angle are detected by a mass spectrometer. In this fashion, the angular distribution of the scattered products can be determined. Moreover, since the incident molecular beam is chopped at a well-defined frequency, usually between 100 to 5000 Hz, the flight time of the scattered molecules can be measured. Thus, from the flight time, the residence time of the reacting molecules on the surface, i.e., the minimum residence time necessary to detect reaction products, can be determined in the range of 10^{-6} to 10^{-2} second.[13] This is in contrast with the steady-state-transport method of reaction-product analysis which gives time-averaged reaction product concentration.

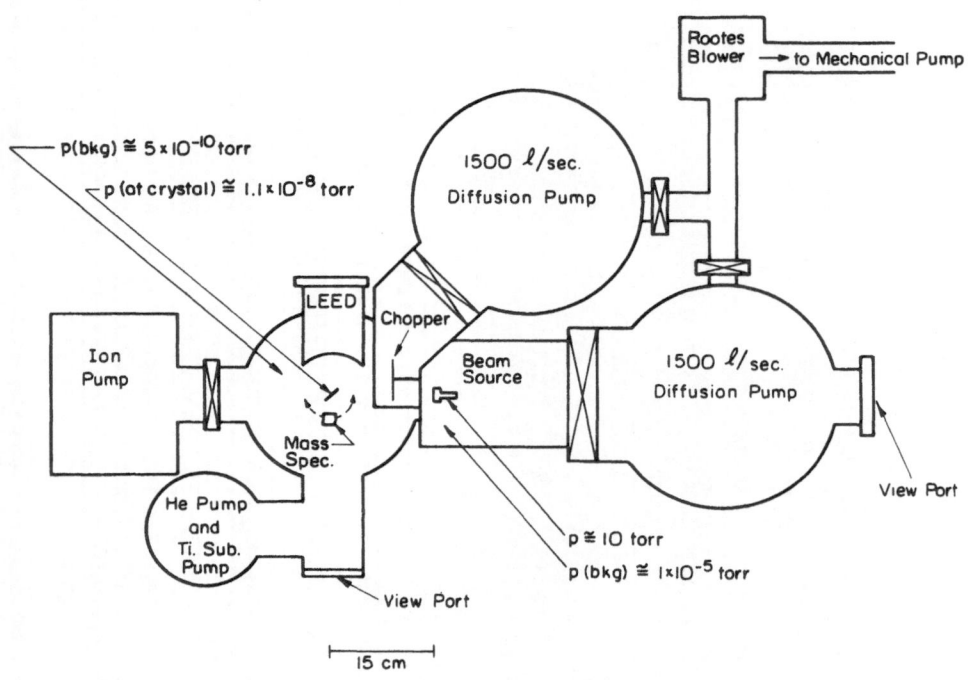

Fig. 6. Molecular-beam surface scattering apparatus.

2.1 Kinetics of H_2-D_2 Exchange on Platinum Crystal Surfaces

Using molecular beams, the HD product concentration and angular distribution were monitored from the Pt(111), the Pt-[9(111)x(111)], and the Pt[5(111)x(111)] surfaces.[13] The chopping frequency of the incident beam was varied in such a way that product molecules that formed in a period of 10^{-2} second or less were detectable, while those formed in times longer than 10^{-2} second were not. Under these conditions, in the temperature range of 300 to 600°C, no HD molecules could be detected from the (111) platinum surface, while quite large HD concentrations were observed from the two stepped surfaces. These observations indicate reaction probabilities of the order of 10^{-1} on the stepped surfaces and less than 10^{-5} on the (111) face. The reaction rate was first order in hydrogen or deuterium pressure; and from the temperature dependence of the HD signal amplitude, an activation energy of 4.5 kcal has been deduced for this reaction. The HD product had cosine angular distribution, indicating complete thermal accomodation with the platinum surface prior to desorption.

Atomic steps on the platinum surfaces are essential in dissociating hydrogen and deuterium. (Without the presence of a large concentration of atomic steps, the probability of the exchange reaction is very low.) In the presence of atomic steps, the hydrogen molecules dissociate with low activation energy and the surface is able to store a large concentration of atoms. This way, the surface converts the bimolecular reaction between H_2 and D_2, which is improbable because of the large energy of dissociation of the molecules (~103 kcal), to an atom molecule reaction (H + D_2) at low activation energy. The atoms react with the incident molecules by a two-branch mechanism. At low temperatures (<600°K), the rate-limiting step appears to be the diffusion of the molecule on the surface to a site where the hydrogen deuteride can be formed. At high temperatures, the reaction between an adsorbed atom and a molecule incident at an atomic step competes with the diffusion-controlled low-temperature branch.[13] These reactions take place on the stepped surfaces in residence times of less than 10^{-2} second, which appears to be too short for the reaction to occur on the low-Miller-Index (111) crystal face.

Summary. Atomic steps play a controlling role in dissociating hydrogen molecules on platinum surfaces.

2.2 Adsorption of Hydrocarbons on the Low-Miller-Index and Stepped Platinum Surfaces

The adsorption of hydrocarbons on the (111), (100), and stepped platinum surfaces has been studied by LEED and work-function-change measurements in the surface temperature range 20 to 300°C.[14,15] Over 40 organic molecules were investigated using the low-Miller-Index surfaces at low pressures in the range of 10^{-9} to 10^{-7} Torr. The observed work-function change is always negative, which indicates that the organic molecules that adsorb are net electron donors to the metal. Table I lists the work-function changes and the surface structures that were obtained on adsorption of the various hydrocarbons. Both the work-function change and the LEED patterns indicate that on these low-Miller-Index platinum surfaces the organic molecules are stable, i.e., do not undergo dehydrogenation or chemical rearrangements readily in the temperature range of 20 to 300°C. Following adsorption, reorientation of the molecules in the adsorbed layer is necessary to form ordered structures. Molecules that have higher rotational symmetry or have only small-size substituents on the benzene rings exhibit better ordering. The adsorbed layers are more ordered on the Pt(111) crystal face than on the Pt(100) face. Both the diffraction and the work-function-change data indicate that substituted benzenes chemisorb with their benzene ring parallel to the surface and interact with the metal via the π electrons in the benzene ring. Benzene itself undergoes reorientation on the platinum surface with increasing exposure which indicates a change from predominantly π bonding to σ bonding as the surface coverage in-

Table I. Work-Function Changes and Structural Information for Adsorption of Organic Compounds on the Pt(111) and Pt(100)–(5 x 1) Surfaces

Adsorbate	Temp, °C	Pt(111) Press., Torr	WFC, volts	Adsorbate Diffraction Features or Surface Structure	Pt(100)-(5 x 1) Press., Torr	WFC, volts	Substrate Structure After Adsorption	Adsorbate Diffraction Features or Surface Structures
Acetylene	20	1×10^{-8}	-1.5	(2 x 2)	4×10^{-7}	-1.65	(1 x 1)	$(\sqrt{2} \times \sqrt{2})$R45°
	20	1×10^{-8} (10 min)	-1.65	Disordered				
	150	4×10^{-7}	-1.8	Disordered	4×10^{-7}	-1.7	(1 x 1)	$(\sqrt{2} \times \sqrt{2})$R45°
Aniline	20	1×10^{-8}	-1.8	Streaks at 1/3 order; Diffuse (1/2 0) features	1×10^{-8}	-1.75	(1 x 1)	Disordered
Benzene	20	4×10^{-7}	-1.8	Poorly ordered	3×10^{-7}	-1.6	(1 x 1)	Diffuse, ringlike, 1/2 order streak
	20	4×10^{-7} (5 min)	-1.4	$\begin{vmatrix} 4 & -2 \\ 0 & 4 \end{vmatrix}$				
	20	4×10^{-7} (40 min)	-0.7	$\begin{vmatrix} 4 & -2 \\ 0 & 5 \end{vmatrix}$	3×10^{-7} (2 hr)	-1.3	(1 x 1)	Diffuse 1/2 order streak
Biphenyl	20	2×10^{-9}	-1.85	Very poorly ordered	2×10^{-9}	-1.8	(1 x 1)	Disordered
n-Butylbenzene	20	8×10^{-9}	-1.5	Disordered	8×10^{-9}	-1.5	(1 x 1)	Disordered
t-Butylbenzene	20	5×10^{-8}	-1.7	Disordered	5×10^{-8}	-1.75	(1 x 1)	Disordered
Cyanobenzene	20	1×10^{-8}	-1.6	Diffuse (1/3 0) features	1×10^{-8}	-1.5	Faint (5 x 1)	Disordered
1,3-Cyclohexadiene	20	2×10^{-8}	-1.75	Poorly ordered	2×10^{-8}	-1.7	(1 x 1)	Diffuse 1/2 order streak
	20	2×10^{-8} (1 hr)	-1.3	$\begin{vmatrix} 4 & -2 \\ 0 & 4 \end{vmatrix}$	2×10^{-8} (1 hr)	-1.6	(1 x 1)	Diffuse 1/2 order streak
	20	3×10^{-7} (5 hr)	-0.8	$\begin{vmatrix} 4 & -2 \\ 0 & 5 \end{vmatrix}$	2×10^{-8} (5 hr)	-1.4	(1 x 1)	Diffuse 1/2 order streak
Cyclohexane	20	6×10^{-9}	-1.2	(1 x 1) low background	6×10^{-9}	-0.75	(5 x 1)	Low background
	20	4×10^{-7}	-0.7	Very poorly ordered	4×10^{-7}	-0.4	(1 x 1)	Diffuse, streaked, (2 x 1) pattern
	150	4×10^{-7}	-1.1	Apparent (2 x 2)	4×10^{-7}	-1.2	(1 x 1)	Streaked (2 x 1) pattern
	300	4×10^{-7}	-1.4	Disordered	4×10^{-7}	-1.5	(1 x 1)	Disordered
Cyclohexene	20	6×10^{-7}	-1.7	$\begin{vmatrix} 2 & 2 \\ 4 & -2 \end{vmatrix}$	6×10^{-7}	-1.6	(1 x 1)	Diffuse (1/2 0) features
	150	6×10^{-7}	-1.6	Apparent (2 x 2)	6×10^{-7}	-1.5	(1 x 1)	Streaked (2 x 1) pattern
Cyclopentane	20	7×10^{-9}	-0.95	(1 x 1) low background	7×10^{-9}	-0.4	(5 x 1)	Low background
	20	4×10^{-7}	-0.7	Disordered	4×10^{-7}	-0.3	(1 x 1)	Diffuse features at 1/2 order
Cyclopentene	20	–	–	–	2×10^{-7}	-1.4	(1 x 1)	Diffuse, streaked, (1/2 0) features
2,6-Dimethyl-pyridine	20	4×10^{-8}	-1.6	Diffuse 1/3.2, 2/3.2 order streaks	4×10^{-8}	-1.5	Faint (5 x 1)	Disordered
3,5-Dimethyl-pyridine	20	6×10^{-8}	-2.3	Diffuse 1/2 order streak	6×10^{-8}	-2.2	(1 x 1)	Disordered

Table I. (Continued)

Adsorbate	Temp, °C	Pt(111) Work-Function Change Press., Torr	WFC, volts	Adsorbate Diffraction Features or Surface Structure	Pt(100–(5 × 1)) Work-Function Change Press., Torr	WFC, volts	Substrate Structure After Adsorption	Adsorbate Diffraction Features or Surface Structures
Ethylene	20	1 x 10⁻⁸	-1.5	Diffuse (1/2 0) features	1 x 10⁻⁸	-1.2	(1 x 1)	(√2 x √2)R45°
	250	1 x 10⁻⁸	-1.7	Disordered	1 x 10⁻⁸	-1.5	(1 x 1)	Disordered
Graphitic Overlayer	950		-1.1	Ringlike diffraction features		-1.0	(1 x 1)	Ringlike diffraction features
n-Hexane	20	5 x 10⁻⁸	-1.1	Disordered	5 x 10⁻⁸	-0.8	(1 x 1)	Disordered
	20	5 x 10⁻⁸ (5 hr)	-0.9	Disordered	5 x 10⁻⁸ (5 hr)	-0.6	(1 x 1)	Disordered
	250	5 x 10⁻⁸	-1.5	Disordered	5 x 10⁻⁸	-1.2	(1 x 1)	Disordered
Isoquinoline	20	6 x 10⁻⁸	-1.9	Diffuse (1/3 0) and (2/3 0) features	6 x 10⁻⁸	-2.1	(1 x 1)	Disordered
Mesitylene	20	4 x 10⁻⁸	-1.7	Streaks at 1/3.4 order diffuse (2/3.4 0) features	4 x 10⁻⁸	-1.7	(5 x 1)	1/3 order streaks
	20	4 x 10⁻⁷	-1.35	Disordered	4 x 10⁻⁷	-1.2	(1 x 1)	Disordered
2-Methyl-naphthalene	20	6 x 10⁻⁸	-2.0	Very poorly ordered	4 x 10⁻⁹	-1.6	Faint (5 x 1)	Disordered
Naphthalene	20	9 x 10⁻⁹	-1.95	Apparent (3 x 1)	9 x 10⁻⁹	-1.7	(1 x 1)	Disordered
	150	9 x 10⁻⁹	-2.0	(6 x 6)	9 x 10⁻⁹	-1.65	(1 x 1)	Disordered
Nitrobenzene	20	9 x 10⁻⁹	-1.5	Diffuse (1/3 0) features (pattern electron-beam sensitive)	9 x 10⁻⁹	-1.4	(1 x 1)	Disordered
Piperidine	20	8 x 10⁻⁸	-2.1	Disordered	8 x 10⁻⁸	-2.05	Faint (5 x 1)	Disordered
Propylene	20	2 x 10⁻⁸	-1.3	(2 x 2) (pattern electron-beam sensitive)	2 x 10⁻⁸	-1.2	(1 x 1)	1/2 order streaks (pattern electron-beam sensitive)
Pyridine	20	1 x 10⁻⁸	-2.7	Diffuse (1/2 0) features	1 x 10⁻⁸	-2.4	(1 x 1)	Disordered
	250	1 x 10⁻⁸	-1.7	Well-defined streaks at 1/3, 2/3, 3/3 order	1 x 10⁻⁸	–	(1 x 1)	(√2 x √2)R45°
Pyrrole	20	6 x 10⁻⁸	-1.45	Diffuse (1/2 0) features (pattern electron-beam sensitive)	6 x 10⁻⁸	-1.6	(1 x 1)	Diffuse (1/2 0) features
Quinoline	20	3 x 10⁻⁸	-1.45	Diffuse 1/3 order streaks	3 x 10⁻⁸ (6 min)		(5 x 1)	Diffuse 1/3 order streaks
					3 x 10⁻⁸ (14 min)	-1.7	(1 x 1)	Disordered
Styrene	20	6 x 10⁻⁸	-1.7	Streaks at 1/3 order	6 x 10⁻⁸	-1.65	(1 x 1)	Very poorly order
Toluene	20	1 x 10⁻⁹	-1.7	Streaks at 1/3 order	1 x 10⁻⁹	-1.55	(5 x 1)	Streaks at 1/3 order
	150	1 x 10⁻⁹	-1.65	(4 x 2)	1 x 10⁻⁹	-1.5	(1 x 1)	Disordered
m-Xylene	20	1 x 10⁻⁸	-1.8	Streaks at 1/2.6 order	1 x 10⁻⁸	-1.65	(5 x 1)	Streaks at 1/3 order

creases. The large work-function change on adsorption of pyridine (−2.5 V) and the adsorption characteristics of substituted pyridines indicate that the nitrogen participates in the adsorbate-substrate bond.

The chemisorption of hydrocarbons has entirely different characteristics on stepped platinum surfaces.[8] While the hydrocarbon molecules remain largely intact on the low-Miller-Index surface in the temperature range 20 to 200°C, they undergo chemical reactions readily, dehydrogenation, and/or decomposition even at 20°C. Products of the chemisorption are partially dehydrogenated carbonaceous deposits whose characteristics depend on the surface structure of the stepped platinum surfaces, the type of hydrocarbon chemisorbed, the rate of adsorption, and the surface temperature. The distinctly different chemisorption characteristics of the various stepped platinum surfaces, described in detail elsewhere, have been explained by considering the interplay of four competing processes[8]: (1) dehydrogenation, (2) decomposition of the organic molecules, (3) the nucleation and the growth of ordered carbonaceous surface structures, and (4) the rearrangement of the platinum substrate by faceting. The effect of increased partial pressure of hydrogen over the surface is to slow down the rate of decomposition so that processes (1) and (3) may predominate on a stable, stepped crystal surface.

Summary. Atomic steps and other microstructures, due to platinum atoms in various states of coordination present in steps, control the rates of breaking C-H and C-C bonds. In the absence of large concentrations of steps, the adsorbed hydrocarbon molecules remain intact below 200°C so that their surface crystallography can readily be studied.

2.3 Comparison of Reaction Rates for the Cyclopropane Ring Opening on Platinum Crystals and Supported Platinum Catalysts

There is a gap between chemisorption and surface-reaction studies performed in high vacuum on single-crystal surfaces and those carried out at 1 atm or higher pressures on highly dispersed supported catalysts. The purpose of our work is to bridge the gap between these two fundamental areas of catalytic research.[10] The objective was to measure reaction rates on well-defined single-crystal surfaces both at high pressures (1 atm) and at low pressures after preparation in ultrahigh vacuum in the range of 10^{-4} to 10^{-8} Torr within the same apparatus. The hydrogenolysis of cyclopropane was studied at 1 atm total pressure on a stepped, single-crystal surface of platinum. The hydrogenolysis of cyclopropane was chosen as the test reaction because of the considerable amount of data and experience which has been collected in various laboratories. The rate is relatively high at room temperature on supported platinum catalysts and only one product, propane, is formed on platinum catalysts below 150°C which simplifies the analysis of the results. Table II summarizes the results obtained and compares our results on stepped, single-crystal surfaces at atmospheric pressures with those of others obtained using supported platinum catalysts.[11] It appears that at 1 atm pressure the platinum stepped single crystal behaves very much like a highly dispersed supported platinum catalyst for the cyclopropane hydrogenolysis. This observation supports the contention that well-defined crystal surfaces are excellent models for polycrystalline supported metal catalysts. It also tends to verify Boudart's hypothesis that the cyclopropane hydrogenolysis is an example of a structure-insensitive reaction. The initial specific reaction rates that were reproducible within 10 percent were within a factor of 2 of published values for this reaction on highly dispersed platinum catalysts. The activation energies observed for this reaction, in addition to the turnover number (number of molecules formed per platinum atom per minute), are close enough on the various platinum surfaces to call the agreement excellent.

Table II. Comparison of Initial Specific Rate Data for the Cyclopropane Hydrogenolysis on Platinum Catalysts[11]

Date Source	Type of Catalyst	Calculated Specific Reaction Rate @ P^{*}_{CP} = 135 Torr and T = 75 °C	
		$\left(\dfrac{\text{Moles } C_3H_8}{\text{Min/Cm}^2 \text{ Pt}}\right)$	$\left(\dfrac{\text{Molecules } C_3H_8}{\text{Min/Pt Site}}\right)$
Present study	Run 10A	2.1×10^{-6}	
	Run 12A	1.8×10^{-6}	
	Run 15	1.8×10^{-6}	
	Run 16	2.1×10^{-6}	
	Average	1.95×10^{-6}	812
Hegedus	0.04 wt % Pt on η-Al$_2$O$_3$	7.7×10^{-7} based on 100% Pt dispersion	410
Boudart et al.	0.3% and 2.0% Pt on η-Al$_2$O$_3$	8.9×10^{-7}	480
and			
Dougharty	0.3% and 0.6% Pt on γ-Al$_2$O$_3$	2.5×10^{-6}	1340

Summary. Single-crystal platinum surfaces with well-defined structures are realistic models for dispersed, supported-metal-catalyst particles. Thus, the information obtained on crystal surfaces can be utilized to explain the reactivity of supported metal catalysts.

2.4 Reactions of n-Heptane on Platinum Surfaces of Varying Atomic Structure

n-Heptane may undergo a variety of chemical reactions on platinum surfaces, including hydrogenolysis, dehydrocyclization, and isomerization:

It has been well documented in the literature[1-3] that, depending on the catalyst preparation, the ratio of products that form in these competing reactions can be widely varied. The turnover numbers have been measured and were in the range of 10^{-3} for hydrogenolysis and isomerization and 10^{-4} for dehydrocyclization. We have studied these various reactions in our steady-state-flux, low-pressure system[17] using a mass-spectrometer detector and hydrogen to an n-heptane ratio of 5:1. In one study we used stepped surfaces with (111) terrace and (100) step orientation, which differed only by the width of their terraces, from 4 to 10 atoms wide. The rates of hydrogenolysis and isomerization reactions appear to increase somewhat with increasing step density without exhibiting marked structural sensitivity. As long as atomic steps are present in fairly large concentrations, these reactions occur readily on single-crystal surfaces at low pressures with turnover numbers comparable to those reported on platinum powders at high pressures.[18] Low-energy electron diffraction and Auger electron spectroscopy studies have revealed that the platinum surface is covered with a layer of disordered carbonaceous deposit while these reactions take place at low pressures. The presence of this deposit does not hinder the catalyzed surface reactions in any way. While the overall rate of the hydrogenolysis reaction shows minimal surface-structure sensitivity, when the turnover numbers are compared for the various stepped platinum surfaces the product distribution is markedly dependent on the atomic structure of platinum. With polycrystalline platinum foils cleaned by high-temperature oxidation (approximately $1400°C$, 10^{-7} Torr oxygen), methane is the predominant hydrogenolysis product of n-heptane at $350°C$. Using the Pt(S)-[6(111)x(100)] surface, the ratio of methane to ethane to propane of 5:3:2 was observed at $350°C$. Thus, the nature of C-C bond breaking depends very much on surface structure. These results indicate a similar mechanism for the hydrogenolysis and isomerization reactions to that suggested by Touroude and Gault.[19] The presence of carbonaceous residues on the metal surface was also observed by Merta and Porec[20] during cyclopropane hydrogenolysis and by Sinfelt[21] during the hydrogenolysis of ethane.

The dehydrocyclization reaction to form toluene exhibited a behavior very different from the other two competing reactions. Toluene formation is detectable only if the stepped platinum surface is covered with an ordered layer of carbonaceous deposit. It appears that this complex reaction will take place only in the presence of such a template, that is, a partially dehydrogenated, ordered carbonaceous deposit. The ordered deposit, in turn, forms only on stepped platinum surfaces with the appropriate atomic structure, terrace width and orientation, and step orientation. These chemical reactions do not take place on the (111) crystal face of platinum. Within minutes (using the same reaction conditions as that for stepped surfaces), the low-Miller-Index surface is covered with a layer of graphitic carbon that poisons the catalytic activity. If this ordered layer fails to form on account of the presence of surface impurities or high-temperature oxidation of platinum or some other reason related to surface preparation, the dehydrocyclization reaction does not occur.

Summary. The catalytic activity of platinum in hydrocarbon reactions is controlled by platinum atoms surrounded by fewer neighbors than the atoms in the close-packed, low-Miller-Index crystal planes.

The dehydrocyclization of n-heptane to toluene is detectable at low pressures (10^{-4} Torr) only if the surface is covered with a layer of ordered carbonaceous deposits. While the competing hydrogenolysis and isomerization reactions do not require the presence of such ordered structures, the product distributions of these two reactions, just as in dehydrocyclization, are markedly dependent on the surface structure of platinum.

DISCUSSION

The various studies using platinum surfaces that were reviewed all indicate that the key to the reactivity of platinum is the atomic step structure. Since the steps occur with well-defined periodicity, the atomic structure of these active sites can readily be studied by diffraction. Platinum atoms in steps can exist in several different atomic environments which are distinguishable by the number of nearest neighbors, and by the various nearest-neighbor bond lengths and bond angles. Changing such a microstructure by suitable surface preparation can drastically alter the relative rates of competing hydrocarbon reactions and the product distribution. Thus, platinum catalysts can be tailored to yield desired products by manipulation of the surface structure. The marked differences in the reactivity of the various surface sites indicate differences in the strengths of the chemical bonds (Pt-C, Pt-H) from site to site. It is likely that such differences in chemical bonding are due to charge density variations for platinum atoms in various surface sites. The largest difference in charge density is perhaps between platinum atoms in a step and in a terrace.

Since platinum atoms may be placed in several different atomic environments on a given surface, each catalyst particle can be multifunctional and can catalyze several simple reactions simultaneously. It is our hope that using stepped crystal surfaces of well-defined atomic structure will allow the various catalytic functions of the surface microstructures to be identified and isolated.

A significant body of data is available that indicates the presence of partially dehydrogenated carbonaceous residues (ordered or disordered) on the metal surface during hydrocarbon reactions at low or high pressures and in the presence of excess hydrogen. More detailed scrutiny of the mechanism of metal catalysis of hydrocarbons must include studies of the electronic structure of the adsorbed carbonaceous layer.

ACKNOWLEDGMENT

This work was done under the auspices of the U.S. Atomic Energy Commission through the Inorganic Materials Research Division of the Lawrence Berkeley Laboratory. Partial support was received from the Office of Naval Research of the National Science Foundation and the Petroleum Research Fund.

REFERENCES

1. Anderson, J. R., and Avery, N. R., *J. Catalysis*, **5**, 446 (1966).

2. Baron, Y., et al., *J. Catalysis*, **5**, 428 (1966).
3. Shepard, F. E., and Rooney, J. J., *J. Catalysis*, **3**, 129 (1964).
4. Somorjai, G. A., *Catalysis Rev.*, **7**, 87 (1972).
5. Somorjai, G. A., *J. Catalysis*, **27**, 453 (1972).
6. Henzler, M., *Surface Sci.*, **19**, 159 (1970).
7. Lang, B., Joyner, R. W., and Somorjai, G. A., *Surface Sci.*, **30**, 440 (1972).
8. Baron, K., Blakely, D., and Somorjai, G. A., *Surface Sci.*, **41**, 45 (1974).
9. Lang, B., Joyner, R. W., and Somorjai, G. A., *Proc. Roy. Soc.*, **A331**, 335 (1972).
10. Kahn, D., Petersen, G. E., and Somorjai, G. A., *J. Catalysis* (1974).
11. Kahn, D., Ph.D. Thesis, University of California, Berkeley (1973).
12. Bernasek, S. L., Siekhaus, W. J., and Somorjai, G. A., *Phys. Rev. Letters*, **30**, 1202 (1973).
13. Bernasek, S. L., and Somorjai, G. A., *J. Chem. Phys.*, **62**, 3149 (1975).
14. Gland, J. L., and Somorjai, G. A., *Surface Sci.*, **41**, 387 (1974).
15. Gland, J. L., and Somorjai, G. A., *Surface Sci.*, **38**, 157 (1973).
16. Ciapetta, F. G., and Wallace, D. N., *Catalysis Rev.*, **5**, 67 (1971).
17. Lang, B., Joyner, R. W., and Somorjai, G. A., *J. Catalysis*, **27**, 405 (1972).
18. Carter, J., Cusumano, L., and Sinfelt, J. H., *J. Catalysis*, **20**, 223 (1971).
19. Touroude, R., and Gault, F. G., *J. Catalysis*, **32**, 279 (1974).
20. Merta, R., and Porec, V., *J. Catalysis*, **17**, 79 (1969).
21. Sinfelt, J. H., *J. Catalysis*, **27**, 468 (1972).

DISCUSSION on Paper by G. A. Somorjai

MENZEL: I should like to mention some work I did quite a long time ago, which may be connected with Professor Somorjai's results. The ortho-para-hydrogen conversion was studied on platinum filaments as a function of pretreating [Menzel, D., and Riehert, L., *Ber. Bunsenges. Phys. Chem.*, (1962)]. While the energy of activation on filaments untreated or treated in hydrogen was 18 kcal/mole, pretreatment in oxygen at 1200°C yielded activities longer by 10^4 and an activation energy of 4.5 kcal/mole. The latter is interestingly identical with that reported by Professor Somorjai for the H_2-D_2 reaction on stepped platinum surfaces. Our results were originally interpreted as being due to oxygen present on the surface, but it appears possible that the steps were really responsible (the mentioned treatment leads to rapid evaporation of PtO_2, which certainly roughens the surface). Interestingly, admixture of gold while lowering the preexponential gave about the same activation energy as pretreatment with oxygen.

PLUMMER: In your experiments on H_2-D_2 exchange, is it possible that the steps are not the active sites for exchange (at high temperature)? Could the steps act as a source of atoms to diffuse across the plane, where these single atoms on top of a plane are the active sites for exchange? At low temperatures there would be no atoms diffusing across the planes and the H_2-D_2 exchange would be independent of step density. At higher temperatures, the steps act as a source of platinum atoms and the exchange rate would depend upon the step density. This could be checked by cooling a Pt(111) crystal to 77°K, depositing platinum atoms onto the crystal, and then performing the H_2-D_2 exchange.

SOMORJAI: It is entirely possible that platinum adatoms are even more effective in dissociating H_2 or D_2 molecules than are atoms in steps if lowered coordination number is important in such a process. It would be important to test this by your proposed studies. There is a possibility, however, that, at low temperatures, the increased surface concentrations of the reacting species may affect the mechanism as well and might make it more difficult to ascertain the effect of adatoms.

RHODIN: This concerns four questions on the more detailed definition of the nature of the microstructure at stepped platinum surfaces. It seems important in the structure analyses of the stepped surface to define as precisely as possible, the step height and to a lesser extent, the terrace width. The questions are:

(1) Do you have any idea of the degree of uncertainty in the step-height determinations? It seems that the kinematic scattering approach developed by Rhead and Perdereau, and also by Henschler, depends very much on a simple, idealized diffraction model. This can describe only approximately the real surfaces which you study.

(2) What effort has been made to determine what fraction of the total sample surface is not in the simple ordered crystallography assumed in your model?

(3) What fraction of the total ordered surface is represented by the step-terrace structure specifically defined ($\pm$ 1 Å) in the schematic diagram of a crystal cut off at the simple vicinal surface?

(4) As regards the remarkable crystallographic stability of the initially prepared stepped surfaces during subsequent reactions, in terms of the Wulff-Herring concepts of surface energy and crystallography, can you explain how so many of the different sectioning orientations reported all tend to be effectively stabilized by each of the different organic reactions that you have studied? Is it possible that all the surface structures are essentially in a metastable state which may revert to a more stable configuration if heated to sufficiently high temperatures?

SOMORJAI: To answer your questions in order:

(1) The high-Miller-Index surfaces that have ordered step structure are cut within 1/2 degree of the desired orientation. The single kinematic models that you have mentioned have been adequate to determine both the average terrace width and step height and we have used none other.

Any useful improvements on these models that permit determination of the atomic step structure in greater detail would be most welcome.

(2) Estimation of that fraction of the surface that is disordered in a LEED experiment is always difficult. The scattering of helium beams can be used to determine surface roughness and has been found useful in our work on stepped surfaces. However, much of our information on surface order comes from the inspection of the LEED pattern, just as in the case of studies on low-Miller-Index surfaces. The diffraction beam size and its intensity with respect to the background intensity is utilized to estimate surface order. Changes of diffraction features due to ordering upon heating after ion bombardment also gives a great deal of information. In general, the quality of the diffraction patterns from stepped surfaces is just as good as the diffraction pattern from the low-Miller-Index face of the same metal.

(3) It is clear that the diffraction pattern yields *average* terrace spacing which of course hides the broad distribution of terrace width that is possible as indicated by the treatment of Park and Houston. The schematic diagram I have shown simply indicates the average terrace width and the step height that can be deduced from the analysis of the diffraction pattern.

(4) The stepped platinum surfaces are remarkably stable; their behavior upon heating to high temperature has been described in our various publications. They are readily regenerated after ion bombardment or after removal of surface impurities or adsorbates. Various possible reasons for this stability have been discussed in a paper by Schwoebel, Mykura, and others. Although the stepped surfaces must have higher free energies than the low-Miller-Index surfaces, their stability appears to be assured both by thermodynamic (Cabrera) as well as kinetic (Schwoebel) reasons. There is a definite need to collect a great deal of experimental data on the stability of stepped surfaces as a function of temperature and in the presence of adsorbates. Then we can develop a more complete theory to explain the remarkable stability of stepped surfaces.

BOND: Professor Somorjai's observation that dehydrocyclization of n-heptane occurs only in the presence of an ordered carbon deposit may mean that its rate on a clean surface is slow compared with the rate of its disruptive chemisorption with low-coordination-number atoms present at steps, a process which precedes carbon formation. Thus, when no further carbon deposition can occur, dehydrocyclization may be observed. Now, as I have explained in my introductory lecture, hydrogenolysis probably goes via multiply bonded intermediates for whose chemisorption (in the case of platinum) low-coordination edge atoms are needed. Thus, hydrogenolysis can compete with carbon formation, and by implication should be able to proceed in its absence, as is observed.

SOMORJAI: I agree that this can certainly be one possible explanation of our findings.

THEORY OF ENERGY AND MOMENTUM EXCHANGE BETWEEN NEUTRAL GAS ATOMS AND METAL SURFACES

E. Drauglis and J. I. Kaplan*

Battelle, Columbus Laboratories
Columbus, Ohio

ABSTRACT

In recent years molecular beam techniques have been developed to the point where it is now almost routine to obtain reproducible data on the scattering of thermal beams of inert gas atoms by solid surfaces. This improvement in experimental technique has motivated several theoretical treatments of gas-surface scattering. However, there is as yet no theory that is completely adequate. Recent work at Battelle has led to a new quantum-mechanical theory which can serve as a logical starting point for a useful theory of gas-surface scattering. This theory, which is a quantum-mechanical generalization of Landau's theory of thermal accommodation, uses first-order, time-dependent, perturbation theory in which all of the higher order terms in the Born series are included. A simple closed-form ex-

*Present Address: Indiana University and Purdue University at Indianapolis, Indianapolis, Indiana.

pression for $P(\Delta\omega)$, the probability of energy loss of a neutral particle scattering from a surface was derived. Numerical evaluations of this expression for helium incident on tungsten indicate that a single-phonon theory is not sufficient to obtain quantitative agreement with a multiphonon theory.

In this paper, the formalism for the calculation of $P(\Delta\omega)$ is reviewed and its adaptation to the calculation of momentum exchange is described.

1 INTRODUCTION

The concept of using thermal beams of neutral gas atoms as a tool for the study of the properties of the surfaces of solids has been of high interest for a long time. The analogous experiments involving crossed molecular beams of well-defined species have been performed successfully for many years. These experiments have been very useful not only in determining the pair potential for the interaction between nonreactive species but also in obtaining values for the cross sections for both reactive and nonreactive scattering. However, molecular beam techniques have not yet been as successful in providing information on surface properties, partly because of technological difficulties and partly because of insufficient theoretical development. Most of the technological difficulties either have been or are in the process of being overcome. It is therefore now appropriate for theoreticians to apply themselves to the problems of the interpretation of the experiments.

Although the techniques have not been perfected to the point where they can provide detailed information about surface properties, two general observations have been made as a result of modern experimental work. Perhaps the most fundamental of these is the absence of diffraction of helium beams in scattering from metals even though the techniques are sufficiently sensitive to detect diffraction by nonmetallic surfaces.[1-6] The other general observation is that inelastic scattering structure may be discerned away from the elastic peaks. It is this latter observation which gives encouragement to the idea that information on surface properties may be obtained by appropriate analysis of the inelastic scattering of inert gas atoms from surfaces. As an example, such an analysis may provide information about the surface phonon modes.[7]

Over the years, considerable effort has been expended on the theory of the interactions of gases with solids. Most prominent in this area have been the classical mechanical theories of F. O. Goodman[8-19], J. D. McClure[20-23], R. A. Oman[24-28], L. M. Raff[29-32], and the MIT group, R. M. Logan, R. E. Stickney, and J. C. Keck[33-36]; the quantum-mechanical theories of N. Cabrera, V. Celli, F. O. Goodman[37-40], J. L. Beeby[41-44], and G. Wolken[45-46]; and semiclassical work of J. Doll[47]. Much good work has been done but there is still no theory capable of correlating the data obtained from beam-scattering experiments with the properties of the solid from which the beam is being scattered. Several models aimed only at explaining (or predicting) intensities of scattered beams as a function of scattering angle have been proposed. Many of these, such as the cube models of Logan, Keck, and Stickney, are successful in predicting scattering intensity but are too simplified to provide much connection between the microscopic parameters of the solid and its scattering properties. Other models such as the quantum-mechanical models of Cabrera[37] el al. and Wolken[45] are much more realistic but necessitate rather difficult and cumbersome numerical computation and are thus of limited utility at present. However, the authors feel that the most serious fault in presently existing theories is their failure to include multiphonon processes. Furthermore, we feel that even starting with approximate models, one should include multiphonon events rather than attempt to improve upon elaborate single-phonon models. In work performed at Battelle-Columbus under the sponsorship of AFOSR, this procedure was employed successfully in the development of a new quantum-mechanical treatment of gas-surface scattering. This treatment has the advantage of straightforward generalization to permit application to more realistic models and is in principle capable of dealing with complicated systems of practical interest as, for instance, problems in heterogeneous catalysis. This model and recently published results[48] obtained from it are described in some detail in the following section.

2 A NEW QUANTUM-MECHANICAL APPROACH TO GAS SURFACE SCATTERING

2.1 Formulation

Recently, a new approach to the problem of calculating the loss of energy of gas atoms scattered from a surface at normal incidence was developed by the authors.[48] This approach can be described as a quantum-mechanical generalization of Landau's classical theory of thermal accommodation.[49] Since this model has already been described in detail in Ref. 48, only a brief discussion is given here.

In developing the model it is assumed that the dominant interaction V_o between incident gas atoms and the solid depends only on the distance z between the gas atom and the surface. In addition to V_o, there are other interactions with the individual surface atoms. The total of such interactions is given by

$$V_1 = \sum_i V_i(r_i) \, u_i \tag{1}$$

in which r_i is the distance between the gas atom and the equilibrium position of the atom i of the lattice and u_i is the displacement from the equilibrium position of atom i. The V_1 may be assumed to be sufficiently small that the trajectory of the incoming gas atom is only slightly modified. It is further assumed that the lattice is harmonic and thus capable of description in terms of phonon normal modes. These modes are represented by means of creation and annihilation operators, as in the following expression for the Hamiltonian for the solid H_c.

$$H_c = \sum_k \hbar\omega_k(a_k^+ a_k + 1/2) \tag{2}$$

where ω_k is the resonant frequency of mode k and a_k^+ creates and a_k annihilates one quantum of this mode (i.e., one phonon). The displacement u_i can be expressed as a linear combination of the normal modes:

$$u_i = \sum_k A_{ik}(a_k^+ + a_k) \tag{3}$$

where the A_{ik} are obtained from the normal-mode analysis of the lattice of interest. Substitution of Eq. (3) in Eq. (1) then yields an expression for V_1 in terms of the A_{ik}, a_k^+, and a_k.

The next step is to express V_1 as a function of time t by solving the classical equation of motion of the gas atom as acted upon by V_o alone. The following expression for t is obtained:

$$t = \left(\frac{m}{2}\right)^{1/2} \int_{z_o}^{z} \frac{dz'}{\sqrt{E - V_o(z') - K_x}} \tag{4}$$

where E is the energy of the gas atom far from the solid and z_o is the distance of closest approach. Here K_x is the kinetic energy associated with motion in the x direction and is a constant of the motion since there is no net force in the x direction. Solution of Eq. (4) for z and substitution of the result in the expression for V_1 gives $V_1(t)$.

$$V_1(t) = \sum_k (a_k^+ + a_k) G_k(t) \tag{5}$$

The Hamiltonian for the system crystal-plus-gas atom is then

$$H = \sum_k \hbar\omega_k(a_k^+ a_k + 1/2) + \sum_k (a_k^+ + a_k) G_k(t) \tag{6}$$

The scattering solutions for this Hamiltonian can be found exactly.[50] If one defines

$$I_k = \int_{-\infty}^{\infty} G_k(t)e^{-i\omega_k t} dt \tag{7}$$

one can obtain the state $\widetilde{\Psi}_F(t)$, which is given the following equation:

$$\widetilde{\Psi}_F(t) = \exp\left[-\frac{i}{\hbar} \sum_k (I_k a_k + I^*_k a^\dagger_k)\right]\Bigg| \widetilde{\Psi}_I(t_o)> \qquad (8)$$

where $\widetilde{\Psi}_I(t_o)$ is the initial state at the t_o.

The transition probability W_{nI} to a particular final state $\widetilde{\Psi}_n(\infty)$ at $t = \infty$ is given by

$$W_{nI} = \left|<\widetilde{\Psi}_F(\infty)\,\Big|\,\widetilde{\Psi}_n(\infty)>\right|^2 \qquad (9)$$

The probability that the scattered gas atom loses energy $\Delta E = E_n - E_I$ in the scattering is given by

$$P(\Delta E) = \frac{1}{2\pi}\int_{-\infty}^{\infty} <\widetilde{\Psi}_I|u^+(t)\,u(0)|\widetilde{\Psi}_I> e^{i(\Delta E/\hbar)t}dt$$

$$= \frac{1}{2\pi}\int_{-\infty}^{\infty} P(t)\,e^{i(\Delta E/\hbar)t}\,dt \qquad (10)$$

where

$$u(t) = \exp\left[-\frac{i}{\hbar}\sum_k \left(I_k a_k\,e^{-i\omega_k t} + I^*_k a^\dagger_k\,e^{i\omega_k t}\right)\right] \qquad (11)$$

Using well-known techniques developed originally for Mössbauer-effect calculations[51] gives

$$\widetilde{P}(t) = \exp\sum \left|\frac{I_k}{\hbar}\right|^2 \left[(2n_k + 1)(-1 + \cos \omega_k t) - i\sin \omega_k t\right] = \exp R(t) \quad (12)$$

where $n_k = (\exp \hbar\omega_k/kT - 1)^{-1}$ and T is the temperature of the lattice. The average energy loss may then be computed by means of

$$\Delta E = \int \Delta E\,P(\Delta E)\,d(\Delta E) \qquad (13)$$

A procedure similar to that of Allen and Feuer[52] can then be used to obtain the thermal accommodation coefficient from Eq. (13).

2.2 Application of the Model to Helium
Impinging on Tungsten

Calculation. Upon completion of the formal development described in the previous section, it was decided to attempt to evaluate the theory by calculation of $P(\Delta\omega)$ for a specific system. Because there is much extremely reliable experimental data available on the scattering of noble gases from metal surfaces[53-56], a typical example of such a system—helium impinging on tungsten—was chosen for initial evaluation. In order to make the task tractable, several additional assumptions were introduced into the calculation. The first of these is that the interaction potential could be sensibly approximated by a Morse potential, i.e.,

$$V_o = D\left\{\exp[-2K(z\text{-}b)] - 2\exp[-K(z\text{-}b)]\right\}. \tag{14}$$

If this potential is used it is possible to obtain an analytic expression for I_k. This is then inserted into Eq. (12) to obtain $R(t)$. In calculating this quantity, it was assumed that the solid can be treated as a Debye solid having continuous single-mode bulk phonons of frequency ω_k with a cutoff frequency of ω_{max}. The summation in Eq. 12 is then replaced by integration over ω. The result of this integration (obtained numerically) was then substituted into Eq. (15) to obtain $\widetilde{P}(t)$. It was found to be more convenient to use $\Delta\omega = E/\hbar$ rather than ΔE)

$$P(\Delta\omega) = \frac{1}{2\pi} \int_{-\infty}^{\infty} e^{i\Delta\omega t} \exp[R(t)]\, dt \tag{15}$$

The strength of the singular elastic peak (Mössbauer peak) was also evaluated by calculating $P(\Delta\omega)$ for $\Delta\omega = 0$. The probability of exchange of a single phonon was also determined by the relationship.

$$P_1(\Delta\omega) = \frac{1}{2\pi} \int_{-\infty}^{\infty} e^{i\Delta\omega t} \sum \left|\frac{I_k}{\hbar}\right|^2 \left[(2n+1)\cos\omega_k t - i\sin\omega_k t\right] dt \tag{16}$$

Details of this calculation can be found in Ref. (48).

The results of the calculation performed on helium scattering from a tungsten surface are shown in Fig. 1. Both $P(\Delta\omega)$ and $P_1(\Delta\omega)$ are shown for $T = 190°K$, $T = 380°K$, and $T = 760°K$ respectively for the following set of parameters:

$$D = 2.78 \times 10^{-14} \text{ erg}$$

$$K = 0.608 \times 10^8 \text{ cm}^{-1}$$

$$E/D = 2, 4$$

$$\omega_{max} = 5 \times 10^{13} \text{ sec}^{-1}$$

$$M_{He} = 6.6 \times 10^{-24} \text{ g}$$

$$M_W = 3.0 \times 10^{-22} \text{ g}$$

The value for ω_{max} was obtained from bulk specific-heat data and corresponds to a Debye temperature of $380°K$.

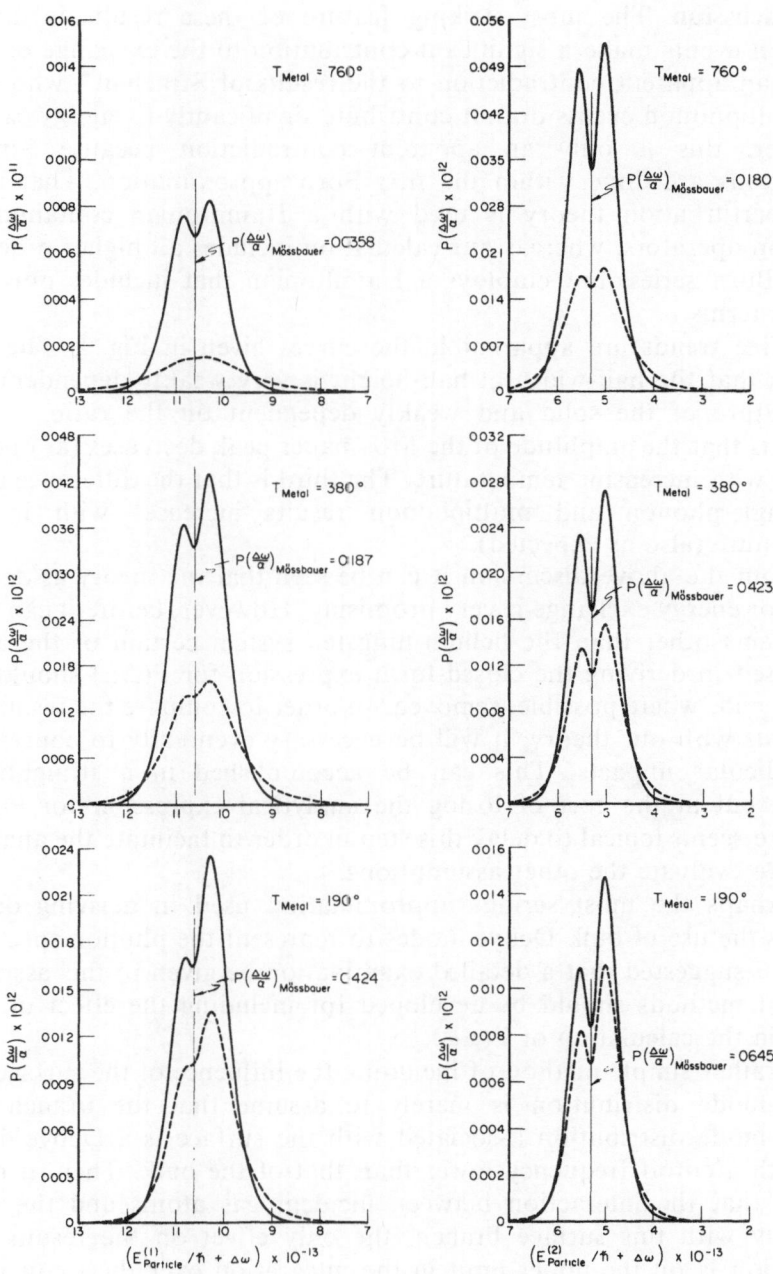

Fig. 1. $P(\Delta\omega/\alpha)$ versus $\Delta\omega$ for three temperatures. $E_1/\hbar = 5.29 \times 10^{13}$ sec^{-1} and $E_2/\hbar = 10.58 \times 10^{13}$ sec^{-1}. $P(\Delta\omega/\alpha)$, the dashed curve, was obtained from the multiphonon calculation and $P_1(\Delta\omega/\alpha)$, the solid curve, was obtained from the single-phonon calculation. The vertical line indicates the Mössbauer line. The parameter is a scaling parameter and is proportional to the square root of the energy E divided by the mass of the gas atom.

Discussion. The most striking feature of these results is that multiphonon events make a significant contribution to the exchange of energy. This is an apparent contradiction to the results of Strachan[48] who showed that multiphonon events do not contribute significantly to energy exchange. However, this is only an apparent contradiction because Strachan's calculations are done within the first Born approximation. That is, first-order perturbation theory is used with a Hamiltonian containing multiphonon operators, whereas our calculation includes all higher order terms in the Born series, but employs a Hamiltonian that includes only single-phonon terms.

Three trends are apparent in the curves given in Fig. 1. The first of these is that the half-width at half-height is very weakly dependent on the temperature of the solid and weakly dependent on the ratio E/D. The second is that the amplitude of the Mössbauer peak decreases (as one would expect) with increasing temperature. The third is that the difference between the single-phonon and multiphonon results increases with increasing temperature (also as expected).

From the above discussion it can be seen that the theory as developed so far for energy exchange is very promising. However, before it can be used on systems other than the helium-tungsten system certain of the assumptions used in deriving the closed-form expression for $P(\Delta\omega)$ should be examined and, where possible, removed. In order to compare the results of experiments with our theory, it will be necessary eventually to consider non-perpendicular impacts. This can be accomplished in a straightforward manner but at the cost of losing the analytical expression for $P(\Delta\omega)$. It therefore seems logical to delay this step in order to facilitate the analysis required to evaluate the other assumptions.

Perhaps the most serious approximation used in deriving our final result is the use of bulk Debye modes to represent the phonon spectrum. It has been suggested that a detailed examination be given to this assumption and that methods should be developed for including the effect of surface modes in the calculation of $P(\Delta\omega)$.

A rather simple method of including the influence of the surface on the normal-mode distribution is merely to assume that the branch of the normal-mode distribution associated with the surface is a Debye distribution with a cutoff frequency lower than that of the bulk. Then, if one can assume that the interaction between incident gas atoms and the solid is primarily with this surface branch, the only effect on the results of our calculation is on the upper limit in the integration over the frequency distribution. However, in general, one would not expect the assumption that the bulk modes can be neglected to be valid, particularly in light of Manson's work on the theory of thermal accommodation.[40] In this theory, three modes are considered—bulk modes, surface (or Rayleigh) modes, and mixed modes. In applying the theory to the calculation of the thermal-accommodation coefficient for helium incident on tungsten, Manson found that all three modes contributed significantly to the final result. However, above 200°K the contributions from each were found to be independent of temperature so that an unambiguous separation of the individual contributions of each would be quite unlikely. This may indicate that one could use a less elaborate treatment than that of Manson, such as taking a weighted average of two Debye distributions. However, this point will have to be checked further. In particular, it will be necessary to be more careful in treating the phonon distribution in the calculation of momentum exchange than in the corresponding energy-exchange calculation.

2.3 Extension of the Theory to the Calculation of Momentum Transfer

In many experimental molecular beam studies of the scattering of gas beams by metallic surfaces, the data are displayed in the form of plots of intensity versus scattering angle. In order to apply the present theory to the interpretation of such data, it is necessary to adapt it to the computation of momentum transfer. How this may be done is outlined below.

Initially we consider the angular distribution of atoms that scatter from an incoming trajectory which is perpendicular to the surface. This case has the advantage of being simple, since only a single angle is needed to define the scattering state. Because the change in energy $\hbar\Delta\omega$ and change in z component of momentum Δp^z define the scattering angle via the relation

$$\Delta p^z = (2m\hbar\Delta\omega + p_o^2)^{1/2} \cos\theta - p_o \tag{17}$$

where p_o is the initial momentum, we need calculate only

$$P(\Delta\omega, \Delta p^z) = \sum_n |\langle \Psi_I(-\infty)|\Psi_n(\infty)\rangle|^2 \, \delta(\hbar\Delta\omega - E_n + E_I)$$

$$\delta(\Delta p^z - p_n^z + p_I^z) \tag{18}$$

$$= (2\pi)^{-2} \int_{-\infty}^{\infty} dt \int_{-\infty}^{\infty} d\tau \exp(i\Delta\omega t + i\Delta p^z \tau) \exp[L_t(t,\tau) + L_{\iota}(t,\tau)] \tag{19}$$

$$L_{t(\iota)}(t,\tau) = \sum_k \hbar^{-2}|I_k|^2 \cos^2\theta_k \left[(2n_k + 1)(-1 + \cos\Delta_{k,t(\iota)}) - i\sin\Delta_{k,t(\iota)}\right] \tag{20}$$

$$v_{t(\iota)} = \text{velocity}_{\text{transverse(longitudinal)}}$$

$$\Delta_{k, t(\iota)} = \omega_k \left[t + (\hbar \tau \cos \theta_k)/v_{t(\iota)} \right] \tag{21}$$

Integrating $P(\Delta p^z, \Delta \omega)$ over Δp^z and $\Delta \omega$ gives

$$1 = \int_{-\infty}^{\infty} d(\Delta \omega) \int_{-p_o}^{-p_o} - (2m\hbar\Delta\omega + p_o^2)^{1/2} P(\Delta p^z, \Delta \omega) \, d\Delta p^z \tag{22}$$

which is just the normalization condition. Making a change of variable from Δp^z to $\mu = \cos \theta$ permits Eq. (22) to be written as

$$1 = \int_o^1 d\mu \int_{-\infty}^{\infty} (2m\hbar\Delta\omega + p_o^2)^{1/2}$$

$$P \left[\Delta\omega, \Delta p_z = \mu(2m\hbar\Delta\omega + p_o^2)^{1/2} - p_o \right] d(\Delta\omega) \tag{23}$$

This then identifies $P(\mu)$ as

$$\widetilde{P}(\mu) = \int_{-\infty}^{\infty} (2m\Delta\omega + p_o^2)^{1/2}$$

$$P \left[\Delta\omega, \Delta p^z = \mu(2m\hbar\Delta\omega + p_o^2)^{1/2} - p_o \right] d(\Delta\omega) \tag{24}$$

It is seen that the evaluation of Eq. (24) is considerably more complicated than the corresponding expression for $P(\Delta\omega)$, that is, Eq. (15). Nevertheless, preliminary computer calculations [in which an incorrect form of Eq. (24) was used] indicate that Eq. (24) can be evaluated with a reasonable expenditure of computer time.

ACKNOWLEDGMENT

The authors wish to express their gratitude to the United States Air Force Office of Scientific Research, which supported part of the work under Research Grant No. AFOSR-72-2354. They also wish to thank Dr. M. L. Glasser for his assistance in the analytical evaluation of some of the key integrals and Dr. R. P. Kenan for his contributions to the numerical analysis and computer programming necessary to obtain the results given in Fig. 1.

REFERENCES

1. Palmer, R. L., O'Keefe, D. R., Saltsburg, H., and Smith, J. N., Jr., *J. Vacuum Sci. Technol.*, **7**, 90 (1970).

2. Weinberg, W. H., and Merrill, R. P., *Phys. Rev. Letters*, **25**, 1198 (1970).
3. Tendulkar, D. V., and Stickney, R. E., *Surface Sci.*, **27**, 516 (1971).
4. O'Keefe, D. R., Palmer, R. L., and Smith, J. N., Jr., *J. Chem. Phys.*, **55**, 4572 (1971).
5. Smith, J. N., Jr., O'Keefe, D. R., Saltsburg, H., and Palmer, R. L., *J. Chem. Phys.*, **50**, 4667 (1969), *ibid.*, **52**, 315 (1970).
6. Beeby, J. L., *J. Phys. C: Solid State Phys.*, **5**, 2098 (1972).
7. Cabrera, N., in *Molecular Processes on Solid Surfaces*, E. Drauglis et al. (Eds.), McGraw-Hill, New York (1969), p. 169.
8. Goodman, F. O., *J. Phys. Chem. Solids*, **23**, 1269 (1962).
9. Goodman, F. O., *J. Phys. Chem. Solids*, **23**, 1491 (1962).
10. Goodman, F. O., *J. Phys. Chem. Solids*, **24**, 1451 (1963).
11. Goodman, F. O., *J. Phys. Chem. Solids*, **26**, 85 (1965).
12. Goodman, F. O., in *Rarefield Gas Dynamics*, 4th Symposium, Vol. II, J. H. de Leeuw (Ed.), Academic Press, New York (1966), p. 366.
13. Goodman, F. O., in *Rarefield Gas Dynamics*, 5th Symposium, Vol. I, C. L. Brundin (Ed.), Academic Press, New York (1957), p. 35.
14. Goodman, F. O., *Surface Sci.*, **7**, 391 (1967).
15. Goodman, F. O., in *The Structure and Chemistry of Solid Surfaces*, G. A. Somorjai (Ed.), John Wiley, New York (1969).
16. Goodman, F. O., in *Rarefield Gas Dynamics*, 6th Symposium, Vol. II, L. Trilling and H. Y. Wachman (Eds.), Academic Press, Inc., New York (1969), p. 1105.
17. Goodman, F. O., *Surface Sci.*, **19**, 93 (1970).
18. Goodman, F. O., *Surface Sci.*, **30**, 1 (1972).
19. Goodman, F. O., *J. Vacuum Sci. Technol.*, **9**, 812 (1972).
20. McClure, J. D., *J. Chem. Phys.*, **51**, 1687 (1969).
21. McClure, J. D., *J. Chem. Phys.*, **52**, 2712 (1970).
22. McClure, J. D., *J. Chem. Phys.*, **57**, 2810 (1972).
23. McClure, J. D., *J. Chem. Phys.*, **57**, 2823 (1972).
24. Bogan, R. A., and Li, C. H., in *Rarefield Gas Dynamics*, 4th Symposium, Vol. II, *op. cit.*, p. 396.
25. Oman, R. A., in *Rarefield Gas Dynamics*, 5th Symposium, Vol. I, *op. cit.*, p. 83.
26. Oman, R. A., *J. Chem. Phys.*, **48**, 3919 (1968).
27. Oman, R. A., in *Rarefield Gas Dynamics*, 6th Symposium, Vol. II, *op. cit.*, p. 1331.
28. Calia, V. S., and Oman, R. A., *J. Chem. Phys.*, **52**, 6184 (1970).
29. Raff, L. M., Lorenzen, J., and McCoy, B. C., *J. Chem. Phys.*, **46**, 4265 (1967).
30. Lorenzen, J., and Raff, L. M., *J. Chem. Phys.*, **49**, 1165 (1968).
31. Lorenzen, J. L., and Raff, L. M., *J. Chem. Phys.*, **52**, 1133 (1970).
32. Lorenzen, J., and Raff, L. M., *J. Chem. Phys.*, **52**, 6134 (1970).
33. Logan, R. M., and Stickney, R. E., *J. Chem. Phys.*, **44**, 195 (1966).
34. Logan, R. M., Keck, J. C., and Stickney, R. E., in *Rarefield Gas Dynamics*, 5th Symposium, Vol. I, *op. cit.*, p 49.
35. Logan, R. M., and Keck, J. C., *J. Chem. Phys.*, **49**, 860 (1968).
36. Stickney, R. E., in *The Structure and Chemistry of Solid Surfaces*, G. A. Somorjai (Ed.), John Wiley, New York (1969).
37. Cabrera, N., Celli, V., and Manson, J. R., *Phys. Rev. Letters*, **22**, 346 (1969).
38. Cabrera, N., Celli, V., Goodman, F. O., and Manson, J. R., *Surface Sci.*, **19**, 67 (1970).
39. Manson, J. R., and Celli, V., *Surface Sci.*, **24**, 495 (1971).
40. Manson, J. R., *J. Chem. Phys.*, **57**, 4504 (1971).
41. Beeby, J. L., and Dobrzynski, L., *J. Phys., C: Solid State Physics*, **4**, 1269 (1971).
42. Beeby, J. L., *J. Phys., C: Solid State Physics*, **5**, 3438 (1972).
43. Beeby, J. L., *J. Phys., C: Solid State Physics*, **5**, 3457 (1972).

44. Beeby, J. L., *J. Phys., C: Solid State Physics*, **6**, 1229 (1973).
45. Wolken, G., Jr., *J. Chem. Phys.*, **58**, 3047 (1973).
46. Wolken, G., Jr., *J. Chem. Phys.*, **59**, 1159 (1973).
47. Doll, J. D., *Chem. Phys.*, **3**, 257 (1974).
48. Kaplan, J. I., and Drauglis, E., *Surface Sci.*, **36**, 1 (1973).
49. Landau, L., *Phys. Zeit. Sowjet*, **8**, 489 (1935).
50. ter Har, D., *Selected Problems in Quantum Mechanics*, Academic Press, Inc., New York (1964), pp. 152–154.
51. Visscher, W. M., *Ann. Phys.*, **9**, 194 (1960).
52. Allen, R. T., and Feuer, P., *J. Chem. Phys.*, **43**, 4500 (1965).
53. Smith, J. N., and Saltsburg, H., in *Fundamentals of Gas-Surface Interactions*, H. Saltsburg, J. N. Smith, and M. Rogers (Eds.), Academic Press, Inc., New York (1967), p. 370.
54. Smith, D. L., and Merrill, R. P., *J. Chem. Phys.*, **53**, 1594 (1970).
55. Weinberg, W. H., and Merrill, R. P., *J. Chem. Phys.*, **56**, 2882 (1972).
56. Stoll, A. G., Smith, D. L., and Merrill, R. P., *J. Chem. Phys.*, **54**, 163 (1971).
57. Drauglis, E., in *Molecular Processes on Solid Surfaces*, E. Drauglis, et al. (Eds.), McGraw-Hill, New York (1969), p. 367.

DISCUSSION on Paper by E. Drauglis

MENZEL: Does your model of momentum transfer include internal conversion to bound states ("selective adsorption")?

DRAUGLIS: No, because of the necessary assumption that the amount of energy transferred is small in comparison with the kinetic energy of the gas particle before collision with the surface.

MENZEL: You check your calculations of energy transfer rather indirectly (Debye-Weller factor of scattering). Would it not be better to do direct calculations of thermal-accommodation coefficients, for which accurate measurements are available?

DRAUGLIS: Our model provides a simple way of calculating thermal-accommodation coefficients. All that we need to do to obtain values to compare with experimental values is evaluate P ($\Delta\omega$) for a distribution of energies of the incident gas particles. In attempting to obtain values for TAC for comparison with experiment, we must select the proper distribution (Maxwellian etc.) and then integrate P($\Delta\omega$) over this distribution. We have not yet performed such calculations because of lack of computer time.

KOHN: In your calculations did you allow for the role of surface vibrational modes?

DRAUGLIS: No, we did not. In future work we hope to look into ways of including such modes.

REACTIONS CATALYZED NEAR SECOND-ORDER PHASE TRANSITIONS OF THE SUBSTRATE*

H. Suhl

University of California, San Diego
La Jolla, California

Although the theory of chemical rate constants, in the form given it by Eyring around 1930, serves rather well in many situations, there is a quite substantial number of reactions in which it seems to fail. The first penetrating examination of its validity, together with an outline of how to proceed when it fails, was given by Kramers[1] in 1940, and there has been no really major advance in the theory since. Kramers' results are summarized as follows.

He views a chemical reaction as a Brownian motion in the space of the relative nuclear coordinates, this motion being propelled by the random forces exerted on the atoms by the heat bath. That heat bath includes *all* degrees of freedom of the system that are not directly involved in the specification of the reaction, such as inert molecules of a carrier gas, internal vibrations of the reactants themselves, photons, etc. The random forces have two counteracting effects: they cause an adequate resupply of reactants in

*Work partially supported by the National Science Foundation DMR-03838.

the Maxwell velocity distribution tail which gets depleted by "escape" of reactants over the reaction barrier, but they also produce a friction that tends to impede that escape. When the strength of the random forces (as measured by a certain time average of their autocorrelation function) is such that these two tendencies are in balance, the Kramers theory reduces to the older Eyring theory (the so-called absolute rate theory, ART). The rate is then at its maximum. It goes to zero as the strength of the random forces goes to zero because of inadequate replenishment of the tail, and it goes also to zero when the strength becomes very large (because of too much friction). In all cases the rate constant can be written:

$$K = v \exp - \Delta F / kT$$

where ΔF is the difference of free energies of the "activated complex" at the top of the barrier and at the originating minimum or valley. (Minimum in the case of isomeric or decomposition reactions, valley in the case of reactions involving initially widely separated reactants.)

For a detailed theory of the prefactor v see Refs. 1–3. Except in the ART range where it is totally independent of the dynamics, v can be related to various electromagnetic response functions of the heat bath (the substrate in the case of catalysis). These response functions have anomalies near a phase transition of the heat bath, and therefore in the range in which ART fails, the prefactor v has a corresponding anomaly. This may be at the basis of the widely scattered but substantial evidence for anomalies in the rate of catalyzed reactions near phase transitions of the catalyst. (For a summary of these effects see the review paper by Voorhoeve.[4])

There is yet another anomaly that should occur near a phase transition of a catalyst, an anomaly that has to do with the exponent rather than the prefactor in the Arrhenius rate formula, and that should therefore appear whother ART applies or not. A phase transition is characterized by an order parameter (such as the magnetization in ferromagnets or the electric polarization in ferroelectrics) to which the degrees of freedom of the reactants on its surface are coupled to a greater or lesser extent. Since the order parameter varies with position as well as time, particularly near the phase transition, the coupling energy at any instant is likewise a function of the relative positions of the reactants. The time variation of the order parameter field may likewise be regarded as a Brownian motion, and thus it is necessary to regard the reaction as a coupled Brownian motion of the reactant atoms and the order parameter field. The detailed theory of the effect is given in Ref. 5.

If it is supposed that most of the order-parameter fluctuations become very long lived near T_c (less rapidly varying than the motion of the chemicals), the effect may simply be regarded as a random modulation of the barrier itself, and thus a modulation of the Eyring rate. The time average of the rate (which is the quantity measured) is the same as the ensemble average over a probability distribution exp $(-F_{L.G.}/kT)$, where $F_{L.G.}$ is the Landau-Ginsberg free energy expression for the fluctuating order parameter field. This average has an anomalous temperature dependence around the transition temperature which is easily computed in the quadratic (molecular field) approximation to $F_{L.G.}$. Far from T_c, the effect of the fluctuations is simply to give some additional activation *entropy*, as one would expect. (Those fluctuations that are rapid compared with the motion of the reactants give, in addition to a contribution to the activation entropy, a contribution to the friction coefficient of the reactants.)

In addition to the anomaly near T_c, the coupling to the order parameter must also give a difference in the asymptotic slopes of the Arrhenius plot well above T_c and below T_c. This is simply because the coupling free energy well below T_c has a nonfluctuating part because of the finite value of the average order parameter, while well above T_c it does not (except for the entropic contribution which does not change the slope of the Arrhenius plot). It is possible to establish a relation between that change in slope and the size of the anomaly near T_c, because both effects come from the same coupling free energy. For the magnetic case, the most important term in that coupling energy is

$$F_c = \sum_{ij} \sum_{\alpha\beta} A^{ij}_{\alpha\beta} M_\alpha(R_i) M_\beta(R_j)$$

where $M_\alpha(R)$ are the cartesian components of the magnetization field at the positions R_i of the various reactants and products and $A^{ij}_{\alpha\beta}$ is the coupling strength. Consider the simple case of a single adparticle, the reaction being desorption. Also, neglect the inevitable anisotropy of A at the surface. Let R_A^z be the normal distance of the adsorption well from the surface, and R_B^z the normal distance of the desorption barrier (see Fig. 1), both measured relative to the point at which the magnetization density goes "effectively" to zero. Then (see Ref. 5) the temperature sensitive part of the change in the activation barrier is

$$\frac{A k_E T}{\pi c} \left[\left\{ \exp\left(-2|R_B^z|\sqrt{\frac{a}{c}}\right) \right\} / R_b^z - \left\{ \exp\left(-2|R_A^z|\sqrt{\frac{a}{c}}\right) \right\} / R_A + 1/R_A - 1/R_B \right]$$

where a and c are the parameters of the L.G. free energy:

$$F_{L.G.} = \int \left[\tfrac{1}{2}c(\nabla M)^2 + \tfrac{1}{2} a M^2 + \tfrac{1}{4}b M^4 \right] \mathcal{D}M$$

and where a varies as $(T-T_c)$ near T_c. The correlation length is $\ell = c/a$, for $T > T_c$. The above expression holds only so long as ℓ is at least several lattice spacings. The size of the anomaly may be measured by the vertical distance between the straight line that fits the Arrhenius plot far from T_c and a

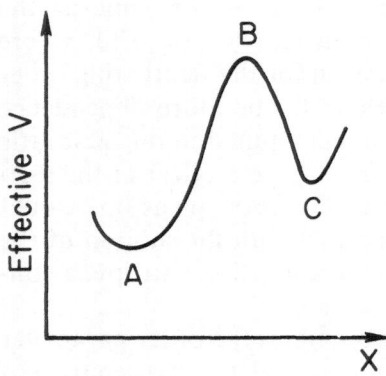

Fig. 1. Potential barrier and reaction path.

parallel straight line drawn through the point $[T_c, \log \mathbf{K}(T_c)]$ where $\mathbf{K}(T_c)$ is the rate at T_c. This anomaly is evidently

$$\frac{A}{4\pi c}\left[\left\{\exp\left(-2|R_B^z|/\ell_s\right)-1\right\}/R_B^z - \left\{\exp\left(-2|R_A^z|/\ell_s\right)-1\right\}/R_A\right]$$

where ℓ_s is the correlation length remote from T_c. On the other hand, the change in *slope* of the plot going from well below to well above T_c is

$$A[M^2(R_B) - M^2(R_A)]$$

where $M(R)$ is the mean magnetization in the ferromagnetic regime at R. To find $M(R)$ in terms of the bulk magnetization M_s within mean field approximation, we must solve:

$$\frac{d^2M}{dz^2} - \frac{a}{c}M - \frac{b}{c}M^3 = 0$$

subject to the boundary condition that $M = 0$ at $z = 0$ and $M = M_s$ at $-\infty$, when $a/c = (1-T_c/T)/\ell^2 < 0$. The solution of this equation may be written:

$$M = M_s \tanh\sqrt{\frac{|a|}{c}}\,|z| \qquad \begin{array}{l} z < 0 \\ a < 0 \end{array}$$
$$= 0 \text{ otherwise.}$$

Therefore the change in the asymptotic slope of the Arrhenius plot is:

$$M_s^2 A[\tanh^2 R_B^z/\ell_s - \tanh^2 R_A^z/\ell_s]$$

and in the present units, M_s is of order unity. For the case $R_B = 0$, $R_A = 2\ell_s$, this is virtually equal to $-AM_s$ (i.e., the slope above T_c is greater than that below*). In the experiments of Measor and Afzulpurkar[6] on the initial oxidation rate of iron, the difference in activation energies below and above T_c is -0.5 volt, so that $A \sim -0.5$ volt, which suggests very intimate coupling of the oxygen ions to the order parameter. This would also account for the large anomaly that is observed.

Figure 2 shows the variation of log $\overline{K}$ in the immediate vicinity of T_c for the case in which $A/\triangle V = \frac{1}{2}$, and in which the exchange stiffness of surface spins is the same as in the bulk. Here $\triangle V$ is the barrier in the absence of coupling to M. It has been assumed in this example that the barrier peak is outside the range of the magnetization, a case which can be obtained by simply setting $R_B = 0$ in the preceding formula. (However, for $T-T_c \sim T_c$, $\sqrt{a/c}$ must be replaced by $1/\ell_s$). For an adatom whose spin is exchange coupled to the spin density of the conduction electrons of a ferromagnetic metal substrate, A is on the order of .05 volts; hence, if $\triangle V \sim$ 1 volt, the above case of $A/\triangle V = 0.5$ would seem unrealistic. However, to obtain the effective value of $A/\triangle V$, the calculated value should be multiplied by $c_{bulk}/c_{surface}$, i.e., the ratio of bulk to surface stiffness. This may well bring $(A/\triangle V)_{effective}$ closer to 0.5. Note that in Fig. 2 there is no trace of the cusp near T_c reported in Ref. 6. A possible explanation of this cusp is that the quartic terms change a linear dependence in $T-T_c$ to one on a fractional power of $T-T_c$. However, another very strong effect may be responsible for

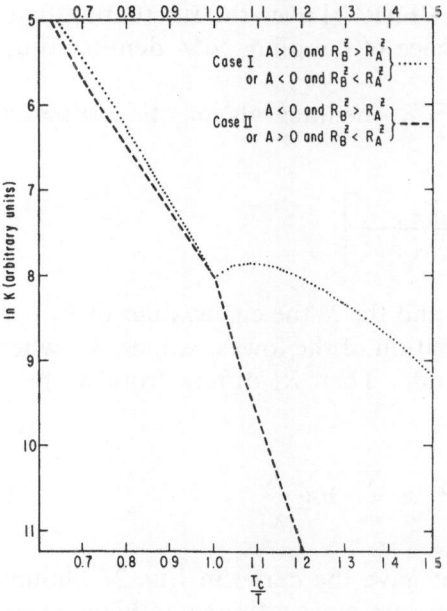

Fig. 2. Anomaly in reaction rate near T_c for two simple cases.

---

*If A is positive and $R_A^z < R_B^z$, or if A is negative and $R_B^z < R_A^z$. The case $R_B^z < R_A^z$ corresponds to thermally activated diffusion from a chemisorbed state to the interior of the solid.

the apparent cusp. The constant A is in some cases negative, at least in second-order perturbation theory of the adatom-spin density coupling. This means that *locally*, at R, the term $M^2 a(\frac{T}{T_c} - 1)$ in the L.G. free energy is replaced by $M^2 \left\{ a\left(\frac{T}{T_c} - 1\right) - A \right\}$ in the total free energy ($F_{L.G.}$ + F_c). This is equivalent to a *locally* higher curie temperature $T'_c = T_c(1 + \frac{A}{a})$.

Of course, this local effect does not cause long-range order for $T > T_c$; however, if A is large enough ($\geqslant 2\pi k T_c$), the *local* magnetization fluctuations go critical a little above T_c. A localized fluctuation mode splits off the usual plane-wave-fluctuation spectrum of $F_{L.G.}$. Up to this point we have studied only the anomaly near T_c due to the "scattering" of the usual modes from the local perturbation. No singularity was obtained because of the averaging over all wave numbers. However, the localized mode, which arises above a critical coupling strength, does lead to singular behavior, essentially because it is all "concentrated" near the adatom.

For simplicity we again neglect anisotropy of the coupling, and again take R_B so far outside the substrate that $F_c(R_B)$ is zero. Then

$$\overline{K} = \overline{K}_0 \int \exp\left[-\frac{F_{L.G.} + F_c(R_A)}{k_B T} \right] \mathcal{D}M \ / \ \int \exp\left[-\frac{\beta F_{L.G.}}{k_B T} \right] \mathcal{D}M$$

This is the average of the rate $\overline{K}_0 \exp\left[-\beta F_c(R_A) \right]$ over the fluctuating field. $\overline{K}_0 \sim \exp(-\beta \Delta V)$ is the rate in the absence of coupling. $\mathcal{D}M$ denotes functional integration.

In the quadratic approximation to $F_{L.G.}$, the integrals may be evaluated, with the result:

$$\overline{K} = \overline{K}_0 \left[\frac{\lambda_1 \lambda_2 \dots}{\lambda'_1 \lambda'_2 \dots} \right]$$

where the λ'_i are the eigenvalues of $F_{L.G.}$ and the λ_i the eigenvalues of $F_{L.G.}$ + F_c. As we shall see, $\lambda'_i \doteq \lambda_i$, with the exception of the lowest values, λ_1, when $A < 0$ and sufficiently large in magnitude. Then λ'_1 differs from λ_1 by a finite amount. We have

$$\log \frac{\overline{K}}{\overline{K}_0} = \tfrac{1}{2} \log \frac{\lambda_1}{\lambda'_1} + \tfrac{1}{2} \sum_{i=2} \log \frac{\lambda_i}{\lambda'_i} \tag{1}$$

The sum on the right-hand side is what gave the curve in Fig. 2. Though each λ'_i differs from λ_i infinitesimally, the sum gives a finite combined effect of all the scatterings. To find the leading term $\tfrac{1}{2} \log \frac{\lambda_1}{\lambda'_1}$ we must find the "bound-state" eigenvalue, if any, of

$$\int \left[\tfrac{1}{2}c(\nabla M)^2 + \tfrac{1}{2}aM^2 \, dV + A\delta(r-R_A)M^2(R_A) \right] \delta M$$

Let

$$M(r) = \sum_q m_q \sin (q_z z c^{iq} T^r T) \qquad z < 0$$

$$= 0 \qquad z > 0$$

where q_z, q_T are the components of $\mathbf{q}$ normal to and in the surface, respectively. Then the last integral is

$$V \sum \tfrac{1}{2}(cq^2+a)|m_q|^2 + \Omega A \Big(\sum_{q'} m_{q'} \sin q'_z R_A^z\Big)^2$$

where Ω is an atomic volume, and the eigenvalues are given by

$$\lambda' m_q = (cq^2+a)m + \frac{A}{V} \sin R_A^z \Big[\sum m_{q'} \cdot \sin(q'_z R_A^z)\Big]$$

or

$$1 = \frac{A}{V} \sum_q \frac{\sin^2 q_z R_A^z}{cq^2 + a - \lambda'}$$

Let $c = \epsilon \ell^2$, where ϵ is the effective exchange energy and ℓ the lattice spacing. Then

$$1 = - (A/8\pi^3 \epsilon \ell^2) \sum_q \frac{\sin^2 q_z R_A}{q^2 + \dfrac{a-\lambda'}{\epsilon}} \, d\mathbf{q} \qquad (2)$$

The integral diverges at the upper limit, and this must be taken into account properly; otherwise the well-known fact that in three dimensions a bound state exists only above a minimum coupling strength is violated. The upper cutoff is of order $1/\ell$, because at $q \sim 1/\ell$, terms like $\int(\nabla^2 M^2)$ in $F_{L.G.}$ (which would ensure convergence) come into play. With such a cutoff, the integral can be evaluated. To simplify matters we sacrifice information regarding the R_A^z dependence, and replace $\sin^2 q_z R_A^z$ by its average, $\tfrac{1}{2}$. Equation (2) then becomes:

$$1 + \frac{4\pi^2 \epsilon \ell^3}{A} = \ell\alpha \, \text{arc} \, \tan(\ell\alpha)^{-1} \qquad (3)$$

where $\alpha = (a - \lambda')/\epsilon$. This equation has a root $\ell\alpha$ only if $A < 0$ and

$$1 - \frac{4\pi^2 \epsilon \ell^3}{|A|} > 0$$

or

$$|A| > 4\pi^2 \epsilon \ell^3$$

Now $|A| = \epsilon' \ell^3$, where ϵ' is the effective spin coupling energy of the impurity to the substrate electrons, evaluated in second or higher order perturbation theory. Thus, for a bound state,

$$\epsilon' > 4\pi^2 \epsilon$$

(for an impurity far in the interior of the solid, the condition will be $\epsilon' > 2\pi^2 \epsilon$).

We have $\alpha\ell = (a - \lambda'\ell/\epsilon$, and $a \approx \epsilon\left(\frac{T}{T_c} - 1\right)$. Hence, if $\alpha_1\ell = z_1$, a root of the characteristic Eq. (3), we have

$$\lambda'\ell = \epsilon\left(\frac{T}{T_c} - 1 - z_1\right)$$

This compares with the lowest root $\lambda_1\ell = \epsilon\left(\frac{T}{T_c} - 1\right)$ in the absence of coupling. Thus, according to Eq. (1), the extra anomaly in the rate due to the bound state is:

$$\tfrac{1}{2} \log\left[\left(\frac{T}{T_c} - 1\right) \Big/ \left(\frac{T}{T_c} - 1 - z_1\right)\right]$$

For $T < T_c(1+z_1)$, the problem is more complicated. A "giant" local moment $M(r)$ then forms around the adatom (to be discussed in another publication) and it is necessary to calculate the fluctuations about $M(r)$. In the case of a pure, uniform ferromagnet this presents no problem. The deviation m about M obeys the same statistics for $T < T_c$ as for $T > T_c$, but with $T-T_c$ replaced by T_c-T. In the present case, this is not so. Although $T-T_c(1+z_1)$ is still replaced by $T_c(1+z_1)-T$, the coefficient of this expression in the L.G. free energy becomes position dependent. Still, it is likely that a large part of the effect is accounted for simply by placing absolute value bars around the argument of the logarithm.

For $A > 0$, we speculate that a similar logarithmic infinity arises, but *below* the bulk curie temperature. With $A > 0$, the local value of T_c is lowered; hence, for sufficiently large A, one expects an island of spin *disorder* to arise somewhat below T_c. However, the theory becomes much more difficult since it is then necessary to consider deviations, not from the uniform magnetization, but from a magnetization that satisfies the equation

$$\nabla^2 M - |a| M^2 + M^3 + AM(0) = 0$$

subject to $M = |a|/b$ far in the interior of the solid.

REFERENCES

1. Kramers, H. A., *Physica,* **7,** 284 (1940).
2. d'Agliano, E. G., Schaich, W. L., Kumar, P., and Suhl, H., Nobel Symposium 24, *Medicine and Natural Sciences, Collective Properties of Physical Systems,* B. Lundqvist and S. Lundqvist (Eds.), Academic Press, New York (1973), pp. 200–208.
3. d'Agliano, E. G., Kumar, P., Schaich, W., and Suhl, H., *Phys. Rev.,* **B11,** 2122 (1975).
4. Voorhoeve, R.J.H., *Proceedings at 19th Conference on Magnetism,* Boston, Massachusetts, November 1973.
5. Suhl, H., *Phys. Rev.,* **B11,** 2011 (1975).
6. Measor, J. C., and Afzulpurkar, K. K., *Phil. Mag.,* **10**(2), 817 (1974).

ISOTHERMAL DESORPTION
MEASUREMENTS

D. Menzel

Physik-Department, Technische Universität München
München, West Germany

ABSTRACT

The advantages of isothermal desorption measurements for the unambiguous determination of desorption-rate parameters are briefly reviewed. Two examples of such measurements for systems exhibiting unusual kinetic behavior are discussed. The desorption of oxygen from silver (110) proceeds according to zeroth order over a considerable fraction of coverage, which is interpreted in terms of desorption from the ends of chains of oxygen atoms. A weakly bound hydrogen state on tungsten precovered with CO desorbs according to first order but with an extremely small preexponential; possible reasons for this behavior are discussed. The importance of desorption kinetics and of coadsorbed states to catalysis is stressed.

1 INTRODUCTION

Desorption is an important step in heterogeneous catalysis. There are many cases of catalytic reactions in which desorption of either products or poisoning species determine the rate of the overall reaction; even where this is not the case, desorption is often involved in the overall kinetics in a decisive way. Therefore, the understanding of this comparatively simple dynamic process is not only a test case for the investigation of more complex surface kinetics, but should be of direct value for our understanding of catalysis. Unfortunately, the test of existing concepts is considerably impeded by the fact that reliable determinations of all parameters in the general Polanyi-Wigner equation[1]

$$R_d = - dN/dt = k_o^{(m)} \cdot N^m \cdot \exp(-E/RT) \qquad (1)$$

namely, formal order m, activation energy E, and preexponential factor or prefactor k_o (which all can contain a dependence on the coverage N) are difficult. In particular, the value of the prefactor, which is of prime importance for the desorption mechanism and the energetic coupling to the solid, usually contains a very large uncertainty.[1]

Most measurements to date have been performed using a special version of temperature-programmed desorption, the so-called flash desorption technique.[2] In this technique, the temperature of the substrate is varied in a programmed way (usually linearly with time), and the variation of gas pressure in the system, which is caused by desorption, is recorded. For large pumping speeds, the increase of gas density (or pressure) over the background value is proportional to the rate of desorption at any time. The occurence of maxima in the rate-versus-temperature curve is usually taken as an indication of the existence of distinct binding modes (see, however, ref. 3); the reaction order and activation energy are derived from the peak locations and their movement with variation of initial coverage and heating rate. The deduction of preexponentials makes a curve analysis necessary, which often leads to a large uncertainty. A number of experimental difficulties arise from the use of gas pressure as the parameter connected to desorption[4], and this could therefore be avoided by following another parameter connected with coverage. The main difficulties in the interpretation arise from the necessity of curve-shape analysis to extract all information, and from the fact that the three kinetic parameters are usually not determined independently. A very instructive example of the ambiguities which can arise from this has been given recently for a not too complicated case, the two-peak desorption spectrum of nitrogen on tungsten, by Pisani et al.[5] Even where reaction orders and activation energies are determined unambiguously (for instance, for clear first order with constant E), the value of the corresponding preexponential is usually quite ambiguous; for instance, peak broadening can be caused by a number of experimental factors, a coverage-dependent activation energy, *or* a small preexponential. While many measurements are compatible with "normal" prefactors, i.e., $\sim 10^{13}$ sec^{-1} for first order and 10^{-2} cm^2 sec^{-1} for second order[1], other values often cannot be excluded. Another difficulty arises for desorption not occuring in one step. As the different parallel or consecutive steps will have different temperature dependences, the state of the layer for a certain pair of coverage and temperature values will be not unique, but will depend on the history up to that point. This has also been noted very recently by Shanabarger[6] for the special case of desorption via a precursor state.

The standard procedure of general chemical kinetics, i.e., the isothermal measurement of concentration change to obtain reaction orders and rate constants for a certain temperature, followed by repetition at different temperatures to derive activation energies and prefactors, seems to be of advantage in cases where reaction orders and/or prefactors are difficult to obtain by temperature-programmed measurements. This has been discussed before by Pétermann.[7] The elucidation of multistep desorption mechanisms should also be easier at constant temperatures, because of the constancy of the various exponential terms. Few isothermal desorption experiments have been carried out, however. This is probably due to the fact that step functions in temperature or pressure, which are necessary to produce a non-equilibrium situation in the adsorption layer and to clearly define the starting point of an isothermal run, are difficult to produce. One method, which in essence uses a pressure jump, is the residence-time measurement using molecular beams.[8] So far, it has been used only for the investigation of adsorption of species, for which either desorption occurs as ions, or the detection is possible by surface ionization, or is easy by electron impact ionization, since the beam can be frozen out after one passage through the ion source. Temperature-jump desorption measurements on well-defined surfaces have been carried out, for instance, by Kohrt and Gomer[9] for CO and oxygen on W(110).

We have conducted isothermal desorption measurements in some systems showing peculiar behavior, which would have been difficult to elucidate with temperature-programmed measurements, and which might be of more general interest to surface reactions also. Two of these are discussed in the following. The basic concept was to follow the desorption process not by detecting the desorbing flux, i.e., monitoring the product, but by measuring a parameter which is connected to the surface coverage in a known way, i.e., monitoring the remaining reactant concentration. The two parameters used were the work-function change, measured with a vibrating capacitor, and the ESD ion current[10] measured mass spectrometrically. In both cases it had been ascertained that the connection between signal and coverage was linear; temperature jump procedures were used.

2 DESORPTION OF OXYGEN FROM SILVER (110)

The desorption of oxygen from the Ag(110) face was followed by monitoring the $\Delta\phi$-change. LEED investigation had shown that adsorption of oxygen on this face first causes streaks normal to the [100] direction, which then coalesce to form (7x1), (6x1), (5x1) ..... (2x1) patterns, one after the other, with increasing coverage.[11] This probably means that chains of oxygen atoms along the "long" [100] distance of this face are formed, starting at very low coverage, which try to keep maximum distance between each other in the "short" [1$\bar{1}$0] direction. The interesting combination of attractive interactions in the long direction with repulsive interactions in the short directions is not discussed here; we just mention that oxygen seems to cause such anisotropic interactions more often [see, for example, the (4x1)-structure of oxygen on the W(100) face[12], and the (2x1) structures on the Ru(001) face[13]]. The work-function increase as measured by a vibrating capacitor was about 0.9 eV for the saturated (2x1) layer. Plotting $\Delta\phi$ against the coverage obtained from the respective LEED structures (assuming the structure to cover the whole surface homogeneously) yielded a good straight line, which suggests that the dipole moment per adsorbed oxygen atom is constant. Temperature-programmed desorption has been done from such layers, following the change in coverage by monitoring $\Delta\phi$ instead of the gas pressure (which is particularly prone to disturbances for oxygen). Typical results are shown in Fig. 1. It is seen that the desorption curves move to higher temperatures with increasing coverage, showing that the formal order is below one. The common leading edge of the peaks suggests zeroth order. The desorption spectra can be analyzed with all ambiguities mentioned; this is not done here, however, but it is shown that additional information can be obtained from isothermal desorption measurements.

Isothermal-desorption runs for different temperatures are shown in Fig. 2. As the crystal used had a rather large heat capacity, it took about 5 seconds to obtain the constant temperature desired even using an automatic temperature controller with a rather high initial heating rate; only those parts of the curves for which the temperature was truly constant have been plotted. The use of a differentiating device made possible a direct plot of $d\Delta\phi/dt$ versus $\Delta\phi$, i.e., rate of desorption versus coverage, as shown in Fig. 2. It can be seen that for low temperatures, the rate of desorption stays approximately constant over a considerable portion of the range at intermediate coverages. This zeroth order of desorption can be qualitatively explained consistent with the LEED results mentioned. The anisotropic interactions discussed above mean that atoms in the middle of a chain are more strongly bound than those at the end. Desorption will then occur from the ends of the chains for a large portion of coverage. If the mean number of chains stays approximately constant in this range with only their lengths

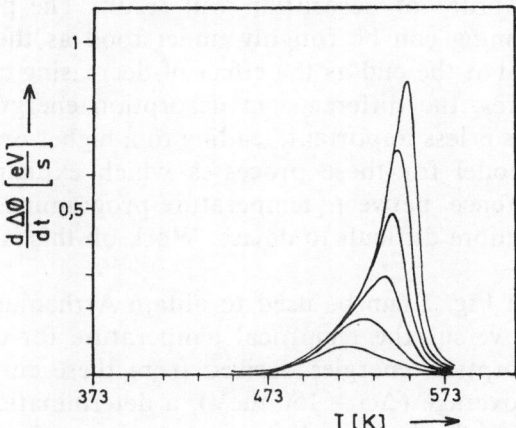

Fig. 1 Temperature-programmed desorption of oxygen from Ag(110), followed by measurement of the work-function change for different initial coverages.

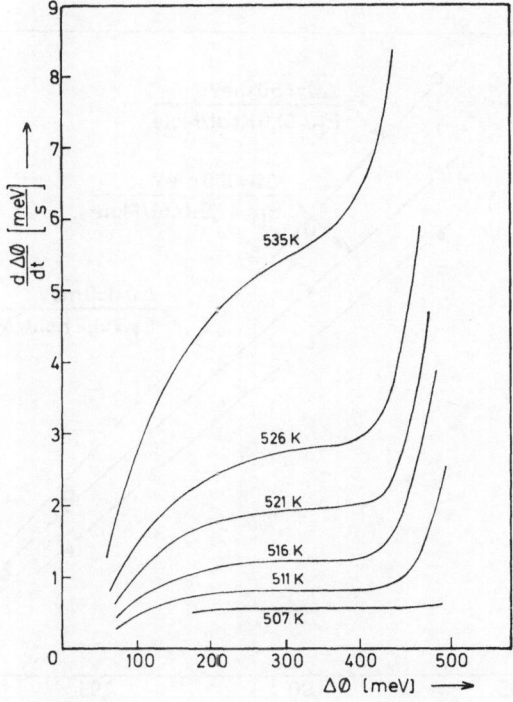

Fig. 2. Isothermal desorption of oxygen from Ag(110) at various temperatures. Abscissa and ordinate give directly the coverage and the rate of desorption, respectively, in relative units. The heating-up region has been discarded.

shrinking, zeroth order of desorption will result. The period with higher rates at the beginning can be roughly understood as the use-up of single atoms, and the tail at the end as the effect of decreasing chain numbers. At higher temperatures, the difference in desorption energy of atoms at the ends and in chains is less important, leading to a higher order of desorption. A quantitative model for these processes which explains all details, including the difference between temperature-programmed and isothermal measurements, is more difficult to devise. Work on this is in progress at the moment.

The curves in Fig. 2 can be used to obtain Arrhenius plots directly by plotting the rates versus the reciprocal temperature for constant coverage (Fig. 3). The desorption energies derived from these curves are shown in Fig. 4. For low coverage ($\Delta\phi = 150$ meV), a determination of the isosteric adsorption energy yielded about 40 kcal/mole, in good agreement with the desorption energy. At higher coverages, the activation energy increases with increasing coverage, which can again be understood qualitatively by the attractive interactions between oxygen atoms. At low coverages, the preexponential of desorption is about 10^{28} cm^{-2} sec^{-1}, which is a normal value for zeroth order.

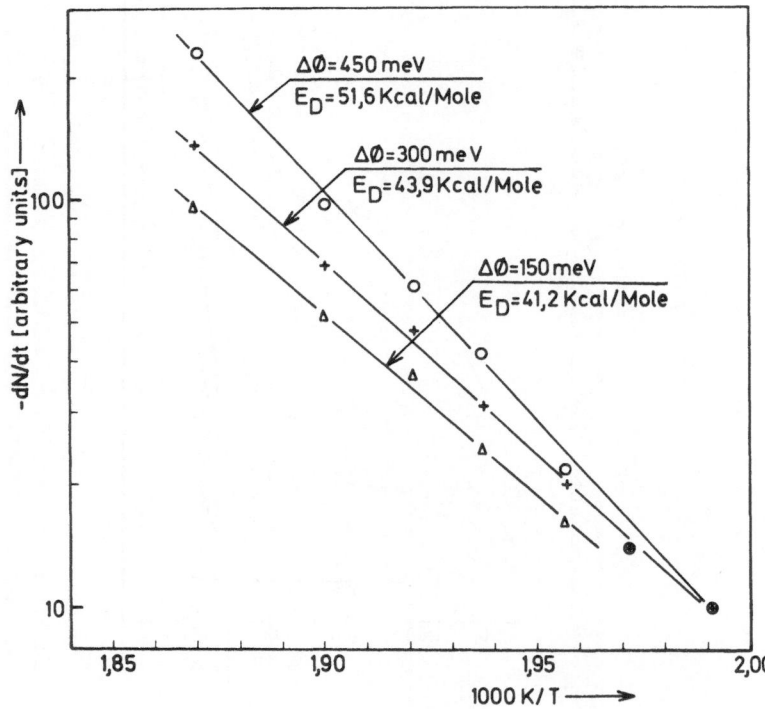

Fig. 3. Arrhenius plots of desorption rates obtained from the runs of Fig. 2 at various constant coverages.

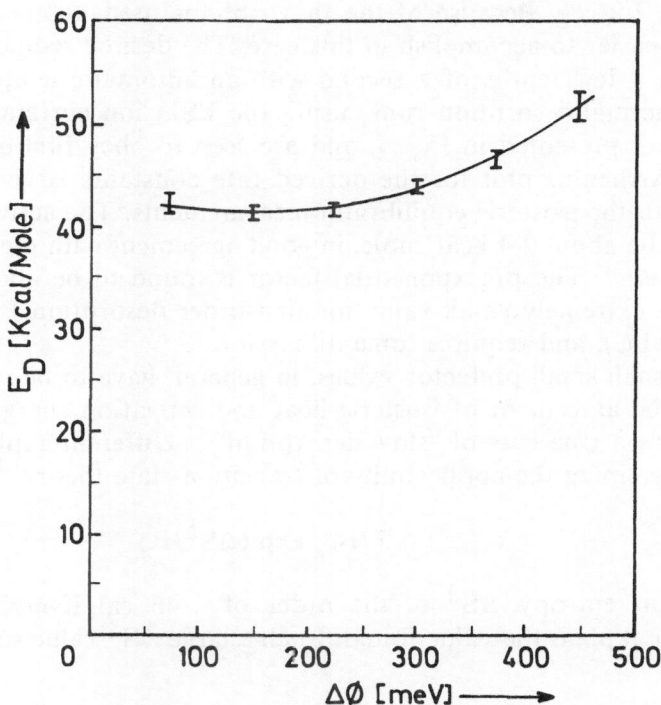

Fig. 4. Desorption energy as a function of coverage obtained from plots like those of Fig. 3 (only some experimental points are shown).

3 DESORPTION OF HYDROGEN ON TUNGSTEN PRECOVERED WITH CO

On tungsten surfaces precovered with CO or oxygen, a weakly bound hydrogen state can be adsorbed, which has been termed κ hydrogen.[14] The kinetic experiments to be mentioned here have been conducted on a CO-covered polycrystalline tungsten ribbon, which had been shown to consist mainly of (100) facets; very little difference in the properties of κ hydrogen has been found on (100) and (110) single-crystal faces.[15] On oxygen layers, κ hydrogen was found to be bound somewhat more strongly.

The κ state is signified by a very large ESD cross section.[1,10] While the β_2 and β_1 hydrogen states have total desorption cross sections of $\sim 10^{-19}$ and $< 10^{-21}$ cm^2 at 100 eV on clean W(100) and of $< 1.4 \cdot 10^{-18}$ and $\sim 2 \cdot 10^{-18}$ cm^2 on W(110), κ hydrogen has a total ESD cross section of $\sim 1 \cdot 10^{-16}$ cm^2 at 100 eV. Therefore, even very small amounts of this species can be detected and monitored accurately by ESD. The binding energy was found to be very small (~ 7.5 to 8 kcal/mole) by isosteric measurements (using the ESD ion current and low current mode as the defining parameter for isosteric conditions, see Fig. 6). Because of the thin ribbons used, temperature jumps were much easier to accomplish in this case. The desired temperatures were obtained in a few tenths of a second with an automatic temperature control.[16] Isothermal-desorption runs, using the ESD ion current as coverage indicator, are presented in Fig. 5, and are seen to obey first-order desorption. The Arrhenius plot for the derived rate constants is given in Fig. 6, together with the isosteric equilibrium measurements. The activation energy is found to be about 7.4 kcal/mole, in good agreement with the isosteric adsorption energy. The preexponential factor is found to be about 10^3 sec^{-1}, which is an extremely small value for first-order desorption ($\sim 10^{-10}$ of the "normal" value), and requires some discussion.

While such small prefactor values, in general, have to be regarded with suspicion, the agreement of isosteric heat and activation energy shows that we have here a true case of "slow desorption".[17] Different explanations are possible. Assuming the applicability of transition-state theory[1,18], so that

$$k_o = r \cdot (kT/h) \cdot \exp(\Delta S^{\ddagger}/R) \qquad (2)$$

an activation entropy $\Delta S^{\ddagger}$ of the order of -46 cal/K-mole would be necessary to explain the value found. Such a large $\Delta S^{\ddagger}$ value seems difficult

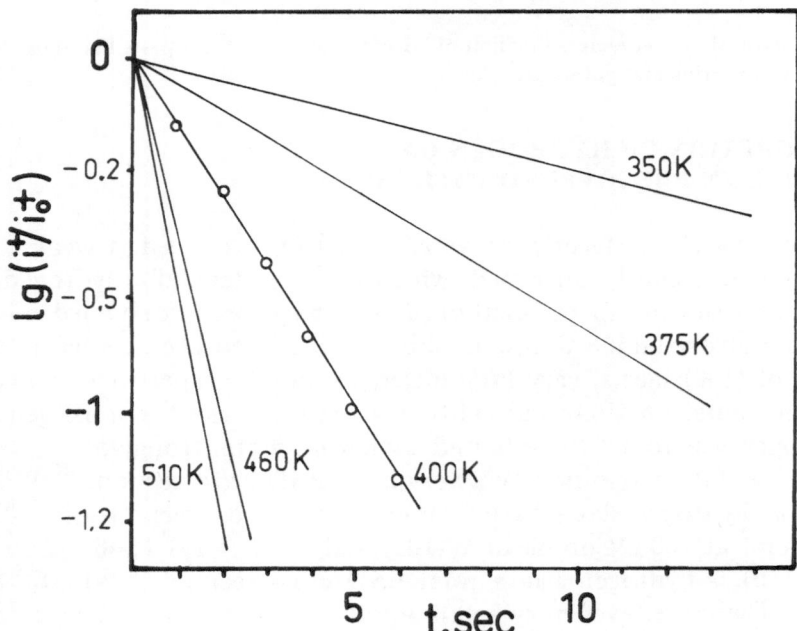

Fig. 5. First-order plots of isothermal desorption measurements of κ hydrogen on CO/W using the ESD ion current of H$^+$ as coverage indicator.[14]

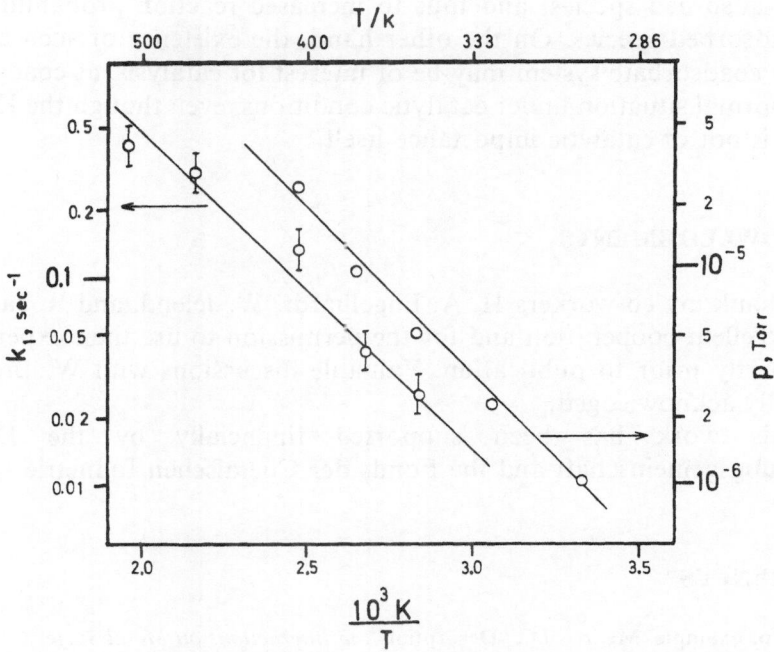

Fig. 6. (a) Arrhenius plot of rates of Fig. 5 and (b) isosteric heat for κ hydrogen on CO/W.[14]

to understand in this case. Simple desorption should not lead to such a large value; a more complicated transition state involving rearrangement of many atoms is very unlikely for such a weakly bound state. Another possibility would be a very small transmission coefficient r. Several other possibilities have been discussed[14] (curve crossing, immediate redissociation if the state is atomic, tunneling), none of which is very convincing either. A final possibility would be the nonapplicability of transition-state theory. If the energetic coupling between adsorbate and substrate is very small, the Eyring treatment breaks down[19] and the preexponential is drastically decreased. This case can also be treated as a reaction sequence, with the energy transfer as the rate-determining step. It seems possible that the κ hydrogen, sitting on top of the adsorbed CO, is so far away from the metal that coupling by electrons is negligible (this would be in line with the very large ESD cross section[10]); coupling by phonons could also be drastically decreased because of the low mass (probably leading to a vibration frequency outside the phonon band of the metal) and the long distance of the adsorbate.

Even if the last possibility is not the correct explanation for the present case, such a mechanism could well have implications for catalytic reactions in general. A drastic decrease in the prefactor for desorption by weak coupling to the substrate heat bath could lead to long surface dwell times even for weakly adsorbed species, and thus to increased reaction probabilities with other adsorbed species. On the other hand, the existence of such an effect for this coadsorbate system may be of interest for catalysis, as coadsorption is the normal situation under catalytic conditions, even though the H-CO-W system is not of catalytic importance itself.

ACKNOWLEDGMENTS

I thank my co-workers H. A. Engelhardt, W. Jelend, and R. Jaeger for their excellent cooperation and for the permission to use their experimental data, partly prior to publication. Valuable discussions with W. Brenig are gratefully acknowledged.

This work has been supported financially by the Deutsche Forschungsgemeinschaft and the Fonds der Chemischen Industrie.

REFERENCES

1. See for example: Menzel, D., "Desorption", in *Interactions on Metal Surfaces* (Topics in *Applied Physics*, Vol. 4), R. Gomer (Ed.), Springer, Berlin (1975).
2. Redhead, P. A., *Vacuum,* 12, 203 (1962); Ehrlich, G., *Adv. Catalysis,* 14, 256 (1963).
3. Toya, T., *J. Vacuum Sci. Technol.,* 9, 890 (1972); Adams, D. L., *Surface Sci.,* 42, 12 (1974).
4. McCarroll, B., *J. Appl. Phys.,* 40, 1 (1969).
5. Pisani, C., Rabino, G., and Ricca, F., *Surface Sci.,* 41, 277 (1974).
6. Shanabarger, M. R., *Surface Sci.,* 44, 297 (1974).
7. Pétermann, L. A., *Progress in Surface Science,* Pergamon, Oxford (1973), Vol. 3/1, p. 1.
8. See for example: Perel, J., Vernon, R. H., and Daley, H. L., *J. Appl. Phys.,* 36, 2157 (1965); Hudson, J. B., and Sandejas, J. S., *J. Vacuum Sci. Technol.,* 4, 230 (1967).
9. Kohrt, C., and Gomer, R., *J. Chem. Phys.,* 52, 3283 (1970); *Surface Sci.,* 24, 77 (1971).
10. See for example: Menzel, D., *Angew. Chem. Intern. Ed.,* 9, 255 (1970); Madey, T. E., and Yates, J. T., *J. Vacuum Sci. Technol.,* 8, 525 (1971); Menzel, D., *Surface Sci.,* 47, 370 (1975).
11. Engelhardt, H. A., Bradshaw, A. M., and Menzel, D., *Surface Sci.,* 40, 410 (1973); Engelhardt, H. A., and Menzel, D., to be published.
12. Bradshaw, A. M., Menzel, D., and Steinkilberg, M., *Jap. J. Appl. Phys.,* Suppl. 2, Part 2, 841 (1974), and work cited therein.
13. Madey, T. E., Engelhardt, H. A., and Menzel, D., *Surface Sci.,* 48, 304 (1975).
14. Jelend, W., and Menzel, D., *Surface Sci.,* 42, 485 (1974).
15. Jaeger, R., and Menzel, D., unpublished.
16. Menzel, D., *Ber. Bunsenges. Phys. Chem.,* 72, 591 (1968).
17. Pétermann, L. A., *Adsorption—Desportion Phenomena,* F. Ricca (Ed.), Academic Press, London—New York (1972), p. 227.
18. Glasstone, S., Laidler, K. J., and Eyring, H., *The Theory of Rate Processes,* McGraw-Hill, New York—London (1941).
19. Kramers, H. A., *Physica,* 7, 284 (1940); d'Agliano, E. G., Schaich, W. L., Kumar, P., and Suhl, H., Nobel Symposium 24, Academic Press, New York (1973), p. 200.

DISCUSSION on Paper by D. Menzel

EMMETT: What was the temperature range covered on your work on hydrogen adsorption? For small-particle discussion, it may be of interest at this point to call attention to a method that has been used for obtaining pore iron crystals that are uniformly about 100 Å. The work to which I referred was carried out by a graduate student at the University of California at Santa Barbara in 1966. The student's name was Chond. His procedure involved the vaporizing of the iron pellets in the helium stream and joining this stream with one consisting of heated helium. By choosing the concentrations and temperature correctly, it is found possible to produce a deposit of uniform size (iron), about 100 Å.

MENZEL: Adsorption took place between 100 and 300 K, and desorption was essentially complete at about 450 K.

BLOCK: (1) In FEM studies the oxygen penetration into the silver lattice is fairly well established at temperatures above 200 to 250°C. It is questionable whether this is a first-order reaction in respect to the oxygen coverage. A correction of the desorption rate of oxygen will probably be necessary in a nonlinear coverage dependence.

(2) Could you comment on the stability of the Ag(110) surface at higher temperatures? Is it stabilized by oxygen adsorption?

(3) Would the assumption of an activated complex during desorption in connection with a first-order desorption kinetics imply desorption of nonequilibrium gas molecules?

MENZEL: (1) There is no penetration into the lattice at low coverage, as shown by the equality of isoteric heats and desorption energy. At higher coverages some of the desorption may actually occur into the solid and not into the vacuum, and we are considering this possibility in the detailed mechanistic analysis mentioned. However, the problem of the formal order, which was the main point here, will be there for both paths.

(2) The (110) face was found to be stable both clean and with adsorbed oxygen up to at least 300°C.

(3) An activated complex of desorption does not imply activated adsorption, i.e., the existence of a saddle point in potential energy. Thus, even for a smoothly varying potential curve, the transition-state theory is applicable, as the activated complex does not have to be a well-defined species. If one wants to picture this statement physically, one can say that the "activated complex" corresponds to the point of no return of the particle leaving the surface. Maybe a theoretician can comment on the formal side.

GRIMLEY: Considering desorption from the standpoint of the quantum theory of reactions, the initial state is a mixture of states characterized by a density matrix. If the energy distribution in the initial state is canonical, the transition rate to a final state in the channel describing desolved atoms can be written in the form $(kT/\hbar)\exp(\Delta F^{\ddagger}/kT)$ familiar in the absolute-reaction-rate theory. However, it is not required that $\Delta F^{\ddagger}$ involves an integration over energy of the energy derivative of the transition rate from all states of given energy in the entrance channel. For the actual definition, see, for example, Levine's book on *Molecular Rate Processes.*

ERTL: (1) With the system O/Ag(110) you were following the removal of oxygen from the surface by a technique which probes the surface concentration. Have you any evidence that this is exclusively due to desorption and that no interference with dissolution of oxygen in the bulk takes place?

(2) According to your model for the interaction between adsorbed oxygen atoms on Ag(110) one would expect an increase of the binding energies at rather low coverage, but in fact this is observed only at higher coverages. Can you comment on this effect?

MENZEL: (1) As already mentioned in reply to Block, we are pretty sure that solution does not play a role at low coverages, but not so sure about its role at higher coverages. I may add, however, that our field emission measurements of oxygen adsorption of silver films on tungsten tips [Wüstner, W., and Menzel, D., *Thin Solid Films*, **24**, 211 (1974)] have shown quite conclusively that oxygen is removed from a smooth silver layer by heating to about 500°K without any penetration.

(2) The increase of binding energy with coverage starts at about one-tenth of a monolayer, i.e., at rather low coverages, in agreement with expectations.

SCHWAB: May I ask whether you have a special idea as to the nature of the binary surface complex in coadsorption?

MENZEL: All we really know is that the hydrogen must be located at a rather large distance from the metal surface. This is shown by the very large ESD cross section and, if the interpretation of the small preexponential of desorption as being due to small coupling is correct, by this latter feature also. Also, the hydrogen is somehow connected with the oxygen atom. Apart from that we cannot give any detailed idea at present.

MASON: Most chemists would believe that there will be little covalence in oxygen linear chains having more than four or so members. The oxygen-oxygen interactions must be small compared with substrate-oxygen interactions. Given that the substrate is determining the geometry of the chains, does the possibility exist of cyclic polymers on different substrates?

MENZEL: The attractive interactions between the oxygen atoms are indeed very weak and are therefore most probably due to indirect interactions. In all similar cases, so far only linear chains have been seen; this may be caused by the rectangular faces used [for oxygen on Ag(111), no ordered oxygen structures are formed].

SOMORJAI: Could you discuss the various models which you have indicated that you have available to explain the low preexponential factors that were observed.

MENZEL: As already pointed out in the text, three possibilities appear for the low preexponential factors:

(1) A large negative value of $\Delta S^{\ddagger}$, the entropy of activation. This would imply an activated complex for desorption with a much more complicated structure or a very small probability of formation. For the present example, this does not seem to be very likely, although not impossible.

(2) A very small value of the transmission coefficient, for which there could be several reasons. A tunneling effect can be excluded because of the good Arrhenius behavior. Nor does desorption by crossing from one potential curve to another with different symmetry seem to be very likely in this case.

(3) As Kramers has shown (ref. 19), the Eyring treatment breaks down if the energetic coupling between the reactive mode (here the movement leading to desorption) and the heat bath (here the solid) is either very strong or very weak, i.e., the characteristic time of energy transfer is either much shorter or much longer than the characteristic times of the system (e.g., the vibration frequency). As the hydrogen species seems to be rather decoupled from the substrate, as shown by its very high ESD cross section, the weak coupling case appears to be more probable here. We cannot conclusively ascribe the low preexponential to such an effect at present, however.

EHRLICH: would hesitate to account for the very low preexponential in the desorption of hydrogen in the κ state in terms of restricted energy transfer. Even with the light rare gases, which are only bound to a surface with an energy of a few kcal/mole, there is quite a finite thermalization in collisions from the gas phase. This is further enhanced by the presence of surface layers. Although the details of the atomic events in desorption are slightly different, it is difficult to see how energy transfer to a particle held in as deep a well as κ hydrogen, and interacting with a layer of intermediate mass, would be limiting.

ANTONINI: You have explained your results of isothermal-desorption experiments of hydrogen adsorbed on polycristalline tungsten by electron-stimulated desorption technique (ESD)—first-order reaction for desorption; activation energy, 7.4 kcal/mole; preexponential factor, 10^3 sec^{-1}—in terms of the existence of a molecular, weakly bound hydrogen adsorption state and a true "slow desorption".

In fact you have neglected to account for possible readsorption during your desorption experiments.

Similar numerical results have been obtained for H_2 on nickel by ESD (ref. 17) and analysis of isothermal-adsorption curves [Antonini, J. F., *Helv. Phys. Acta* (December, 1974)] when readsorption is neglected. However, when readsorption in the 10^{-10} to 10^{-9} Torr range is taken into account, the interpretation of the same isothermal experiments results in a perfect second-order reaction for the desorption reaction, with the appropriate activation energy 14 to 15 kcal/mole and preexponential factor 10^{12} to 10^{13} sec^{-1}.

Later results of Pétermann which do not figure in the reference you mention (ref. 17) have shown that for nickel and Pd/Ni alloys the hydrogen phase which exhibits a large ESD cross section is not a "surface phase" and is replenished by bulk diffusion.

MENZEL: Readsorption certainly did not play a role in our experiments. This is borne out by the fact that the sticking coefficient of κ-hydrogen is so small ($\sim 3.10^{-4}$ at zero coverage, strongly decreasing with coverage) that under the conditions of the desorption experiments (background pressure about 10^{-10} Torr, time a few to a few tenths of seconds) the number of readsorbed hydrogen molecules is absolutely negligible. Furthermore, the excellent agreement of isosteric heat and activation energy of desorption which were determined independently, would not be understandable if readsorption were important. The latter agreement also shows that the very small preexponential found is not an artifact, and that no such effects as dissolution (which would not be expected for W anyway) can have played a role. Furthermore, the very large ESD cross section ($>10^{-16}$ cm^2 at 100 eV) shows that the species concerned is quite decoupled electronically from the metal, and must therefore sit at a considerable distance from it. This could easily lead to weak vibrational coupling also, especially in view of the small mass of H. Nevertheless, weak energetic coupling is only one of the possible reasons for the small preexponentials; another would certainly be a very large negative activation entropy due to special configurational requirements for desorption to occur. For details see ref. 14.

CONDENSATION KINETICS AND MECHANISMS*

L. D. Schmidt

*Department of Chemical Engineering
and Materials Science
University of Minnesota
Minneapolis, Minnesota*

ABSTRACT

Representative results on sticking coefficients of chemisorbing gases on metals are summarized and discussed in terms of various models of condensation. The dependences on solid material, crystal plane, gas and surface temperature, and adsorbate mass have been measured with isotropic fluxes for a number of chemisorption systems. Coverage dependences of s generally agree with models of direct or precursor intermediate processes, but values of initial sticking coefficients can be interpreted only by considering the mechanism of energy transfer. Recent results using molecular beams to determine s as a function of angle of incidence are summarized. Predictions of several models of energy transfer are discussed and compared with the results of these experiments.

1 INTRODUCTION

Catalytic reactions on surfaces generally require that reactant molecules first adsorb. Rates of condensation, described through the sticking coefficient or probability of condensation s, are observed to vary widely and as yet unpredictably between different solids and between different crystal planes of a given solid.[1,2] In this paper we review some of the experimental data on sticking coefficients and their coverage and temperature dependences. It will be apparent that, while coverage dependences can generally be fit by various models, such measurements do not generally permit one to decide on mechanisms of energy transfer because they are averaged over incident angles. Results of molecular beam measurements of the variation of s with angle of incidence are described and compared with predictions of various models.

2 STICKING COEFFICIENTS

Sticking coefficients are easy to estimate in a variety of experiments[1,2], but measurements are highly susceptible to systematic errors on the order of a factor of two. This is because of the difficulty in measuring absolute pressures and pumping speeds in most ultrahigh-vacuum systems. In the desorption method one measures coverage n versus time at constant pressure (for example by flash desorption or work function change); then s is given by the expression

$$s = \frac{\sqrt{2\pi mkT_g}}{P} \frac{dn}{dt} \tag{1}$$

Absolute measurement of s requires absolute values of P and n. The coverage dependence s(n) requires data of sufficient accuracy that slopes can be obtained as a function of coverage. The latter is troubled by crystal cooling during exposure and by pressure variations. In the adsorption method, one measures pumping by the adsorbing surface by following the pressure P(t) as it deviates from the steady-state value P_0 when the surface is saturated. Then s is obtained from the expression

$$s = \frac{(2\pi mkT_g)^{1/2}}{ART_g\tau} \left(\frac{P_0}{p}\right) - 1 \tag{2}$$

and n from the expression

$$n = \frac{V}{ART_g\tau} \int_0^t (P_0 - P) \, dt' \tag{3}$$

Note that absolute pressures are not required to obtain s. However, the pumping time constant τ must be known for absolute values. This method appears from our experience[2] to yield more accurate values of both s (because absolute pressures need not be known) and coverage dependences (because slopes need not be measured). The adsorption method requires a sufficiently low pumping speed that a significant fraction of the pumping is caused by the surface, and it requires that the pumping speed be known accurately. Madey[3,4] has developed a method for determining s which uses a molecular beam from a calibrated leak and therefore does not require absolute values of pressure or pumping speed.

Representative data on initial sticking coefficients s_0 and coverage dependences of s on single crystal planes of transition metals are shown in Table I and Fig. 1.[1-19] We caution, however, that there are several discrepancies of at least a factor of two in s in the literature. We believe that values shown are much more reliable than this, and, in any case, comparisons should be accurate because they were obtained in similar vacuum systems and sometimes simultaneously in the same system. There is a consensus from several laboratories that on tungsten, s for H_2 is considerably less than unity[1,4,5], that s for N_2 is ~0.4 on the (100) plane[9] and much lower (perhaps zero) on the (110) plane[10], and that s for O_2 on the (100) plane is nearly unity[4,11]. Coverage dependences for N_2 and O_2 on W(100) have been reproduced in several laboratories, but most of the others are unchecked, and two systems [O_2 on Pt(111) and O_2 on W(110)] have been the subject of controversies[13-15,18,19], as is discussed later.

Table I. Sticking Coefficients

Adsorbate	Substrate	s_0[a]	Coverage Dependence	References
H_2	W(100)	0.11, 0.50	Precursor	1,4,5
	W(110)	0.07	$1-\theta$	1,5
	W(111)	0.24	Precursor	1,5
	Mo(110)	0.10	$(1-\theta)^2$	6
	Ta(100)	0.18	Precursor	7
D_2	W(100)	0.25, 0.50	Precursor	1,4
	W(110)	0.7	$(1-\theta)$	8
N_2	W(100)	0.41	Precursor	9
	W(110)	0.004	$(1-\theta)^2$	10
	Mo(110)	0.09	$(1-\theta)$	6
CO	W(100)	0.49	Precursor	11
	W(110)	0.55	Precursor	12
O_2	W(100)	1.0	$\sim 1-\theta$	4,10
	W(110)	0.34 ($T_s=30^\circ$K)	Precursor	13
		Small	$\sim\theta^{1/2}$ for $\theta<0.2$	14
		0.25	Increases slightly	15
	W(211)	0.8	Precursor	14
CO_2	W(100)	0.05	Precursor	11
NH_3	W(100)	~1.0		15
H_2O	W(100)	~1.0		17
O_2	Pt(111)	~0.1	$\exp(-m\theta)$	18,19

(a) All values are for gas and surface at 300 K except when indicated.

Coverages in Fig. 1 are given as θ, the fraction of the saturation coverage at $T_s = 300°$K, and in Table I coverage dependences are given for the most tightly bound state. There may, however, be additional states with $s < 10^{-3}$ which are not populated in these experiments. In any case, the coverage dependences can be interpreted only if one or at most two binding states are being populated.

The variations of s_0 with gas and surface temperatures T_g and T_s are displayed in Fig. 2 for gases on (110) and (100) planes of tungsten.[2,9,11-13] Measurement of s at different T_s is fairly straightforward if crystal heating

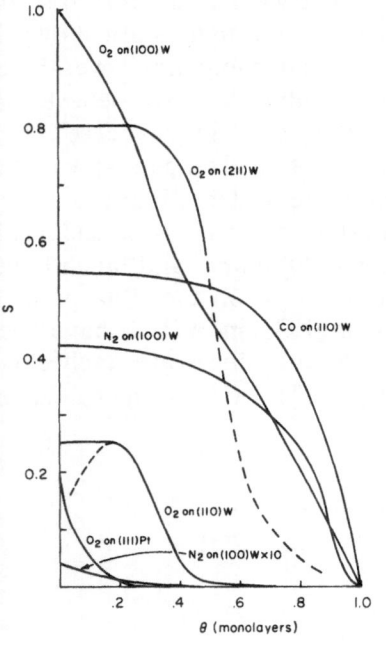

Fig. 1. Sticking coefficient s versus fraction of saturation coverage θ for chemisorbed gases on single-crystal planes of transition metals.

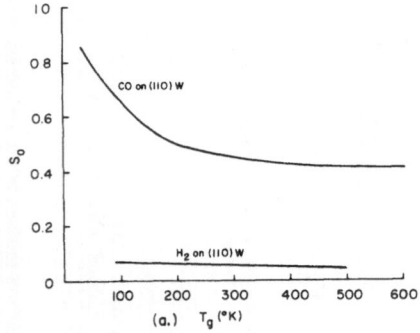

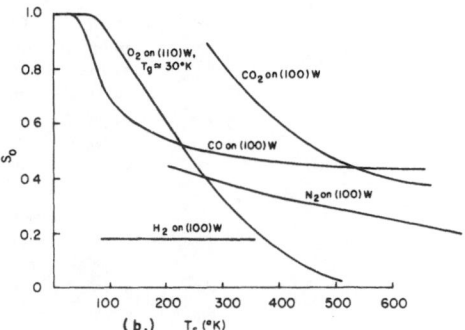

Fig. 2. Initial sticking coefficients s_0 versus gas and surface temperature; except where indicated the other temperature was $300°$K.

does not interfere. In all of these experiments, the gas was at $300°K$, a temperature fixed by wall collisions in the vacuum system.

The dependence of s on T_g is more difficult to obtain, because it requires either an independently heated molecular beam or the ability to vary the temperature of the entire vacuum system. Data shown[13,20] were obtained from Knudsen effusion molecular beam sources. Temperatures are those of the reservoirs, and should be multiplied by a factor of two if T_g is to be interpreted as that corresponding to an isotropic flux. Data are representative of normal incidence[13] for CO on W(110) and are averaged over incident angle for H_2 on W(110)[20], as discussed later.

The dependence of s on adsorbate mass can be measured readily for hydrogen because H_2 and D_2 differ in mass by a factor of two. The results in Table I show that $s_{D_2}/s_{H_2} \simeq 1.4$ on W(100)[8] and Ta(100)[7], but $s_{D_2} \simeq s_{H_2}$ on W(110)[8].

Our discussion of these results is concerned with causes of the observed coverage dependences of s and the variations of s_0 with metal, crystal plane, adsorbate mass, T_s, and T_g. The first of these, $s(\theta)$, can be explained fairly well in most cases by simple models, but the other dependences require consideration of the mechanisms of energy transfer and the dependence of s on angle of incidence.

3 CONDENSATION KINETICS

The simplest description of the coverage dependence of s assumes that it depends on the probability $g(\theta)$ of finding suitable vacant sites

$$s = s_0 \, g(\theta). \tag{4}$$

For single-site adsorption, $g(\theta) = 1-\theta$, and if two uncorrelated sites are required, $g(\theta) = (1-\theta)^2$. The latter is expected for the simplest type of dissociative adsorption, but $g(\theta)$ will be more complicated if the adsorbate is immobile, requires correlated pairs of sites, or exists as islands or patches. This is the "direct" or site-occupation model first suggested by Langmuir.[21] It is the form of $s(\theta)$ usually assumed in adsorption-isotherm calculations, but the data in Table I and Fig. 1 show that s is seldom proportional to $1-\theta$ or to $(1-\theta)^2$ for these chemisorption systems.

In fact, s is frequently observed to be nearly independent of θ. Models developed some time ago[22,23] to explain the constancy of s (from data on polycrystalline surfaces) considered condensation to proceed not in a single-step process but rather through an intermediate adsorbed state or precursor. The simplest version of this idea, developed by Becker[22] and by Ehrlich[23], considered condensation to proceed via the steps

$$A_g \underset{k_d{}^*}{\overset{s^*}{\rightleftharpoons}} A^* \xrightarrow{\ k_a g(\theta)\ } A_s \tag{5}$$

where A_g, A^*, and A_s represent the adsorbate in gas, precursor, and chemisorbed states, respectively. The rate constants S^* and $k_d{}^*$ apply to condensation and desorption of the precursor state, and $k_a g(\theta)$ is the rate constant for conversion from the precursor state into the chemisorbed state. The conversion rate is assumed to be proportional to $g(\theta)$, the probability of having suitable vacant sites in the chemisorbed state. If the coverage in A^* is always small, one can apply the pseudo-steady-state assumption on its density n^* to obtain

$$s = \frac{dn_s}{dt}\Big/ \text{flux} = k_a g(\theta)\, n^* = \frac{s_0(1+K)}{1 + \dfrac{K}{g(\theta)}} \tag{6}$$

In this expression, $K = k_d{}^*/k_a$ and $s_0 = s^*/(1+K)$.

A similar model developed by Kisliuk[24] is similar to that implied by Eq. (5), except that if the site visited by the molecule in the precursor state is occupied by a chemisorbed species, the molecule is allowed to diffuse over the surface with rate constant k_{diff} to another site. This may be sketched as shown below:

$$\tag{7}$$

The coverage dependence predicted by this model is easily formulated[12,13,33] by considering probabilities of adsorption P_{a_i} on successive sites, $i=1,2,\ldots$ Then s is given by the expression

$$s = \sum_{i=1}^{\infty} P_{a_i} = \frac{s_0}{1 + \dfrac{K'\theta}{g(\theta)}} \tag{8}$$

with K' being a ratio of conversion and desorption probabilities. The only difference between this expression and Eq. (6) is the θ which occurs in the second term in the denominator of Eq. (8). In fact it can be shown that Eqs. (6) and (8) give identical coverage dependences if $g(\theta) = 1-\theta$.

Kohrt and Gomer[12,13] considered a modification of the Kisliuk model which includes the possibility that the molecule in the precursor state may not be able to chemisorb even if the chemisorbed site associated with it is empty. This could arise, for example, if an adjacent site must also be vacant for chemisorption to occur. A probabilistic argument, similar to that just sketched, yields

$$s = \frac{s_0}{1 + \frac{K'}{g(\theta)} + \frac{K''[1-\theta-g(\theta)]}{g(\theta)}} \tag{9}$$

where K'' is another grouping of rate constants or probabilities.

These precursor models give very similar coverage dependences, and it is questionable whether any experimental results are sufficiently accurate to decide between them. Even if there is only one major binding state [probable for N_2 on W(100) and for CO and O_2 on W(110)], the presence of binding states associated with defects or other crystal planes can alter $s(\theta)$ slightly. We showed[9] that $s(\theta)$ for N_2 on W(100) could be fit better by Eq. (6) than by Eq. (8) and that $g(\theta) = (1-\theta)^2$ gave better agreement than did $g(\theta) = 1-\theta$. Kohrt and Gomer[12,13] analyzed their results for CO and O_2 on W(100) using Eqs. (8) and (9), but did not critically test the expressions.

Values of K from Eq. (6) have been determined as a function of T_s for N_2 on W(100). It was shown[1,9] that K is of the form

$$K \simeq \frac{k_a}{k_d{}^*} \simeq \frac{A_d{}^*}{A_a} \exp[-(E_d{}^*-E_a)/RT] \tag{10}$$

as expected if K is a ratio of conversion to desorption-rate constants. The activation energy in K is positive, and this shows that the energy barrier between the precursor and chemisorbed states, E_a in Fig. 3a, is less than the heat of adsorption in the precursor state, $E_d{}^*$.

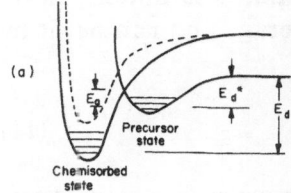
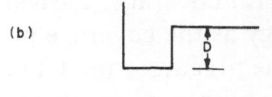
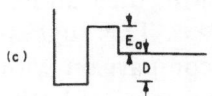
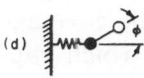

Fig. 3. Potentials and models in condensation: (a) potential perpendicular to surface for chemisorbed and precursor state, (b) square well potential, (c) square well potential with barrier, (d) model used for quantized oscillator calculations.

Adsorption of N_2, CO, and H_2 on W(100) and W(110) agrees quite well with coverage dependences predicted by direct adsorption or by the various precursor models.

Next we consider adsorption of O_2 on tungsten and platinum. On W(110), the early results of Kohrt and Gomer[13], assumed to obey a Kisliuk model, can be reconciled with the more complex behavior found by Tracy and Blakely[14] and by Hopkins et al.[15] if it is assumed that Kohrt and Gomer were observing only the lower coverage state. Tracy and Blakely reported that s increased with coverage for $\theta < 0.2$. They found by LEED that oxygen adsorbed as islands, and showed that the observed behavior could be explained if O_2 has a much higher sticking coefficient on oxygen-covered patches than on bare surface regions. Hopkins et al. agreed qualitatively with the results of Tracy and Blakely, but reported only a slight increase in s with θ at low coverages.

LEED shows that O_2 also exists in patches on W(211) just as it does on the (110) plane, but in this case, s is independent of θ for $\theta < 0.35$. The difference appears to be associated with the size of the islands and the relative sticking coefficients on bare and oxygen-covered surfaces.[14] On W(110), O_2 can only condense on the latter and must then diffuse to the edges of the islands to chemisorb. If this process is diffusion limited, Tracy and Blakely showed that $s \sim \theta^{1/2}$, in good agreement with their experimental results. On W(211) the constancy of s implies that molecules can condense both on bare and oxygen-covered patches and that diffusion across the islands is rapid. (If condensation occurred only on bare regions, $s \sim 1-\theta$, a dependence also predicted by site-occupation kinetics with no islands.)

On Pt(111), Bonzel and Ku[18] found a variation in s for O_2 which is different from any reported previously. They found that s is initially fairly high, $s_0 \sim 0.1$, but that it decreases rapidly with coverage and can be fit by the expression

$$s = s_0 \exp(-14.1\theta) \tag{11}$$

They also explained their results by assuming a precursor intermediate, Fig. 3a. If the heat of adsorption decreases strongly with coverage (dashed curve in Fig. 3a), the curves will cross at a higher energy as the coverage increases because E_a will be larger. Now if one assumes the precursor model of Eq. (6) and lets K vary by having E_a decrease linearly with θ, then an expression of the form of Eq. (11) will be obtained for $K \gg 1$. This suggestion, while quite plausible, needs to be further tested by comparison with heat of adsorption data on Pt(111); the results are consistent with the measured heat of adsorption on polycrystalline platinum, but this may be fortuitous. An alternative and less interesting explanation of this coverage dependence is that there are several binding states of different s which populate sequentially as the total coverage θ increases to by chance fit Eq. (11).

The examples we have considered exhibit all possible coverage dependences of s: decreasing as $1-\theta$, $(1-\theta)^2$, $\exp(-m\theta)$; independent of θ; and even increasing with θ. The models which have been proposed can be brought into satisfactory agreement with most results with suitable choice of parameters, but there is little detailed information on mechanisms of the processes. One perhaps improbable feature of all precursor models is that, for s to be initially independent of θ, the sticking coefficient into the precursor state s* must be *totally independent* of whether the surface is clean or adsorbate covered. In fact much of the data where precursor models are assumed may simply involve the linear decrease (more slowly than as $1-\theta$) associated with different values of s* on clean and adsorbate-covered regions.

In the early literature, one finds frequent reference to "activated adsorption". This would occur if the gas molecule had to overcome a potential energy barrier ($E_a > E_d*$ in Fig. 3a) and if the molecule did not need to become thermally accommodated in a precursor state before crossing the barrier. We know of no examples of adsorption on metals which indicate activated adsorption. This would be seen as an increase in s with increasing gas temperature, a seldom measured quantity as indicated in Fig. 2. On the other hand, some investigators have reported increases in s with increasing surface temperature, for example, in oxide formation. This is not activated adsorption in the sense of requiring a potential barrier; various sequences of processes after accommodation can yield to increase in s with T_s.

4 ENERGY TRANSFER

We have thus far considered various models of kinetics of condensation and their coverage dependence but have said little of the values of the initial sticking coefficients s_0 and their dependences on metal, crystal plane, T_g, T_s, and adsorbate mass. As indicated in Figs. 1 and 2 and Table I these vary widely for the systems shown. For N_2, s is lower by a factor of ~100 on the (110) plane than on the (100) plane, for O_2 by a factor of ~5, and for H_2 by a factor of ~3. Between corresponding planes of bcc transition metals, differences of at least a factor of 2 are usually found, but no obvious trends with solid properties are evident. On W(100) there is a mass dependence in s for hydrogen[8], but apparently there is no difference in s between H_2 and D_2 on W(110). Figure 2 shows that dependences on T_s and T_g vary considerably with the gas, plane, and temperature range.

Mechanisms of condensation can be understood only by considering energy transfer between the gas and the solid because condensation requires (at least) that the gas molecule transfer its kinetic energy to the solid upon impact. We have recently carried out molecular beam experiments in which the dependence of s on angle of incidence with respect to the surface was examined.[20] This eliminates one of the averages incorporated in sticking coefficients measured with isotropic fluxes and provides further clues as to the possible modes of energy transfer.

Some experimental results for several gases on the (100) and (110) planes of tungsten are shown in Fig. 4. Curves are drawn through data points and are reported as initial sticking coefficients s_0 at angle ϕ divided by s_0 at normal incidence. They were obtained in a simple molecular beam system in which a uniform and collimated beam from a capillary impinged on a single-crystal surface which could be rotated with respect to the beam. The system pumping speed was high enough that adsorption from the background could be subtracted from that from the beam.

The trend for most systems is clear: s is generally only weakly dependent on ϕ and increases slightly toward grazing incidence. For H_2 and D_2 on W(110), however s decreases strongly toward grazing incidence. Dependences on T_s and T_g have been examined, and the variation of s_0 (angle averaged) with T_g is shown in Fig. 2.

Energy transfer to solids by phonon generation has been considered by many investigators.[25-33] These have included one- and three-dimensional classical and semiclassical models[25-28,33] and also models where impulsive interactions and one-dimensional interaction potentials were assumed[29-32].

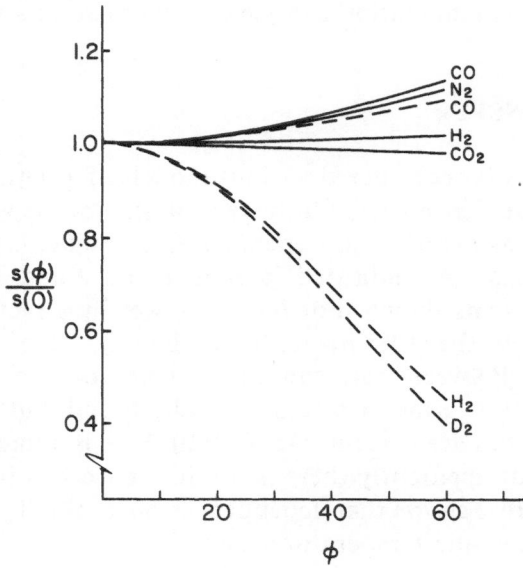

Fig. 4. Sticking coefficients s(ϕ) versus angle of incidence with respect to surface normal ϕ for several gases on W(110) and W(100), All systems except H_2 and D_2 on W(110) exhibit slight increases with increasing ϕ.

The former are too cumbersome to permit use of real material parameters or finite surface temperature, while in the latter, these can be included as adjustable parameters. We have extended[20] these so-called hard and soft cube models, originally developed to explain beam-scattering experiments, to include an attractive potential which is assumed to be a one-dimensional square well, Fig. 3b. The criterion for trapping of a gas molecule in this well is simply that after collision the normal component of the velocity of the molecule v_n be too small to escape the barrier, $v_n < (2E_d*/m)^{1/2}$. In this model $s(\phi)$ is given by the expression

$$s(\phi) = \frac{\int_0^\infty dv \int_{w_m}^\infty (v \cos \phi) \, dw(v-w) \, v^2 \exp\left(-\frac{mv^2}{2kT_g}\right) \exp\left(-\frac{Mw^2}{2kT_s}\right)}{\int_0^\infty dv \int_{-\infty}^v dw(v-w) \, v^2 \exp\left(-\frac{mv^2}{2kT_g}\right) \exp\left(-\frac{Mw^2}{2kT_s}\right)} \quad (12)$$

where w is the velocity of the surface atom, and m and M are the masses of gas and surface atoms, respectively. This expression can be integrated numerically. It predicts that s always increases toward grazing incidence because the energy which must be transferred for condensation to occur (that associated with the normal component of the velocity) is smaller. This agrees quite well with $s(\phi)$ data in Fig. 4 for reasonable choices of parameters for all gases except H_2 and D_2 on W(110).

The potential E_d* in Eq. (12) is assumed to be that for adsorption in the precursor state, Fig. 3a, and s is that into the precursor, s*. These are consistent with coverage dependences described previously in the limits of low coverage for appropriate values of K, K', or K''. The surface temperature is accounted for by assuming a Maxwellian velocity distribution of surface atoms, and therefore the solid is characterized only through the mass of the atom M and T_s.

This model and all similar classical phonon models appear to be incapable of predicting a decrease in s toward grazing incidence, because s should increase when the amount of energy which needs to be transferred is smaller. For H_2 and D_2 on the (110) plane, there appears to be some sort of a threshold in which a minimum perpendicular component of velocity is required. We have considered two possible models[20] in which a threshold occurs: a classical potential barrier and a quantized surface oscillator.

A potential barrier between the gas and the chemisorbed state could exist for dissociative adsorption if the potential curves for the dissociated and undissociated species were such that $E_a > E_d*$ in Fig. 3a. To consider this situation we have modified the one-dimensional square well model to include a barrier of height E_a, indicated in Fig. 3c. As expected, this model can yield sticking coefficients which decrease toward grazing incidence because the perpendicular component of the velocity of the inpinging molecule decreases as ϕ increases. Sticking coefficients can be computed for this model from an expression similar to Eq. (12), except that the limits of integration are altered by the presence of the barrier. In the limits of $E_a \gg kT_g$ and large E_d, the model predicts

$$s(\phi) = (1+E_a/RT_g \cos^2\phi) \exp(-E_a/RT_g \cos^2\phi) \qquad (13)$$

so that s should decrease with increasing ϕ or T_g.

Energy transfer to the lattice requires vibrational excitation of the solid, and these modes may be quantized. This may arise because, while there is a quasi-continuum of phonon modes in a solid, the time of the interaction is so short that only a small number of solid atoms can respond during the collision. We have considered this in the model sketched in Fig. 3d. The gas molecule is assumed to collide with a surface atom connected through a harmonic potential to the rest of the solid. As before, a one-dimensional impulsive potential between gas and surface is assumed. Sticking coefficients are computed from the formula

$$s(\phi) = \frac{\int_0^\infty dv\ v^3 \exp(mv^2/kT_g) \sum_{i=0}^\infty n_i \sum_{j>i}^{\infty\prime} P_{j-i}(\frac{1}{2}mv^2 \cos^2\phi + D)}{\int_0^\infty dv\ v^3 \exp(mv^2/kT_g) \sum_{i=0}^\infty n_i \sum_{j=0}^\infty P_{j-0}(\frac{1}{2}mv^2 \cos^2\phi + D)} \qquad (14)$$

The summation Σ' extends over all states which lead to trapping, and P_{j-i} are transition probabilities of going from the ith to the jth vibrational levels. These were computed according to the ITFITS method developed for vibrational excitation of gases.

To summarize the essential features of these models, all use one-dimensional impulsive potentials and assume that the gas molecule impinges only on a single surface atom. These approximations seem reasonable because of the short duration of the collision for light gas molecules on heavy surface atoms and because the potential experience by a molecule near a close-packed plane of a metal should be smoothed parallel to the surface by interactions with conduction electrons. The models have the advantage that, because of their simplicity, they contain only a few experimental parameters (M_g, M_s, T_g, T_s) and a few adjustable parameters (D, E_a, surface vibrational level spacing) for which reasonable values can be chosen for computations.

Results of calculations will be presented elsewhere, but Fig. 5 shows some representative curves for the three models. It is seen that both models involving thresholds are capable of correctly predicting a decrease in s towards grazing incidence. However, neither simultaneously gives the observed dependences on adsorbate mass and T_s for H_2 and D_2 on W(110) with reasonable choices of parameters. We conclude from these calculations that either mechanism may be operative, but it is difficult to determine

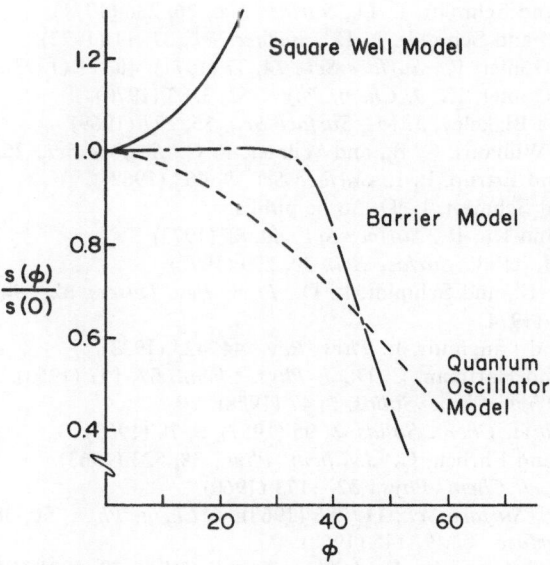

Fig. 5. Calculated $s(\phi)$ curve for square well, barrier, and quantum oscillator. The classical square well correctly predicts $s(\phi)$ for most systems, and the barrier and quantum oscillator models predict decreases in $s(\phi)$ with increasing ϕ, with suitable choices of parameters.

whether lack of quantitative agreement results from incorrect mechanisms or the simplifications used in developing workable models. We have also considered energy transfer involving electronic excitation of metals and internal modes of the gas molecule. These models cannot be formulated in sufficient detail to permit comparison with experimental results. Electronic excitations could result in decreased $s(\phi)$ with increasing ϕ since the transition probabilities should be smaller if the distance of closest approach to the metal decreases.

REFERENCES

1. Schmidt, L. D., in *Adsorption—Desorption Phenomena*, F. Ricca (Ed.), Academic Press, London—New York (1972), p. 391; *Catalysis Rev.*, **9**, 113 (1974).
2. Tamm, P. W., and Schmidt, L. D., *J. Chem. Phys.*, **54**, 4775 (1971); **55**, 4253 (1971).
3. Madey, T. E., *Surface Sci.*, **33**, 355 (1972).
4. Madey, T. E., *Surface Sci.*, **36**, 281 (1973).
5. Barford, B. D., and Rye, R. R., *J. Chem. Phys.*, **60**, 1046 (1974).
6. Mahnig, M., and Schmidt, L. D., *Z. Phys. Chem.*, **80**, 71 (1972).
7. Ko, S. M., and Schmidt, L. D., *Surface Sci.*, **47**, 557 (1975).
8. Ko, S. M., Steinbruchel, C. S., and Schmidt, L. D., *Surface Sci.*, **43**, 521 (1974).
9. Clavenna, L. R., and Schmidt, L. D., *Surface Sci.*, **22**, 365 (1970).

---

*Partially supported by NSF under Grant No. GK16241.

10. Tamm, P. W., and Schmidt, L. D., *Surface Sci.,* **26,** 286 (1971).
11. Clavenna, L. R., and Schmidt, L. D., *Surface Sci.,* **31,** 11 (1972).
12. Kohrt, C., and Gomer, R., *Surface Sci.,* **24,** 77 (1971); **40,** 71 (1973).
13. Kohrt, C., and Gomer, R., *J. Chem. Phys.,* **52,** 3283 (1970).
14. Tracy, J. C., and Blakeley, J. M., *Surface Sci.,* **15,** 257 (1969).
15. Hopkins, B. J., Williams, C. B., and Wilmer, P. C., *Surface Sci.,* **25,** 663 (1971).
16. Anderson, J., and Estrup, P. I., *Surface Sci.,* **9,** 463 (1968).
17. Sanders, G., and Schmidt, L. D., to be published.
18. Bonzel, H. P., and Ku, R., *Surface Sci.,* **40,** 85 (1973).
19. Weinberg, W. H., et al., *Surface Sci.,* **30,** 299 (1972).
20. Steinbruchel, C. S., and Schmidt, L. D., *Phys. Rev. Letters,* **32,** 594 (1974); *Phys. Rev.,* **B10,** 4209, 4215 (1974).
21. Taylor, J. B., and Langmuir, I., *Phys. Rev.,* **44,** 423 (1933).
22. Becker, J. A., and Hartman, C. D., *J. Phys. Chem.,* **57,** 157 (1953).
23. Ehrlich, G., *J. Phys. Chem. Solids,* **5,** 47 (1958).
24. Kisliuk, P., *J. Phys. Chem. Solids,* **3,** 95 (1957); **5,** 78 (1958).
25. McCarroll, B., and Ehrlich, G., *J. Chem. Phys.,* **38,** 523 (1963).
26. Zwanzig, R. W., *J. Chem. Phys.,* **32,** 1173 (1960).
27. Goodman, F. O., *Surface Sci.,* **11,** 283 (1968); *J. Chem. Phys.,* **50,** 3855 (1969).
28. Armand, G., *Surface Sci.,* **9,** 145 (1968).
29. Madix, R. J., and Koros, R. A., *J. Phys. Chem. Solids,* **29,** 1531 (1968).
30. Steinbruchel, C. S., and Schmidt, L. D., *J. Phys. Chem. Solids,* **34,** 1379 (1973).
31. Logan, R. M., and Stickney, R. E., *J. Chem. Phys.,* **44,** 195 (1966).
32. Logan, R. M., and Keck, J. C., *J. Chem. Phys.,* **49,** 860 (1968).
33. Weinberg, W. H., and Merrill, R. P., *J. Vacuum Sci. Technol.,* **8,** 718 (1971).

DISCUSSION on Paper by L. D. Schmidt

KOHN: It seems to me that your model with a single barrier does not allow for a precursor-bound state. Is that correct?

SCHMIDT: That is correct. This would apply to the case where the incident molecule impinges on the barrier without first becoming accommodated in the precursor state. The situation with accommodation could easily be treated, but this would introduce another empirical parameter.

KOHN: I find it hard to accept your suggestion that there is effectively no continuous lattice spectrum because of the short impact time during the process of accommodation. Certainly in a Mössbauer emission, which involves even shorter times, the continuous phonon spectrum is observed.

SCHMIDT: I believe that the fact that the collision time is short compared with the lattice vibrational period requires that equilibrium phonon descriptions are invalid, although the details of this process are complicated. Low-energy phonons require motion of too many atoms to effect the energy transfer from the gas molecule (but these will be involved in subsequent energy equilibration). We believe that our models of an un-coupled surface atom and a surface oscillator coupled to a rigid lattice are reasonable ones, but certainly the problem should be examined in more detail. At least there appears to be a minimum energy which can be ex-cited. The Mössbauer problem should be quite different because of the much higher energy involved, and because of the much longer times (10^{-9} second versus 10^{-13} second) in condensation.

ERTL: Your model used to explain the angular variation of the sticking coefficient contains a small activation barrier. This, however, would give rise to an increase in the sticking coefficient with temperature, which in fact is not observed.

SCHMIDT: That is correct. This is a major objection to the barrier model as applied to our H_2 on W(110) data.

AGENDA DISCUSSION: KINETICS AND TRANSPORT

*Gert Ehrlich**

University of Illinois
Urbana, Illnois

*Richard A. Craig***

Battelle-Columbus
Columbus, Ohio

1 INTRODUCTION

In order to couple closely with the papers presented during the morning session as well as to allow additional time for discussion, which had been quite limited, the agenda was confined to a few principal topics. These were explored with specific reference to the papers presented earlier in the day. A summary of the discussion follows. Although the chronological sequence has not been maintained, an attempt has been made to preserve the flavor of

*Chairman.
**Secretary.

the proceedings, with only minor additions, such as references and occasional summaries.

2 THE ROLE OF DEFECTS IN SURFACE REACTIONS

To initiate the discussion, the Chairman proposed to examine the experimental evidence for the effect of lattice steps upon adsorption and reaction at surfaces. This could be a very important effect, but Somorjai's reports on hydrogen adsorption and exchange [*Phys. Rev. Letters,* **30**, 1202 (1973)] raised interesting questions. The Chairman observed that under the conditions of the experiments, the diffusion distance of hydrogen adatoms amounted to hundreds of lattice spacings. As such, the entire surface could be rapidly populated from steps accidentally present on an otherwise perfect surface, even if molecules were able to dissociate only at steps. If dissociation at the steps were the limiting process, then it appeared that minor changes in the terrace width should not have much of an effect.

Somorjai stated that up to 600°K, (111) faces of platinum with 5- and 8-atom wide terraces were equally reactive; at higher temperatures, the surface with the higher density of steps yielded about twice as much HD as that with the lower density. In the lower temperature regime, the reaction was governed by a Langmuir-Hinshelwood process with an activation energy of 4.5 kcal/mole. Only the flat (111) was unreactive. He emphasized the advantage of molecular beam techniques over other methods which average the product over more than 10^{-2} second; with these methods even the unstepped (111) might show some HD. Somorjai also stressed that his results should not be interpreted as suggesting that all stepped surfaces would be as active as those of platinum. He indicated that similar studies on iridium had not yet been completed, but that stepped gold surfaces showed poor activity in chemisorption compared with platinum; however, oxygen formed an ordered structure on gold above 775°K, the same structure as observed for water layers under similar conditions.

Ertl inquired about the heat of adsorption for hydrogen on platinum, and about the concentration of adsorbed hydrogen atoms under the conditions of Somorjai's experiments. Schmidt indicated that the heat of adsorption had been measured by Rye; it was ≈ 17 kcal/mole on the (111) plane and seemingly higher on others. Ertl suggested that such variations would be significant in the interpretation of Somorjai's experiments. He also expressed doubt that a first-order dependence of reaction rate on pressure necessarily implied a Langmuir-Hinshelwood reaction. In Somorjai's experiments, there was a great excess of hydrogen over deuterium. A deuterium molecule from the beam that struck the surface and dissociated would therefore have little chance of recombining with another deuterium molecule before reacting with an adsorbed hydrogen atom. Under these conditions, the rate of HD formation by recombination in the adsorbed layer would also be first order in the deuterium pressure. Ertl wondered how this possibility had been ruled out in the experiments. Somorjai felt that reaction via surface recombination would go as $p^{1/2}$, but indicated he would consider Ertl's suggestion.

Schmidt also offered an alternative interpretation of Somorjai's experiments. Rye at Cornell [K. E. Lu and R. R. Rye, *Surface Sci.,* **45**, 677 (1974)] measured both heats of desorption and sticking coefficients for hydrogen on different planes of platinum. The sticking coefficient was small on the (111) plane, but the binding energy also varied quite a bit from plane to plane, and seemed lowest on the (111). He inferred from these results that the lower reactivity of flat surfaces in H_2-D_2 exchange might have been due to a lower hydrogen coverage on the (111) planes rather than to a low sticking coefficient.

Somorjai indicated that their measurements of the angular distribution could perhaps be interpreted as showing increased energy transfer or increased residence time on the step surfaces. However, the phase shifts observed experimentally were too small for quantitative determination of the residence time. He agreed there could be an accommodation problem, but this did not follow from the data available.

Johnson inquired whether the possibility of hydrogen diffusing into the metal had been taken into consideration, either for platinum or possibly for palladium. Ertl replied that diffusion into the bulk occurred quite readily for palladium. Adsorbed and absorbed material could be easily distinguished—flash desorption gave two traces, one from the bulk and one from the surface. Work-function measurements also were helpful, because hydrogen in the bulk did not affect the work function. It was of interest that the heat of solution was much lower than the heat of adsorption at the surface. Ertl reported that in his laboratory, such measurements had been made on the (111) and on stepped surfaces [H. Conrad, G. Ertl, and E. E. Latta, *Surface Sci.,* **41**, 435 (1974)]. The heat of adsorption of hydrogen at low coverages, at which steps were just beginning to fill, was about 3 kcal/mole higher than on the (111) plane. This indicated clearly that the binding energy at the steps was higher than on the flat. Similarly, the maximum work-function changes were ~ 20 percent higher on the stepped surface than on the flat. With carbon monoxide, however, no special effect at all was detected for steps, so there seemed to be a considerable chemical specificity operating.

At this point, Boudart expressed surprise at the lack of reactivity of Somorjai's single crystals. In studies of H_2-D_2 exchange over supported platinum, rapid reaction had been found even at liquid-nitrogen temperatures, with turnover times on the order of 1 per second. Boudart wondered whether this shocking difference in activities might stem from a lack of thermal accommodation in the beam experiments. Menzel rose to discuss the role of surface structure in thermal accommodation. He quoted experiments by Messer and Molière [*Z. Angew. Phys.*, **20**, 481 (1966)] which suggested that thermal accommodation was sensitive to the detailed surface structure. Menzel indicated that these experiments were probably influenced by surface contamination. In his own laboratory, thermal accommodation of rare gases had been studied on tungsten filaments, the pretreatment temperature of which had been varied between 1500 and 3000°K [J. Kouptsidis and D. Menzel, *Ber. Bunsenges.*, **71**, 725 (1967)]. This work showed no change in the accommodation coefficient, despite the fact that quenching a filament from 3000°K produces quite a different number of surface defects and steps than does quenching from 1500°K. Menzel concluded from these observations that the surface structure did not affect thermal accommodation of rare gases.

Plummer than proposed that experiments could be constructed to test the role of structural defects in promoting the dissociation of hydrogen into atoms. One possibility was that dissociation actually occurred at individual adatoms, and that the steps just acted as source for such adatoms. To test this, one could deposit metal atoms on an otherwise perfect (111) plane of platinum, and could then check whether this promoted H_2-D_2 exchange. Bassett interjected, however, that at other than cryogenic temperatures this would not be practical. The atoms would just diffuse over the surface until they coalesced into islands, resulting only in an increase in step density.

In a different vein, Grimley pointed out that for a theoretical understanding of heterogeneous reactions, H_2-D_2 exchange was extremely important. It was the one reaction upon which theoreticians would first have to cut their teeth. He therefore inquired whether any information was available about the internal state of hydrogen in Somorjai's incident beam, about the internal state of HD in the reacted beam, or about the dependence of the reaction on the angle of incidence of the beam.

Somorjai's response was negative. His experiments were done using a multichannel source, with a typical distorted cosine distribution. For the internal states, a Boltzmann distribution was expected. He indicated that the angular dependence of the reaction rate had not been measured. He and his co-workers had examined the dependence of the reaction rate on beam temperature in order to look for activation energies and found none. No information was available about the rotational states of HD molecules coming off.

Emmett indicated that Somorjai's results could be interpreted in quite a reasonable way. In the high-temperature range, for which there was no temperature dependence and the pressure was 10^{-8} mm, it was possible that the only hydrogen on the surface was in the crevices where steps met the top surface. If these crevices were pretty well filled with hydrogen and the deuterium from the beam had to strike the hydrogen, then the reaction would be independent of temperature; it would also be directly proportional to the pressure of the deuterium in the beam, and proportional to the number of steps. In the low-temperature region, one still had to assume that steps had something to do with the reaction, but there could be more hydrogen out on the flat surfaces coming from the crevice to begin with. There might be some remaining questions, but the general idea of hydrogen atoms on the surface being struck by deuterium and forming HD seemed reasonable.

The Chairman agreed that this was a reasonable view, but pointed to the need for firm experimental information to substantiate the model. Yates raised the possibility of molecular hydrogen being present on the platinum (111). He quoted results of Weinberg [W. H. Weinberg and R. P. Merrill, *Surface Sci.*, 33, 493 (1972)] who had made bond energy—bond order calculations for hydrogen on this plane. These indicated that in addition to an atomic state, a molecular state might be present, with a binding energy of ~ 17 kcal/mole. Yates inquired whether there was any experimental evidence for molecular hydrogen on this plane. Fischer stated that he had attempted to observe hydrogen on the (100) plane of platinum in photoemission experiments. At pressures as high as 10^{-6} Torr, he had been unable to find any evidence whatsoever for hydrogen, either atomic or molecular.

In response to this, Rhodin reported on studies at Cornell in which hydrogen adsorption on the low-index planes of platinum had been probed by means of field emission electron tunneling [N. J. Dionne and T. N. Rhodin, *Phys. Rev. Letters*, 32, 1311 (1974)]. The electron tunneling spectra were found to be extremely sensitive to any perturbation of the electronic structure at the surface by adsorption, and three conclusions could be drawn immediately: field evaporation produced clean low-index faces of platinum; these surfaces contaminated rapidly at liquid-nitrogen temperature unless extreme precautions were taken; and hydrogen chemisorbed on the low-index faces of platinum at liquid-nitrogen temperature. There was no doubt that hydrogen chemisorbed on the clean (111) face, but at these low temperatures, chemisorption was probably not dissociative.

The Chairman at this point interjected that one of the big problems in surface studies was the characterization of the structure. The presence of fine-scale imperfections could have a big effect on reactions but was difficult to establish. Nevertheless, this was important, both to provide a firm base for the interpretation of experiments in the laboratory, and for the eventual transfer to practical catalysis.

In this regard, Oudar mentioned work his group had done about 10 years ago to characterize the specific activity of steps on the (100) plane of copper for H_2-H_2S mixtures. Adsorption occurred most rapidly on stepped planes. By LEED techniques, Perdereau and Rhead [*Surface Sci.*, 24, 555 (1971)] deduced that there was selective adsorption on the steps. The mean width between steps determined from adsorption experiments was confirmed by LEED. In general, Oudar felt that LEED techniques could establish the connection between the single-crystal low-index planes of the laboratory and the complex surfaces characteristic of real catalytic processes.

Bassett suggested that field-ion microscopy had great potential for characterizing the microstructure of metal surfaces under catalytic conditions. For example, elaborate step structures were observed by field ion microscopy on tungsten surfaces on which hydrocarbon adsorption had taken place at elevated temperatures. Most interesting from the point of view of catalysis was the stability of these carbon-contaminated surfaces. Recently, Drechsler [M. Pichaud and M. Drechsler, *Surface Sci.,* **32**, 341 (1972)] had measured enormously high activation energies for metal atoms diffusing on surfaces contaminated by carbon. There clearly was an important effect of these carbonaceous residues on the stability of step structures. Bassett pointed out that in the field ion microscope these details could be seen and their contributions to surface processes readily characterized. The Chairman added that Bassett's earlier FIM investigations [*Proc. Roy. Soc. (London),* **A286**, 191 (1965)] were quite pertinent as well. In these, the effect of temperature on lattice steps had been studied. For tungsten such steps remained straight up to $\sim 650°$ K, and only then did kinking occur. If one extrapolated these results to materials less refractory than tungsten, surfaces with straight lattice steps were unlikely except at very low temperatures.

The Chairman closed this aspect of the discussion. He noted that in studying the role of defects it was crucial to be able to ascertain their presence or absence on the surface under examination. R. S. Polizzotti at Illinois had been able to devise techniques for establishing the perfection of low-index planes in the field emission microscope. On rhodium, a metal closely related to platinum, Polizzotti (CSL Report R-646, University of Illinois-Urbana, April, 1974) found that chemisorption of hydrogen occurred on the perfect (111) plane at rates comparable to those on rougher planes. There was in these experiments, done at temperatures from 300 down to 80° K, no indication of preferred dissociation at lattice steps. The experimental situation on the role of defects in reactions on the noble metals was thus not clear, and further work to resolve some of the conflicts on platinum would be desirable.

Bond then appealed to theorists for help in understanding the rate of molecular dissociation at flats and steps. Was there any indication from theory that H_2 had difficulty in dissociating on crystal flats, and if so, how did steps assist this process? In reply, Kohn indicated that his calculations of schematized hydrogen atoms located near a highly schematized metal surface showed that dissociation of hydrogen molecules would occur on a perfect surface. Reactivity at steps had not yet been studied theoretically. However, at Irvine, Maradudin, together with a student, was doing a calculation using the density functional formalism to determine the electronic structure near a simple corner. Such calculations were relatively easy, Kohn pointed out, and should lead to some insight into the problem. Smith added that in their approximate calculations, in which H_2 dissociated on a metal modeled on tungsten, one could think of chemisorption in terms of the Hellman-Feynman theorem. According to this theorem, the net force on one of the protons in H_2 is given by the electric field due to all the other electrons and ions in the system. For dissociation to occur, the electronic charge had to be spread out parallel to the surface and had to be diminshed between the protons. In the one-dimensional, self-consistent, field treatment of this problem, this was precisely what happened. The calculations had not yet been done for steps, however.

Stimulated by these remarks, Weber showed pictures of the charge density in the proximity of a step on platinum (100). Charge densities from 8,000 atoms, obtained from relativistic, self-consistent, field calculations, were superimposed to obtain the potentials. It was of interest that the potential surfaces were not really very flat on the scale of energies familiar to chemists. Messmer, however, reminded the audience that Kohn had already pointed out how important it was to get the surface potential self-consistently. Just superimposing atomic potentials could be misleading, as this would fail to account for charge rearrangements that could further change the potentials.

Grimley added that any theoretician with any sort of a model for chemisorption was in a position to study chemisorption at steps and other microstructures; that is, he would be able to produce a potential surface. To go on from there to discuss reactive collisions in such an environment was a problem an order of magnitude harder than just to study chemisorption.

In closing this part of the discussion, the Chairman noted that although the theory was not yet in hand, clear-cut experimental evidence about the lack of reactivity of flats was in fact available. In his thesis at Illinois, Polizzotti had studied not only rhodium, on which there was no indication of preferred reactivity at steps compared to flats, but had also examined tungsten. On the perfect (110) plane of tungsten, hydrogen did not dissociate. This plane could only fill up with hydrogen atoms brought in by diffusion from a rough surface, where dissociation occurred, presumably at steps. The differences in the behavior of rhodium and tungsten emphasized the importance of chemical effects in dictating the relative reactivity of flats and steps.

3 HYDROCARBON REACTIONS

The discussion was turned toward different surface reactions by Kemball. He emphasized the importance of exchange reactions other than H_2-D_2, stressing the exchange of hydrocarbons such as propane, butane, or heptane with deuterium. A great deal could be gained by learning whether the molecule had undergone multiple exchange on reaction, and by determining the pattern of products formed. Kemball indicated that such experiments could illuminate the nature of the intermediates on the surface, the reactivity of those intermediates, and hence their part in mechanisms. Exchange reactions of hydrocarbons on various faces of platinum could yield important information about rates of molecular dissociation under conditions of coverage interesting for catalytic reaction—that is, about reversible rates of adsorption on surfaces covered by substantial amounts of gas.

Kemball also recalled studies, done in his group and in that of Bond, which indicated that for the exchange of molecules like propane and butane, the form of the catalyst was unimportant. It did not seem to matter whether a supported platinum catalyst was used, or whether platinum was in the form of an evaporated film, albeit somewhat dirty (for these experiments were done under a conventional vacuum). There was an unmistakable pattern of multiple exchange products characteristic of platinum. Even in reactions in which the poisoning of the surface was observed by laying down of hydrocarbon species, the exchange reactions still maintained the same characteristic pattern associated with platinum, suggesting some unique chemical specificity.

Block reported that he had been able in field emission experiments to obtain patterns characteristic of individual hydrocarbons; these patterns could be used like fingerprints to identify surface processes. He indicated that for a (100)-oriented platinum emitter exposed for some time to n-butane, his group had found hydrocarbon residues on higher index planes with a molecular weight up to 1000 or more.

Fischer raised a question about the role of carbonaceous overlayers, which had been reported necessary in order for n-heptane to rearrange to toluene on platinum. In the refinery this reaction was carried out on a supported platinum catalyst, a so-called bifunctional catalyst. Hydrogen removal from the n-heptane occurred on the platinum; the remainder of the molecule then moved over to the acidic silica alumina support, where it was ionized and underwent skeletal rearrangement. It seemed possible that the carbonaceous overlayer on a platinum crystal might just play the role of the acidic insulator in the industrial catalyst.

Mason was surprised by Somorjai's results on chemisorption without dissociation of allenes and polycyclic hydrocarbons on the (111) and (100) faces of platinum. It would be worthwhile, he remarked, to see whether the kind of calculation successful for three-dimensional crystals would be capable of predicting the structure of the two-dimensional layers. Furthermore, it could be interesting to do classical crystallographic experiments, involving isomorphous replacement, in LEED studies of compounds like bromobenzene, chlorobenzene, and bromonaphthalene. Questions of resolution parallel to the surface would not enter for adsorbates like para-dihalcbenzenes, as the heavy atoms were about 6.5 Å apart and ought to be quite easily seen. Mason emphasized the promising physics and chemistry of the systems Somorjai had investigated. Many people had been concerned with electron transfer and semiconductivity in molecular crystals, and with the question of exciton transfer. The general interest in generating organic superconductors would also apply *prima facie* to the type of system Somorjai was investigating.

Kemball pointed to experiments he had performed some time ago, us-ing evaporated platinum films of dubious cleanliness, which were quite effective in catalyzing reactions of aromatic molecules at room temperature. This, he argued, made it doubtful that one would get layers of aromatic molecules as suggested by Mason. Aromatics with substituted halogens were extremely reactive, and decomposed at quite low temperatures on all sorts of catalysts. Mason agreed with Kemball, but remarked that Somorjai had not found changes in LEED pattern until the sample was heated to 250°C. Somorjai confirmed Mason's comment, and indicated that Gland, in his laboratory, had studied LEED patterns and work-function changes for a series of 55 organic molecules on the (111) and (100) faces of platinum. Naphthalene and benzene showed no trace of decomposition until the platinum was heated above 200°C; on the other hand, cyclohexane did show a very slow dehydrogenation.

Yates recalled that Demuth and Eastman [*Phys. Rev. Letters,* **32,** 1123 (1974)] had done photoemission experiments on benzene adsorbed on Ni(111) and found evidence for perturbation of π electrons in the adsorbed species. He suggested chemical studies, exchange studies such as Kemball mentioned, as supplements for spectroscopic methods, in order to determine once and for all whether or not undissociated aromatic molecules existed.

In regard to Mason's comment about aromatics chemisorbed on a metal surface, Johnson pointed out the existence of classes of layer-like compounds, for example, metal thiocyanates, with stable metal atoms at the center. These compounds existed for many metals in the periodic table, not just the transition metals, and polymerized. Johnson recalled that Davis (MIT) had made two-dimensional polymers consisting of arrays of transition-metal atoms surrounded by these organic ring systems. Johnson thought it interesting to look for a connection between the coordination of a transition-metal ion in a ring system and the coordination of a ring system with a transition-metal surface of a three-dimensional solid. "After all, coor-dination chemistry is coordination chemistry", Johnson said, "and the same fundamental principles underlie all these things".

4 FACTORS AFFECTING THE CHEMICAL REACTIVITY OF METALS

The Chairman raised the question of why metal surfaces were reactive and useful in catalytic processes. Did this come about because of their ability to break chemical bonds, their efficiency in energy transfer, or some other factor?

Messmer indicated that coordination chemistry might help here. From coordination chemistry, one knew that different coordinations were possible around metal atoms, and that in organo-metallic compounds, ligands were quite labile. He suggested that perhaps these features of transition metals made them good catalysts as opposed to more covalent compounds, where bonding was stronger and more directional. Johnson reinforced Messmer's remarks that transition metals form stable compounds with ligands in which the latter were labile. He felt that it was for these reasons that transition metals made the best catalysts, and pointed out that they also performed catalytic functions in enzymes. Their function there was better understood than their function in the refinery. Johnson suggested that we might take a hint from what was known about metal enzymes; the important properties there were lability and stability.

Bond pointed out that catalysis on metals was really quite a simple and straightforward matter. All the surface had to do was to chemisorb the reactants in the right way with the right energy. Transition metals, specifically the group VIII transition metals, did this par excellence. They had the facility to dissociate molecules like hydrogen, which are rather difficult to dissociate by other means; having performed that act, the transition metals bound the chemisorbed species, but not too strongly, so that they were still reactive. That is what catalysis by metals really is: chemisorbing reactants in the right way and with the right energy.

After Bond had so succinctly stated the problem, Schmidt pointed out that metals were not necessarily the best catalysts. In terms of the dollar amount of catalysts used and the chemical magic that they performed, oxides were really preeminent. Metals did only the easy things, such as chemisorption and recombination of fragments. Complicated reactions, such as partial oxidation and dehydrocyclization occurred on oxides, and therefore, oxides really had considerably more capability than the metals.

Mason observed that in homogeneous catalysis the emphasis in past years had been on the generation of electron-rich species. That is why, on the one hand, there had been emphasis on the early transition elements, such as titanium complexes, and, on the other hand, on rhodium and platinum zero complexes. These were known to activate organic ligands in the sense that when acetylene or ethylene was put on platinum, carbonium ions were strongly stabilized. With that, the scene was set for cyclic ligandization reactions. The emphasis in homogeneous catalysis had been on complexes with those elements whose oxidation states could be changed quite easily. Platinum, for example, had a formal charge of zero, but this could change to platinum +2. There obviously was a built-in flexibility of electron transfer. Mason suggested that a simple answer to the question of why metals, particularly transition metals, were good catalysts might be that in transition metal complexes one could generate a variety of electronic conditions and this variety of conditions could be supplemented by the incoming ligands. Mason also noted the importance of the template: two molecules that had been modified electronically by the transition metal might be brought into closer chemical proximity for the right reaction.

Mason also mentioned recent work by the Bradford group as being interesting in this connection. They had studied carbon monoxide adsorption on the transition series by ESCA and photoelectron spectroscopy. For molybdenum at low temperatures, CO was associatively chemisorbed, as evidenced by the ESCA carbon and oxygen spectra; at room temperature it was dissociated. In contrast, on nickel, the chemisorption of CO at room temperature was associative; it dissociated above 300°C. One could also look at the stretching frequency for adsorbed carbon monoxide; on going through the transition series, this frequency dropped. This trend was no mystery; as the π bonding into carbon monoxide increased, the CO stretching frequency decreased and the CO bond strength increased. It had been found at Bradford that in going across the transition series, the oxygen 1s and carbon 1s ESCA binding energy increased just prior to dissociation. On copper, for example, the 1s level was close to that of carbon monoxide in the gas. For molybdenum (on which CO dissociates), it was 1.5 eV higher. The argument here was that dissociation of CO happened because, on moving across the transition series, the metal increasingly transferred charge; the strength of the metal-carbon bond was increased at the expense of the carbon-oxygen bond.

At this point, the Chairman posed the question whether dissociation on metals was in fact always rapid or whether there were simple molecules for which adsorption was limited by an activation barrier? That is, were there well-documented examples of the old concept of activated adsorption so important in the days of H. S. Taylor.

Johnson cited an example in both homogeneous and heterogeneous catalysis. Zeise's anion will, if illuminated with photons having energies of 3.5 to 4 eV, undergo photoaquation: an ethylene molecule is given off and replaced by water. The interpretation of this exchange was that the photons induce an electron transfer from a deep level, associated with the platinum, to an antibonding level between the platinum d orbitals and the ethylene π level. This excitation destabilizes the bond and allows the ethylene to go off, to be subsequently replaced by a water molecule. A similar thing happens on a platinum surface. Allen Dromowitz, a student at MIT a few years ago, observed that illumination by photons with energies between 3.5 and 4 eV dropped the catalytic activity for the hydrogenation of ethylene on platinum. The interpretation was that ethylene was forced off the surface under illumination. Once again simple model calculations similar to those reported by Messmer for ethylene interacting with platinum allow one to interpret this photochemical effect in the same way as the aquation of Zeise's anion, that is, as an excitation from d levels to antibonding platinum ethylene levels.

Kemball suggested that ethylene was much too complicated and that methane would be a far more suitable choice for looking at adsorption. Ethylene could be chemisorbed, it could be dehydrogenated, it could polymerize, it could also be hydrogenated; consequently, it was not an especially good test molecule. Methane, on the other hand, had a strong carbon-hydrogen bond, stronger than that in the higher hydrocarbons. In fact, an activation energy had been found for its dissociative chemisorption; adsorption of the higher hydrocarbons did not involve as large an activation energy. Kemball suggested that for finding out when a metal was really good at handling hydrocarbons one should first try it on methane.

This point of view was seconded by the Chairman, who pointed out that Charles Stewart at Illinois had carried out a detailed study of the dissociation of methane on clean rhodium surfaces (CSL Report R-638, University of Illinois-Urbana, December, 1973). Stewart had established that it was indeed an activated process and that the barrier could be overcome just by heating the gas. Methane was a simple molecule that could be studied by various powerful techniques, yet provided a nice introduction to hydrocarbon chemistry.

Yates then described recent experiments at the NBS to test dissociation at surfaces. They had used ESCA and flash desorption methods to look at nitrogen, nitric oxide, and carbon monoxide adsorbed on various crystal faces of tungsten. For all three gases, they found distinct states at low temperatures—in the case of CO, the virgin state, in the case of nitrogen, the γ state—all fairly weakly bonded; upon heating, these converted to more stable states on the surface. For γ nitrogen, Ehrlich had shown many years ago that there was the possibility of competition for sites between the dissociated β nitrogen and the molecular γ state [*J. Chem. Phys.*, **34**, 29 (1961)]. Yates stated that there was evidence from a variety of experimental methods that this type of activated surface conversion occurred. The question of the primary act of chemisorption being activated was still undecided.

Emmett noted a number of systems in which chemisorption did not occur at all; nitrogen, for example, did not chemisorb on copper or platinum. In his view, saying that this was because there were no nitrides of copper or platinum was begging the question. It was of interest that there were numerous materials with which nitrogen was almost inert; on others, such as uranium and vanadium, nitrogen chemisorbed too tightly for catalytic purposes.

The Chairman observed that such effects could be governed either by kinetics or thermodynamics. From the limited data available, it appeared that with nitrogen on copper and on platinum, thermodynamics prevented chemisorption. For dissociative chemisorption, thermodynamics required a decrease in free energy. For that, the bond energy between nitrogen and the metal had to equal at least one-half the dissociation energy of the molecule [G. Ehrlich, *J. Chem. Phys.*, **31**, 1111 (1959)]. There was no indication that this requirement was in fact met for copper and platinum interacting with nitrogen. By way of confirmation, Ertl recalled that Lee and Farnsworth [*Surface Sci.*, **3**, 461 (1965)] had observed no interaction between molecular nitrogen and copper; when they dissociated the nitrogen into atoms, however, they found tightly bound species. Smith added that Burkstrand at GM was presently doing a LEED analysis of nitrogen on copper.

Grimley inquired about the experimental situation. In principle, at least, theoreticians knew what to do to discover an activation barrier. He asked if the experimenters knew how to perform experiments to confirm or deny such predictions.

Schmidt indicated that an energy barrier would be evidenced by an increase in sticking coefficient with increasing gas temperature. He stated that only a few such measurements had been made as these were difficult experiments. The two studies cited in his presentation showed the sticking coefficient to be decreasing with increasing temperature. Such experiments should be carefully reviewed, because these kinetics probably involved thermal accommodation followed by chemisorption: the chemisorption step could still have some activation energy. The chairman added that there were really two types of experiments that gave information on whether adsorption was activated. The simplest was done by exposing a surface to a gas and observing the rate of adsorption at one temperature. The temperature of both gas and solid was then raised and the change in the rate of reaction with temperature yielded the activation energy. The next stage of sophistication was to resort to molecular beams, and to change the temperature of the solid and the gas independently. That was a rather more troublesome experiment, but it could be done and, in fact, had been done. In the first type of experiment, Kemball [*Proc. Roy. Soc. (London)*, **A207**, 539 (1951)] long ago found that adsorption of methane on nickel films was an activated process. For the interaction of methane with rhodium, Stewart at Illinois had followed the rate on changing the temperature of the gas alone. He had established where the activation energy entered, as well as the nature of the activation process. Experimenters are on an equivalent footing with the theoreticians: in principle they know how to do the experiments, but by and large prefer not to do the difficult things.

Somorjai indicated that he had looked for a variation in reaction rate with beam temperature in the H_2-D_2 exchange on platinum and had found no activation energy. Such experiments, he noted, were easy once a beam system was available. He cited studies with hydrogen on copper in other laboratories where activation energies had been found. Kemball added that an excellent way for testing activated chemisorption was to study isotope exchange, as had in fact been done for methane with deuterium.

In summary then, although the reasons for the reactivity of metals or the lack thereof were not quantitatively understood, wide variations in activity existed. Despite the generally high ability of metals to dissociate molecules, several simple systems have been clearly identified in which adsorption required an appreciable activation energy. Information about these is still limited; to the extent that hydrogen and hydrocarbons are involved, these processes could be important, not only for understanding adsorption kinetics in general, but for catalytic problems as well.

5 SURFACE TRANSPORT PROCESSES

The discussion moved to the question of surface diffusion, especially of gases on metals. The Chairman asked if this was an area about which enough was known to make sensible predictions. Was the theory understood? Was this a field in which the experimental situation was well in hand? Was it possible to perform calculations with confidence in the results? Were predictions possible about what should happen on going from one material to the next?

Messmer indicated that he believed this to be an important subject. It involved kinetics and potential surfaces. However, these were extremely difficult things to calculate, much more difficult than the study of chemisorption. Calculations of surface diffusion based on first principles were far off, and for the immediate future we would probably have to settle for the use of rather simple models.

Kohn indicated that from the theoretical standpoint, the problem of surface diffusion was twofold: First, one had to know the potential energy surface; then one had to understand the mechanism by which the particle was agitated to move over the surface. Kohn believed that both parts of the problem were simpler for the theoretician than an actual chemical reaction involving interactions between two atoms, and that therefore one might expect progress earlier. On the second aspect, namely, what propelled a particle over the constant energy surface, considerable work was being done by Professor Suhl and his students at La Jolla [E. G. d'Agliano, P. Kumar, W. Schaich, and H. Suhl, *Phys. Rev.*, **B11**, 2122 (1975)]. They were interested in making correlations between the motion of a particle on the surface with the collective modes, electronic as well as lattice modes, that existed within the metal. Professor Nozieres and his students at Grenoble were also doing some work along the same line.

Kohn also mentioned that Professor Gomer at Chicago had devised a novel technique for studying surface diffusion. From the speed of fluctuations in the electron tunnelling current from a tiny portion of a single crystal plane, Professor Gomer was able to deduce the rate of diffusion *Surface Sci.*, **38**, 373 (1973)]. The Chairman noted that it was important to recognize that a great deal had already been learned about the diffusion of gases over metals through the application of field emission methods. Such experiments had been ingeniously carried out by Professor Gomer, and were nicely summarized in his book, *Field Emission and Field Ionization* (Harvard, 1961).

Mason added that inelastic neutron scattering could now provide root-mean-square distances for atoms diffusing over single-crystal surfaces as a function of temperature. Experiments on the motion of hydrogen and deuterium were under way at Grenoble.

Boudart raised a question about a principle of deBoer's, who emphasized that if an adsorbed entity could desorb from a surface, it could undoubtedly move quite freely over the surface. Boudart had the impression that this useful rule had now been violated in some instances and wondered how such deviations should be interpreted.

Grimley indicated that there were at least three levels on which one could consider this question. The first level was to assume that deBoer considered only the energy term, and not the possible preexponential factor in an Arrhenius-like equation. The preexponential which controls desorption could be very different from that for surface diffusion. To address this problem on the second and less trivial level, the potential energy surfaces would have to be known and a trajectory analysis performed to see what happened. Desorption and diffusion were both possible fates of the particle and all one could say on this level was that the activation free energy into a transition state for the exit channel was different from that for the diffusion channel. On the third level, one might observe that to desorb a particle from the surface, a suitable fluctuation in the motions of the substrate was needed. Surface mobility involved a different fluctuation in the motion. The relative probabilities of these fluctuations would determine the preexponential factor governing the validity of deBoer's proposition.

Smith then addressed the question of whether we should expect insulators or metals to have different barriers to surface diffusion. To answer that question, the potential surfaces had to be known. As yet these were not available for either an insulator or a metal, but there were certain facts that could afford insight into the problem. The electronic roughness, the amplitude of the variation in electron density parallel to the surface, was apparently smoother for metals than for insulators. Many years ago, Smoluchowski [*Phys. Rev.,* **60**, 661 (1941)] introduced the concept of spreading and smoothing of the electron density at the surface of a metal. Recent calculations by Appelbaum and Hamann, in which they included the full ion-core potential and went beyond the perturbation approximation, confirmed these ideas: they found that for Na(100), the electron density was indeed quite smooth *Phys. Rev.,* **B6**, 2166 (1972)]. However, when they went on to silicon [*Phys. Rev. Letters,* **32**, 225 (1974)], they found that much to their surprise the spreading and smoothing concept was not at all true. For silicon, the electron density, or equivalently, the potential seen by the electron was rough; that still had to be translated into the potential seen by an atom or molecule, but it could convey some feeling for the situation. Messmer indicated that such calculations had not yet been done for transition metals; there, he suggested, quite a different behavior from that on sodium might be found.

Schmidt observed that if the potential energy surfaces perpendicular and parallel to a solid were known for both atoms and molecules, then one could do everything. However, it seemed obvious from the state of the theory, which was not yet capable of yielding binding energies at adsorption sites, let alone the subtleties of a transition state, that for the foreseeable future we would have to rely on experimental measurements. In view of this, Schmidt suggested that there should be greater emphasis on the theory of energy-exchange processes. The experimentalist was going to need considerable help in the calculation of preexponential terms, that is, in the understanding of the transfer dynamics. Webb commented that a considerable amount of information was becoming available, both from experiment and calculations, on the dynamics of the surface lattice. This should help in understanding the dynamics of atomic events.

Emmett's timely remarks closed the discussion. He observed that, although surface diffusion was important for understanding surface behavior, in real catalysis, transport over the surface was not the limiting factor. Material transfer occurred primarily through the gas phase rather than over the surface, and in the standard discussions of catalytic reactions, transport over the surface was neglected.

Part Six

APPLICATIONS TO CATALYSIS

CLASSICAL METHODS IN CATALYSIS RESEARCH

R. P. Eischens

Texaco Research Center
Beacon, N. Y.

ABSTRACT

Two topics are discussed. The first involves infrared spectra of carbon monoxide on nickel and is related to the question of whether carbon monoxide dissociates on atomically clean nickel. The second is concerned with reduction to metallic nickel on refractory supports. It is concluded that dissociation is not an important factor in the infrared studies and that metallic nickel may be prepared with relative ease on some types of silica but with difficulty on alumina.

1 INTRODUCTION

The preliminary announcement of this symposium implies that the participants are either catalytic chemists or surface physicists. Experimental methods were classified as "classical" or "techniques for studying atomically clean surfaces". It is difficult to define catalytic chemistry. However, for the purpose of this presentation, catalytic chemists are those who are interested in the fundamental aspects of catalysis and who use classical methods in their studies of surface chemistry. In turn, classical methods may be defined as those which utilize vacuum systems in which the gauges cannot read below 10^{-6} Torr.

Surface physicists may not understand why catalytic chemists are content with classical methods when the means to attain better vacuum are available. The attitude of the catalytic chemist toward classical methods is influenced by the assumption that the vacuum system is not a limiting factor in catalysis work. This assumption appears to be valid in almost all catalysis studies. In the petroleum industry, the catalytic chemist is likely to encounter difficulties in situations involving interaction between metal and support which are not related to the vacuum system. A typical problem is illustrated by studies of the reducibility of nickel on different supports.

A second factor is the importance attached to the concept that heterogeneous supported-metal catalysis is simply an aspect of conventional metal-ligand chemistry. Contaminants are studied as potential ligands which might affect the catalytic activity of the metal atom. Thus, the catalytic chemist is interested when high-vacuum work indicates a chemisorption mechanism which is not analogous with known metal-ligand chemistry. Examples of such evidence are rare. One example, to be discussed here, involves the infrared spectrum of carbon monoxide on nickel where some results suggest that carbon monoxide dissociates on atomically clean nickel.

2 CARBON MONOXIDE ON NICKEL

Most infrared studies of adsorption have been carried out in classical vacuum systems. When more rigorous vacuum techniques are used, the results are usually similar to those observed under classical conditions.[1] An apparent exception was encountered in the case of carbon monoxide on nickel. Studies in a high-vacuum system led to the conclusion that carbon monoxide dissociates on atomically clean nickel.[2,3] This conclusion implies that the carbon-oxygen functional groups, detected by infrared, can be found on nickel only when dissociation is retarded by contamination.

Spectra of carbon monoxide chemisorbed on three different types of samples are shown in Fig. 1. Spectrum A is from samples (at 25°C) in which the nickel particles are suspended in oil and are, therefore, covered by a layer of hydrocarbon prior to exposure to carbon monoxide.[4] Spectrum B represents carbon monoxide on metal films (−160°C) prepared under ultrahigh-vacuum conditions and which were atomically clean prior to exposure to carbon monoxide.[2] In this case, bands were not detected during

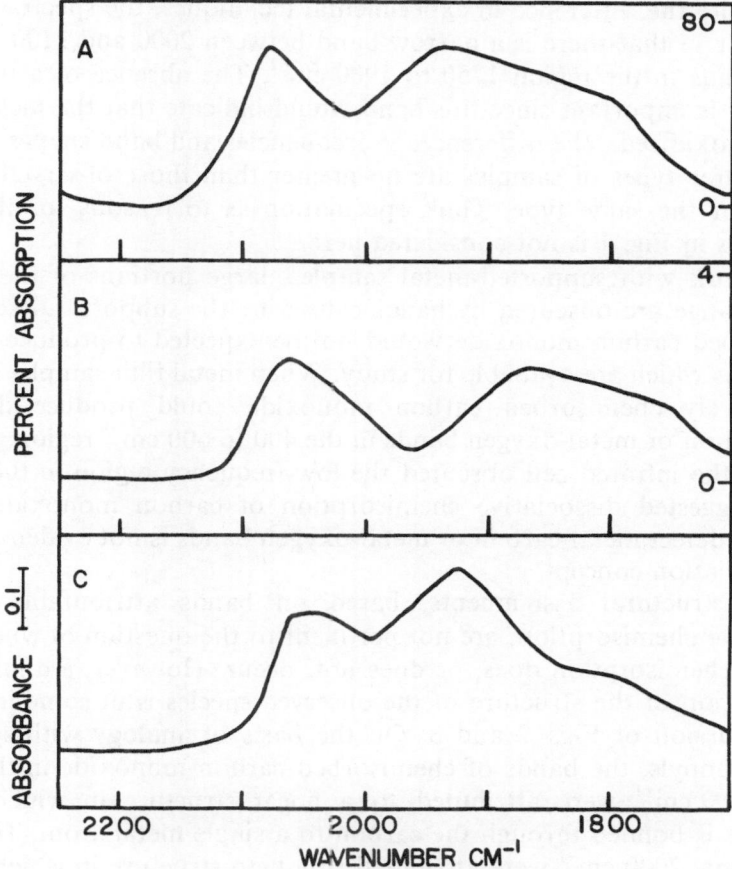

Fig. 1. Carbon monoxide on nickel: A, dispersed in hydrocarbon[4]; B, evaporated film[2]; C, supported on Cabosil.

the initial phases of the dosing while the pressure was increased from 10^{-8} to 10^{-6} Torr. The absence of bands was attributed to the dissociation referred to above. For spectrum C, obtained by previously described methods[5], the sample (30°C) was nickel supported on Cabosil. This type of nickel is presumed to be atomically clean by many catalytic chemists even though this presumption may not be acceptable to surface physicists. This aspect of supported-metal samples is discussed later.

All three spectra represent chemisorbed carbon monoxide at residual pressures of 10^{-5} to 10^{-6} Torr. For spectrum B the experimental pressure was attained by dosing until the pressure increased from the initial 10^{-8} Torr. For spectra A and C the carbon monoxide was initially at a higher pressure (about 0.1 Torr) and was then pumped down. (The spectra in Fig. 1 were redrawn so they could be fitted to a single wave-number scale.)

Despite the difference in experimental techniques, the spectra in Fig. 1 are similar in that there is a narrow band between 2000 and 2100 cm^{-1} and broad bands in the region 1750 to 1900 cm^{-1}. The absence of a band near 2200 cm^{-1} is important since this band would indicate that the nickel atoms had been oxidized. The differences in frequencies and band shapes observed for the three types of samples are no greater than those observed between samples of the same type. Thus, speculation as to reasons for the minor differences in Fig. 1 is not considered here.

In work with supported-metal samples, large portions of the infrared spectral range are obscured by bands caused by the support. Dissociatively chemisorbed carbon monoxide would not be expected to produce bands in the regions which are available for study, When metal film samples are used, dissociatively chemisorbed carbon monoxide could produce detectable metal-carbon or metal-oxygen bands in the 400 to 600 cm^{-1} region. The windows on the infrared cell obscured the low-frequency region in the studies, which suggested dissociative chemisorption of carbon monoxide.[2] Thus, failure to detect metal-carbon or metal-oxygen bands is not evidence against the dissociation concept.

The structural assignments, based on bands attributable to non-dissociative chemisorption, are not pertinent to the question of whether dissociative chemisorption does, or does not, occur. However, a digression to consideration of the structure of the observed species is of some interest to later discussion of Figs. 2 and 3. On the basis of analogy with spectra of metal carbonyls, the bands of chemisorbed carbon monoxide in the region above 2000 cm^{-1} were attributed to a linear structure in which carbon monoxide is bonded through the carbon to a single metal atom. The broad bands below 2000 cm^{-1} were attributed to a keto structure in which the carbon was bridged between two metal atoms. Later carbonyl work showed that in some cases the linear structure could also produce bands below 2000 cm^{-1}. This made it appear that the question of whether the carbon monoxide could be bonded to more than one nickel atom was not resolvable by infrared alone. However, recent infrared work in which the bands below 2000 cm^{-1} decreased as palladium was diluted in Pd-Ag alloys provides strong evidence for the bridge concept.[6] X-ray studies have shown that carbon monoxide can bridge between three metal atoms.[7] This structure produces an infrared band at 1800 cm^{-1}. Carbon monoxide bridged between three or four metal atoms would correspond to chemisorption in an interstitial position. Recent calculations show that the interstitial position would be favored on nickel.[8]

At first thought, it might not be expected that a technique which involves chemisorption on metal particles suspended in oil (spectrum A) would produce spectra similar to those produced on an atomically clean surface. However, the similarity is not unexpected when chemisorption is visualized as simply another aspect of metal-ligand chemistry. Inadvertent

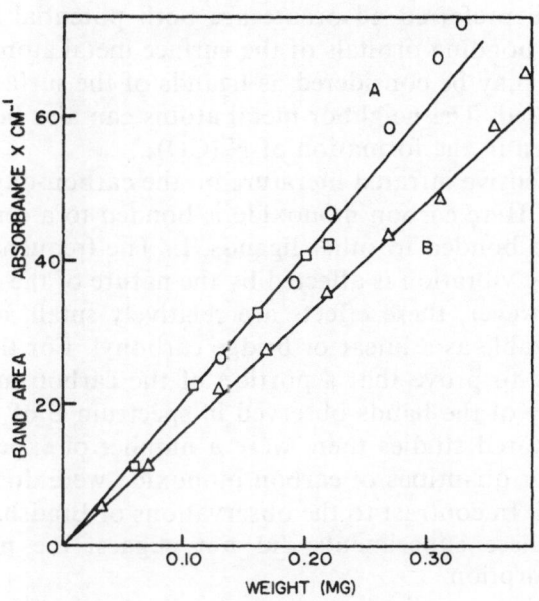

Fig. 2. Total integrated band intensities versus weight of carbon monoxide on nickel-Cabosil.

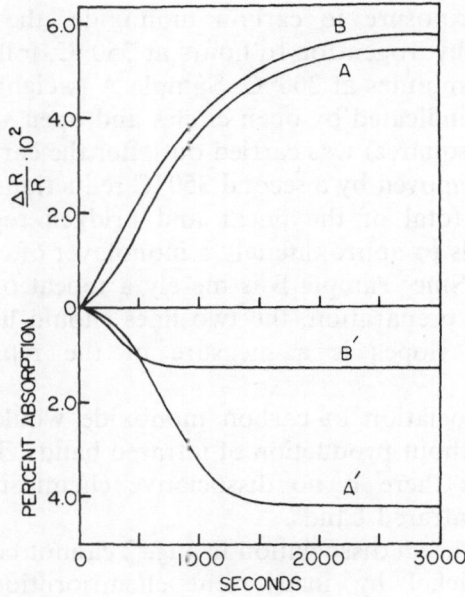

Fig. 3. Carbon monoxide on evaporated nickel film: intensity of bands and resistance of film as function of time of exposure to carbon monoxide.[10]

contaminant and preferred adsorbate are both potential ligands and must compete for the bonding orbitals of the surface metal atom. Even neighboring metal atoms may be considered as ligands of the surface atom on which attention is focused. The neighbor metal atoms can also be displaced by the adsorbate, as seen in the formation of $Ni(CO)_4$.

There is extensive infrared literature on the carbon-oxygen bands of the system L_nMCO. Here carbon monoxide is bonded to a central metal atom, M, which is also bonded to other ligands, L. The frequency of the carbon-oxygen stretching vibration is affected by the nature of the other ligands and by solvents. However, these effects are relatively small, and the spectrum remains recognizable as a linear or bridge carbonyl. For the above reasons, it is not possible to prove that a portion of the carbon monoxide has dissociated by study of the bands observed in spectrum B of Fig. 1.

In early infrared studies there were a number of experiments in which partial monolayer quantities of carbon monoxide were dosed onto Cabosil-supported nickel. In contrast to the observations of Bradshaw and Pritchard (spectrum B)[2], these experiments did not suggest the possibility of dissociative chemisorption.

Infrared-dosing experiments are conveniently carried out in apparatus designed for simultaneous gravimetric and infrared measurements.[9] Fig. 2 shows the integrated band intensities as a function of the weight of carbon monoxide chemisorbed at 30°C. Two sample disks, A and B, were used. These were both nickel on Cabosil, similar to that used for spectrum C of Fig. 1. Prior to exposure to carbon monoxide, the samples had been reduced in flowing hydrogen for 16 hours at 350°C, followed by evacuation at 10^{-6} Torr for 30 minutes at 300°C. Sample A (weight about 100 mg) was used in two trials (indicated by open circles and open squares). The second trial on sample A (squares) was carried out after the carbon monoxide from the first trial was removed by a second 350°C reduction. The bands were integrated over the total of the linear and bridged regions. A weight of 0.35 mg corresponds to approximately a monolayer of carbon monoxide on the nickel surface. Since sample B is merely a repeat of sample A, starting with a new sample preparation, the two lines should have the same slope. The difference in slopes is a measure of the reproducibility of the experiment.

In Fig. 2, dissociation of carbon monoxide would be detected from weight increases without production of infrared bands. The extrapolation to zero indicates that there is no dissociative chemisorption prior to the appearance of the infrared bands.

The failure to detect dissociation in Fig. 2 cannot be attributed to deactivation of the nickel by inadvertent chemisorption of hydrocarbon. Samples prepared under these conditions will avidly chemisorb a monolayer of hydrocarbon (0.2 mg of ethylene) at 30°C. Stratification, in which adsorption is first concentrated on particles near the outer portions of the sample, is also not likely to be a misleading factor in the interpretation of Fig. 2. Stratification should be considered because the lines in Fig. 2 would go through zero if a constant fraction of each dose were dissociated. The possibility of misleading results due to stratification is rejected because gaseous carbon monoxide would tend to go to surfaces which had not been deactivated by a dissociatively adsorbed species.

As discussed earlier, it is not easy to detect contamination by observation of the spectrum of chemisorbed carbon monoxide (except in some cases of contamination by strongly electronegative elements such as oxygen). Thus, the similarity of spectrum B to spectrum A is not evidence for dissociation of carbon monoxide. The conclusion regarding dissociation on atomically clean nickel in spectrum B must be based on the absence of bands after the initial addition of CO.

Nickel was deposited on the walls of the infrared cell during deposition of the nickel films for spectrum B. These walls remain at room temperature while the sample is at −160°C. The nickel on the walls is not necessarily a problem. However, it is a disturbing factor when the conclusion regarding dissociation is based on negative evidence, i.e., the absence of bands. In a subsequent study of carbon monoxide on nickel films, Bradshaw and Vierle measured the conductivity of the film simultaneously with observation of the infrared spectra.[10] This experiment was based on the concept that dissociative adsorption could be detected in the absence of infrared bands because either dissociatively or associatively chemisorbed carbon monoxide would decrease the conductivity of the film. The conductivity measurements would therefore serve a purpose equivalent to the gravimetric measurements of Fig. 2.

In Fig. 3, the change in resistance, $\Delta R/R$, of the film and the intensities of the infrared bands are plotted as a function of time of exposure to 10^{-6} Torr of CO at −170°C. The intensities of the linear species at 2090 cm^{-1}, A′, and the bridged species at 1810 cm^{-1}, B′, were monitored in separate resistance measurements A and B, respectively. It is seen that there are changes in resistance after the infrared bands have reached their maximum. These changes are discussed in detail in the original publication. For present purposes, the significant aspect of Fig. 3 is the absence of changes in resistance prior to appearance of the infrared bands. Thus, the results of the ultrahigh-vacuum experiment illustrated in Fig. 3 are consistent with those shown in Fig. 2. In neither case is there evidence for dissociatively adsorbed carbon monoxide. The difference between the two high-vacuum experiments, Fig. 3 and spectrum B of Fig. 1, awaits further clarification.

The discussion of whether dissociative chemisorption of carbon monoxide occurs on nickel at low temperature has compared classical and high-vacuum techniques in studies of a single system. This system was selected because it is one of the few cases, involving infrared studies, where high-vacuum techniques appeared to give a result different from those obtained by classical methods. Moreover, the difference, if real, might have been attributed to contamination of supported-metal samples and therefore would have implications which extended beyond the question of carbon monoxide dissociation on nickel.

3 SUPPORTED-METAL SAMPLES

The large surface area of the supported metal protects against contamination under classical vacuum conditions. The 100-mg samples used for the work in Fig. 2 have metal areas of about 1 square meter. Samples used for infrared work are small because of special requirements of the infrared methods. Samples which are a hundred, or even a thousand, times larger may be conveniently used in adsorption and catalytic studies. A monolayer of carbon monoxide on the nickel weighs about 0.3 mg for the 100 mg nickel-on-Cabosil sample. At 10^{-6} Torr, a volume of 1 liter of carbon monoxide weighs less than 2×10^{-6} mg. It is apparent that gases at this pressure cannot cause significant contamination unless the pressure is due to an adsorbable gas whose concentration is being replenished by transfer from an essentially inexhaustible source. Such a pseudo-equilibrium might be produced by an external leak. However, this possibility can be excluded by simple static leak tests. More care is required to exclude back diffusion from pumps or inefficient cold traps in which the equilibrium vapor pressure of the potential contaminant is too high. For gravimetric systems, the efficiency of trapping can be determined by observing weight changes as the sample is exposed to the classical vacuum over an extended period of time. The system is satisfactory if weight changes are small and the sample retains the ability to chemisorb potential contaminants such as hydrocarbons or oxygen.

The above discussion should not suggest that the catalytic chemist faces no difficulties in working with supported-metal samples, but instead that the difficulties are associated with the necessary presence of the support and cannot be attributed to inadequacy of classical vacuum systems.

4 REDUCIBILITY OF SUPPORTED NICKEL

A further question is whether metal can be produced if the particles are supported on an oxide. Supported-metal catalysts are often prepared by adding a solution of the metal salt to the support, followed by drying and reduction in hydrogen. The drying step is sometimes severe and involves calcining in air to convert the salt to oxide. Unsupported transition-metal oxides are easily reduced by hydrogen. However, when the transition-metal oxide is associated with an oxide support, the reduction can be difficult. The difficulty may be caused by the formation of a mixed oxide, for example, nickel hydrosilicate in the case of nickel on silica.[11] Bulk mixed oxides have been detected by X-ray. Another theory implies that the trouble is caused by the high degree of dispersion which prevents production of metal nuclei.[12] A high dispersion would imply that many of the transition-metal atoms share oxygens with nonreducible cations (most commonly aluminum or silicon) of the support. Such surface mixed oxides would be expected to be difficult to reduce.

Fig. 4 shows how the nature of the support affects the reducibility of nickel.[13] When Cabosil was used as the support, the reduction was complete at the first determination which came after 6 hours. For alumina, the reduction was only 60 percent complete after 24 hours. Ordinary silica shows essentially complete reduction after about 12 hours. Cabosil is a special form of silica. It is prepared by oxidation of silicon chloride. This type of preparation produces a silica which is in the form of nonporous microspheres. Cabosil is more inert than silica prepared by dehydration of silica gel. In all three cases, the sample was calcined at 538°C and the nickel content would be 7 percent if reduction were complete. In the case of Fig. 4 (and also Fig. 5), the extent of reduction to nickel was determined by magnetic methods. The general nature of the results, shown in Fig. 4, has been confirmed in numerous studies in which other methods of determining the extent of nickel reduction have been used.[14,17]

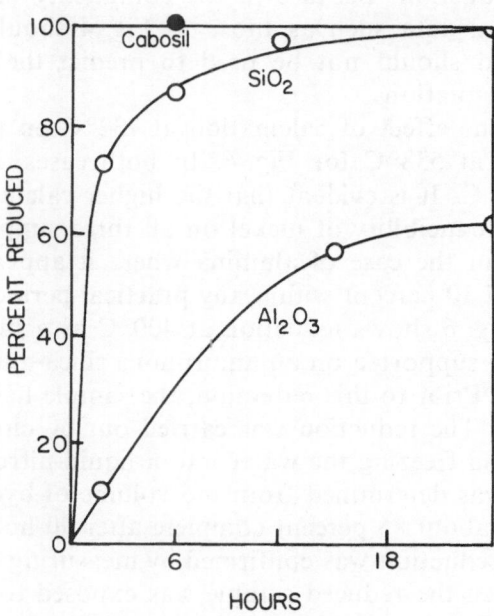

Fig. 4. Effect of support on reducibility of nickel.[13]

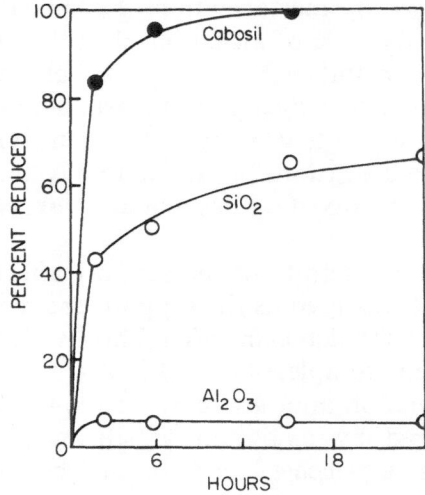

Fig. 5. Effect of calcination temperature on reducibility of supported nickel, calcined at 732°C compared with 538°C for Fig. 4.[13]

Other than the support, there are many factors which affect the reducibility of supported metals. These include the chemical properties of the metal itself, concentration, degree of dispersion, and the treatment of the sample prior to reduction. Because of the complexity of the variables involved in reduction, data such as those in Fig. 4 should be taken as indicating trends and should not be used to predict the precise extent of reduction in other situations.

Fig. 5 shows the effect of calcination at 732°C on the samples which had been calcined at 538°C for Fig. 4. In both cases, the reduction was carried out at 371°C. It is evident that the higher calcination temperature has decreased the reducibility of nickel on all three supports. The effect is most pronounced in the case of alumina where it appears that reduction would never exceed 10 percent within any practical period of time.

Curve A of Fig. 6 shows reduction at 400°C as a function of time for nickel (4.4 percent) supported on an amorphous silica-alumina (13 percent) cracking catalyst.[18] Prior to this reduction, the sample had been calcined at 330°C for 3 hours. The reduction was carried out by circulating hydrogen over the catalyst and freezing the water into a liquid-nitrogen trap. The extent of reduction was determined from the volume of hydrogen consumed. The reduction was about 85 percent complete after 90 hours.

The extent of reduction was confirmed by measuring the volume of oxygen consumed when the reduced sample was exposed to excess oxygen at 400°C. Curve B of Fig. 6 represents a reduction of the reoxidized sample. In this case, 85 percent reduction was attained in a short time. The ease of

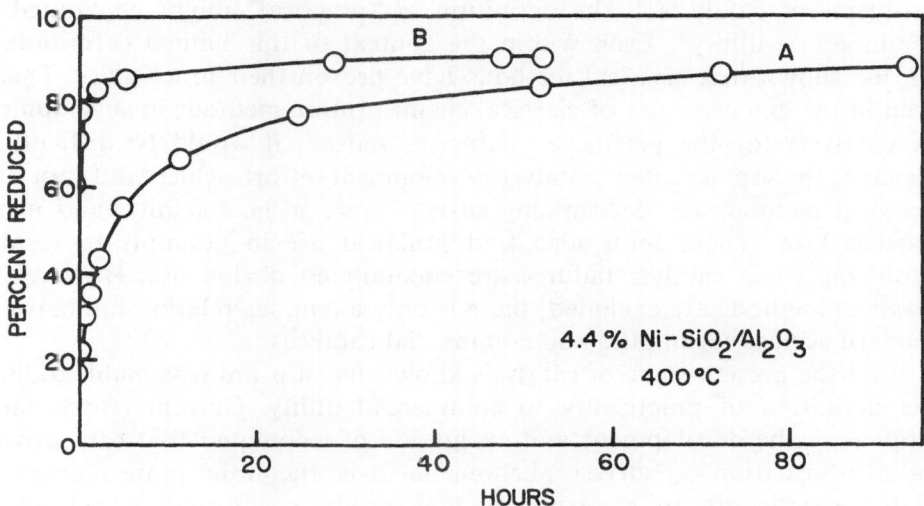

Fig. 6. Reduction of nickel supported on silica-alumina.[18]

reducing reoxidized supported nickel was also observed by Bicek and Kelly.[15] The reduction shown by curve b is similar to what would be expected for unsupported nickel oxide. This suggests that little interaction between the support and the nickel oxide is produced by reoxidation.

Nickel is intermediate with regard to ease of reduction on supports; platinum is easier to reduce and iron more difficult. Thus, the failure to attain complete reduction may be a factor in many experiments. The severity of the problem is mitigated by the tendency of individual particles to be completely reduced or not reduced at all. In most chemisorption situations, nickel aluminate and nickel silicate are inert. The unreduced metal is therefore wasted and the error is inversely proportional to the extent of reduction. For example, particle sizes are often calculated on the basis of chemisorption capacity and the amount of metal thought to be contributing to the surface area. An error equivalent to the 15 percent loss, as in Fig. 6, would not be serious in most static chemisorption measurements. However, in situations where the reduction was only 50 percent, misleading results could be produced. If the unreduced cation has catalytic activity, the difficulties due to incomplete reduction are magnified by kinetic factors. Thus, reducibility of supported metals must be of concern to catalytic chemists.

5 PROGRESS IN PRACTICAL CATALYSIS

The theme of this session of the symposium is: "What progress has been made in applying modern surface techniques to the study of practical problems of catalysis?" The definition of "practical" might be limited to "commercial utility". Even within the context of this limited definition, it can be shown that classical methods have proven their practicality. This is seen in the extensive use of classical chemisorption methods in development of catalysts for the petroleum industry. Indeed, it would be difficult to visualize a sophisticated catalyst-development effort which did not use classical methods for determining surface area, surface acidity, and metal particle size. These tools also find practical use in attempts to resolve problems when catalyst failures are encountered during use. However, if classical methods are excluded, there is only a tenuous relationship between modern surface techniques and commercial catalysis.

At the present stage of catalysis knowledge, it is not reasonable to limit the definition of practicality to commercial utility. Current efforts must emphasize the development and evaluation of techniques that can provide valid information on surface phenomena. It is, therefore, more realistic to assign practicality to experiments that produce information helpful to catalytic chemists. This requirement implies that the information gained in studies of bulk metals or thin films should be transferable to supported metals. The catalytic chemist is confident that there is no obstacle to this transfer, so it is easy for him to accept the information produced in studies of transition metals by surface physicists. Under this second definition, the application of modern techniques has produced significant practical results. Modern methods are especially attractive for studies involving individual crystal faces or characterization of the surfaces of alloys since supported-metal studies encounter serious difficulties in these areas.

REFERENCES

1. Harrod, J. F., Roberts, R. W., and Rissman, E. F., *J. Phys. Chem.*, **71**, 343 (1967).
2. Bradshaw, A. M., and Pritchard, J., *Surface Sci.*, **17**, 373 (1969).
3. McCoy, E. F., and Smart, R.St.C., *ibid.*, **39**, 109 (1933).
4. Blyholder, G., *Proc. Third Internat. Congress on Catalysis* (1964), p. 657.
5. Eischens, R. P., and Pliskin, W. A., *Adv. Catalysis*, **10**, 1 (1958).
6. Soma-Noto, Y., and Sachtler, W.M.H., *J. Catalysis*, **32**, 315 (1974).
7. Corey, E. R., Dahl, L. F., and Beck, W., *J. Am. Chem. Soc.*, **85**, 1202 (1963).
8. Blyholder, G., private communication, 1974.
9. Mertens, F. P., and Eischens, R. P., *Proc. Fourth Materials Symposium,* Berkeley, June 1968, G. A. Somorjai (Ed.), Paper 53.
10. Bradshaw, A. M., and Vierle, O., *Ber. Bunsenges.*, **74**, 633 (1970).
11. Schuit, G.C.A., and vanReijen, L. L., *Adv. Catalysis*, **10**, 243 (1958).
12. Roman, A., and Delmon, B., *J. Catalysis*, **30**, 333 (1973).
13. Sieg, R. P., Constabaris, G., and Linquist, R. H., Am. Chem. Soc. Meeting, New York (September, 1963).
14. Levinson, G. S., Preprints, Division of Petroleum Chemistry, American Chemical Society, Chicago (September, 1967), p. 47.

15. Bicek, E. J., and Kelly, C. J., *ibid.*, p. 57.
16. Beuther, H., and Larson, O. A., *Ind. Eng. Chem. Process Design and Dev.*, **4**, 177 (1965).
17. Swift, H. E., Lutinski, F. E., and Tobin, H. H., *J. Catalysis*, **5**, 285 (1966).
18. Webb, A. N., private communication, 1968.

DISCUSSION on Paper by R. P. Eischens

MASON: The assignment of $\nu_{C=O}$ at ca. 1980 cm^{-1} to a zinc carbonyl species is interesting and perhaps surprising in the light of the vibrational data on upper carbonyls where I believe $\nu_{C=O}$ is 2050 cm^{-1} unless there were special features of the complex such as a very low coordination number of the zinc at or around the surface.

EISCHENS: Dr. Mason's comment pertains to a zinc carbonyl species which is described in *J. Phys. Chem. Solids*, **28**, 2135 (1967). This carbonyl frequency is 60 to 70 cm^{-1} lower than that for carbonyls (linear) of transition metals. The specific frequency of the band is not so surprising as the fact that a zinc carbonyl band was observed, since zinc carbonyl is not known in conventional chemistry. A future adequate explanation of the exact bonding which makes possible this zinc carbonyl will probably explain why the frequency is lower than that for transition-metal carbonyls.

WAGNER: Zinc oxide contaminated with carbon probably contains an excess of metal. Upon treatment with oxygen, the metal excess decreases. This may have an effect on the absorption spectrum.

EISCHENS: Dr. Wagner's observation is correct. Exposure to oxygen increases the transmission of zinc oxide. The increase in transmission is due to a decrease in conductivity caused by a decrease in excess zinc. The unique zinc carbonyl bands of impure zinc oxide are observed, together with the effects due to conductivity changes. The bonding problem involves the question of relating the decrease in excess zinc to formation of zinc carbonyl.

WALTER: Zinc oxides can exhibit photooxidation as well as photoreaction in white light and in the presence of oxygen. Could it be possible that illumination during experimentation affected your results of CO bonding to ZnO exposed to oxygen as chlorine?

EISCHENS: Our spectrometer was fitted with a filter which cuts off radiation above 4300 cm^{-1}. I do not believe that the low-energy infrared radiation is likely to produce photooxidation. Studies of the effect of white light on the impure zinc oxide-oxygen system would be of interest but we have not done this.

SCHWAB: I wonder whether you had any special reason for not mentioning the probability of the spinel formation causing the incomplete reduction of NiO over Al_2O_3. Such spinels form very easily around 100°C and in presence of water even below that temperature.

EISCHENS: I did not intend to exclude spinels in my mention of mixed oxides. Spinels have been detected by X-ray in these systems.

BASSETT: Dr. Eischens, you mentioned in reply to Dr. Walter that before reduction, the supported nickel catalysts represent mixed oxide systems. I therefore wonder if the difference in reduction behavior with different supports is associated with mutual solubility and whether this results in the presence of any aluminum or silicon in the final nickel particles. This would be of interest in relation to the surface conditions, since Professor Gomer's early field emission studies of nickel showed that it was impossible to produce clean nickel surfaces from nickel containing silicon, since the silicon was surface active.

EISCHENS: I do not believe that solubility of silicon or aluminum in reduced nickel is a factor which results in contamination of the nickel surface. The metal-on-support systems are not analogous to Dr. Gomer's experiments in which the large bulk-to-surface ratio could produce extensive surface contamination even when the solubility was low. For nickel-on-silica, and generally for supported metals, it is doubtful whether the silica is reduced to silicon. Bulk contamination of metal particles is not detected in cases which are amenable to X-ray examination. Surface contamination can be excluded on the basis of chemisorption measurements. The assumptions underlying the determination of particle size by chemisorption are vitiated by surface contamination. In practice, a lack of agreement between chemisorption methods and other means of determining particle size (X-ray, electron microscope) is used as a method for studying surface contamination (see, for example, ref. 11).

SOMORJAI: You have mentioned the possibility of mechanistic similarity between homogeneous and heterogeneous catalysis. Have there been studies made to determine the turnover number for the *same* reaction carried out homogeneously and heterogeneously? Don't you think it would be important to carry out such studies to answer this question?

EISCHENS: The relationship between homogeneous and heterogeneous catalysis has been reviewed by Weller and Mills, Halpern, Orchin, and Ballard in *Advances in Catalysis*, Vols. 8, 11, 16, and 23, respectively. These papers include rate data which could be recalculated in terms of turnover numbers; however, I am not aware of any direct comparison of turnover numbers for the same reaction carried out homogeneously and heterogeneously. The homogeneous work includes studies of the effect of the coordinated ligands on reaction rates. Since reaction rates are influenced by the nature of the ligands, there is more than one valid turnover number for a homogeneous reaction.

The fact that many reactions are catalyzed by complexes which contain only a single metal atom shows that the metallic state is not essential to catalysis by metal atoms. This does not mean that factors such as crystal face or defects may not be important in specific cases. A study of the effect of crystal-face exposure is analogous to the study of the effect of ligands in homogeneous systems.

Dr. P. B. Weisz has presented an interesting comment on the relationship of homogeneous and heterogeneous catalysis in his introduction to Vol. 13 of *Advances in Catalysis*. He suggests that there is no "secret of catalysis" and that heterogeneous catalysis involves only conventional chemical reactions carried out on solid surfaces which contribute engineering advantages such as ease of separation of product and temperature flexibility. Many catalytic chemists agree with this view.

EMMETT: What is the source of the nitrogen that you indicate is present in zinc oxide?

EISCHENS: The nitrogen was inadvertently introduced in a preparation of zinc oxalate from zinc nitrate and ammonium oxalate. Professor D. M. Smith of the University of Denver has studied red zinc oxide which contains nitrogen, and he has confirmed that electronegative adsorbents produce the bands which we attribute to nitrogen impurities in the zinc oxide we prepared from the impure zinc oxalate.

WALTER: We, as well as Justi and co-workers, and others, investigated the stabilization of Raney nickel by controlled oxidation and subsequent reduction. It was found that hydrated nickel oxide was difficult to reduce as opposed to nickel oxide. Do you think that your observation of slow reduction of the original nickel oxide and fast reduction of the reoxidized supported nickel may be due to the presence of hydrated nickel oxide in the starting material?

EISCHENS: Some nickel may be present as hydrated nickel oxide, especially in cases where samples are prepared by coprecipitation. However,

I do not believe that hydration of the oxide is a major cause for difficulty in reducing supported nickel because the difficulty is enhanced by high-temperature calcination. Bicek and Kelly (ref. 15) have shown that the easily reduced reoxidized nickel becomes difficult to reduce if the supported nickel oxide is heated to high temperatures prior to the second reduction.

SCHRIEFFER: I believe you mentioned that because a carbon atom poisoning the metal surface neither donates or accepts electrons from a metal atom, it does not affect the bondings of CO to the surface. I would like to note that even though no charge transfer occurs, this does not rule out a change in the local density of states on the surface metal atoms in contact with the carbon and consequently a change in the CO bond strength. I believe the fact that no shift is observed must be understood by including such possible shifts in local density of states.

Is there evidence concerning the change in infrared spectrum when a second metal atom is added to a metal carbonyl molecule?

EISCHENS: There have been extensive infrared studies of carbonyls containing more than one metal atom. I am not aware of a consistent pattern of carbon-oxygen frequency modifications for linear carbonyl groups that can be attributed to metal-metal interactions.

PARTICLE GROWTH IN SUPPORTED CATALYSTS

P. Wynblatt, R. A. Dalla Betta, and N. A. Gjostein

Scientific Research Staff
Ford Motor Company
Dearborn, Michigan

ABSTRACT

Experimentally determined particle growth kinetics are presented for two types of systems: a dispersion of platinum particles in a microporous alumina support, which is typical of commercial catalysts, and a dispersion of platinum particles on a flat, nonporous alumina substrate, which is more amenable to interpretation by means of theoretical models.

Comparison of the experimental results with estimates derived from a number of models for particle growth indicates that, under oxidizing conditions, growth is most likely controlled by facet-inhibited interparticle transport over the substrate surface. Although detailed agreement of the theory with experiments on microporous catalysts is not obtained, qualitative theoretical trends are generally substantiated by experiment.

1 INTRODUCTION

Supported catalyst systems composed of metal particles on nonmetallic substrates have been widely used for many years in the chemical industry. During service, phenomena which tend to decrease the effectiveness of these systems can take place. The present paper will focus on one of these phenomena: namely, the decrease in total active surface area resulting from the growth of the catalytically active supported metal particles or crystallites.

Recently, Wynblatt and Gjostein[1] have proposed a general modelling scheme for the various stages in the formation and growth of supported-metal crystallites. From a review of the available experimental information, they concluded that particle growth can proceed in one of two fairly distinct modes, identified as "inhibited" and "noninhibited" growth, respectively. The noninhibited mode refers to growth which generally obeys either the classical models of growth by interparticle transport[2-4] or the more recent models of growth by particle migration, impingement and coalescence[5,6]. The inhibited mode refers to growth by the same general processes but where the metal crystallites tend to facet, i.e., develop singular surfaces, and thus require periodic nucleation events on their surfaces for continued growth; such inhibition is well known in crystal growth theory.

In this paper we present results derived from experiments using an idealized model system consisting of metal particles supported on a flat ceramic substrate as well as from experiments performed on more conventional supported catalysts, where the metal particles are dispersed throughout a microporous ceramic substrate. Next, we estimate particle growth rates from a number of models describing noninhibited growth and show that these do not agree well with our measurements. Finally, we estimate particle growth rates from a model which includes the inhibiting effects of nucleation and show that such an approach promises to improve agreement with experiment.

2 EXPERIMENTAL PROCEDURE AND RESULTS

2.1 Model Experiments on Flat Supports

Samples for the model experiments were prepared by depositing a thin film of alumina ($\sim$100 Å thick) onto quartz microscope slides, followed by deposition of a platinum film ($\sim$10 Å thick) on the alumina. Both films were produced by radiofrequency (RF) sputtering in argon from suitable targets. The advantages of this type of sample are ease of handling of the quartz-supported alumina film and associated platinum, as well as the feasibility of stripping the thin films from the quartz support (in dilute HF) for observation in the transmission electron microscope.

If a fragment of the duplex alumina-platinum film is placed in an electron microscope, on a hot stage, and heated for a few minutes at 500°C, the platinum film may be observed to "dewet" partially and break up into equiaxed platinum particles having a projected radius of ~70 Å and a structure similar to that shown in Fig. 1a. The structure of the alumina film, as determined by electron diffraction, corresponds to the γ-alumina phase.

Normally, however, the duplex alumina-platinum film was heat treated while still on the quartz support, the time consumed by film breakup being neglected. After heat treatment, the samples were stripped from their quartz supports and the average platinum particle size was determined from analysis of transmission electron photomicrographs. Typical photomicrographs are shown in Fig. 1.

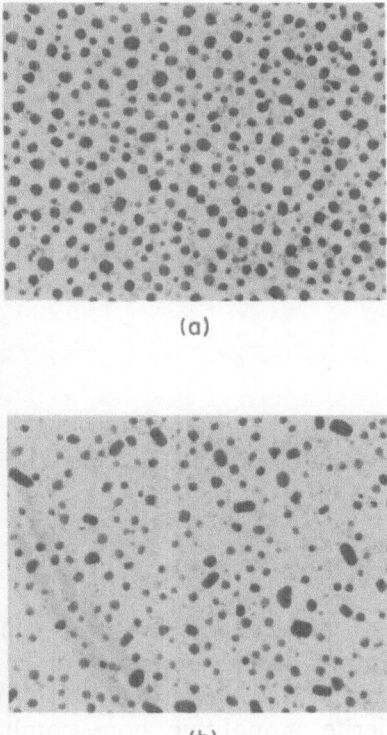

(a)

(b)

Fig. 1. Transmission electron micrograph of flat substrate samples: (a) after heat treatment at 700°C for 3 hours in O_2 at 0.02 atm pressure; (b) after heat treatment at 700°C for 68 hours in O_2 at 0.02 atm pressure.

Previous experiments[7] had shown that if this type of sample is exposed to air in an open furnace at 1000°C, platinum is gradually lost from the sample and disappears entirely in less than 1 hour. That result is consistent with the known volatilization rates of PtO_2[8,9], which is always present on the surface of platinum metal after exposure to an oxygen-bearing environment. In order to prevent this type of platinum loss, samples were heat treated after encapsulation in quartz tubes containing an appropriate atmosphere. Particle growth kinetics were obtained at 700°C for samples heated in two different atmospheres: air and 2 percent O_2 in N_2 at a total pressure of 1 atm. The results obtained are plotted in Fig. 2 and they show that particle growth rates are larger for experiments conducted under higher oxygen partial pressures.

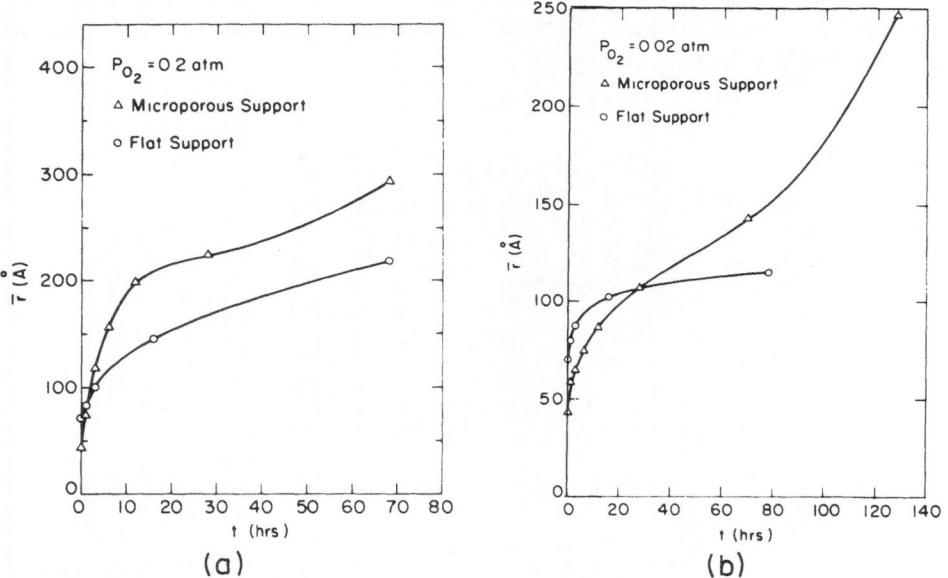

Fig. 2. Plot of average radius, $\bar{r}$, versus time, t, for two types of experiments.

2.2 Experiments on Microporous Supports

The samples for these experiments consisted of platinum deposited on a high-surface-area γ-alumina layer supported on a cordierite honeycomb structure. The γ-alumina was deposited by dipping the honeycomb (American Lava cordierite monolithic honeycomb, 3M Company) in a 30 percent aqueous dispersion of alumina (Alon 30-D colloidal alumina, Cabot Corporation). The honeycomb was dried at 350°C and then heated in air at 1000°C for 200 hours to sinter and stabilize the alumina. The pore size distribution was determined by means of a mercury porosimeter and was found to consist of a single peak at 250 Å diameter, with a width at half height of 70 Å.

Platinum was deposited from an aqueous solution of chloroplatinic acid and after excess solution was blown off from the channels, H_2S was passed through the honeycomb to precipitate the platinum as the sulfide. This was subsequently decomposed by heating in air at 500°C for 5 hours. The alumina and platinum deposition was uniform as determined by total surface area and platinum-surface-area measurements on several samples weighing 2 to 3 grams, which were cut out of the honeycomb.

Platinum surface area was measured by CO adsorption from a flowing hydrogen stream in a manner similar to that described by Gruber.[10] In all cases the results are the average of at least two measurements. The average platinum particle size was calculated assuming cubic particles with five faces exposed to the gas phase and an area of 8.4 Å^2 per surface platinum atom. Particle growth kinetics were obtained at 700°C for samples heated in the same atmospheres as the model samples.

The results obtained here are also plotted in Fig. 2, with the "radius" taken to be half of the cube edge dimension calculated for the particle size. It can be seen that in this case the particle growth rates also depend on oxygen partial pressure. However, the time dependence of the average particle radii inferred from the metal surface area measurements of samples supported on microporous substrates differs both in magnitude and in kind from that obtained by direct measurements on flat substrate samples.

3 MODELS FOR NONINHIBITED PARTICLE GROWTH

There are at least two types of processes by which supported particles can grow: (1) by interparticle transport, analogous to the Ostwald ripening phenomenon; and (2) by particle migration, collision, and coalescence. The results of various theoretical models based on these processes are presented herein, without derivation, and estimates (or at least certain bounds) of particle growth kinetics are obtained.

3.1 Growth by Interparticle Transport

Consider a distribution of partially wetting particles, in the form of spherical segments, in contact with a ceramic substrate, as illustrated schematically in Fig. 3. We note that (when locally equilibrated) each particle will make a contact angle, θ, with the substrate, defined by:

$$\gamma_s = \gamma_{ms} + \gamma_m \cos (\theta) \tag{1}$$

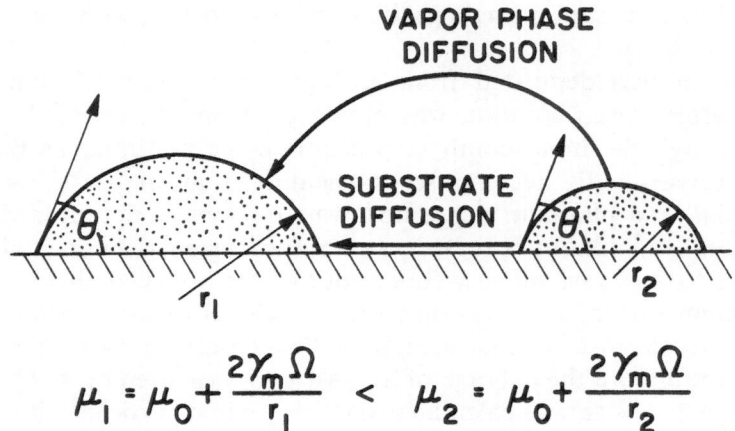

$$\mu_1 = \mu_0 + \frac{2\gamma_m\Omega}{r_1} < \mu_2 = \mu_0 + \frac{2\gamma_m\Omega}{r_2}$$

Fig. 3. Schematic of particles in the shape of spherical segments, making a contact angle θ with a flat substrate.

where γ_s and γ_m are the surface energies of the substrate and metal, respectively, and γ_{ms} is the metal-substrate interfacial energy. Furthermore, the chemical potential, μ, of any given particle within the distribution will depend on its radius of curvature, r, according to the Gibbs-Thompson relation:

$$\mu = \mu_0 + \frac{2\gamma_m\Omega}{r} \tag{2}$$

where Ω is the atomic volume. Thus, larger particles with lower chemical potential will tend to grow at the expense of smaller particles with higher chemical potential, the driving force being the reduction in the total surface energy of the system. Under those conditions, some particles grow and others dissolve away, with a resulting increase in the average particle size of the distribution. There are generally two critical processes that can control the average rate of growth in this type of problem: the transport process itself, commonly referred to as "diffusion control", or the process of attachment and detachment of the diffusing entities from the particles, generally referred to as "interface control". We consider here two modes of transport between particles: diffusion over the substrate surface and diffusion through the vapor phase. (The third possible mode, diffusion through the substrate, does not appear to be a likely process.)

(a) Transport by Diffusion Over the Substrate Surface

Wynblatt and Gjostein[11] have recently derived expressions for particle growth under these conditions using as a basis the classical treatments of Ostwald ripening by Wagner[2] and by Lifshitz and Slyozov[3] and extensions thereof by Chakraverty[4]. We first describe expressions and estimate particle growth rates for the case of growth under reducing conditions. Although these expressions are unsuitable for the present experiments, as they do not account for the effect of oxygen partial pressure on growth rate, shown in Fig. 2, the exercise is instructive in that it clarifies the procedure used to arrive at estimates suitable for the case of growth under oxidizing conditions.

The concentrations and energetics of metal monomers on, and in the vicinity of, a particle are shown schematically in Fig. 4. The upper portion of the figure shows the variation in concentration of metal monomers and the lower portion shows various energy differences associated with certain important monomer transitions. For growth controlled by the diffusion of monomers over the substrate, the average radius at time t is given by:

$$\bar{r}(t) = \bar{r}(0) \left\{ 1 + \frac{K_D t}{[\bar{r}(0)]^4} \right\}^{1/4} \tag{3}$$

where

$$K_D = \frac{27}{64} \frac{D_1 C_s^{eq} \gamma_m \Omega^2}{\alpha_1 \, kT \, \ell n \, (L/r \sin \theta)} \tag{4}$$

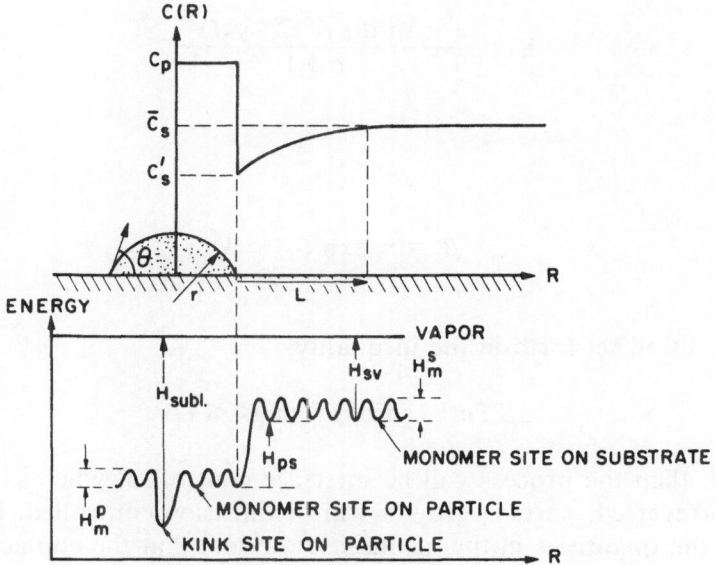

Fig. 4. Above: concentration of monomers on and around a particle. Below: energy changes associated with various important monomer transitions.

D_1 is the diffusivity of monomers over the substrate, C_s^{eq} (atoms/unit area) is the monomer concentration on the substrate which would be in equilibrium with a metal particle of infinite radius, L is the distance from a particle at which the concentration reaches the far-field concentration $\bar{C}_s$ (see Fig. 4), k and T have their usual significance and

$$\alpha_1 = (2\text{-}3 \cos \theta + \cos^3\theta)/4 \tag{5}$$

For present purposes we define D_1 approximately as:

$$D_1 = a^2\nu_s \exp (- H_m^s/kT) \tag{6}$$

where a is the distance between monomer sites on the substrate, ν_s is the vibrational frequency for a monomer on the substrate, and H_m^s is the activation energy for the migration of a monomer over the substrate.*

For "interface" controlled growth, the average particle radius is given by:

$$\bar{r}(t) = \bar{r}(0) \left\{ 1 + \frac{K_1 t}{[\bar{r}(0)]^3} \right\}^{1/3} \tag{7}$$

where

$$K_1 = \frac{4}{9} \frac{a \sin(\theta) \, \beta' \, C_s^{eq} \, \gamma_m\Omega^2}{\alpha_1 kT} \tag{8}$$

and

$$\beta' = \nu_s \exp (-H_m^s/kT) \tag{9}$$

If the condition set forth in the inequality:

$$a\beta' \ell n(L/r \sin\theta) \, r \sin\theta \ll D_1$$

is obeyed, then the process will be interface controlled, whereas if the inequality is reversed, particle growth will be diffusion controlled. In order to compare the quantities in the inequality, we note that the contact angle for

---

*Rigorously, the Gibbs free energy function should be used in these expressions. However, we use the enthalpy function H throughout, which is a reasonable approximation.

platinum on alumina is $\theta \approx 90$ degrees and $L/r \sin\theta$ is approximately 2 or 3. Combining this with the expressions for D_1 and β', Eqs. (6) and (9), we find that:

$$ar\beta' = ar\nu_s \exp (- H_m^s/kT) > D_1 = a^2\nu_s \exp (- H_m^s/kT) \qquad (10)$$

for physically realistic values of r. In this case, therefore, growth by inter-particle transport over the substrate will be diffusion controlled.

Estimation of K_D reduces to an evaluation of $D_1 C_s^{eq}$. C_s^{eq}, the concentration of monomers on the substrate, can be expressed as:

$$C_s^{eq} = C_0 \exp [-(H_{subl.} - H_{sv})/kT] \qquad (11)$$

where C_0 is the concentration of sites on the substrate at which monomer adsorption can take place, $H_{subl.}$ is the heat of sublimation of platinum, and H_{sv} is the heat of adsorption of a platinum monomer on the substrate (see Fig. 4). Thus, from Eqs. (6) and (11), and recognizing that $C_0 = 2/a^2$ (for the basal plane of γ-Al$_2$O$_3$), we obtain:

$$D_1 C_s^{eq} = 2\nu_s \exp [- (H_{subl.} + H_m^s - H_{sv})/kT]$$

and

$$K_D = \frac{27}{64} \frac{2\gamma_m \Omega^2 \nu_s}{\alpha_1 kT \, \ell n \, (L/r \sin\theta)} \exp [-(H_{subl.} + H_m^s - H_{sv})/kT] \qquad (12)$$

$H_{subl.}$ for platinum in 135 kcal/mole[12] and H_{sv} can be estimated[1] from the work of adhesion of platinum on alumina determined by McLean and Hondros[13] to be ~ 9 kcal/mole. H_m^s is not known but must be in the range $0 < H_m^s < H_{sv}$; here we give this quantity its minimum possible value. That will lead to an upper bound for K_D. Taking $\gamma_m = 2100$ ergs/cm^2, $\theta = 81$ degrees, T = 973°K, and $\ell n \, (L/r \sin\theta) \approx 1$, we obtain $K_D = 1 \times 10^{-44}$ cm^4/sec. Finally, inserting K_D into Eq. (3) and taking $r(o) = 70$ Å (the experimental value for flat substrate experiments), we find this theoretical upper bound to be negligibly small, amounting to a growth of only 1 Å in about 10^{18} seconds. It should be emphasized that this low rate stems primarily from the very large heat of sublimation of platinum and from the small heat of adsorption for platinum monomers on the substrate.

We now proceed to estimate the growth rate controlled by diffusion over the substrate, under oxidizing conditions. The surface of platinum in an oxygen-containing environment will have some coverage of chemisorbed oxygen, probably in the form of a surface oxide. At high enough temperature, and when equilibrium between platinum and its surroundings is achieved, there will be significant vapor pressure of PtO_2 in the gas phase, given by:

$$P_{PtO_2} = P_{O_2} K_{eq} \qquad (13)$$

where K_{eq} is the equilibrium constant for the reaction:

$$Pt_{(s)} + O_{2(g)} = PtO_{2(g)} \qquad (14)$$

Under these conditions, the energy required to remove a platinum atom to the vapor phase (albeit in the form of a PtO_2 molecule) will be the enthalpy change for the above reaction, ΔH_{ox} ($\approx$ 42 kcal/mole)[8], which is considerably less than the heat of sublimation of platinum. Thus, we may expect to arrive at more reasonable growth rates. We now assume that the diffusing species on the substrate will be PtO_2 molecules rather than platinum monomers. Accordingly we rewrite

$$C_s^{eq} = \frac{1}{\nu_s' \exp(-H_{sv}'/kT)} \cdot \frac{\chi\, P_{PtO_2}}{(2\pi mkT)^{1/2}} \qquad (15)$$

and

$$D_1 = a^2 \nu_s' \exp(-H_m'^s/kT) \qquad (16)$$

where the primed quantities are the appropriate ones for a PtO_2 molecule, χ is a sticking coefficient for PtO_2 on Al_2O_3 (assumed to be unity), and m is the mass of a PtO_2 molecule. Substituting Eqs. (15), (16), and (13) into Eq. (4) we have

$$K_D = \frac{27}{64} a^2 \frac{P_{O_2} K_{eq}}{(2\pi mkT)^{1/2}} \frac{\gamma_m \Omega^2}{\alpha_1 kT\, \ell n\, (L/r \sin\theta)} \exp\left[-(H_m'^s - H_{sv}')/kT\right] \qquad (17)$$

Unfortunately, there is no way at the present time to estimate $H_m'^s$ and H_{sv}' independently of particle growth experiments. However, the experiments of Huang and Li[14] on the growth of large platinum particles (1500 to 3000 Å) supported on single-crystal sapphire surfaces have shown noninhibited growth kinetics consistent with a fourth-power growth law such as Eq. (3). Furthermore, the kinetics of growth have been found to depend strongly on surface orientation, which indicates that surface diffusion is the controlling mechanism; finally, their growth kinetics have been found to depend on oxygen partial pressure. Their results, therefore, lend themselves well to interpretation by means of Eq. (3), with K_D as expressed in Eq. (17). Taking an intermediate value of K_D from their experiments performed at 900°C: $K_D = 3.3 \times 10^{-24}$ cm^4/sec and comparing with Eq. (17) (using T = 1173°K, a = 2.74 Å, $P_{O_2} = 0.2$ atm, $K_{eq} = 7.92 \times 10^{-8}$)^8, we find $(H'_{sv} - H^\delta_m) = 33.6$ kcal/mole, a result consistent with the known limits for this quantity, namely: $0 < (H'_{sv} - H^\delta_m) < H_{ox}$ (~ 42 kcal/mole). We can use this value of $H'_{sv} - H^\delta_m$ to compute appropriate values of K_D for our experimental conditions, i.e., T = 973°K, $P_{O_2} = 0.2$ and 0.02 atm. The estimated growth kinetics under those conditions are shown in Fig. 5, together with our experimental results, and they can be seen to be considerably higher than the experimental data.

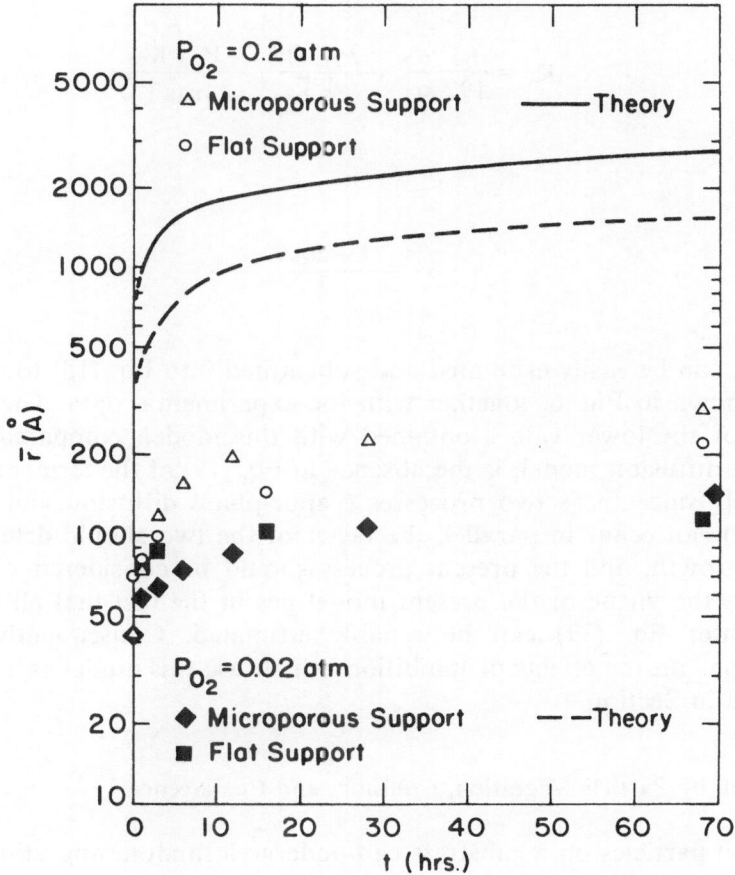

Fig. 5. Comparison of all data with estimates of noninhibited growth by diffusion over the substrate (note logarithmic scale of $\bar{r}$ axis).

(b) Transport by Diffusion Through the Vapor Phase

Wynblatt and Gjostein[11] have also derived an expression for particle growth with interparticle transport occurring through the vapor phase. In this case, the interface process is rate controlling since diffusion through the gas phase is rapid. Again it is assumed that the diffusing species are PtO_2 molecules, the "interface" process being the adsorption and desorption of PtO_2 molecules to and from platinum particles. In this instance, it can be shown that

$$\bar{r}(t) = \bar{r}(0) \left\{ 1 + \frac{K_v t}{[\bar{r}(0)]^2} \right\}^{1/2} \tag{18}$$

where

$$K_v = \frac{64}{192} \frac{\alpha_2}{\alpha_1} \frac{2\gamma_m \Omega^2}{kT} \frac{P_{O_2} K_{eq}}{(2\pi mkT)^{1/2}} \tag{19}$$

and

$$\alpha_2 = \frac{1 - \cos\theta}{2} \tag{20}$$

Thus, K_v can be easily estimated and substituted into Eq. (18) to obtain the curves shown in Fig. 6, together with the experimental data. The principal reason for the lower values obtained with this model, compared with the substrate diffusion model, is the absence in Eq. (19) of the term: $\exp[(H'_{sv} - H_m^{rs})/kT]$. Since these two processes (vapor-phase diffusion and substrate diffusion) can occur in parallel, the faster of the two should determine the rate of growth, and the present process should be considered no further. However, the virtue of the present model lies in the fact that all quantities which enter Eq. (19) can be reliably estimated. Consequently, sample calculations on the effects of inhibition, which use this model as a basis, are presented in Section 4.

3.2 Growth by Particle Migration, Collision, and Coalescence

Small particles on a substrate can undergo a random migration process provided that atomic diffusion over the surface of the particle is rapid, i.e., that the distance $\sqrt{D_s t}$ (where D_s is the surface diffusivity) is comparable with the particle diameter. Monomers (adatoms) diffusing over the particle

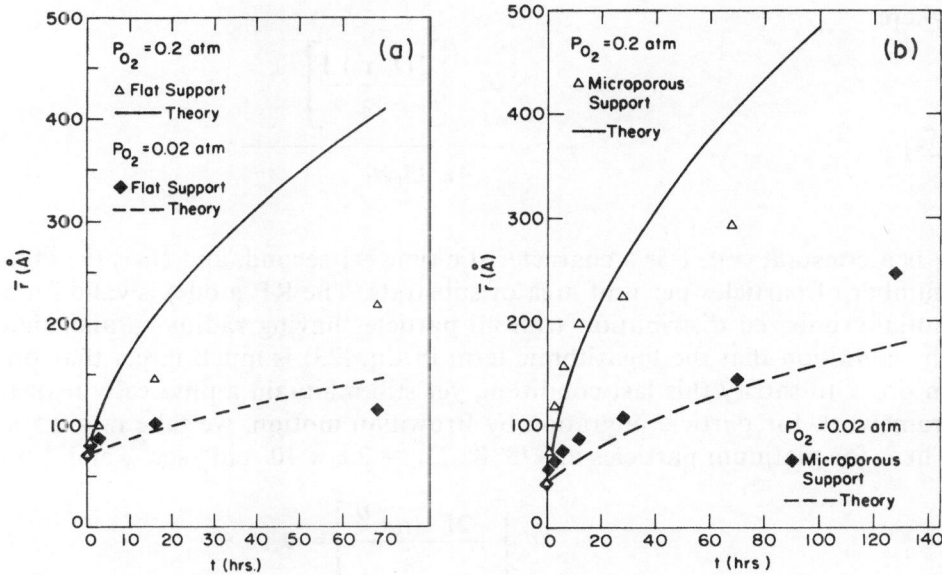

Fig. 6. Comparison of data from (a) flat-support experiments and (b) microporous substrate experiments, with estimates of noninhibited growth by vapor-phase transport.

surface will occasionally accumulate on one side, by random fluctuations, causing the particle to execute Brownian motion over the substrate. As an approximation to the particle diffusivity, D_p, we employ a result obtained by Gruber[15] for the migration of pores in solids by surface diffusion, namely that:

$$D_p = 0.3 \ D_s \left(\frac{a_0}{r} \right)^4 \tag{21}$$

where a_0 is the interatomic spacing of the particle and D_s is the surface diffusivity on the particle.

Recently, Ruckenstein and Pulvermacher[5,6] have developed a binary collision model for migrating particles on a planar substrate in which the rate-controlling step for particle growth can be either particle migration or coalescence. However, estimates of the relative kinetics of the migration and coalescence processes[1] have shown that for conditions of interest in catalyst systems, the particle-migration process should always be rate controlling. If the Ruckenstein and Pulvermacher (RP) model is combined with the expression for D_p given in Eq. (21), it can be shown that

$$\left[\frac{\bar{r}}{r_0} \right]^7 = 1 + q \frac{t}{\tau} \tag{22}$$

where

(23)
$$\tau = \frac{\ell n \left[\dfrac{2D_p(r_0)\ \Gamma}{r_0^2} \right]}{4\pi\ D_p N_0}$$

q is a constant ≈ 4, Γ is a characteristic time ≈ 1 second, and N_0 is the initial number of particles per unit area of substrate. The RP model is valid for an initially unisized distribution with all particles having radius r_0, and under the condition that the logarithmic term in Eq. (23) is much larger than one. In order to satisfy this last condition, yet still maintain a physically realistic framework for particle migration by Brownian motion, we take $r_0 = 2.5\ a_0$. Then, for platinum particles at 973°K, D_s, $= 2.8 \times 10^9$ cm^2/sec[20], and

$$\ell n \left[\frac{2D_p(r_0)}{r_0^2} \theta \right] = 9.1$$

which meets the requirements of the model. With the above values, together with the quantities $a_0 = 2.77$ Å and $N_0 = 10^{12}$ particles/cm^2, we can estimate the growth rate from this model, as shown in Fig. 7. As can be seen from the figure, this model gives a rather slow growth rate when compared with the experimental data. As presented, this model corresponds to reducing conditions. However, in contrast with the previous cases discussed, it is rather more complicated to modify this model to include effects which depend on oxygen partial pressure. The rate-controlling process under reducing conditions is the surface self-diffusion of platinum which has an activation energy of only 30 kcal/mole[20], not leaving much scope for acceleration of the process by an oxidizing environment. However, under oxidizing conditions, it is conceivable that platinum adatoms (monomers) associated with oxygen could migrate over the particle faster than unassociated adatoms, and the concentration of migrating complexes would certainly depend on oxygen partial pressure in a manner similar to that expressed in Eq. (15) for the concentration of complexes on the substrate.

4 A MODEL OF INHIBITED PARTICLE GROWTH

In this section we deal only with the effects of facet inhibition on growth by interparticle transport and, specifically, on growth by vapor-phase transport. We do not discuss inhibition of growth by particle migration, collision, and coalescence for two reasons: (1) we have not been able to

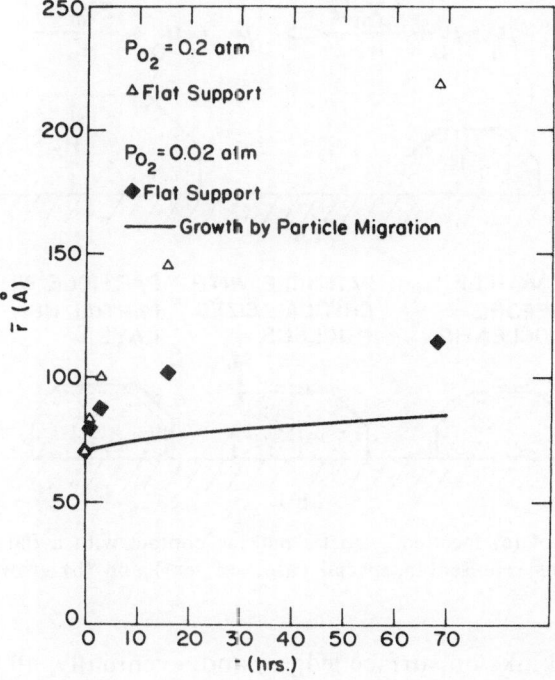

Fig. 7. Comparison of data from flat-support experiments with estimate of noninhibited growth by particle migration, collision, and coalescence.

frame this model in a manner applicable to growth under oxidizing conditions and (2) the growth rate under reducing conditions, shown in Fig. 7, is so slow that inhibition of that mode would not be of interest. In principle, however, one can foresee that particle faceting would inhibit that mode, as has been discussed in general terms previously.[1]

Consider an array of faceted particles lying on a substrate, as shown schematically in Fig. 8a. Since $(\gamma_{ms} - \gamma_s)/\gamma_m \approx 0$ for the case of platinum on Al_2O_3, we can approximate the chemical potential of faceted particles in terms of their heights, H, as*:

$$\mu = \mu_0 + 2\gamma_m \, \Omega/H \tag{24}$$

Thus, there exists in this case a driving force for growth analogous to that obtained for particles shaped in the form of spherical segments. As the larger particles grow, incoming monomers will tend to settle at energetically

---

*In general, $(\gamma_{ms} - \gamma_s)/\gamma_{mi} = H_s/H_i$, where γ_{mi} is the surface energy associated with the ith crystal face and H_s and H_i are the distances from the center of the crystallite to the substrate and the i_{th} crystal face, respectively.[16]

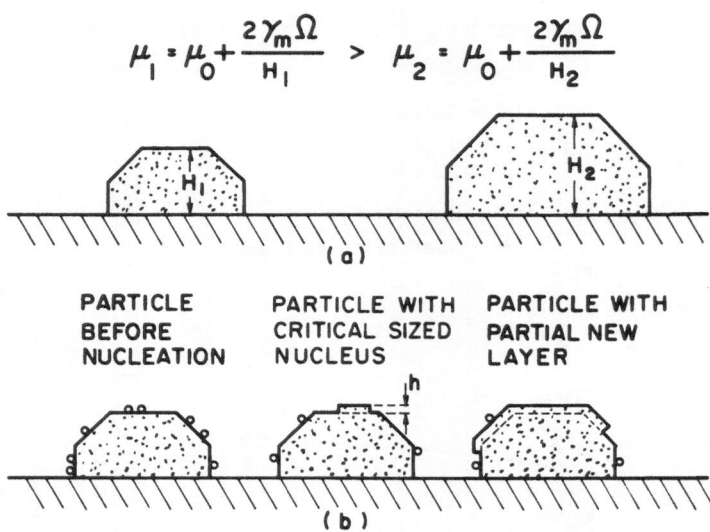

Fig. 8. Schematics of (a) facetted particles making contact with a flat substrate (chemical potential expressions represent a special case, see text) and (b) growth of a particle by nucleation.

favorable sites (kinks on surface ledges) and, eventually, all sites will be consumed, leading to a perfect particle. Thus, faceting can arise as a consequence of growth as well as from a high anisotropy of surface energy. In either case, however, continued growth of the particle must await the nucleation of a new ledge before large numbers of monomers can once again be accommodated on the particle. After nucleation of a new ledge, a whole new monolayer of growth can occur before a new nucleation event is required. The process is illustrated schematically in Fig. 8b. Under those conditions, the rate of change in height of a growing particle may be approximated by:

$$\frac{dH}{dt} = S\dot{N}h \tag{25}$$

where S is the surface area of the particle, N is the nucleation rate (nuclei per unit area per unit time) on the particle surface, and h is the height of a monatomic ledge or nucleus (see Fig. 8b). Equation (25) may be rewritten as[11]

$$\frac{dH}{dt} = \frac{BHH^*}{H\text{-}H^*} \exp\left[\frac{2\gamma_m\,\Omega}{kTH^*} - \frac{h\gamma_m}{2kT}\left(\frac{\epsilon}{h\gamma_m}\right)^2 \frac{HH^*}{(H-H^*)}\right] \tag{26}$$

where

$$B = \frac{4\pi^2 \alpha_2 Z h D_s}{\alpha_0^5} \left(\frac{\epsilon}{h\gamma_m} \right) \tag{27}$$

H* is the particle size within the distribution which neither grows nor shrinks (i.e., the particle size in equilibrium with the far-field conditions), ϵ is the energy associated with the edge of a ledge (energy/unit length), and Z is the Zeldovich factor of nucleation theory.[17]

The nucleation barrier to growth described above does not apply to small dissolving particles, which always possess some atoms in low-energy sites (e.g., corner or edge sites) that are easy to remove. Thus, the expressions governing the dissolution rates of small particles will be the same as those given in Section 3.1, and depend on the prevailing mechanism of interparticle transport. Thus, by appropriately combining Eq. (26) with the models for noninhibited growth one can obtain (numerically) growth kinetics of assemblies of particles under conditions of inhibited growth. At this time, this procedure has been applied only to the case of interparticle transport by vapor phase diffusion[11], because that case presents the least amount of computational difficulty and, furthermore, all the quantities which enter the theoretical expressions are known reasonably well.

Some sample curves for two values of the parameter $\epsilon/h\gamma_m$ are plotted in Fig. 9 along with the data from flat-substrate experiments.

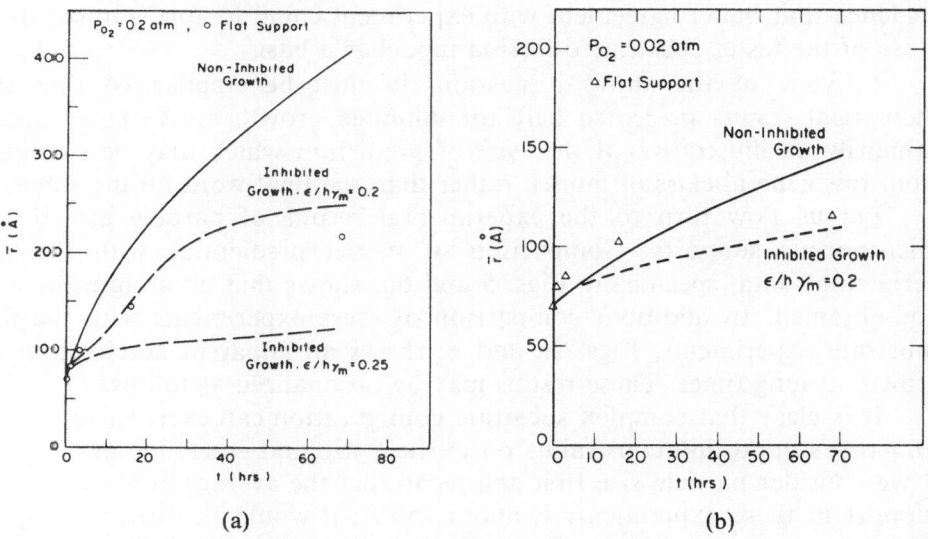

Fig. 9. Comparison of data from flat-support experiments with estimates of inhibited growth by vapor-phase transport.

5 DISCUSSION

Let us first compare the experimental results of particle growth on flat substrates with the various models. Consideration of Fig. 5 indicates that noninhibited growth by diffusion over the substrate is extremely fast compared with experimental data. Furthermore, even the slower growth by vapor-phase transport, Fig. 6a, lies mainly above experiment. The comparison of experimental data with both inhibited and noninhibited vapor-phase transport given in Fig. 9 shows that inhibited growth kinetics are quite sensitive to the parameter $\epsilon/h\gamma_m$ and that reasonable fit is obtained with the data for $\epsilon/h\gamma_m = 0.2$ (at least at high oxygen partial pressure). It is also interesting to note that the model for inhibited growth predicts curves which display inflection points. Unfortunately, the present data are not precise enough to substantiate or reject this feature. Experimental values of $\epsilon/h\gamma_m$, obtained from measurements of the anisotropy of surface energy[18], fall in the range 0.3 to 0.45; thus, the best fit with experiment at $\epsilon/h\gamma_m = 0.2$ is somewhat below the known range for this quantity. This is actually not surprising since, as mentioned in Section 3, it is expected that under non-inhibited growth conditions the mode of interparticle transport would be diffusion over the substrate. In order to obtain good agreement with experiment, using the faster substrate diffusion model as a basis for inhibited growth, one would require greater inhibition than in the case shown here and would therefore expect a larger value of $\epsilon/h\gamma_m$ to be predicted. Furthermore, the fact that the data at short times and low oxygen pressure lie above the curve for noninhibited growth by vapor-phase transport provides further evidence that better agreement with experiment could be forthcoming from a use of the faster substrate diffusion model as a basis.

In view of the above discussion, it must be emphasized that the theoretical results presented here for inhibited growth are to be regarded primarily as illustrative of the type of prediction which may be expected from this general class of model, rather than the final word on the subject.

Let us now turn to the experimental results of particle growth on microporous supports. Comparison of model predictions with the experimental data, specifically Figs. 5 and 6b, shows that good agreement is not obtained. In addition, comparison of these experiments with the flat substrate experiments, Figs. 2a and b, shows an apparent acceleration of growth at long times. These results may be rationalized as follows.

It is clear that complex substrate configuration can exert various (and sometimes opposing) constraints on particle size and accessible metal area. If we consider particle size first and recall that the average pore size of the support in these experiments is about 250 Å, it would be difficult to conceive of an average particle size much in excess of that limit. The pore size therefore imposes spatial constraints on particle size and may be thought of as a stabilizing influence. On the other hand, as a particle grows and fills up the pore it occupies in the support, an increasing fraction of the particle surface will come into contact with the walls of the pore, forming a metal-substrate interface. The possible evolution of particle shape in some simple pore geometries is illustrated schematically in Fig. 10. Thus, the particle surface area accessible to the gas phase will decrease sharply and the apparent particle size inferred from that surface area (assuming cuboids with five out of six faces exposed to the gas) will be grossly overestimated. This latter effect certainly contributes to the difficulties experienced here in reconciling theory and experiment.

In spite of the complex nature of microporous catalysts, substantial progress has been achieved here in understanding and predicting their behavior. On the basis of the study of particle growth in our idealized flat substrate system, it has been possible to develop theoretical models which identify the important parameters controlling growth. For example, if we examine expressions such as Eqs. (12) and (17), which describe the rate constants for growth by substrate diffusion under reducing and oxidizing conditions, respectively, we can draw a number of general conclusions about particle growth which will be applicable at least qualitatively to microporous catalysts.

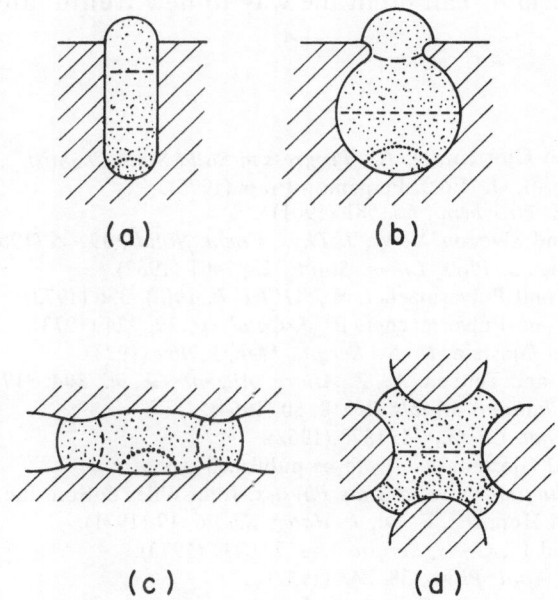

Fig. 10. Schematic showing stages of particle growth in some simple pore geometries. . . . initial stage; - - - intermediate stages; ———— final shape.

1. Growth under oxidizing conditions will be considerably faster than under reducing conditions. This has been borne out by Somorjai[19] in earlier experiments on microporous systems.

2. Under oxidizing conditions, the growth rate will increase with increasing oxidation potential, as is confirmed by the present experiments.

3. The effect of temperature on growth can be estimated from the explicit form of the equations.

4. Somewhat more complicated manipulation of the equations can demonstrate that increased wetting of the substrate by particles (i.e., a decrease in θ) will slow down growth.[1] This could be achieved to some extent by appropriate doping of either particles or substrate. The models therefore indicate possible ways of increasing the resistance of the system to particle growth.

5. Since growth under reducing conditions is controlled in a major way by the heat of sublimation of the metal component, one may predict the relative stabilities of noble-metal catalysts under those conditions as follows: Pd < Rh = Pt < Ru < Ir < Os.

6. Similarly, growth under oxidizing conditions is strongly dependent on the enthalpy change of the reaction $M(s) + xO_2(g) = MO_{2x}$ (g). Under this assumption, we may predict the stability series as Os < Ru < Ir < Pt < Pd = Rh.

7. Finally, lack of information about some of the parameters which enter the rate constants can point the way to new fruitful areas of study.

REFERENCES

1. Wynblatt, P., and Gjostein, N. A., *Progress in Solid State Chemistry,* **9**, 21, McCaldin, J. O., and Somorjai, G. (Eds), Pergamon Press (1975).
2. Wagner, C., *Z. Electrochem.,* **65**, 581 (1961).
3. Lifshitz, I. M., and Slyozov, V. V., *J. Phys. Chem. Solids,* **19**, 35 (1961).
4. Chakraverty, B. K., *J. Phys. Chem. Solids,* **28**, 2401 (1967).
5. Ruckenstein, E., and Pulvermacher, B., *AIChE J.,* **19**(2), 356 (1973).
6. Ruckenstein, E., and Pulvermacher, B., *J. Catalysis,* **29**; 224 (1973).
7. Wynblatt, P., and Gjostein, N. A., *Scripta Met.,* **9**, 969 (1973).
8. Schafer, Von H., and Tebben, A., *Z. Anorg. Allgem. Chem.,* **304**; 317 (1960).
9. Chaston, J. C., *Platinum Metals Rev.,***8**, 50 (1964).
10. Gruber, H. L., *Anal. Chem.,* **34**, 1828 (1962).
11. Wynblatt, P., and Gjostein, N. A., to be published.
12. Kittel, C., *Introduction to Solid State Physics*, John Wiley & Son, Inc., New York p. 99.
13. McLean, M., and Hondros, E. D., *J. Mater. Sci.,* **6**, 19 (1971).
14. Huang, F. H., and Li, C.-Y., *Scripta Met.,*7, 1239 (1973).
15. Gruber, E. E., *J. Appl. Phys.,* **38**, 243 (1967).
16. Winterbottom, W. L., *Surfaces and Interfaces,* Vol. I, Syracuse University Press, Syracuse, New York, (1967), p. 133.
17. Russell, K. C., *Phase Transformations,* ASM, Metals Park, Cleveland, Ohio (1969), p. 219.

18. McLean, M., and Mykura, H., *Surface Sci.,* 5 466 (1966).
19. Somorjai, G. A., *Prog. Anal. Chem.,*1, 101 (1968).
20. Gjostein, N. A., *Surfaces and Interfaces,* Vol. I, Syracuse University Press, Syracuse, New York (1967), p. 271.

DISCUSSION on Paper by P. Wynblatt

BOUDART: We have not found evidence in our work for your model of platinum particles sitting like a geodesic dome on a flat part of the support. On the contrary, all the evidence we have suggests that platinum particles 5 nm or less in diameter in a hydrogen atmosphere are irregular spheres with little contact with the support. If this is so, the interesting question there is: Are dispersion forces adequate to anchor the metallic particles on the support? Have you made estimates of the resulting energy of adhesion?

WYNBLATT: The interpretation of transmission electron micrographs of microporous catalysts is quite difficult, as was evident from the photomicrograph given by Dr. Eischens in the preceding paper. But even if a metal particle residing in an isolated pore could be readily observed, local substrate curvature could substantially modify the equilibrium shape of the particle from that of a simple spherical segment to more complex shapes, as I discussed in my paper in connection with particles growing in pores. However, in our model catalysts consisting of particles supported on a flat substrate, we have found the particles to be reasonably hemispherical (both by tilting experiments in the electron microscope and by replica shadowing), in agreement with our postulate. It should be pointed out that these experiments were performed on particles 12 nm or more in diameter, and that this general picture will become less meaningful in the limit of small atom clusters. The implications of our observations with respect to the work of adhesion is that this quantity has much the same value for 12-nm particles as for those in the macroscopic system (see ref. 13).

SCHMIDT: Oxygen could affect the sintering rate through alterations in the properties of the alumina as well as through those of platinum and through formation of volatile oxides. Professor Somorjai showed several years ago that the oxygen content of α-Al_2O_3 surfaces could be altered by varying the O_2 pressure. Have you measured contact angles and surface diffusion coefficients of platinum on alumina? If not, how do you estimate diffusion coefficients?

WYNBLATT: We have not made detailed measurements of the wetting of alumina by platinum. However, as mentioned in my reply to Professor Boudart's question, we have found that the platinum particles in our flat substrate experiments are reasonably hemispherical, in good agreement with macroscopic experiments performed under oxidizing conditions (see ref. 13). Although there have been no direct measurements of the surface diffusion coefficient of platinum over alumina, we have inferred a value for this quantity from the work of Huang and Li (see ref. 14), as described in the text of the paper.

BOND: Have you measured particle size distributions? If so how does the shape of the distribution change when the average size increases?

WYNBLATT: The initial particle size distribution in our flat substrate samples is a broad gaussian, with the maximum particle size being about twice as large as the mean. However, as growth proceeds, this distribution develops a long tail towards larger particle sizes. This result is in qualitative agreement with our calculated particle size distributions for facet-inhibited vapor phase transport.

BOND: I am a little surprised that you think the conjunction of atoms composing an incomplete outer layer on a particle constitutes the barrier to growth. Does this relate to Bassett's observations that pairwise interactions between metal atoms on a metal surface are somewhat weak?

WYNBLATT: The concept of a nucleation barrier impeding the growth of facetted particles is not new and has been generally recognized since the classical work of Burton, Cabrera, and Frank [*Phil. Trans. Roy. Soc.,* **A243**, 299 (1950)]. However, this does not involve the weakening of the pairwise interaction between adatoms reported by Bassett, which surprised me as much as it did many other participants in this conference.

JAFFEE: The change of rank in particle size stability from reducing conditions, where osmium is highest, to oxidizing conditions, where rhodium is highest, is consistent with experience with platinum-alloy bushings for production of glass fibers. Excessive oxidation of platinum and loss by PtO_2 evaporation may be effectively inhibited by alloying the platinum with 10 to 20 percent rhodium. This suggests a practical method of inhibiting particle growth in supported platinum catalysts.

In the event that it has not been brought out in the paper, what is the reason for the change in stability rank from reducing conditions where heat of sublimation prevails to the rank found under oxidizing conditions?

WYNBLATT: Thank you for your comment. This consititutes yet another example of technological problems brought about by the volatility of PtO_2. With respect to your question, I mentioned in my presentation that the stability ranking of noble metals, under oxidizing conditions, is based on the relative values of the enthalpy for the reaction:

$$Metal_{(solid)} + O_{2(gas)} \longrightarrow Metal\ Oxide_{(vapor)}$$

This criterion for oxidizing conditions produces a different stability series from the one for reducing conditions which is based on the relative magnitudes of the heats of sublimation.

HAIDINGER: Are you aware of the work on particle diffusion (Brownian movement) carried out by Prof. Kern of the University of Marseille? He studied a system in ultrahigh vacuum, so that in his case vapor-phase diffusion can be ruled out. He put forward interesting models including the charging of small particles owing to charge transfer with the support.

WYNBLATT: I am aware of Prof. Kern's work and agree with you that it is very interesting and relevant to our work. However, we have not looked carefully at particle charging resulting from charge transfer with the substrate because, if anything, this would tend to further inhibit the already slow process of particle migration, collision, and coalescence.

OUDAR: Did you try to use another substrate? Maybe it is the best way to confirm your particle growth mechanism in the presence of oxygen.

WYNBLATT: We have not tried other substrates yet, but we hope to do so in the future.

IMELIK: If your model concerning the porous system is correct (filling up of the micropores by the metal) the BET surface area should decrease and the pore size distribution should be changed even for the bigger pores which are not filled. So, simple analysis of the N_2 adsorption isotherms may give some information about the particle growth. A much better approach, however, is to study the system with the low angle X-ray scattering which gives more precise data since we can directly determine the metal particle size and distribution.

WYNBLATT: Even if the growing particles do fill up the pores in which they reside, there is not enough metal around to fill more than 3 to 5 percent of the pores. Thus, it might be difficult to measure the resulting small change in BET surface of the support. On the other hand, I do agree with you that we need some form of direct measurement of particle size and distribution.

EMMETT: I am a little puzzled at your detailed reference to the way in which a capillary could fill up with metallic platinum. If the vapor pressure of a platinum particle is greater for small particles than for large particles, then eventually all of the material would empty out of pores into large crystals on the surface of the support. Under these conditions, the filling of capillaries would be a transient stage in the process.

WYNBLATT: As I tried to explain my paper, the chemical potential (and hence the vapor pressure) of a metal particle depends on the radius of curvature of the metal-vapor interface, rather than on the size of the particle. If one assumes a contact angle of 90 degrees (which is quite close to the measured contact angle of platinum on alumina), it can be shown that particles in pores tend to be stable with respect to infinitely large particles on a flat surface, as illustrated schematically for a variety of plausible pore geometries in Fig. 10. However, the conclusion that particles in pores are more stable than infinite-size particles on a flat surface is not generally true, but depends on both pore geometry and the magnitude of the contact angle θ. If we consider for simplicity the case of a particle in a cylindrical pore, then the particle will be stable with respect to an infinite-size particle for 0 degrees $\leqslant \theta <$ 90 degrees, but unstable for 90 degrees $< \theta \leqslant$ 180 degrees.

A NEW METHOD TO ELUCIDATE THE MECHANISM OF HETEROGENEOUS CATALYSIS BY MEANS OF MICROWAVE SPECTROSCOPIC TECHNIQUES

Toshihiko Kondo

Sagami Chemical Research Center
Nishi-Ohnuma, Sagamihara, Kanagawa, Japan

Kenzi Tamaru

Department of Chemistry
The University of Tokyo, Hongo, Bunkyo-ku
Tokyo, Japan

ABSTRACT

A new method is proposed to identify the intermediates of the hydrogen-deuterium exchange reaction of propene, namely, the microwave spectroscopic techniques which permit determination of the distribution of geometrical isomers of propene-d_1 and -d_2 formed during the course of the reaction. This makes it feasible to identify many reaction intermediates, i.e., 1-propyl, 2-propyl, 1-propenyl, 2-propenyl, σ-allyl, and π-allyl intermediates, as well as the concerted mechanism.

1 INTRODUCTION

In the reaction between hydrogen and oxygen

$$2H_2 + O_2 = 2H_2O$$

we know much about the properties of the hydrogen molecule, such as its binding energy, vibrational frequencies, electronic states, and so forth. It is also the case for the oxygen molecule. This information, however, has almost nothing to do with the elucidation of the mechanism of the H_2-O_2 reaction or its rate of reaction. Recently, many varieties of new physical techniques have been developed which have been revealed to be effective in studying chemisorption on solid surfaces. These chemisorption studies, however, are not necessarily directly associated with the elucidation of mechanisms of heterogeneous catalysis, as in the case of the study of molecular properties of hydrogen and oxygen in elucidating the mechanism of H_2-O_2 reaction.

The chemisorption during the course of reaction cannot be estimated from the adsorption separately measured for each of the reactants and products. It is dependent not only upon the interaction among the surface species, but also upon the position of the rate-determining step, because the chemical potential of the reaction intermediates is largely determined by the position of the rate-determining step.

The properties of the catalytic surface, for instance, work function, in its working state are not the same as those of the clean surface. Under the reaction conditions, some of the reactants, products, intermediates, and others are usually adsorbed on the surface, thus considerably influencing the properties of the sites where the reaction takes place. It is accordingly necessary to study the properties of the catalyst surface in its working state and the adsorption, species, structure, and amounts. The reactivity of each of the surface species under the reaction conditions has to be compared with the rate of the overall reaction, as was first emphasized by one of the authors.[1] The existence of certain adsorbed species does not prove that it is a reaction intermediate through which the overall reaction proceeds. The most abundant adsorbed species is not necessarily the reaction intermediate. The dynamic behavior of each adsorbed species should be examined under the working conditions to identify the reaction intermediates.

Another approach to identification of the reaction intermediate may be realized by examining the dynamic behavior of reaction by means of other methods, such as microwave techniques. Since microwave spectroscopy has extremely high resolution and high sensitivity, it can be applied to the studies of reaction mechanisms through the quantitative analyses of gaseous molecules with permanent electric dipole moments.[2] Propene-d_1 (C_3H_5D) has four geometrical isomers, $CH_2DCH=CH_2$, $CH_3CD=CH_2$, cis-$CH_3CH=CHD$, and trans-$CH_3CH=CHD$, and propene-

d_2 has seven isomers. The rotational transitions of these isomers are observed at the frequencies completely separated from each other. Accordingly, the amount of each of the geometrical isomers of propene-d_1 and -d_2 can be determined by microwave spectroscopy, even in a mixture of all possible isomers $(d_0, d_1, d_2, \cdots d_6)^3$, which is difficult with other methods.

Although some people, including the authors, have studied the mechanism of the catalytic hydrogenation and exchange reactions of propene over various transition metals and their complexes by means of microwave spectroscopic techniques, their results are generally not enough to identify the reaction intermediates, since they did not study the variation of the distributions of propene-d_1 and -d_2 isomers during the course of the reaction.[4]

In this paper the authors propose a new method of identifying the reaction intermediate, i.e., by following the change in the distributions of not only propene-d_1 but also propene-d_2 isomers during the course of the hydrogen-deuterium exchange reaction of propene by means of microwave spectroscopy.

2 THEORY

The hydrogen exchange reaction between propene and a substance containing deuterium (such as, for instance, D, D^+, and D^-) takes place in various ways, but at least one of the hydrogen atoms of the propene molecule should be dissociated prior to, after, or simultaneously with the deuterium-atom addition to propene; these are called the dissociative, associative, or concerted mechanisms, respectively.

In the associative mechanism, there are two intermediates, namely, 1-propyl and 2-propyl species, the half-hydrogenated state of olefin. In the dissociative mechanism, there are many possible intermediates: the 1-propenyl, 2-propenyl, σ-allyl, and π-allyl species. If the hydrogen exchange proceeds by multiple exchange processes, or if the rate of the adsorption-desorption process of propene is slower than that of the exchange process, the propene containing more than one deuterium atom would be produced even at the initial stage of the reaction. In our experiments, mass spectrometric measurements of the deuteropropene formed have demonstrated no such multiple exchange, but rather the stepwise exchange reaction.

Let us consider the distribution of propene-d_1 and -d_2 isomers formed during the course of the reaction via these possible intermediates.

1. *1-Propyl Intermediate.* If the 1-propyl species is the reaction intermediate of the exchange reaction, the propene-d_1 species formed is only propene-2-d_1, deuterium appearing at the central carbon atom of propene. In this case no double-bond shift isomerization of olefin takes place.

$$CH_2=CH-CH_3 \xrightarrow{+D} \underset{*}{CH_2-CHD-CH_3} \xrightarrow{-H} CH_2=CD-CH_3$$

2. *2-Propyl Intermediate.* If a deuterium atom or ion is added to propene to form 2-propyl intermediate, both end carbons become equivalent and the propene-d_1 isomer, which is formed by dissociating one of the five hydrogen atoms of the two methyl groups, is 60 percent propene-3-d_1 and 40 percent propene-1-d_1 species, if the secondary kinetic isotope effect is negligible. Under similar considerations, the distribution of propene-d_2 isomers should be 10 percent propene-$1,1$-d_2 60 percent propene-$1,3$-d_2, and 30 percent propene-$3,3$-d_2. These compositions of propene-d_1 and -d_2 isomers stay constant throughout the reaction, which is the characteristic feature for the 2-propyl intermediate. The double-bond migration also takes place with the exchange reaction.

$$CH_2=CH-CH_3 \xrightarrow{+D} \underset{*}{CH_2D-CH-CH_3} \xrightarrow{-H} \begin{cases} CHD=CH-CH_3 & 40\% \\ \\ CH_2D-CH=CH_2 & 60\% \end{cases}$$

$$CHD=CH-CH_3 \xrightarrow{+D} \underset{*}{CHD_2-CH-CH_3} \xrightarrow{-H} \begin{cases} CD_2=CH-CH_3 & 10\% \\ \\ CHD_2-CH=CH_2 & 30\% \end{cases}$$

$$CH_2D-CH=CH_2 \xrightarrow{+D} \underset{*}{CH_2D-CH-CH_2D} \xrightarrow{-H} CHD=CH-CH_2D \quad 60\%$$

3. *1-Propenyl Intermediate.* In the dissociative mechanism, as propene has four distinguishable hydrogen atoms in its molecule, four different σ-reaction intermediates are possible. In the case of the 1-propenyl intermediate, (Z)-propene-1-d_1 and (E)-propene-1-d_1 are formed through (Z)-1-propenyl and (E)-1-propenyl intermediates, respectively. No olefin isomerizations take place through this mechanism.

$$\underset{\text{H}}{\overset{\text{H}}{>}}\text{C=C}\underset{\text{H}}{\overset{\text{CH}_3}{<}} \xrightarrow{\ -\text{H}\ } \underset{*}{\overset{\text{H}}{>}}\text{C=C}\underset{\text{H}}{\overset{\text{CH}_3}{<}} \xrightarrow{\ +\text{D}\ } \underset{\text{D}}{\overset{\text{H}}{>}}\text{C=C}\underset{\text{H}}{\overset{\text{CH}_3}{<}}$$

4. *2-Propenyl Intermediate.* Propene-2-d_1 is the only product formed through 2-propenyl intermediate mechanism. The distribution of propene-d_1 isomers is the same as that of the 1-propyl intermediate mechanism, although its kinetics are generally different. However, no cis-trans isomerization of olefin takes place through 2-propenyl intermediate, whereas it occurs with the exchange reaction through 1-propyl intermediate. Accordingly, if (Z)-propene-1-d_1 instead of propene-d_0 is the reactant, (Z)-propene-$1,2$-d_2 should be obtained through 2-propenyl intermediate and (E)-propene-$1,2$-d_2 should be produced through the 1-propyl intermediate mechanism.

$$\underset{\text{H}}{\overset{\text{H}}{>}}\text{C=C}\underset{\text{H}}{\overset{\text{CH}_3}{<}} \xrightarrow{\ -\text{H}\ } \underset{*}{\overset{\text{H}}{>}}\text{C=C}\underset{\text{H}}{\overset{\text{CH}_3}{<}} \xrightarrow{\ +\text{D}\ } \underset{\text{H}}{\overset{\text{H}}{>}}\text{C=C}\underset{\text{D}}{\overset{\text{CH}_3}{<}}$$

5. *σ-Allyl Intermediate.* Propene-3-d_1 and propene-$3,3$-d_2 are formed only through σ-allyl intermediate as mono- and dideuteropropenes, respectively. No deuteropropene containing more than three deuterium atoms should appear from this mechanism.

$$CH_2\text{=}CH\text{-}CH_3 \xrightarrow{\ -\text{H}\ } \underset{*}{CH_2\text{=}CH\text{-}CH_2} \xrightarrow{\ +\text{D}\ } CH_2\text{=}CH\text{-}CH_2D$$

$$CH_2\text{=}CH\text{-}CH_2D \xrightarrow{\ -\text{H}\ } \underset{*}{CH_2\text{=}CH\text{-}CHD} \xrightarrow{\ +\text{D}\ } CH_2\text{=}CH\text{-}CHD_2$$

If the intramolecular double-bond migration of σ-allyl intermediate is much faster than the desorption process, the distributions of propene-d_1 and -d_2 isomers should be the same as those for the π-allyl intermediate mechanism. In this case, the intermediate may be called dynamic σ-allyl intermediate.

$$\underset{*}{CH_2\text{=}CH\text{-}CH_2} \rightleftharpoons \underset{*}{CH_2\text{-}CH\text{=}CH_2}$$

6. *π-Allyl Intermediate.* When π-allyl intermediate is formed by dissociating one of the three hydrogen atoms of the methyl group of propene, both of its end carbons become equivalent. Propene-d_1 isomers are formed by adding a deuterium atom or ion to one of the two end carbons of the π-allyl species, and, consequently, the propene-d_1 isomer initially produced is only propene-3-d_1, while that for propene-d_2 is 50 percent propene-1,3-d_2 and 50 percent propene-3,3-d_2. According to this mechanism, since propene has five hydrogen atoms which participate in the exchange reaction, the statistical distribution of propene-d_1 and -d_2 isomers in the exchange equilibrium is the same as that for the 2-propyl intermediate mechanism.

7. *Concerted Mechanism.* In this mechanism, one of the hydrogen atoms of the methyl group of propene is dissociated, a deuterium atom or ion being simultaneously added to the methylene group of the propene. The initial product is, accordingly, only propene-3-d_1. The propene-d_2 species initially formed is accordingly 100 percent propene-1,3-d_2 species. Propene-1-d_1, propene-3,3-d_2, and propene-1,1-d_2 are subsequently produced by repeating the process.

The changes in the compositions of those isomers for each of the mechanisms are schematically summarized in Fig. 1, which illustrates the various distributions of propene-d_1 and -d_2 isomers with the reaction time for the possible reaction mechanisms. Of course, varieties of modifications of each of the mechanisms are theoretically possible, but the figure gives a sort of classification of these possibilities on the basis of the distributions of d_1 and d_2 species.

From Fig. 1 it is clear that the distributions of propene-d_1 isomers are the same both in the π-allyl intermediate mechanism and in the concerted mechanism, but from the distributions of propene-d_2 isomers one can easily tell which mechanisms are operative. It is also to be noted here that the changes in the distributions of d_1 and d_2 isomers with time give the first crucial evidence for the concerted mechanism.

3 EXPERIMENTAL RESULTS

3.1 2-Propyl Intermediate

The hydrogen exchange reaction takes place between propene and deuterophosphoric acid (D_3PO_4) at room temperature, and is a reaction similar to the double-bond migration of olefin by phosphoric acid. The

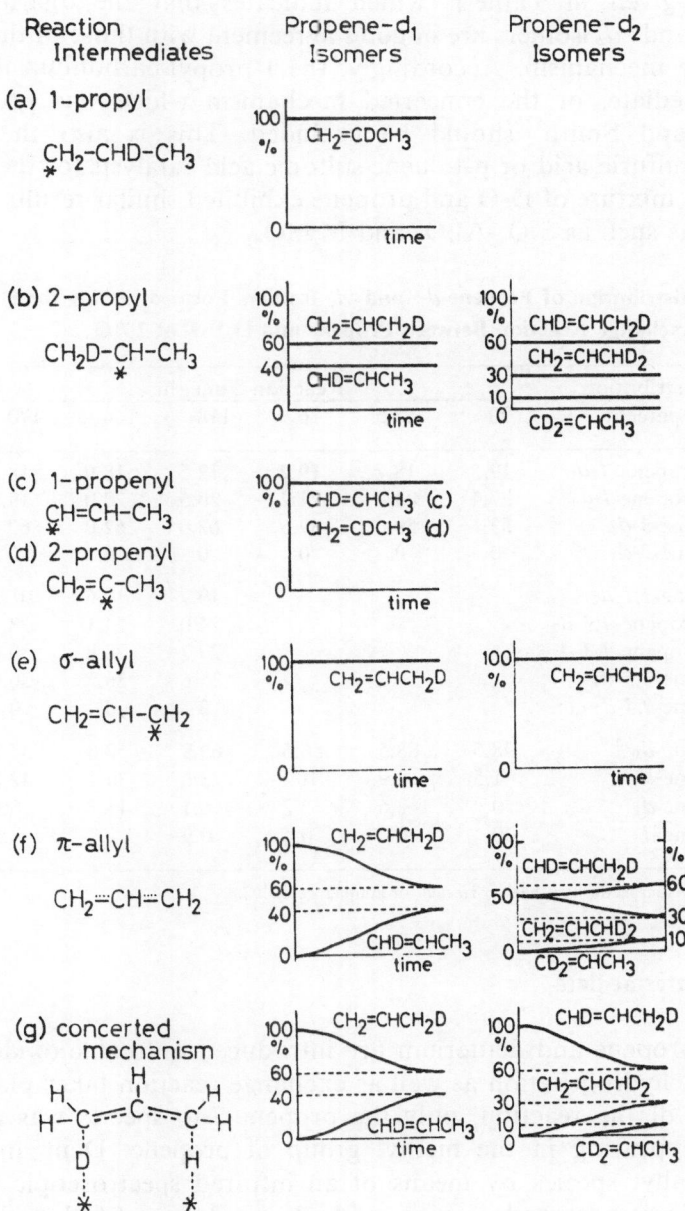

Fig. 1. Possible reaction intermediates of the hydrogen-deuterium exchange reaction of propene and the change in the distribution of propene-d_1 and -d_2 species with time.

results are given in Table I, which indicates that the distributions of propene-d_1 and -d_2 isomers are in good agreement with those of the 2-propyl intermediate mechanism. Accordingly, the 1-propyl carbonium ion, the π-allyl intermediate, or the concerted mechanism which was proposed by Turkevich and Smith[5] should be excluded. This is also the case for deuterated sulfuric acid or p-toluene sulfonic acid catalysts for the exchange reaction.[6] A mixture of D_2O and propene exhibited similar results over solid acid catalysts such as SiO_2-Al_2O_3 and $MgSO_4$.

Table I. **Distribution of Propene-d_1 and -d_2 Isomers Formed During the Hydrogen Exchange Reaction Between Propene and D_3PO_4 at 28°C**

Distribution, percent	Reaction Time, hr					
	20	68.5	76	118	144	170
(Z)-Propene-1-d_1	19.5	18.7	19.2	17.5	18.0	18.4
(E)-Propene-1-d_1	27.4	21.7	21.2	20.5	20.0	19.6
Propene-3-d_1	53.1	59.6	59.6	62.0	62.0	62.0
Propene-2-d_1	0	0	0	0	0	0
Propene-1,1-d_2				10.7	10.6	10.3
(Z)-Propene-1,3-d_2				24.0	24.1	23.1
(E)-Propene-1,3-d_2				29.7	28.8	30.6
Propene-3,3-d_2				35.6	36.5	36.0
Propene-2,3-d_2 etc.				0	0	0
Propene-d_0[a]	98.5	88.5	86.5	69.9	52.3	37.2
Propene-d_1	1.5	8.9	10.8	23.0	34.1	42.0
Propene-d_2	0	2.6	2.2	6.1	11.1	16.9
Propene-d_3	0	0	0.4	0.9	2.2	3.5

(a) All mass spectra have been corrected for abundance of ^{13}C.

3.2 π-Allyl Intermediate

When propene and deuterium are introduced onto zinc oxide at room temperature, hydrogenation as well as exchange reaction takes place. In the initial stage of the reaction, only the propene-3-d_1 species was produced, deuterium appearing in the methyl group of propene. Dent and Kokes[7] observed π-allyl species by means of an infrared spectroscopic technique when propene is adsorbed on zinc oxide, but it does not follow that the π-allyl species would be the reaction intermediate for the hydrogen exchange reaction of propene with deuterium gas.

The formation of propene-3-d_1 species suggests the possibility of π-allyl intermediate and the concerted mechanism. To distinguish the two mechanisms the deuterium distribution in the propene-d_2 was examined as shown in Fig. 2, which demonstrates π-allyl intermediate.[8] The same

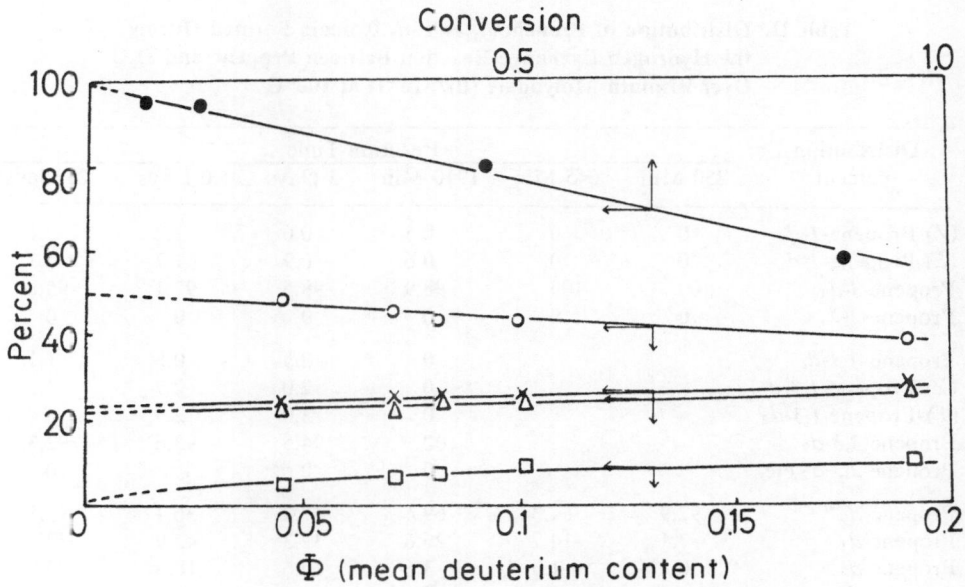

Fig. 2. The deuterium distribution in propene-d_2 and propane-d_2 molecules over ZnO (10 g); $P_{C_3H_6} = P_{D_2} = 12.5$ cm Hg at room temperature. O : propene-$3,3$-d_2; △ : propene-cis-$1,3$-d_2;

×: propene-trans-$1,3$-d_2; □: propene-$1,1$-d_2; ●: propane-$1,2$-d_2; $\Phi = \sum_{i=1}^{6} id_i \Big/ 6 \sum_{i=0}^{6} d_i$ (mean

deuterium content). Conversion = propane/(propene + propane).

propene-D_2 reaction on a stoichiometric lamellar compound of graphite and potassium, $C_{24}K$, occurred via π-allyl intermediate[9], while propene over DMSO(d_6)-BuOK(+BuOH) gave a deuterium distribution of propene similar to π-allyl intermediate mechanism.

3.3 σ-Allyl Intermediate

The next example is a σ-allyl intermediate, which is involved in the reaction of propene and D_2O vapor over bismuth molybdate catalyst. The bismuth molybdate catalyst is a selective catalyst for the oxidation of propene to acrolein, and it is, accordingly, interesting to examine the adsorbed state of propene on such catalysts. When a mixture of propene and D_2O was circulated over bismuth molybdate at 100°C, the results of the microwave and the mass analyses were as given in Table II. The main parts of the propene-d_1 and -d_2 isomers are propene-3-d_1 and propene-$3,3$-d_2, respectively, throughout the course of the reaction. Consequently, the reaction proceeds through σ-allyl intermediate.

When the reaction was carried out at higher temperature (300°C), the results changed to some extent, as shown in Table III. The distributions of

Table II. Distribution of Propene-d_1 and -d_2 Isomers Formed During the Hydrogen Exchange Reaction Between Propene and D_2O Over Bismuth Molybdate (Bi/Mo=1) at 100°C

Distribution, percent	Reaction Time					
	250 Min	645 Min	1910 Min	3 Days	6 Days	23 Days
(Z)-Propene-1-d_1	0	0	0.5	0.6	1.2	2.1
(E)-Propene-1-d_1	0	0	0.6	0.9	1.7	2.1
Propene-3-d_1	100	100	98.9	98.5	97.1	95.8
Propene-2-d_1	0	0	0	0	0	0
Propene-$1,1$-d_2			0	0.5	0.8	1.1
(Z)-Propene-$1,3$-d_2			0	2.0	2.7	3.2
(E)-Propene-$1,3$-d_2			0	3.0	2.9	3.4
Propene-$3,3$-d_2			100	94.5	93.6	92.3
Propene-$2,3$-d_2 etc.			0	0	0	0
Propene-d_0[a]	92.9	84.3	69.4	47.9	36.1	15.5
Propene-d_1	7.1	14.7	26.6	39.3	42.9	39.3
Propene-d_2	0	1.0	3.6	10.6	18.4	33.8
Propene-d_3	0	0	0.2	2.2	2.6	10.8

(a) All mass spectra have been corrected for abundance of ^{13}C.

Table III. Distribution of Propene-d_1 and -d_2 Isomers Formed During the Hydrogen Exchange Reaction Between Propene and D_2O Over Bismuth Molybdate (Bi/Mo=1) at 300°C

Distribution, percent	Reaction Time					
	10 Min	25 Min	45 Min	185 Min	1872 Min	9 Days
(Z)-Propene-1-d_1	3.6	3.7	4.8	7.2	10.4	13.7
(E)-Propene-1-d_1	4.8	4.6	4.9	6.8	10.3	15.4
Propene-3-d_1	91.6	91.7	90.3	86.0	79.3	70.9
Propene-2-d_1	0	0	0	0	0	0
Propene-$1,1$-d_2				0	5.1	6.6
(Z)-Propene-$1,3$-d_2				13.4	23.0	24.3
(E)-Propene-$1,3$-d_2				17.1	25.0	26.4
Propene-$3,3$-d_2				69.5	46.9	42.7
Propene-$2,3$-d_2 etc.				0	0	0
Propene-d_0[a]	97.8	94.9	93.4	87.6	81.7	70.3
Propene-d_1	2.2	4.9	6.4	10.9	16.5	24.4
Propene-d_2	0	0.2	0.2	1.3	1.5	4.4
Propene-d_3	0	0	0	0.2	0.3	0.8

(a) All mass spectra have been corrected for abundance of ^{13}C.

propene-d_1 isomers are similar to those of π-allyl intermediate (or concerted mechanism), but the initial distribution of propene-d_2 isomers is between those of σ-allyl and π-allyl intermediates, which may be interpreted to exhibit dynamic σ-allyl intermediates, since we know some examples of dynamic σ-allyl species which have been reported in various systems, for instance, allylmagnesium bromide, C_3H_5MgBr[10], and diallylzinc $(C_3H_5)_2Zn$[11]; in these systems the species is σ-allyl at lower temperatures and dynamic σ-allyl at higher temperatures.

3.4 1-Propenyl Intermediate

When a mixture of deuterium and propene was introduced over a γ-alumina at room temperature, the d_1 species formed in the very initial stage of the reaction consisted of only cis-1-d_1 and trans-1-d_1 propene, but this was followed by 3-d_1 species formation through the double-bond migration of the propene molecule.[12]

The reaction intermediates in other catalytic systems were also studied. It was reported that the deuterium-propene exchange reaction proceeds through 1-propyl and 2-propyl intermediates over stoichiometric metal phthalocyanine-alkali metal electron donor-acceptor complexes.[13] The contribution of the two intermediates markedly depends upon the stoichiometry of the complexes, as shown in Table IV.

Table IV. Propene-Deuterium Exchange Reaction Over Ni-Pc^{x-} xNa$^+$ and Co-Pc^{x-} xNa$^+$ (180° C. Pc = Phthalocyanine = 0.32 g); $P_{C_3H_6}$ = 13.0 cm Hg

Complexes	Electron Configuration	Rate Φ/Hr^{-1}	Reactivity of Each Hydrogen Bonded to		
			C_1	C_2	C_3
Ni-Pc^{4-} 4Na$^+$	$d^8 + \pi^4$	8.3 x 10^{-2}	0.15	0.79	0.06
Ni-Pc^{3-} 3Na$^+$	$d^8 + \pi^3$	5.8 x 10^{-3}	0.20	0.70	0.10
Ni-Pc^{2-} 2Na$^+$	$d^8 + \pi^2$	2.1 x 10^{-4}	0.23	0.54	0.23
Ni-Pc$^-$ Na$^+$	$d^8 + \pi$	5.8 x 10^{-5}	0.25	0.59	0.16
Co-Pc^{5-} 5Na$^+$	$d^8 + \pi^4$	1.9 x 10^{-2}	0.41	0.05	0.54
Co-Pc^{4-} 4Na$^+$	$d^8 + \pi^3$	1.8 x 10^{-4}	0.34	0.41	0.25
Co-Pc^{3-} 3Na$^+$	$d^8 + \pi^2$	4.5 x 10^{-5}	0.21	0.72	0.07
Co-Pc$^-$ Na$^+$	d^8				

4 CONCLUSION

The microwave spectroscopic method has been shown to be effective for identifying the reaction intermediates of the hydrogen-deuterium exchange reaction of propene or adsorbed state of propene over catalyst surfaces.

REFERENCES

1. Tamaru, K., *Bull. Chem. Soc. Japan,* **31,** 647 (1958); *Advances in Catalysis,* Vol. XV Academic Press, New York (1964), p. 65.
2. Townes, C. H., and Schawlow, A. L., *Microwave Spectroscopy,* McGraw-Hill, New York (1955).
3. Morino, Y., and Hirota, E., *J. Chem. Soc. Japan,* **85,** 535 (1964); Scharpen, L. H., Rauskolb, R. F., and Tolman, C. A., *Anal. Chem.,* **44,** 2010 (1972).
4. See for example, Hirota, K., Hironaka, Y., and Hirota, E., *Tetrahedron Lett.,* p. 1645 (1964); Hirota, K., and Hironaka, Y., *Bull. Chem. Soc. Japan, 39, 2638 (1966); J. Catalysis,* **4,** 602 (1965), Sakurai, Y., Kaneda, Y., Kondo, S., Hirota, E., Onishi, T., and Tamaru, K., *Trans. Faraday Soc.,* **67,** 3275 (1971); Hirabayashi, K., Saito, S., and Yasumori, I., *J. Chem. Soc. Faraday Trans. I,* **68,** 978 (1972); Tolman, C. A., and Schalrpen, L. H., *J. Chem. Soc. Dalton,* 584 (1973).
5. Turkevich, J., and Smith, R. K.. *J. Chem. Phys.,* **16,** 466 (1948).
6. Sakurai, Y., Kaneda, Y., Kondo, S., Hirota, E., Onishi, T., and Tamaru, K., *Trans. Faraday Soc.,* **67,** 3275 (1971); Kondo, T., Ichikawa, M., Saito, S., and Tamaru, K., *J. Phys. Chem.,* **77,** 299 (1973).
7. Dent, A. L., and Kokes, R. J., *J. Am. Chem. Soc.,* **92,** 6709 (1970).
8. Naito, S., Kondo, T., Ichikawa, M., and Tamaru, K., *J. Phys. Chem.,* **76,** 2184 (1972).
9. Kondo, T., Ichikawa, M., Saito, S., and Tamaru, K., *Bull. Chem. Soc. Japan,* **45,** 1580 (1972).
10. Nordland, J. E., and Roberts, J. D., *J. Am. Chem. Soc.,* **81,** 1769 (1959).
11. Wilke, G., et al., *Angew. Chem., Int. Ed. Engl.,* **5,** 151 (1966).
12. Sakurai, Y., Onishi, T., and Tamaru, K., *Trans. Faraday Soc.,* **67,** 3094 (1971); *Bull. Chem. Soc. Japan,* **45,** 980 (1972).
13. Naito, S., Ichikawa, M., Saito, S., and Tamaru, K., *J. Chem. Soc. Faraday Trans. I,* **69,** 685 (1973).

DISCUSSION on Paper by K. Tamaru

YATES: Could you tell us the amount of gaseous material needed to do these microwave analyses of deuterium distribution in organic molecules?

TAMARU: In our experiments, the pressure of the gas sample is 0.02 Torr at dry-ice temperature and its volume is 600 to 700 ml. We can detect any isomers containing more than 0.1 percent of the sample.

FISCHER: If I understand correctly, microwave spectroscopy couples to the rotation of molecules. Can it observe molecules adsorbed on metal surfaces? Can it observe reaction intermediates adsorbed on the insulator in bifunctional catalysts? Can it measure molecules in the gas phase in dispersed metal catalysts?

TAMARU: To observe molecules adsorbed on metal surfaces by means of microwave spectroscopy it is necessary for an appreciable amount of adsorbed species to have the same discrete rotational energy levels in the microwave region with uniform geometrical configurations in spite of the perturbation from solid surfaces, which seems to be not probable (not impossible, however) in many cases. In the gas phase, the metal atoms combined with gas molecules may be studied by microwave spectroscopy in some special cases such as that of metal carbonyl, but, I am afraid, the preparation of a system of uniformly dispersed metal catalyst with the same numbers of metal atoms as well as adsorbed molecules with the same geometrical configuration would be very difficult.

KEMBALL: We have investigated the exchange of propylene recently on a number of oxides using the microwave technique and there is good general agreement with the results reported by Prof. Tamaru. But there is one case where caution is required in the interpretation of the results. The distribution of products corresponding to a 2-propyl or carbonium ion intermediate is exactly the same as the equilibrium distribution for the scrambling of deuterium atoms between the five terminal positions of the molecule. Such a distribution is found for reactions of propylene with deuterium on magnesium oxide at $220°K$. It is unlikely that carbonium ions are involved under these conditions. A preferred explanation is that during residence on the catalyst, there is a rapid (perhaps intramolecular) scrambling of the terminal atoms of the molecule—a suggestion which is supported by the high activity of magnesium oxide for the isomerization of but-1-ene at low temperatures.

TAMARU: In the case of the 2-propyl intermediate we have to be always careful on your point. We have to examine the rate of isomerization of olefins and that of the deuterium exchange reaction to check the point.

SCHRIEFFER: It appears that the experiments you described involving phthalocyanine charge-transfer complexes are particularly significant for the theorists, since as I understand, it is the π-electron structure that is primarily involved in the reaction. This π structure is likely to be easier to treat theoretically than the more conventional α-orbitals in transition-metal catalysis. The observed variation of activity with electron transfer should be a valuable guide in constructing a theoretical treatment of this reaction. Am I correct in assuming that the π-structure dominates the reaction mechanism?

TAMARU: Correct. When D_2 is introduced onto the film of electron donor-acceptor complexes of various metal phthalocyanines with alkali metals, HD is readily formed. The rate of the HD formation is dependent only upon the number of electrons donated to the π-conjugated system of the ligand, irrespective of the kind of central metal ion of the phtalocyanines [Naito, S., Ichikawa, M., and Tamaru, K., *J. Chem. Soc., Faraday Trans. I,***68**, 1451 (1972)].

WALTER: In what phase were the phthalocyanine complexes used in your experiments? (Solid, dissolved?)

TAMARU: The stoichiometric electron donor-acceptor complexes of phthalocyanine with alkali metals were prepared in solution, but catalytic reactions were mainly studied over the films of these complexes after the solvent was evaporated. (In some cases, however, they were studied in solution.)

EISCHENS: You have determined the reaction intermediates in reactions catalyzed by phthalocyanines. Is it now possible to examine the adsorbed species by methods such as infrared to learn whether the spectroscopic methods provide a valid observation of the intermediates?

TAMARU: It is possible, as you know, to study the adsorbed species by infrared methods. However, even if we observe an adsorbed species, it does not follow that it is a reaction intermediate through which the overall reaction proceeds. It is necessary to study the dynamic behavior of adsorbed species to identify the real reaction intermediate. Such dynamic treatment of the adsorbed species in its working state is not easy for many catalytic reactions and, accordingly, this microwave spectroscopy method is important in this sense, as it deals with the direct dynamic behavior of the reacting molecule.

EXPERIMENTAL RESULTS RELATING TO THE INFLUENCE OF HOT ELECTRONS (ELECTRICAL CURRENT FLOW THROUGH THE CATALYST) ON CATALYZED REACTIONS

W. Haidinger

Technicum Cantonal de Bienne
Bienne, Switzerland

J. Figar

Batteile
Geneva Research Centre
Geneva, Switzerland

ABSTRACT

It is shown experimentally with various chemical reactions that hot electrons (electrons having some energy excess above the Fermi level) flowing from inside the catalyst towards its surface may influence the catalytic activity and the selectivity. To produce a hot-electron flow, a metal-insulator-metal (M-I-M) thin-layer device was used, and the catalytic behavior of the metal top layer, exposed to the reacting gases, was studied. Reactions with different mechanisms, such as decomposition of formic acid, ammonia, and tert-butyl chloride, and oxidation of propylene, showed an unusual behavior during the passage of the hot-electron flow through the catalyst. A substantial increase in overall reaction rates was observed for all reactions, and an interesting variation of selectivity for propylene oxidation. Principal experimental results are given and discussed.

1 INTRODUCTION

Our experimental study, which will be referred to here, is based on the leading idea that *momentum and energy exchange may be assumed to occur* between free electrons moving inside the catalyst and adsorbed species on the surface. Such energy exchange would *influence the rate of* elementary processes on the surface which are involved in the catalyzed reaction, and would, in this way, influence the rate and/or the pathway of this catalyzed reaction.

By free electrons inside the catalyst, we mean electrons having some excess energy (energy above the Fermi level) and consider these electrons as an "electron beam" which moves *from inside the catalyst towards its surface.* An analogical situation to this would be an electron beam bombarding a surface (with preadsorbed species) under high vacuum; this would lead to the well-known electron impact desorption. The essential difference, however, is that in our case ("bombardment" of the surface from inside) we have transformed a complex physical instrument (electron-impact desorption requires UHV installation) into a *true catalyst* that can be employed in any ordinary catalytic reactor, i.e., a catalyst the activity and selectivity of which can be regulated, independently of reaction temperature and reactant concentrations, by means of the applied voltage.

One of the most interesting features of such a setup is the possibility of studying the interaction between the electron beam and adsorbed species in *any* desired working condition outside the solid surface, meaning at *any* desired gas pressure and temperature. Let us remind you of the famous "pressure gap" in understanding catalytic processes, which has been repeatedly mentioned during this colloquium and could possibly be bridged in this way.

Now, let us suppose we have the catalyst with the chemisorbed complexes (intermediates of a catalyzed reaction) on its surface, and an electron beam comes from the inside of the catalyst body towards this surface. An energy transfer (energy and momentum exchange) from these electrons to the chemisorbed species may occur, and a variation in the rate of the catalyzed reaction (in which these species are involved) is to be expected.

With this general picture in mind, we had to deal with two principal problems before starting the experimental work:

1. How can we create an electron beam coming from inside the catalyst, arriving at the active surface, and interacting with the adsorbed species?

2. How can we prove that the assumed catalytic effect of those electrons does occur, and which catalyzed reactions are the most suitable for such a proof?

Concerning the first problem, we used metal-insulator-metal (M-I-M) devices of relatively large areas, composed principally as shown in Fig. 1.

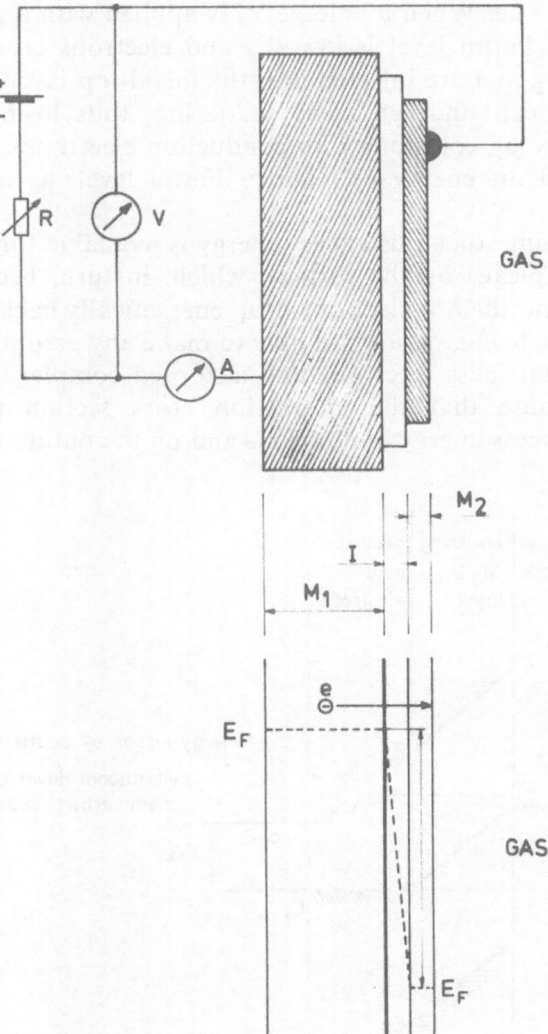

Fig. 1. Sandwich structure of a catalytic device and d.c. supply circuit (schematical); M_1, M_2 ... metal layers; I ... insulator layer.

Each device is deposited on a thick nickel disk by covering the nickel surface with a thin layer of Al₂O₃ or SiO₂, on which is deposited a thin layer of gold.

Although large-area junctions are usually not employed for the production of hot electrons, in order to prevent pinhole effects, etc., we used M-I-M devices, each of which had an active area of about 1 cm². For the purpose of measuring a sizable effect on reaction rate in a reactor of relatively great volume, five such devices were switched in parallel.

A corresponding band diagram (Fig. 2) shows a possible hot-electron path in such a device. When a voltage V_1 is applied with a positive bias on the top layer, its Fermi level is lowered and electrons cross the insulator layer by tunnelling and are injected into the metal top layer. It must be admitted that they can undergo some scattering, thus losing part of their energy (principally by collisions with conduction electrons). They reach the surface then with an energy eV_2 above Fermi level, as indicated in the Fig. 2.

Now, we assume that this excess energy is available for transfer to the chemisorbed complexes on the surface, which, in turn, become activated. Owing to this event, the hot electrons drop energetically back into the Fermi level. For the time being, we are not able to make any assumption about the interaction between these electrons and adsorbed complexes, with one exception: we suppose that the interaction cross section in this process depends on the excess energy of electrons and on the nature of the adsorbed

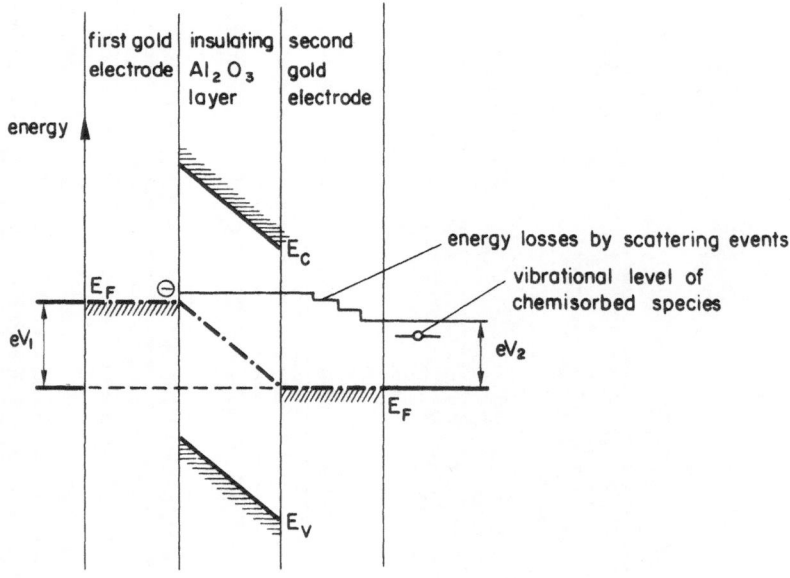

Fig. 2. Band diagram illustrating a possible hot-electron path.

complex. Thus, a reaction with a simple elementary step on the surface is expected to be accelerated because of activation of the chemisorbed intermediate which should be manifested by an increase in reaction rate (enhancement of activity). For a complex reaction, where several elementary steps take place on the surface with different intermediates involved, only one of the intermediates is expected to be preferentially activated and, in this way, only one of the possible reactions accelerated by hot electrons of appropriate energy (selectivity of catalyst is changed).

The electron transport in the mentioned M-I-M device involved many complex problems, theoretical as well as practical. Instead of tackling these problems, we decided, first, to find out whether or not the predicted effect does occur in experiment.

So we temporarily put aside the knowledge of the energy exchange mechanism, the energy distribution, and other properties of the hot-electron beam. Instead, care was taken with the chemical part of the experiments in order to insure that significant conclusions could be drawn from the results.

With these limitations in mind, the aim of our work may be defined more precisely on the basis of two questions:

1. Is there a variation of catalytic activity which can be attributed to an interaction of hot electrons with adsorbed species?

2. If so, is there any influence on the selectivity for complex reactions, as would be expected on the basis of different interaction cross sections of different intermediates?

To obtain the answers to these questions, we carried out activity and selectivity measurements on the reactions summarized in Table I.

The main results of these measurements are described in more detail in the following.

Table I. Reactions Employed in Activity and Selective Measurements

Reaction	Reactor Type	Measured Property	Catalytic Property
Formic acid decomposition	Static	Reaction rate, dp/dt	Activity
NH_3 decomposition	Static	Reaction rate, dp/dt	Activity
Propylene oxidation	Flow	Product composition, GCH	Activity, selectivity
Tert-butyl chloride decomposition	Flow	Product composition, GCH	Activity

2 EXPERIMENTS ON FORMIC ACID DECOMPOSITION

Formic acid decomposition was chosen as the first reaction. This reaction, with a well-known and relatively simple mechanism, is reported to proceed on the gold surface almost entirely as dehydrogenation:

$$HCOOH \rightarrow H_2 + CO_2$$

The reactor used for this study was described elsewhere.[1]

The typical experimental-record curve is shown in Fig. 3. Measurement of the pressure increase is divided into three subsequent time intervals characterized by current off—current on—current off cycles. With this experimental scheme, we sought to prevent aging or poisoning effects on the measured reaction rate.

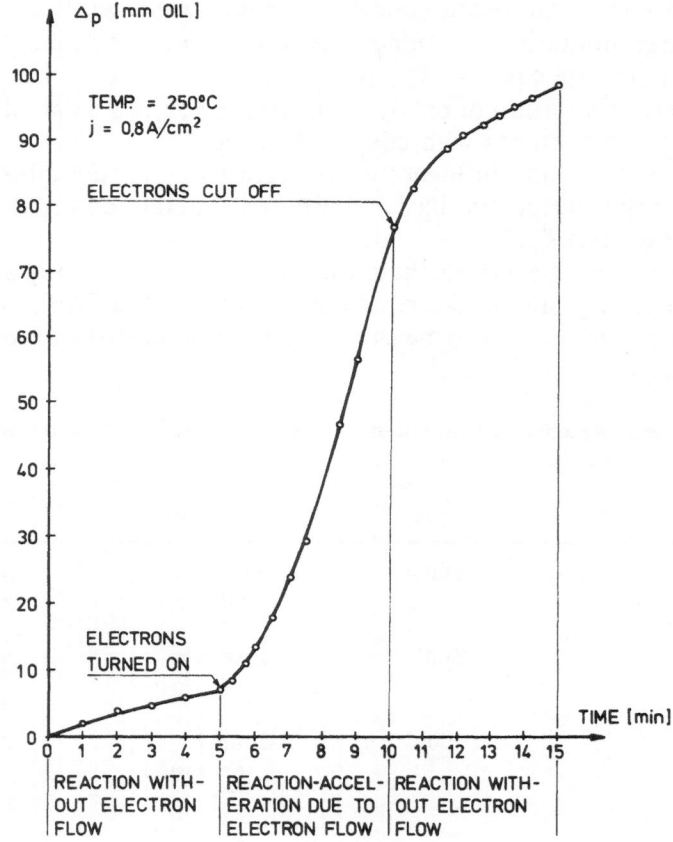

Fig. 3. Pressure difference Δp versus time plot showing the increase of the HCOOH decomposition rate during hot-electron flow.

Closer examination of the curve reveals that the first and the last parts are almost parallel, which is to be expected. The middle part has two tails, which are more pronounced at smaller concentration of formic acid. To explain these tails, we assume that a rearrangement of the surface population must take place at the moment of sudden change in catalytic activity of the surface.

In fact, steady-state concentrations of reactants and reaction products are established on the catalyst surface during the reaction, with constant activity. Since the activity is suddenly changed, new steady-state concentrations must be established, which are determined by the rates of the elementary processes involved in the overall reaction (diffusion, adsorption, surface reaction, desorption). Let us consider the situation at the moment of switching off the current. There must be an excess of reaction products in the surface population, so that before the steady-state corresponding to low activity is reestablished, the products must get away from the surface. This can be described formally as diffusion from a planar source of limited capacity into a semi-infinite medium. Solution of the differential diffusion equations, for this case, can be approximated, for small time values, by a linear function concentration versus square root of time. Figure 4 shows that this is indeed the case for the second tail. The case of the first tail is, of course, the reverse situation, which means diffusion into a limited volume from semi-infinite media, and can no longer be approximated in this simple way.

If you look at the reaction rate line in the first current-off period, you see that it is slightly curved. This behavior does not correspond to the zero-order kinetics. We suppose that an autopoisoning is taking place, owing for instance to the possible side reactions where formaldehyde is formed

$$2 \ HCOOH \rightarrow HCHO + CO_2 + H_2O$$

During the current-on periods, this was never observed. So we suppose that the electron action is the cause here. This is indicated also by the following observation: the initial activity of the gold surface in the first experiments decreased successively from experiment to experiment. When we included a period of activating current passage through the catalyst in vacuum, the original activity was restored and reproducible results were obtained.

A series of experiments was performed with different catalytic devices and under different reaction conditions. The results seemed promising, for an increase of the decomposition rate by as much as 100-fold was observed.

At this point the question has to be asked whether it is really the influence of hot electrons, and not another trivial effect, which increases the activity of the catalyst. The results of an experiment giving answers to this question are shown in Fig. 5.

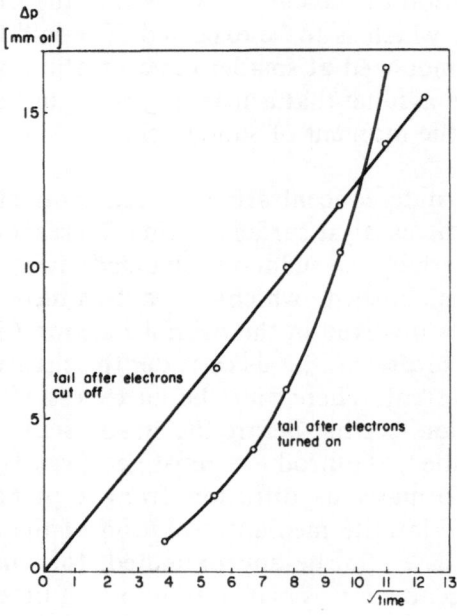

Fig. 4. Pressure difference versus square root of time plot for the first and second tail.

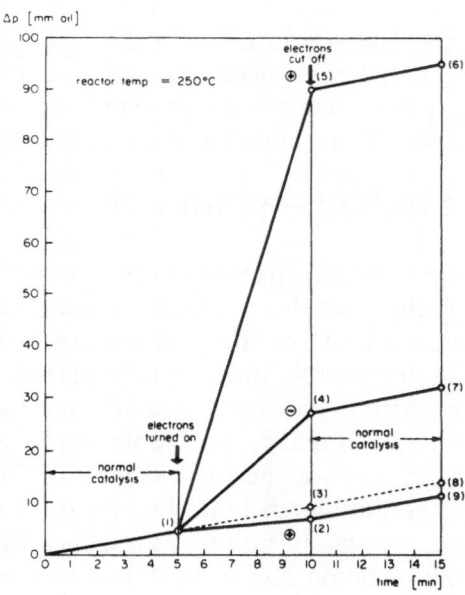

Fig. 5. Influence of the bias voltage polarity and of the structure of the catalytic device on HCOOH decomposition rate: Curve 0-1-5-6 ... top layer positive; Curve 0-1-4-7 ... top layer negative; Curve 0-1-2-9 ... device without insulator layer.

All the experiments represented here were performed at the same temperature and formic acid concentration, the first and second being performed even on the same catalytic device. The first curve corresponds to the experiment where the active surface was positively biased, so that the electron flow was directed towards the surface. In the second experiment, the surface was negatively biased. Electron flow is, consequently, directed away from the surface. There is still an increase in reaction rate, though much smaller, being about 20 to 25 percent of the reaction-rate increase observed in the first experiment. This behavior was found repeatedly with almost all of the investigated catalytic devices. We do not know exactly why an increase is still observed in the second case, but we assume that the presence of hot holes as energy carriers might be the cause.

The third experiment was carried out with a device where the gold layer was directly evaporated on the nickel support, without an intermediate insulator layer. There is thus no appreciable shift of the Fermi level, and hot electrons cannot participate in this case. Almost the same electric power was applied to the device as in the first and second experiments. No increase in reaction rate was found; on the contrary, a slight decrease was observed. From these results, we concluded that the presence of hot electrons is responsible for the increase in reaction rate.

So far, the experimental results obtained correspond well with the results we expected on the basis of the simplified model of energy transfer. The situation is, however, not so simple, as we discovered from two experiments, the results of which are shown in Figs. 6 and 7.

The Arrhenius plot, shown in Fig. 6, was obtained with a catalytic device where the insulator layer was Al_2O_3. A similar plot, shown in Fig. 7, was obtained with devices where the insulator was SiO_2. The straight lines, corresponding to reaction rates at different intensities of activating current, were calculated by the least-squares method. What is rather surprising is that the typical pattern of the compensation effect is found. This goes so far that above the temperature of the intersection point, the reaction is slowed down at the moment when the activating current is switched on.

What would instead be expected, if the simple "bombardment" mechanism were valid, is no compensation or anticompensation behavior; in fact, the quantity of molecules decomposed because of the electron action should be added to the quantity of molecules decomposed because of the "natural" activity of the gold surface. The reaction rate without activation current should therefore be the smallest reaction rate at a given temperature. This, however, is not the case. It is worthwhile noting here that the existence of the compensation effect completely rules out a trivial thermal effect by Joule heating of the catalyst surface (as discussed in greater detail in Ref. 1).

We are not sure that the known hypotheses proposed for the explanation of the compensation effect for conventional catalysts can be applied to our case. Also, we do not yet know whether the same behavior is to be

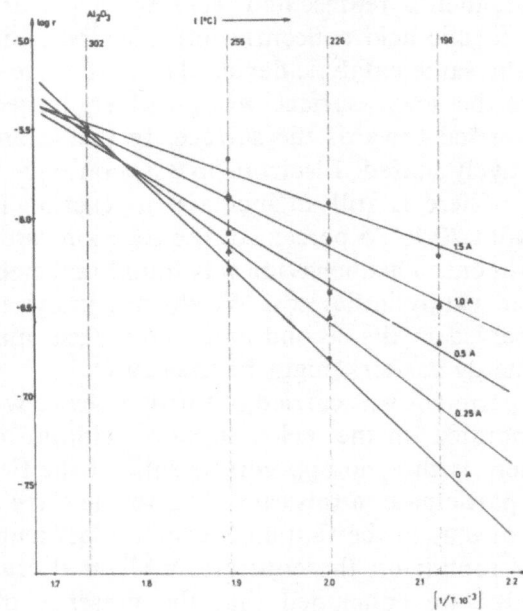

Fig. 6. Arrhenius plot showing compensation effect for a device with an Al_2O_3 insulator layer.

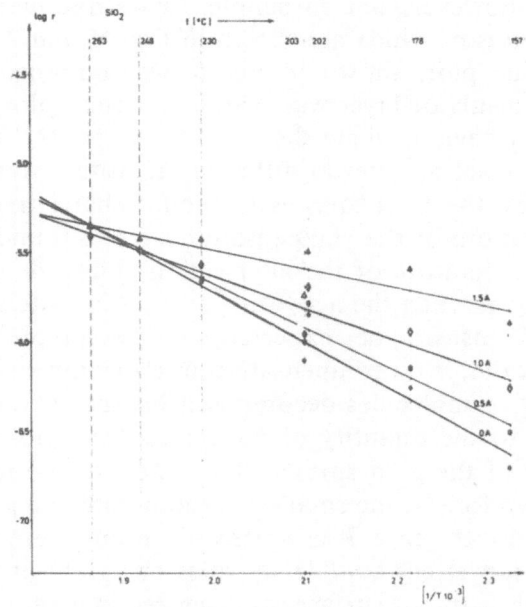

Fig. 7. Arrhenius plot showing compensation effect for a device with an SiO_2 insulator layer.

found for each type of our catalytic device, and larger scale experimental work must be undertaken to this end. What is already clear is that different devices have different temperatures of intersection (compare Figs. 6 and 7).

What we may imagine is that the hot-electron action diminishes the surface coverage, the nonreacted molecules being desorbed with different rates at different temperatures. But this assumption must be tested further.

Some other possible explanations can be taken into consideration. One, proposed by Kemball, is based on the compensating relation between heat and entropy of adsorption. This would explain qualitatively the experimental results, if the change in heat and entropy of adsorption due to hot-electron action would follow formally the same compensating relation as observed with usual catalysts.

Another possible explanation is that we cannot ignore the fact that a kind of dynamic heterogeneity is created on the surface by the action of electrons. This can be due to the energy distribution of electrons, as well as to the heterogeneity of the insulating layer (preferential ways for hot-electron passage may not be neglected). Therefore, the approach proposed originally by Constable could be the possible explanation for the compensation effect. More experimental results are needed in order to decide what the right explanation is.

Two more interesting characteristics were deduced from these experiments. The first is the turnover numbers as a function of the current intensity and temperature, as shown in Figs. 8 and 9.

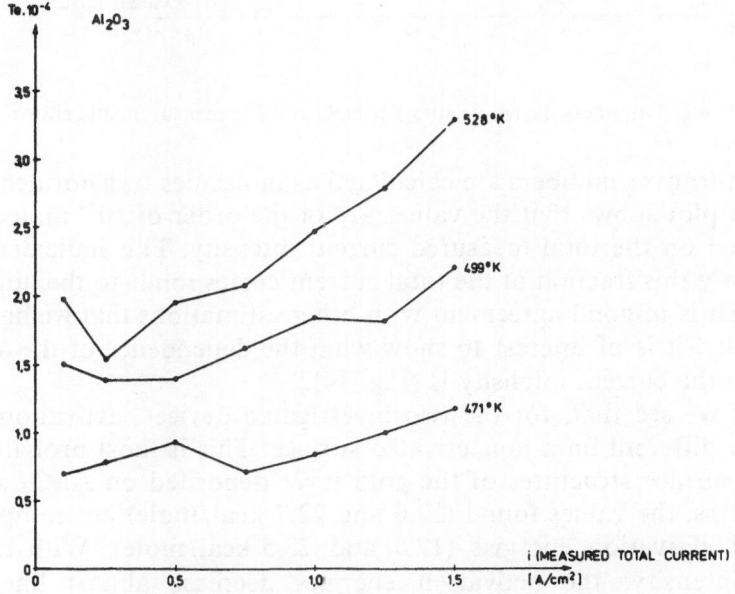

Fig. 8. Turnover numbers Te (molecules per electron) for measurements shown in Fig. 6.

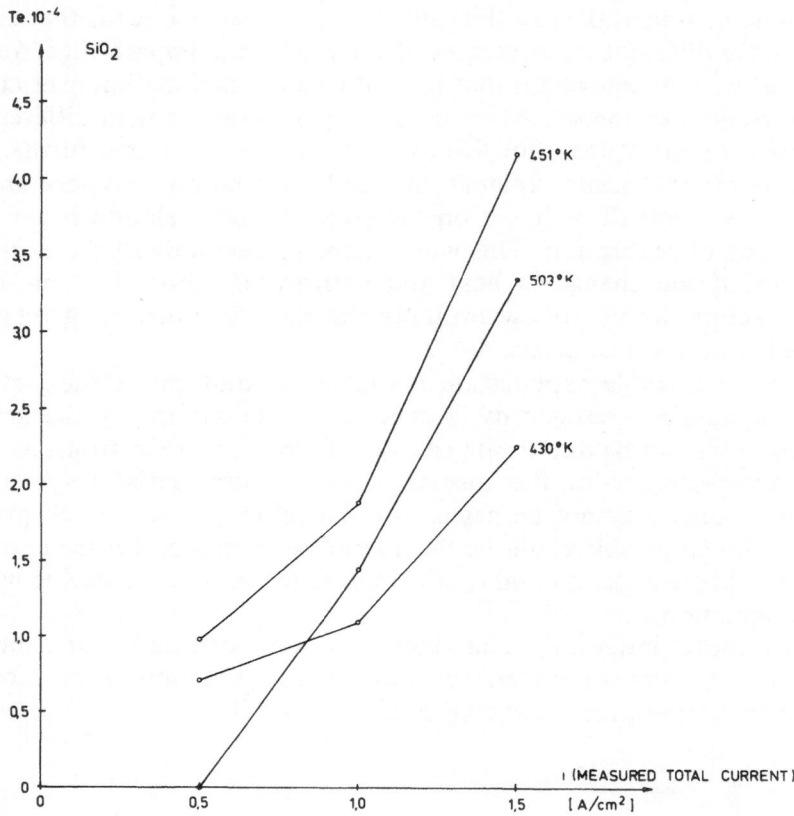

Fit. 9. Turnover numbers Te (molecules per electron) for measurements shown in Fig. 7.

The turnover numbers are calculated as molecules transformed per electron. The plot shows that the values are of the order of 10^{-4} molecule/electron, based on the total measured current intensity. This indicates that approximately this fraction of the total current corresponds to the hot-electron flux, which is in good agreement with other estimations that we have made.

Finally, it is of interest to show what the dependence of the activation energy on the current intensity is (Fig. 10).

First we see that, for the two investigated devices, activation energies are rather different on a nonactivated surface. This is most probably due to different surface structures of the gold layer deposited on Al_2O_3 and SiO_2. Nevertheless, the values found (12.6 and 22.7 kcal/mole) are comparable to published activation energies (12.5 and 23.5 kcal/mole). With increasing current intensity, the activation energies decrease almost linearly and parallel for both types of devices.

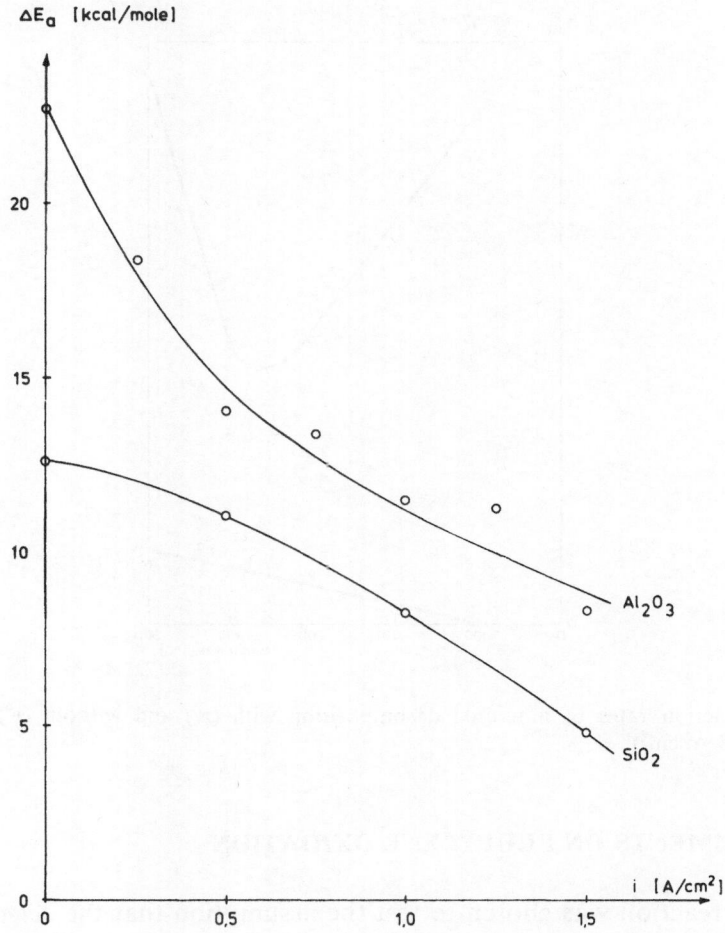

Fig. 10. Activation energy ΔEa for HCOOH decomposition, calculated from the plots Fig. 6 (Al_2O_3) and Fig. 7 (SiO_2).

3 EXPERIMENTS ON NH_3 DECOMPOSITION

One series of experiments was performed in order to see whether our catalytic devices behave in a similar manner in NH_3 decomposition. The same apparatus was used. The results are shown in Fig. 11.

What we can say is only that an increase in decomposition rate is observed. We do not have, at present, any explanation concerning the curious shape of the curve. Also, we drew only one conclusion from this experiment, namely, that the same type of catalytic device can accelerate the rates of reactions having very different reaction mechanisms.

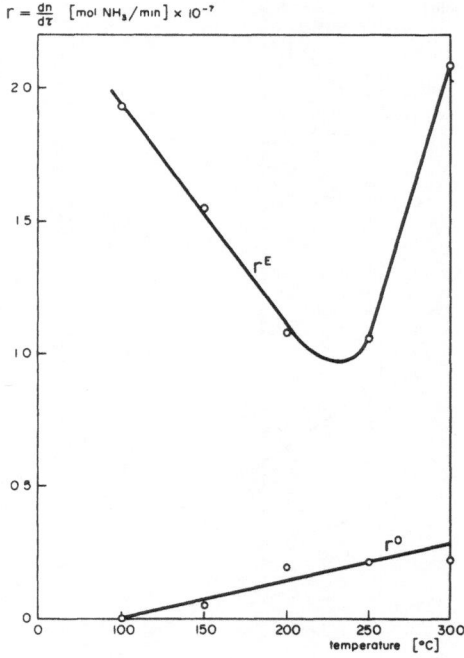

Fig. 11. Reaction rates of ammonia decomposition with (r^E) and without (r^o) activating current (1.75 A/cm^2).

4 EXPERIMENTS ON PROPYLENE OXIDATION

This reaction was chosen to test the assumption that the *selectivity* can be changed by the hot-electron action. In fact, necessary conditions for such an effect are fulfilled for this reaction because at least two parallel reactions go on at the surface of silver or gold, propylene oxide and acrolein being the first stable products that can be found. It is also known that different kinds (at least three) of chemisorbed complexes of oxygen participate in different partial reactions of propylene oxidation on Ib-group metals. Without going into details, we can conclude that this is the situation where the influence of hot electrons on the selectivity can be tested as mentioned in the introduction. If the different chemisorbed species have different interaction cross sections, one or another of them would be activated at an appropriate hot-electron energy, and this would be seen by the increase in production rate of one or another stable intermediate product.

To do the experiments, catalytic devices with a top layer of gold were tested with Al$_2$O$_3$ as insulating layer. The flow reactor used for these experiments has been described elsewhere.[2]

The results of one of the most interesting experiments are summarized in Fig. 12. The production rates are represented here for the main reaction products, namely, propylene oxide, formic acid, CO_2, and H_2O, as a function of electron current density. The principal features are as follows: (1) variation of production rates between $0-1$ A/cm^2 is small; and (2) between 1 and 1.5 A/cm^2, a very significant increase in the total oxidation rate is observed; (3) at the same time, the production rate of different products vary; that is, for propylene oxide, the rate increases by the factor 73.6, for CO_2 it increases by the factor 10.4, and for acrolein, it decreases to zero.

We can therefore conclude that the hot electrons influence, in this case, both the activity and the selectivity of gold for propylene oxidation.

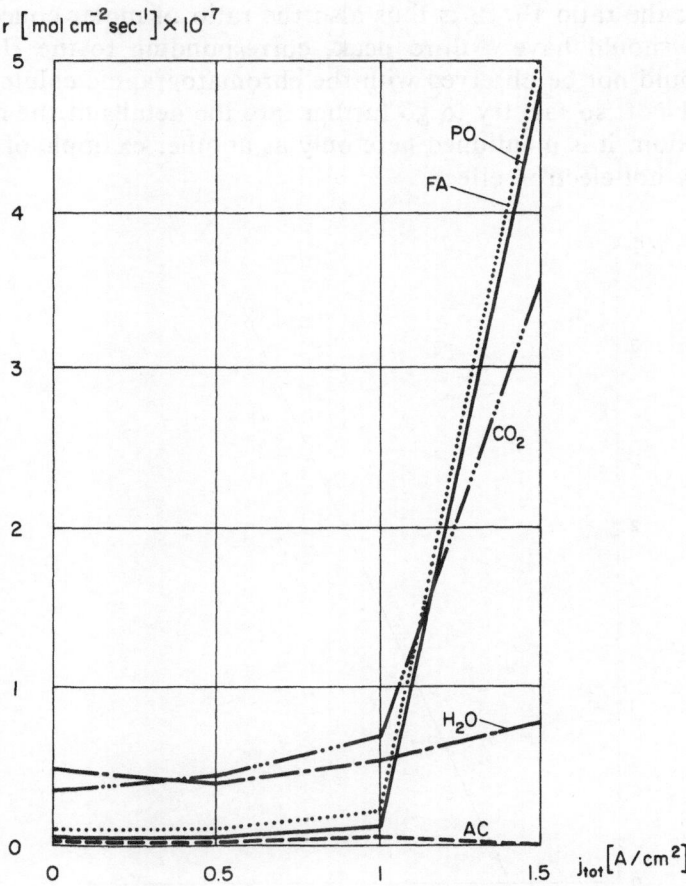

Fig. 12. Production-rate increase of principal oxidation products during propylene oxidation as a function of the activating current density: PO, propylene oxide; AC, acrolein; FA, formic acid.

5 EXPERIMENTS WITH BUTYL CHLORIDE DECOMPOSITION

We would like to mention, very briefly, another reaction which we found to be catalyzed by our catalytic devices, namely, the decomposition of tert-butyl chloride. It was found in the apparatus for propylene oxidation that tert-butyl chloride carried by the propylene and oxygen mixture was decomposed by the hot-electron action at 150°C, where otherwise no decomposition took place on the nonactivated gold surface of our catalyst. The results are shown in Fig. 13.

The ratio P_1/P_2 is plotted as a function of inverse current density and shows a fairly linear dependence. P_1, P_2 are the surface areas of the peaks in the gas chromatogram, corresponding to tert-butyl chloride and isobutylene, respectively; the ratio P_1/P_2 is thus also the ratio of molar concentrations. In fact, we should have a third peak, corresponding to the HCl, which, however, could not be observed with the chromatographic column used.

We did not, so far, try to go further into the details of the mechanism of this reaction; it is mentioned here only as another example of a reaction catalyzed by hot-electron effect.

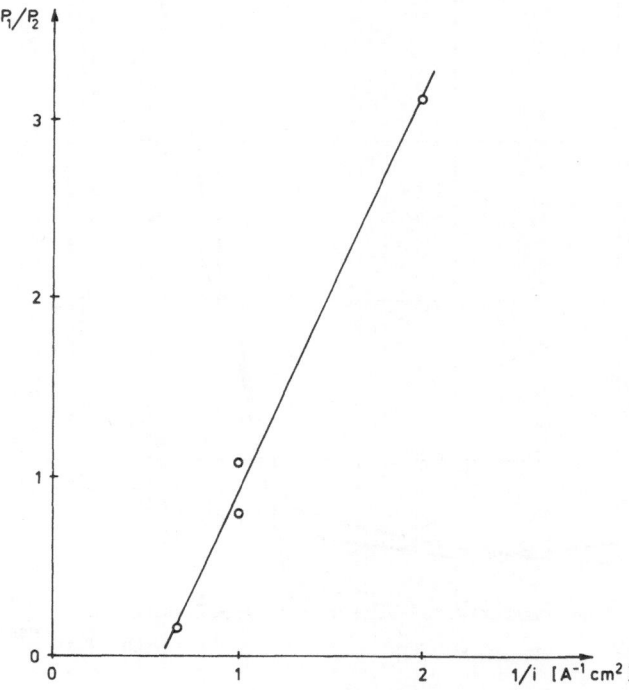

Fig. 13. Decomposition of tert-butyl chloride. P_1/P_2 is molar concentration ratio of tert-butyl chloride and isobutylene.

CONCLUSIONS

The described experimental work shows sufficient evidence that hot electrons flowing towards the catalyzing surface influence significantly its activity and selectivity with respect to several reactions with very different reaction mechanisms. This is what we were looking for. The catalytic behavior of a hot-electron-activated catalyst shows that the originally supposed, very simplified model must be modified, and even that other, quite different mechanisms involving hot-electron flow cannot be excluded at present. The goal of this preliminary research program was, at least in our opinion, fulfilled.

REFERENCES

1. Haidinger, W., and Figar, J., *Chem. Phys. Letters*, **11**, 545 (1971).
2. Figar, J., and Haidinger, W., *Chem. Phys. Letters*, **19**, 564 (1973).

DISCUSSION on Paper by W. Haidinger

KEMBALL: I wonder whether it is correct to describe these interesting results as true catalysis. The low values for the turnover numbers suggest that a minute fraction of the electrons function as reactants or perhaps merely as a source of some of the energy needed for reaction at the surface. There is no evidence that the electrons initiate chain processes such as those found when radicals are introduced into many types of homogeneous reactions.

HAIDINGER: There is obviously no reason why the catalytic effect we observed should not be considered as heterogeneous catalysis. The hot-electron beam should not be considered only as a valid supply of activation heat, as it also represents a direct supply of free energy of activation.

FIGAR: Phenomenologically speaking, our system conforms to the classic definition of a catalyst, because neither gold nor hot electrons appear in reaction products. The activating electron current flows in a closed circuit which has no electrical communication with the reaction space. From the point of view of the reactants, which are not aware of what is going on below the catalyst surface, I would say this is a true catalyst. It must be assumed that energy transfer takes place on all kinds of heterogeneous catalysts, and perhaps, in our case, we use this energy supply in a more efficient way.

IMELIK: Your catalytic disk is made up in such a way that a part of Al_2O_3 is in direct contact with the reactants. Therefore, the variation of the thickness of the metal layer does not prove that alumina is not activated by the current. It would be interesting to determine the selectivity of the decomposition of HCOOH, since when dealing with oxides, the reaction $HCCOH \rightarrow H_2O + CO$ predominates, contrary to what is observed for metals which give essentially the dehydrogenation reaction.

FIGAR: We cannot exclude the fact that a part of alumina layer which is not covered by the gold layer participates as a catalyst in the reaction. Nevertheless, gas chromatographic analysis of the reaction product of HCOOH decomposition was performed in several experiments, and *no* significant variation of CO concentration (as a product of dehydration reaction) was observed, with or without activation by electron flow. (Even if Al_2O_3 would participate in the reaction, its influence could be ignored when compared with the catalytic influence of the glass walls of the reactor.)

SMITH: This is a fascinating idea. First I would like to point out that a possible relationship exists between your work and that of Professor Honig of the University of Arizona. He has noted a change in exo-electron emission with variation in catalytic activity. Second, have you measured the temperature of your gold surface? I am wondering whether the "hot" electrons you are injecting into the gold are not affecting the catalytic rate by increasing the temperature.

FIGAR: Thank you very much for your comment and the very useful suggestion about the exo-electron emission. We shall certainly consider it in our future work.

The problem of surface temperature of our catalyst was intriguing indeed, and we have examined it quite carefully. Now, direct measurement of temperature on thin layers is a difficult task, and therefore a series of experiments was undertaken, the results of which *cannot* be explained by temperature increase. The most convincing are: (1) the variation of decomposition rate of formic acid, when the top layer was biased positively or negatively with the same power dissipated in the catalyst (Fig. 5); (2) decrease of decomposition rate due to activating current passage at high temperatures (above a critical value, see Arrhenius plots in Figs. 6 and 7); and (3) change of selectivity for propylene oxidation (Fig. 12). All these proofs are indirect, but as they all denote the same conclusion, we are now fairly sure that we are not dealing with trivial surface heating effects.

BLOCK: In a recent publication by N. Sato and M. Seo [*J. Catalysis*, **24**, 224 (1972)], it has been reported that propylene oxide formation and exo-electron emission from silver were proportional to one another. This was a most surprising result. The mechanism proposed is related to negatively charged intermediates which may decay partially by electron emission. Perhaps this observation may throw some light on your findings.

FIGAR: Thank you very much for your comment.

YATES:
1. From your experiments can you determine whether the chemical processes during current flow occur in the oxide film or within the gold film?
2. Do you have any estimate of the effect of electron energy on the chemical process? For electron-impact desorption processes on surfaces, threshold energies usually range from ~ 5 to 20 eV, depending upon the particular process being studied. Furthermore, cross sections for electron-impact induced ionization processes maximize at energies near 100 eV. Since your bias voltages are in the range of 1 to 2 eV, I find it difficult to believe that you are dealing with the production of electronically excited surface species.
3. One of your arguments which states that Joule heating is not involved is based on the extrapolation of your Arrhenius plots (at various current densities) to high temperature, where a crossing of the straight lines is deduced. However, judging from the scatter of the data points, which are all at temperatures below this crossing point, I am not certain that the crossing is in fact a valid conclusion. Are there other experimental indications that firmly exclude the possibility of Joule heating?

HAIDINGER:
1. This can be done only indirectly by analyzing the reaction products. It is known that metal surfaces favor dehydrogenation and that oxide surfaces favor dehydration of formic acid. However, electron microscopy showed that the thin gold layer was very dense, and if one looks at the total area of alumina and gold layers, only a very small fraction of the alumina layer is exposed to the gases, which means that the presence of alumina may be neglected in the reaction process. Furthermore, as pointed out in a previous comment, gas chromatography confirms this assumption.
2. Indeed there must be a difference in the interaction mechanism between the electron-impact desorption and the interaction when hot electrons are used coming from the inside of the catalyst. However, we

have put forward the analogy between both phenomena in order to give an indication of how our work is classified. It is well known that scattering cross sections rise up very steeply by decreasing the electron energy in the range of 1 eV or below. But in order to give a better reply to your first question, it should be considered that the chemical process takes place via some intermediate states. In most chemical reactions, the breaking or at least excitation of a certain covalent bond is the elementary rate-determining step. Whereas most of the other degrees of freedom of the system have an almost continuous spectrum, quantum effects cannot be neglected.

A reason may be given why an electron beam, having a relatively low excess energy, may be an available source of energy if one considers the absolute rate theory with the condition that $h\omega_i \ll 1$, where ω_i is the frequency of vibration along the reaction coordinate. Infrared spectroscopy gives the following values for the ratio $h\omega_i/kT$ (at room temperature) for some important processes.

Process	$h\omega_i/kt$
O_2 dissociation	7.9
HC dissociation	15.0
$C = C$ double bond in a chain	8.0
$C - C$ single bond in a chain	7.0

The fact that there are very few excitations of the these high ratios available at normal temperatures makes the breaking process of these bonds a very slow one. From estimations about the actual excess energy, we found that by applying 1 volt across the device, the energies of the hot electrons would be centered around 0.1 eV (cited in the Fig. 2 as V_2) corresponding to a ratio

$$\frac{eV_2}{kT} \simeq 4$$

Perhaps this gives better information about the amount of transferred energy by hot electrons compared with that by thermal excitations (citations from collaboration with Dr. Bergmann of Battelle-Geneva).

FIGAR: The reaction-rate decrease was not extrapolated from the straight lines in the Arrhenius plot, but was directly measured in the case of the catalyst with Al_2O_3 layer. There is a distinct decrease of slope on the Δp v time record (which was not shown here) at the moment when the activating current was switched on. See comment in reply to Dr. Smith's questions for other indications excluding the effect of Joule heating.

HAIDINGER: At the very beginning of this work we devoted a great number of experiments to proving that the observed activity enhancement is not a result of trivial Joule heating. Although the kind of compensation effect shown by Dr. Figar is the strongest evidence, there are many other experimental indications against the Joule heating effect. These are summarized in Ref. 1, but to mention only one, for the formic acid decomposition, we found a rate at 270°C (reactor temperature) for normal catalysis on gold (current off) which was equal to the decomposition rate with hot-electron flow (current on) at 20°C. If this difference is due only to a Joule heating effect, an electric power of about 7 kW would be necessary to maintain the temperature gradient in our setup. The effective power consumption was, however, about 0.1 watt.

FISCHER: In your studies, the variation of the thickness of the top gold layer is an excellent tool for separating real hot-electron effects from others. It does not change significantly any other property of the device, but it will effectively cut off the arrival *at* the surface of hot electrons. In case of experimental difficulties, take a good catalytic device and redeposit gold after the first study.

FIGAR: Thank you for your comment. We shall certainly take your suggestion into consideration in our future experimental work.

KOHN: Could you please tell us what you know about the energy of the hot electrons entering the gold foil? I was surprised that the mean free path was as long as 800 Å.

HAIDINGER: We did not undertake any work in this field. The value of 800 Å mentioned in our paper is in fact an attenuation length and was taken from the paper of C.R. Crowell et al., *Phys. Rev.,* **127**, 2006 (1962). In this paper, the relation between attenuation length and electron mean free path due to electron-phonon and electron-electron interaction is discussed. The mean free paths are about half the attenuation length in gold.

Instead of the mean free path, it is better to consider the attenuation length λ as it may be related to both the electron-phonon and the electron-electron mean free path. Values of about 800 Å to 1000 Å are reported in literature for hot electrons having about 1 eV excess energy which are injected into a gold layer.

There are other authors [e.g., Thomas, H., *Z. Physik.*, **147**, 395 (1957)] who report much larger attenuation lengths.

KOHN: I would like to mention work by Shanaberger which correlates resistance changes with changes in chemisorption.

HAIDINGER: Thank you very much for your comment. There might be some correlation between the resistivity change in chemisorption and our work. We are familiar with some of these works (for instance Suhrmann's review article) on this topic. However, in resistivity measurements only the conduction current (Fermi electrons) can be responsible for any effect on chemisorption. This may therefore be related to the curve in Fig. 5 showing a decrease in the decomposition rate of formic acid when the conduction current only (absence of the sandwiched oxide layer) crosses the catalyst.

Part Seven

CONCLUDING AGENDA
DISCUSSION—CRITICAL ISSUES

CONCLUDING AGENDA DISCUSSION: CRITICAL ISSUES (EXPERIMENTAL)

T. E. Fischer*

J. R. Smith**

This Colloquium has brought together two disciplines that are widely different in their traditions and in their methods and whose cooperation could bring about progress in heterogeneous catalysis that neither discipline would be likely to achieve alone. Heterogeneous catalysis has a long and distinguished history of industrial accomplishment and scientific research. Its predominantly physicochemical methods have already produced a wealth of information on and much understanding of catalyzed reactions (e.g., reaction mechanisms, energetics, and kinetics). Information is also available about the catalysts themselves (e.g., their state of dispersion, adsorption properties, and correlations of catalytic activity with position in the periodic table). Research can be performed under conditions very similar to those in industrial applications.

*Chairman.

**Secretary.

In the last two decades, surface physics has experienced such an extensive development of measuring techniques and theoretical capabilities that it can be considered a new science. The objects of its research include the fundamental processes at the atomic and electronic level that are involved in catalysis. The information obtained is often very detailed. In its present stage of development, surface physics requires highly simplified and controlled conditions, hence its predilection for very low gas pressures and plane surfaces of single crystals. These characteristics of the two fields point to complementary activities with limited overlap. The broad picture of catalysis, the first investigation and ordering of a new situation, and the development of new catalysts for particular applications will probably remain the province of "traditional" catalysis research, and so will, to a great extent, the formulation of the most important questions on the frontier of catalysis science. Many of these questions, particularly those that pertain to the solid state properties of the catalyst itself, may, however, not find an answer except through the novel investigative tools of surface physics. Examples of such questions follow:

- What is the chemical composition of the active catalyst surface?

- What influence does the configuration of surface atoms have on a catalytic activity (i.e., what constitutes active centers)?

- What are the properties of a solid that make it a good catalyst for a given reaction?

It appears, then, that surface physics will have more of an impact on the frontier of the science of catalysis than on the development of the novel catalysts needed in industry.

Considering this situation, the critical issues appear to be the following:

1. What are the most important questions the answer to which would push back the frontiers of an understanding of catalysis?

2. How is knowledge acquired from highly simplified and controlled surface physics experiments (i.e., plane single crystal surfaces and very low pressures) relevant to catalysis occurring under vastly different conditions (i.e., high dispersion of the catalyst and high gas pressures)?

The Chairman began the discussion by requesting that Ertl recapitulate the information about surfaces and molecules provided by modern experimental techniques of surface science.

Ertl first considered the question of structure or geometric arrangement of the atoms on the solid surface. LEED is in a stage where we can learn structures from it for simple systems. One might supplement these structural analyses with insight obtained from chemical information in the manner that Mason suggested. The second topic in surface analysis is chemical identification of species in the surface. We now have two powerful techniques for this, AES and ESCA (as shown by Dr. Yates). With these new techniques we can now quantify the role of poisons, i.e., we can ask the question of what is the necessary concentration of a poison on the surface to significantly influence the course of a reaction. The third subject is the use of infrared spectroscopy to determine the molecular structure and the influence of the surface on the molecule as exemplified by Eischens' measurement of the shift of the carbonyl bands in the presence of the substrate. It is now possible to perform infrared spectroscopic analysis on single crystals even, in some cases, with a single reflection of the infrared. Finally, ultraviolet photoelectron spectroscopy can be used to quantify the change in the electronic structure of the molecules due to chemisorption.

Fischer then asked Kemball in what direction would he suggest that surface science move to become more useful to catalytic science?

Kemball responded that it seems very important to him to realize that there are two gaps in the spectrum of work on surface science. First of all, we have, on the one hand, a gap between surface physics and surface chemical physics and, on the other hand, the studies on fundamental catalysis carried out in universities and in industry using conventional vacuum techniques. The second gap is between the latter type of research and the problems that have to be faced by the chemist and chemical engineer in industry who has to deal with actual catalytic processes. We are chiefly concerned with the first of these gaps, though we must not forget that the second exists. He commented that Boudart is right when he says that the exchange of hydrogen with deuterium is a trivial reaction. It can be carried out easily in the laboratory and the results have little relevance when viewed from the standpoint of the industrial catalytic chemist. However, he agreed that Grimley is also right when he describes the same reaction as far from trivial, because he looks at it from the viewpoint of the theorist. This example illustrates the existence of the two gaps.

Kemball stressed the danger of overstating the relevance of specific surface-physics investigations to catalysis. He noted that although he realized that in many cases the claim may arise from genuine aspirations if it is overstated there is a danger or loss of scientific credibility that will harm all of us. He asked those concerned with the theory or the physics of surfaces not to underestimate the amount of basic information that is known to the catalytic chemist. This whole Colloquium could have been planned the other way around, with the chemist setting out in some detail the basic concepts established and then asking the physicists to choose the areas that seem to them of potential interest in bridging the gap.

Kemball then proffered a few suggestions, perhaps mainly for some shift of emphasis in order to create some tenuous bridges between the work of the surface physicists and catalytic chemists. We must be content with limited but well-defined objectives. These suggestions are:

1. Shift the emphasis of some of the work on individual faces of single

crystals away from adsorption and spectroscopic measurements toward the determination of catalytic behavior, i.e., rates and selectivity. Two of the papers that have indicated a real contribution toward bridging the gap are those of Ertl and of Somorjai, both of whom made kinetic measurements on reactions. Chemisorption and spectroscopic studies necessarily tell us about the more strongly adsorbed species, while much of catalysis may depend on more labile and much less abundant surface species.

2. More emphasis should be put on studies of reactions of polyatomic molecules and somewhat less on the reactions of diatomic molecules. There would be great value in having more data about reactions of polyatomic molecules on individual crystal faces. The point made earlier about the value of exchange reactions is particularly relevant to polyatomic species.

3. More emphasis should be put on the use of defined surfaces, but which are deliberately covered with known amounts of poisons or deposits in order to obtain results more directly comparable with "real" systems. Poisons can be helpful in indicating not only the number but also the extent of catalytic active sites.

4. Many of the important industrial catalysts are oxides or sulfides and not metals. Emphasis is needed on the development of methods for studying fundamental processes on well-defined oxide surfaces. The control of the hydroxyl population of the surface will not be easy but it will be vitally important.

Somorjai suggested that it would be very helpful to the surface scientist to read the catalytic literature. He admitted that it may be confusing at first, but pointed out that often it has been helpful to him, especially where there is no fundamental literature on the subject available.

Schwab commented that looking back to the last decade of catalysis research, one can state that the most important progress in that field had been made by taking loans from solid state physics. However, this involved the necessity of drawing conclusions on catalytic behavior from a knowledge of bulk properties of the solid state rather than from knowledge of surface properties. This has brought considerable progress, however. He suggested that he need only mention the concept of the electronic factor. Only recently has the surface scientist achieved the means to definitively investigate surface properties in a direct manner. With this new sophistication, surface science can be very useful to catalysis science. However, the surface scientist should not restrict himself to metal surfaces but should, with equal intensity, study surfaces of semiconductors, oxides, salts, etc., which are equally important as catalysts. As an example, Scwab mentioned the two kinds of basic planes of zinc oxide, one occupied by Zn^{+2} and the other by O^{-2}, which have different chemical and catalytic properties.

Jaffee interjected at this point that the organizing committee was well aware of the importance of semiconductors and other types of catalysts but chose to focus on transition-metal catalysts, in order that the subject be narrow enough that all the viewpoints could be brought together in a Colloquium.

Yates asked any of those present who worked for companies that were actually using catalysts to make chemicals, or otherwise had close experience with practical catalysis, whether they recalled any examples of fundamental research having impact on the world of real catalysis.

Emmett responded that he could recall one example of an individual who transferred a result from very basic research into commercial application. He referred to the discovery of Otto Beeck. Beeck's rhodium was 3 orders of magnitude more active than nickel when prepared in high vacuum as thin films. Emmett stated that this finding was translated into commercial patents using rhodium as a hydrogenation catalyst.

Fischer added that very often new concepts and phenomena are uncovered through surface science that are helpful to catalytic chemists in an indirect way. He recalled an example in his own laboratory where a physical phenomenon being studied by the surface science group there led to the initiation of a new line of catalyst research by a catalytic chemist in the laboratory.

Wynblatt described the following contribution of surface science: he mentioned a technique for studying reactions on alloy surfaces which was developed by H. P. Bonzel and himself, and which was described in *Surface Science* about a year ago. This technique has the advantage that it allows the study of a whole binary alloy system with a single sample. The sample consists of a fairly thick foil of metal A which is coated with a thin film (100 Å or less) of metal B. One can run a reaction on the foil surface (i.e., pure metal B). The foil may then be annealed so as to cause interdiffusion and bring some concentration of metal A to the surface. At this point, the reaction of interest may be run once again on this surface of new composition. By alternate interdiffusion anneals, and activity measurements, it is quite simple to determine activity as a function of surface composition, all the way from pure B to almost pure A.

Bond commented that in connection with the question posed by Yates, he would not like it to be thought that activity is the only parameter of practical importance in chemical processes. The formation of by-products, heat and mass flow characteristics, mechanical stability, and resistance to poisoning are also vital considerations. These are not, however, problems whose solution surface physicists can usefully contribute to at the moment; they must remain for some time the concern of chemists and chemical engineers.

Ertl remarked that an obvious question that can be asked is: How many metal atoms are needed for the reaction? Is it a single metal atom, a cluster of metal atoms or an even larger configuration? We would, first of all, like to know the change in the electronic structure of one component of the alloy as the percentage of the other component increases. An assumption that has been made, for example, is that a nickel atom in a copper alloy has the electronic properties of pure nickel. The validity of that assumption can perhaps best be answered by Ehrenreich. Ertl noted that they have done experiments with Pd-Ag alloys, measuring the variation of the adsorption energy of chemisorbed species as a function of their coverage on the alloys and as a function of the alloy fraction. He also commented that they found that only a few percent of silver in the palladium would change the binding properties of chemisorbed gases markedly. That is, for pure palladium, the adsorbed layer was nearly uniform. However, when a few percent of silver was added, it was found that the adsorbed layer was heterogeneous. He offered the tentative explanation that, in fact, chemisorption involves ensembles of a few surface atoms whose compositions and therefore adsorption properties are altered by alloying.

Fischer asked Ehrenreich if it is possible for alloying to produce electronic properties different from the sum of the properties of the constituents, even in those cases where the rigid band model does not apply. For example, would the properties of nickel atoms in copper-nickel alloys not be different from those in bulk nickel?

Ehrenreich's answer was "yes". He then added: This can be seen by looking at ESCA data taken by Wertheim and collaborators on Ni-Cu alloys. As nickel is added to copper a nickel virtual level is formed at first. The width of that nickel level then grows with increasing nickel concentration, and the copper peak declines. Finally, on the pure-nickel side, the virtual nickel level has grown into a nickel d band and the copper peak has gone away. So this is not at all the kind of behavior that one would expect from a rigid band model.

Bond then asked Ehrenreich whether there are any binary systems for which the rigid band model is applicable, explaining that he asked this because it would be nice to check out this simple band picture against catalytic activity changes.

Ehrenreich responded that the rigid band model is valid provided the difference in electronic potential between the constituents of the alloy is sufficiently small. Apparently, in some semiconductor systems like Si-Ge alloys, this is the case. However, in anything involving d electrons, this does not seem to be so.

Kohn commented that the rigid band model at best has validity only for average properties of the system, such as the total electronic specific heat. Since the catalytic processes are highly localized, this model is especially unsuitable for them. He stated, that he however, believes that the band theory in its general form of a theory of spatially delocalized eigenfunctions is still our best bet for catalytic phenomena involving metals.

The Chairman then raised the question of the pressure gap. He recalled that it is common practice for the surface scientist to do his experiments at pressures many orders of magnitude lower than those encountered in many industrial applications. He asked the Colloquium members for comments on cases where the differences are qualitative and thus prevent transfer of information.

Schmidt noted that there is evidence from work in their laboratory that platinum wires may remain free of carbon and similar contaminants, when used as a catalyst in an oxidizing atmosphere. This comes from AES examination of platinum after heating in O_2 at 1 Torr to 1 atmosphere of pressure, and from NH_3 oxidation kinetics. It appears that in high O_2 pressures, reaction equilibrium with carbon and sulfur produce volatile oxides which continually remove these species. This implies that conventional catalyst preparation techniques may in fact produce essentially clean metal surfaces. Perhaps oxidation catalysis may be a far more fruitful area of study in elucidating mechanisms of catalytic reactions than are hydrocarbon reactions.

Walter added that from the free energy of adsorption, the equilibrium pressure in the gas phase p can be calculated by

$$\Delta F = RT \ln p/p_0$$

(p_0 = pressure in the standard state). Inserting numerical values of ΔF (chemisorption), p is found to be far below the lowest pressures obtainable in present-day vacuum chambers. Thus, one can rather safely assume chemisorption and catalysis observed in investigations necessitating UHV (LEED, Auger, ESCA, etc.) to correspond to situations in the high-pressure regime, as long as physisorbed layers do not affect the surface chemistry. An example of the latter case is the chemisorption of water on platinum. There, the more physisorbed water is present on the platinum surface, the faster is the chemisorption.

Schrieffer asked if there had been progress with differentially pumped systems in which electrons are emitted into vacuum above a layer of high pressure.

Fischer mentioned the work of R.T.K. Baker, who has built a differential pumping system into an electron microscope. Baker has produced a film called "Catalysis at Work" which shows reactions occurring at pressures much higher than usually permissible in the electron microscope.

Somorjai stated that the correlation between the reactivity of the surface and its atomic structure and composition can readily be studied at low pressures in the range of 10^{-5} to 10^{-4} Torr [*J. Catalysis,* **27,** 405 (1972)]. Electron scattering techniques like LEED and AES can be utilized at these pressures while the reactant and product distributions in the gas phase are monitored by a mass spectrometer.

However, most of the industrial catalysts are used at pressures of $\sim 10^3$ Torr or higher and, thus, there is a pressure difference of over seven orders of magnitude between the low-pressure catalytic studies and the pressure range utilized by industrial catalytic processes. If our aim is to apply the understanding of catalytic reaction mechanisms gained from low-pressure studies on the atomic scale to practical catalyst systems, we must be able to correlate the kinetics and mechanism of the surface reaction over this pressure range.

It is possible that the reaction mechanism remains unchanged upon changing the total pressure by seven orders of magnitude and that the turn-over number may change linearly or with some power of the reactant pressures. On the other hand, marked changes in the surface concentration of bulk solubility of the reactants with increasing pressure may completely change the mechanism of the surface tension. For this latter case, low- and high-pressure studies cannot be correlated at all until the nature of the pressure-dependent change in mechanism is fully investigated.

Somorjai noted that they have studied the kinetics of catalytic reactions at high pressures ($\sim 10^3$ Torr) on single-crystal platinum surfaces using a gas chromatograph as a detector [*J. Catalysis,* **34,** 291 (1974)] and successfully compared the rates of cyclopropane-ring opening with those reported on polydispersed platinum catalysts. The same apparatus can be used to monitor the rate of hydrocarbon reactions in the range of 1 to 1000 Torr partial pressures of the reactants. At the low-pressure end, the total pressure may be varied in the range of 10^{-6} to 10^{-4} Torr. Bridging the gap in the pressure range of 10^{-3} to 10^{-1} Torr, however, requires modification of the apparatus presently used.

Studies of hydrocarbon reactions over the whole pressure range using the same catalyst surface should determine to what extent low- and high-pressure reaction studies can be correlated. Such investigations are presently feasible and should be of great value in deciphering the atomic details of heterogeneous catalysis.

Oudar mentioned some of their attempts to fill the gap between the low- and high-pressure catalytic experiments. He noted that if they are doing some experiments, for example, with sulfur on single crystals of gold, silver, or iron, they obtain only the first stages of adsorption at low pressure of H_2S. They find that the pressure of H_2S has to be increased considerably in order to reach the maximum coverage. (This maximum coverage was determined from adsorption isotherms taken during high-pressure experiments using a mixture of H_2 and H_2S.) On gold and iron, they observe also that a controlled amount of H_2 in H_2S favors the adsorption of sulfur by some catalytic effect. He also noted that they occasionally obtain a surface coverage which cannot be reached by a low pressure experiment by introducing, for example, a controlled amount of sulfur into the bulk of the material and then populating the surface via segregation by heating.

Fischer noted that infrared spectroscopy might be usable in these high-pressure experiments, and said that he was very interested in Ertl's comments that infrared could be used on flat surfaces. He noted also that differential optical reflectivity of flat surfaces in the visible and ultraviolet ranges had been developed to such a sensitivity that fractions of a monolayer can now be observed. He pointed out that optical spectroscopy may lead to the following simplifications: (1) relaxation effects may be minimized and (2) high pressure gases may not hinder the experiment.

Fischer asked what the sensitivity of such infrared experiments on flat surfaces is, as concerns coverage of adsorbed gases. Ertl replied that at present we could see CO, but that further work is needed before we know whether we can see such things as CH stretching frequencies.

Yates concurred, noting that they have shown that single-reflection infrared spectroscopic studies of CO chemisorbed on an atomically clean tungsten surface had a practical limit of detection of several percent of a monolayer for the high extinction coefficient CO species. He commented that this work was done with conventional infrared spectroscopic equipment, and stated that he shares Ertl's pessimism about the utility of this method for the study of species which have a low extinction coefficient.

Drauglis commented that Fourier transform spectroscopy potentially offers a method for looking at the CH stretching, bending, and other modes in less than monolayer quantities on clean flat surfaces. However, in order to do this, one must scan for periods ranging from 15 minutes to 1 hour or more (a scan takes usually ½ to 1 second). Therefore, the method is limited to static measurements and cannot be used when it is desired to study kinetics. Furthermore, Eischens made the statement a few days ago (in a private conversation) that Fourier transform spectroscopy has not been as sensitive as theory says it should be. Drauglis suggested that perhaps Eischens would like to comment on this point.

Eischens stated that Fourier transform spectroscopy has potential advantages of one or two orders of magnitude in sensitivity. However, this theoretical advantage has not produced results in the more difficult systems such as hydrogen on supported platinum or hydrocarbons on flat surfaces. He commented that he could not cite examples in adsorbed molecule studies where Fourier transform spectroscopy has produced spectra which are significantly better than those attainable with conventional spectrometers. He added that direct comparison of Fourier transform and conventional spectroscopy on a single system would be worthwhile in order to determine whether the potential advantages of Fourier transform spectroscopy can be realized in practice.

Eischens went on to say that an additional problem in reflection work on single crystals is the expense of moderately large (about 1 cm^2) single crystals, and that they have not been able to obtain suitable single crystals for some of the most interesting metals such as platinum and nickel.

The Chairman then asked Emmett to conclude the session by giving his evaluation of this conference and offering his advice on the interaction of surface physics and catalysis.

Emmett felt that such conferences as we were having were the very best for promoting conversation between the various disciplines attempting to understand catalysis. He felt that there was especially good communication between surface experimentalists and catalytic chemists. On the question of what are the areas in which surface scientists can contribute to the understanding of catalytic chemists, Emmett noted that Kemball had covered most of his list already. However, he suggested that an important item that has not been discussed at length was that of the study of the effect of promoters, and he gave the example of the role of promoters in ammonia synthesis. He noted that it might be particularly interesting to study the interaction of poisons with stepped surfaces, as produced by Professor Somorjai. He closed by suggesting that surface scientists should attempt to introduce simple catalytic reactions into their experimental chambers.

CONCLUDING AGENDA DISCUSSION: CRITICAL ISSUES (THEORY)

T. B. Grimley*

Donnan Laboratories
University of Liverpool
Liverpool, England

J. R. Smith**

Physics Department, Research Laboratories
General Motors Technical Center
Warren, Michigan

1 INTRODUCTION

Before the participants assembled in Gstaad, the Chairman circulated the following agenda:
    a. The fundamental problem of heterogeneous catalysis
    b. The adiabatic approximation in catalysis

*Chairman.
**Secretary.

c. The potential energy hypersurfaces
d. Elementary reactions
e. The role of surface spectroscopy in catalysis
f. Calculations on clusters.

At the Colloquium there was time in Session VII for discussions on only the first five items, but item f had already received some discussion in Session III: Theory of Chemisorption.

The *ab initio* theoretical study of even simple heterogeneous reactions such as the H_2/D_2 exchange on Pt(111), or the CO/O_2 reaction on Pd(111) evidently involves so many unsolved and difficult problems that one may question the value of even formulating the theory of such molecular rate processes at all. But it is only by undertaking this formulation that we are able, on the one hand, to bring to the attention of experimental workers in catalysis important theoretical developments that have occurred in other fields, and on the other, to point to the areas where either the formal theory is insufficiently developed or present theoretical methods are inadequate to deal with our problems.

2 THE FUNDAMENTAL THEORETICAL PROBLEM

The fundamental theoretical problem is to calculate for a heterogeneous reaction the rate of reactive transitions out of an initial state $|i>$ to a final state $|f>$, the states of the catalyst being included in the specifications $|i>$ and $|f>$. To introduce the discussion, the Chairman reviewed briefly the theoretical chemist's approach to this problem.[1]

First we need a solution $|i^+>$, say, of Schrödinger's equation

$$(H_0 + V - E)\,|i^+> = 0$$

which satisfies the boundary conditions, and which reduces to $|i>$ when the mutual interactions V of all reactants (including the catalyst) vanish. This solution is expressed formally by the Lippman-Schwinger[2] equation

$$|i^+> = |i> + G_0^+ V|i^+>$$

$$= |i> + G^+ V|i>$$

where the Green operators are defined by

$$G_0^+(E) = (E + i0 - H_0)^{-1}$$

$$G^+(E) = (E + i0 - H)^{-1}$$

$$= G_0^+(E) + G_0^+(E)VG^+(E).$$

The amplitude of the final state $|f\rangle$ in the exact state $|i^+\rangle$ is $\langle f|i^+\rangle$, and consequently the rate of change of the probability to observe the final state $|f\rangle$ due to reactive collisions from the initial state $|i\rangle$ is

$$R_{fi} = \frac{d}{dt}|\langle f|i^+\rangle|^2$$

This is the required transition rate. Using Lippman's[3] generalization of Ehrenfest's[4] theorem we can write

$$R_{fi} = h^{-1}|\langle f|V|i^+\rangle|^2 \, \delta(E_f - E_i) \tag{1}$$

or, if we introduce the T-matrix (reaction matrix) defined by

$$T = V + VG^+V = V + VG_0^+T$$

then

$$R_{fi} = h^{-1}|\langle f|T|i\rangle|^2 \, \delta(E_f - E_i) \tag{2}$$

Equations (1) and (2) are correct whether we use the initial state interaction and T-matrix, or the final state quantities.

In general the initial state is a mixture characterized by a density matrix. If it is drawn from a canonical ensemble, then R_f, the transition rate to $|f\rangle$ from all states in the entrance channel i, is given by

$$R_f = kT \int_0^\infty dE \, \exp(-E/kT)dR_i^i(E)/dE \tag{3}$$

Here $R_i^i(E)$ is the transition rate to $|f\rangle$ from all states of energy E in the entrance channel i. We can write Eq. (3) in the form

$$R_f = (kT/h) \exp(-\Delta F^\ddagger/kT) \tag{4}$$

if we define $\Delta F^\ddagger$, the "free energy of activation from the entrance channel i" by equating the right hand sides of Eqs. (3) and (4).

Of course, the working concepts used by experimental chemists (mechanisms, rate-determining steps, transition states, and so on) are not to be found in the formal theory. These concepts emerge only when simplifying approximations are made.

3 THE ADIABATIC APPROXIMATION

The application of the adiabatic approximation to the electronic motion, leading to the concept of potential energy surfaces, is familiar in the study of reactions in the homogeneous gas phase. We first calculate the surfaces $E_m(\mathbf{R})$ by solving Schrödinger's equation for the motion of the electrons in the fixed potentials of the nuclei:

$$[H_e(\mathbf{p}) + H_{el}(\mathbf{p},\mathbf{R})]\Psi_m(\mathbf{p},\mathbf{R}) = E_m(\mathbf{R})\Psi_m(\mathbf{p},\mathbf{R}) \tag{5}$$

Then the motion of the nuclei is described by the system of coupled equations

$$[H_i(\mathbf{R}) + E_m(\mathbf{R})]\Phi_m(\mathbf{R}) + \sum_n A_{nm}(\mathbf{R})\Phi_n(\mathbf{R}) = E\Phi_m(\mathbf{R}) \tag{6}$$

Neglecting the couplings A_{nm} gives the usual adiabatic equations for the nuclear motion.

The Chairman asked what heterogeneous processes were known to involve nonadiabatic behavior, and Boudart mentioned the slow rate of transition of NO on alumina from a paramagnetic to a diamagnetic state. This process was discovered by Solbakken[5], who interpreted the experimental observation that the rate decreases with increasing temperature, in terms of a nonadiabatic process. Solbakken's observations have been confirmed recently by one of Boudart's students[6], and the process has been shown to involve a disproportionation of adsorbed NO molecules into N_2O and NO_2. The nonadiabatic behavior which may be encountered here is associated with the familiar crossing of electronically adiabatic potential energy curves determined from Eq. (5); if E_n and E_m cross, and A_{nm} in Eq. (6) does not vanish, nonadiabatic behavior is possible.

Schrieffer thought that nonadiabatic behavior would be at least as important for metals as for insulators, and for a different reason, namely, because of the high degeneracy of the electron gas. He believed that adiabatic *versus* nonadiabatic behavior should be viewed rather as in the electron-phonon interaction problem. That is, the Hamiltonian is divided into an adiabatic part and a nonadiabatic part treated in terms of the basis states of the adiabatic Hamiltonian. Then the nonadiabatic terms may well turn out to be as important in reaction dynamics calculations as the adiabatic potential energy surfaces. Kohn pointed out that many catalytic reactions, by their very nature, would involve nonadiabatic transitions between initial and final states which are eigenstates of an adiabatic Hamiltonian, and that therefore all the interest really centers on the nonadiabatic terms in the Hamiltonian.

Newns mentioned Toulouse's[7] explanation of the well-known absence of diffraction patterns when helium atoms are scattered from metal surfaces at very low temperatures. Toulouse suggests that large numbers of electron-hole pairs are generated by the atom-solid interaction, thus destroying momentum conservation in the atomic beam. If this explanation is correct, it means that, in these scattering experiments, the helium-metal interaction cannot be calculated adiabatically, i.e., the concept of an electronically adiabatic, potential energy surface is not very useful here. However, Kohn pointed out that there is a difficulty in reconciling Toulouse's explanation with the fact that there is, in these experiments, a well-defined sharp specular reflection. This specular peak is difficult to understand if there is a loss of momentum conservation through the creation of a large number of electron-hole pairs, each of which has a small momentum.

The Colloquium reached no firm conclusion on the status of the adiabatic approximation. Some reactions, like the atomization of H_2 for example, might be pictured as taking place on a single adiabatic potential energy surface. But as Schrieffer had pointed out, for metal catalysts, the Schrödinger equation [Eq. (5)] will have, in general, a quasi-continuous spectrum of potential energy surfaces $E_m(\mathbf{R})$, so that nonadiabatic behavior (transitions from one to another) might be the norm. However, Schrieffer also remarked that, essentially because of the Pauli Exclusion Principle, the number of electronic degrees of freedom excited might be rather small. In this case it seems that the situation would be similar, in some respects, to that encountered in treating the phonon spectrum of the metal, and the adiabatic approximation would be expected to hold to a very high approximation. What is needed of course is for the coupling operators A_{nm} between neighboring states in the quasi-continuous spectrum to tend to zero more rapidly than their energy separation. The characteristic feature of metals—their quasi-continuous spectrum of electronic excitations—would not then give rise to new possibilities for nonadiabatic behavior in catalysts. Evidently, the definitive treatment of this problem is an important matter.

4 POTENTIAL ENERGY SURFACES

The calculation of electronically adiabatic, potential energy surfaces is the task of chemisorption theory, but so far little progress has been made, and earlier in the Colloquium a number of members expressed pessimism about our ability to make these calculations in the foreseeable future. Experience with such calculations for reactions in the homogeneous gas phase shows that the task is formidable, but there is an opinion building up among students of molecular quantum mechanics that where reactants and products are closed shell systems, the ordinary Hartree-Fock approximation can, in certain circumstances, give meaningful potential energy surfaces. An example of such a system is the reaction between HCl and NH_3 treated some years ago by Clementi[8] using the IBMOL IV Hartree-Fock program. However, Messmer pointed out that Clementi calculated only a potential energy curve, not a surface*, and that the computing time required was about 1 week. He believed therefore that *ab initio* calculations of potential energy surfaces for reactions of catalytic interest are many years away, although simple model calculations could, and should, be done sooner.

Derouane claimed to be less pessimistic than Messmer, but then went on to explain why the ordinary Hartree-Fock scheme would not be adequate for calculating the activation energy for a simple process like dissociative chemisorption. The initial state is the admolecule A-A and the substrate S. The final state has two A-S molecules, so an A-A bond is broken and two A-S bonds are made. The ordinary Hartree-Fock approximation might give reliable information on the initial and final states which are located at minima in the potential energy surface, but if the transition-state complex is near the dissociation limit, the calculated activation energy will not have much meaning because of the well-known failure of the ordinary Hartree-Fock scheme to describe open-shell dissociation products correctly.

Schrieffer numbered himself among the more optimistic members of the Colloquium, although he agreed with Messmer that *ab initio* Hartree-Fock calculations by the LCAO method seem very unlikely to provide answers with reasonable computing times. He believed however that the problem could be approached by treating the adsorbates and the first (perhaps the second also) coordination shells of substrate atoms surrounding the adsorption sites by relatively accurate techniques ($X\alpha$ for example) and the remaining substrate atoms more approximately (by tight-binding methods for example)**. The muffin-tin potentials of the $X\alpha$ scheme are known to give unreliable total energies, but recent developments suggest that by adding "empty spheres", or by treating non-muffin-tin potentials in first-order perturbation theory[11], one can obtain rather accurate total energies, at least for small molecules. Schrieffer thought that more accurate methods of treating correlation effects might be necessary, but he was hopeful that at least the broad features of a potential energy surface could be determined. At this point Ehrlich commented that, for the experimentalist, calculations revealing qualitative trends resulting from changes in either surface geometry or chemical constitution are important. Quantitative agreement with the actual measured quantities, by itself, is not very interesting. The Chairman thought that there was no conflict between Schrieffer and Ehrlich on this matter, and went on to suggest that molecular dynamics on even poor potential energy surfaces might be useful in giving us some idea of the features determining the course of a reaction.

---

*Chairman's note: this is not quite fair. Clementi calculated five potential energy curves for a hydrogen atom moving between nitrogen and chlorine in the linear geometry H_3N—H--Cl because he used five different N-Cl distances. He also made two computations to assess the importance of rearrangements of the NH_3 hydrogens when NH_4Cl is formed.

**Chairman's note: Grimley and Pisani have proposed[9] and used[10] a scheme of this sort to calculate potential energy curves for chemisorption. See also Session III.

Another member who counted himself among the optimists on the question of calculating potential energy surfaces was Kohn who, while again agreeing with Messmer that full-scale Hartree-Fock calculations were not feasible, believes that, using simplified methods (the density functional or Xα methods for instance), a calculation of the potential energy surfaces for a single atom on a metal surface should soon be possible. There are two features that in Kohn's view make this problem simpler than the calculation of potential energy surfaces for gas phase reactions between molecules: (1) the substrate atoms are rather firmly held at fixed positions, (2) the adatom is confined to the vicinity of only a two-dimensional surface. Of course, the problem of calculating potential energy surfaces for a complex chemical reaction on a solid surface is clearly very much harder.

Ehrlich welcomed Kohn's interest in potential energy surfaces for a single atom migrating over a single-crystal surface, and remarked on the growing body of experimental work in the field which provides much material to compare with the theory. Moreover present experimental indications are that the dynamics of the atomic motion are quite simple, so the potential energy surfaces themselves should prove informative.

The Secretary was interested to hear the views of Schrieffer and Kohn, but thought that our best course of action at this time would be to look for a generalization of the Woodward-Hoffmann rules. A generalization is necessary because, at a catalyst surface, there is generally a continuum of eigenstates available. Smith reminded the Colloquium members that Schrieffer had pointed to the local density of states as the proper quantity to focus on. This quantity might also allow us to take account of symmetry effects in reactions at surfaces, and certainly the determination of the local densities of states at a bare catalyst surface should be much easier than the determination of the potential energy surfaces for a complex reaction on the catalyst.

It is evident from the above discussion that, with some reservations, the theoreticians at the Colloquium thought that meaningful calculations of potential energy surfaces for simple heterogeneous processes could be made using simplified models whose predictive power had, however, already been tested in other fields (see Section 5 below). The existence of this view must mean that such calculations will soon be attempted, probably with models which aim at self-consistency only over a surface complex, but use Green-function techniques (Dyson's equation) to draw the rest of the catalyst approximately into the local self-consistency problem. Because of the well-known failure of the ordinary Hartree-Fock model to describe open-shell systems correctly, if this model is used to treat the surface complex, it will be important to choose a reaction where the chemically interesting reaction path involves only closed-shell species. No such restriction need be imposed if the Xα scheme is used to treat the local self-consistency problem; open-shell dissociation products are handled properly in this scheme.

In this discussion, the Colloquium moved closer to the working concepts used by catalytic chemists. To postulate a *mechanism* is to postulate a reaction coordinate on one or more electronically adiabatic, potential energy surfaces, and to propose a certain transition state is to propose the location on this coordinate of the critical stationary point. This then is the meeting ground of catalytic chemists and theoreticians working on chemisorption.

5 ELEMENTARY REACTIONS

We learn from chemical kineticists to regard complex reactions as proceeding *via* a sequence of elementary steps, one of which controls the rate. In the ammonia synthesis for example, the chemisorption of N_2 is believed to be the slow step (see Emmett's lecture). However, there are simpler heterogeneous processes than this, atomic and molecular beam scattering for example, and the Chairman invited Dr. Wolken, Jr. to review the present state of the theory of elastic and inelastic scattering of gas molecules from solid surfaces. His response follows:

I would like to point out one area of gas-solid collisions for which the dynamical problem is rather well under control. As Dr. Messmer pointed out, the potential energy surface is only part of the problem, there remains the problem of how the particles move under the influence of the assumed forces. I wish to turn the process round, and attack the dynamical problem first. The hope is that if one can treat the dynamics reliably, one can tell which parts of the potential affect the scattering, and therefore, which regions of the potential most warrant *ab initio* treatment. For the elastic scattering (diffraction) of closed shell atoms or molecules from solids, which we may treat as scattering from an infinite, static potential, periodic in the plane of the surface (e.g., scattering from alkali halide crystals), the calculation of the intensities in the various diffraction spots can be done with good accuracy. We expand the exact wave function:

$$\psi(\mathbf{R,z}) = \sum_{m,n=0,\pm 1,\pm 2,\ldots} f_{mn}(\mathbf{z})\exp\left\{i(\mathbf{K}+\mathbf{G}_{mn}\cdot\mathbf{R})\right\}$$

where $\mathbf{K}$ stands for the x and y components of the propagation vector (momentum) of the incident particle, $\mathbf{G}_{mn}$ are surface reciprocal lattice vectors, and $\mathbf{R}$ is the two-dimentional (x,y) position vector. This expansion is substituted into the Schrödinger equation to obtain a set of coupled second-order ordinary differential equations. These are solved by well-known numerical techniques. Effects such as bound-state resonances, states diffracted into the surface (closed channels) and, if a suitable potential can be estimated, scattering from multiple planes of surface atoms are all treated without further approximations beyond those already mentioned. One of the several conclusions we[12] were able to draw from these studies is that, for diffraction, the potential in the region of the classical turning point is most important. The long-range behavior of both the flat, and the oscillatory parts of the potential does not seem to matter very much.

The treatment can be extended to molecule-solid inelastic scattering in which energy is transferred not to the solid, but between translational and internal modes of the gas molecule.[13] These inelastic events have recently been observed by the group at Genoa [Professor Boato's group (Chairman)] scattering H_2 from cold LiF surfaces. The theory of this kind of collision, designed for comparison with the experimental results, is being developed.

In the short discussion that followed, Ehrlich emphasized that, for the understanding of chemical events at surfaces, information about inelastic collisions in which internal energy (rotational energy for example) in the gas molecule is transferred to the solid is likely to be of considerable importance. Kohn commented that the calculation of the scattered intensities, even for elastic scattering, requires a knowledge of the dynamics of the interacting atom-lattice system, and he wondered, therefore, in what sense the dynamical problem could be said to be "rather well under control". Wolken replied that, for processes such as the scattering of closed-shell atoms from alkali halide crystals, it does not appear to be a bad approximation to replace the dynamical model by a model with a static potential. However, because of Debye-Waller effects, this static potential is really an effective potential which is different at different temperatures. (Chairman's note: for a review of atomic and molecular-beam scattering from surfaces prior to March, 1973, the forthcoming lecture notes of Course LVIII of the 1973 Enrico Fermi International School may be consulted. See also Somorjai and Brumbach.[14])

To move the discussion nearer to reactive scattering, the Chairman mentioned a simple process, the recombination of hydrogen atoms on a catalyst surface, and distinguished two extreme situations. In one, the molecules (or atoms) spent a long time on the surface, came into thermal equilibrium with it, and so lost all memory of their states in the incident beams; in the other, they spent only a short time on the surface, and the desorbing reaction products would then be in states controlled to some extent by the states of the reactants in the incident beams. Schrieffer said that in his view, there is nothing in the problem of reactive scattering that is not already in the chemisorption problem. That is, as long as the molecules which are reactively scattered are in their electronic ground states, the potential energy surface, in the Born-Oppenheimer sense, can be calculated.

He said that he and J. Davenport were considering the dissociation of H_2 on Ni(100) using a combination of $X\alpha$, and simple LCAO techniques.

Next the Chairman asked the question: "What other elementary processes are simple enough for theoreticians to think about, and are at the same time catalytically interesting?" Ertl, Schmidt, Tamaru, and Bond responded. Ertl said that we must choose reactions for which we have a feeling about the mechanisms, and about the geometries of the collisions, the oxidation of CO for example, with the oxygen atom adsorbed, and the CO molecule coming down from the gas giving the linear OCO—metal geometry. This would be a one-dimensional collision problem, but Ertl did not think there were many reactions of this type, although the decomposition of N_2O might be another example. The molecule is linear, and one should consider the case where it comes down oriented perpendicular to the surface, the oxygen atom sticks, and the N_2 is reflected back. The H_2/D_2 exchange reaction may be more complicated than it looks Ertl said. The nature of the transition state for the reaction in the homogeneous gas phase is certainly not clear.

Schmidt responded to the question as follows:

I have worked on two problems which are important in catalysis, but for which theoretical limitations prevent detailed calculations. The first concerns the rate of dissipation of the energy released in the catalytic reaction. This can control the reaction channels available, and also the local temperature of a supported metal catalyst. How is the energy transferred from electronically excited intermediates, and how does this energy ultimately dissipate itself into phonons and excited conduction electrons? For supported catalysts, there are temperature differences between the catalyst particles and the support, over distances which are of the order of the electron mean-free path, so conventional treatments are useless. The question of the thermal conductivity at a metal surface also arises, and there is some experimental work which seems to indicate that the thermal conductivity near the surface is far less than the bulk value. A high degree of disorder near the surface may account for this.

Second, the possibility of a chemisorbed species existing one or more layers below the surface (but not really dissolved in the bulk) appears to be important in many gas-solid systems. This possibility should be investigated by theoreticians; so far, most of us have assumed that adsorbates must exist outside the surface layer.

Tamaru commented as follows:

A simple catalytic reaction, which theoretical people may be interested in, and which I studied some years ago with Professors Boudart and Taylor, is the decomposition of gaseous GeH_4 on a germanium surface. During the decomposition, the germanium surface is constantly renewed by the deposition of fresh germanium atoms, and the catalytic system contains only two elements, germanium and hydrogen. The reaction is zero order in GeH_4 and

H_2, and during the decomposition, practically all the germanium surface is covered with chemisorbed hydrogen.* Hydrogen desorption from the fully covered surface is the rate-determining step. Accordingly, if we could treat theoretically the rate of hydrogen desorption from a fully covered germanium surface, we could compare it with the experimental results, as we know experimentally the rate and the activation energy for the process. The adsorption isotherm and the heat of adsorption of hydrogen on germanium are also known.

Bond thought that only reactions of hydrogen should be studied theoretically in the first instance, and he included in these the hydrogen-ion discharge reaction at an electrode during electrolysis.

The Chairman remarked that theoreticians will need all possible help in their studies of these elementary processes, and Webb pointed out that there are available some data on mean square amplitudes of vibration, and on thermal expansion coefficients at solid surfaces, either clean, or with adsorbed overlayers. He wondered if these data would be useful to theoreticians in their computations of potential energy surfaces, and other quantities of interest in catalysis. Kohn replied that they would be useful because, to make theoretical calculations of surface energies, work functions, and so on, one needs to know where the surface atoms are and how tightly they are held.

Taking advantage of the opportunity presented by the unique structure of this Colloquium, the Chairman invited the experimentalists present to say where *they* thought that theoreticians skilled in solid state and molecular quantum mechanics could help with the problems of heterogeneous catalysis. Boudart and Bond responded. Boudart said:

Is it not more logical for physicists, who are materials scientists, to give us more help on catalysts, that is, on catalytic materials rather than catalytic reactions? Here traditional expertise in solid state physics can bring help in perhaps a more straightforward manner. I list several specific instances where new theoretical approaches might be most useful:

1. We would like very much to have something besides Pauling's percentage d-band character to describe transition metals. It has been very useful to us as a rough correlating parameter, and we have little else. Other examples of useful ideas which provide correlating parameters are the concept of variable valence proposed a few years ago by Thor Rhodin, and the concept of metal electronegativity due to Pauling.

2. We would like to know both the atomic arrangement, and the electronic structure of small particles. We are interested in the formation of multimetallic clusters, in the stability of clusters, in super-paramagnetism, and finally in the heat flow in small particles.

---

*Tamaru, K., and Boudart, M., *Advan. in Catalysis,* **IX,** 699 (1957).

3. If you were to suggest that new compounds might exist which would be catalytically active, we would try to make them.

Ehrenreich responded at once to these remarks by saying that, on the basis of his work on the heats of formation of bulk metal hydrides (see Session III), there is not, unfortunately, anything very simple to say about the factors affecting the binding. The heat of formation is determined by a complicated interplay of a low-lying (bonding) hydrogen impurity band, d band lowering, and Fermi energy shifts, which vary from material to material. It is possible to build up a body of experience, both for pure metals and for alloys, based on similar calculations, which may suggest directions to be pursued in the search for both new catalytic materials and hydrogen storage media. It is also possible that systematics having some predictive qualities will emerge from such a study, but that remains to be seen. Ehrenreich suggested that since in many cases there exists a correlation between bulk and surface properties, it may be simpler to pursue such a study for the bulk, since bulk electronic properties are easier to calculate. Furthermore, it does not seem clear whether the catalytic process involving hydrogen is better visualized in terms of a model with a single atom located on a perfectly clean surface, with an atom located just beneath the surface, or with an atom in a surface vacancy. If either of the last two is more relevant, a bulk description of the heat of formation may be the more appropriate.

Bond responded to the Chairman's invitation as follows:

Catalytic chemists would be greatly helped by a clear and simple exposition of the electronic structure of metal surfaces. I find the notion of surface valency of assistance, but I do not know how valency varies with coordination number, or with Group number in the Periodic Table. The chemisorption bond may of course involve the participation of a variety of molecular orbitals, either filled or empty, provided by surface metal atoms. What we would like to know therefore is: How can the electronic structure of metal surfaces be described in quantitative MO language? Given this we might well be able to understand how the strength of the chemisorption bond and catalytic activity are affected by electronic structure. At the moment we still have to use Pauling's theory of metals if we want to attempt quantitative correlations.

From these responses it seems clear that experimental chemists are asking primarily for theoretical work on surface crystallography, and surface electronic structure. Regarding the latter, it may be helpful to emphasize that theoreticians are most likely to present the results of their calculations on surface electronic structure in the form of a local density of states $\rho(\mathbf{r},\epsilon)$ varying both with position and energy, and defined by

$$\rho(\mathbf{r},\epsilon) = \sum_{\mu} |\psi_{\mu}(\mathbf{r})|^2 \delta(\epsilon-\epsilon_{\mu})$$

where $\psi_{\mu}(\mathbf{r})$ are the molecular orbitals of the system (see Schrieffer's paper in Session III). This function contains all the information chemists traditionally expect of a theory of electronic structure. At given energy ϵ, the dependence of ρ on position outside a surface tells us about the directional properties of both occupied and vacant orbitals of this energy which can be used for forming the chemisorption bond. The dependence on ϵ is present because of the continuous spectrum of orbitals at solid surfaces; if a discrete spectrum also exists, $\rho(\mathbf{r},\epsilon)$ has δ-function singularities in its energy dependence, and in molecular quantum theory, $\rho(\mathbf{r},\epsilon)$ has *only* such singularities. At single-crystal surfaces, the matrix of $\rho(\mathbf{r},\epsilon)$ in the representation afforded by the atomic orbitals ϕ_i of a surface atom [$\rho(\mathbf{r},\epsilon)$ has the periodicity of the surface net of course] seems likely to provide experimental workers with much of the information they are seeking. The diagonal elements of ρ

$$<i|\rho(\mathbf{r},\epsilon)|i> = \tilde{\rho}_{ii}(\epsilon) = \sum_{\mu} |\int \phi_i^* \psi_\mu d\mathbf{r}|^2 \delta(\epsilon - \epsilon_\mu)$$

are the gross densities of states on the atomic orbital ϕ_i, and

$$\int_{-\infty}^{\epsilon_1} d\epsilon\, \rho_{ii}(\epsilon)$$

is the quantity known in quantum chemistry as the gross population (occupancy) of the orbital ϕ_i. These densities of states are simply the generalizations of the familiar concepts of MO theory required to deal with the case where the MO energy levels ϵ_μ may contain a continuous spectrum. It remains an open question whether this sort of information will be very useful in understanding chemisorption and catalysis; it should however help us to discern some of the symmetry factors controlling the reaction mechanism.

6 THE ROLE OF SURFACE SPECTROSCOPY

Among the more important recent developments in surface science are the surface-sensitive spectroscopic techniques: ultraviolet photoelectron spectroscopy (UPS), X-ray photoelectron spectroscopy (XPS or ESCA), ion-neutralization spectroscopy (INS), field emission spectroscopy (FES), and Auger spectroscopy. The question naturally arises: what contributions have these new spectroscopies made to our understanding of heterogeneous processes, and what contributions are they expected to make?

Hagstrum noted that the Colloquium expected that a knowledge of the local density of states would be useful in trying to understand some of the fundamentals of catalysis (see Section 5), and then went on to say that surface spectroscopic techniques (INS and UPS) provide some information, although they do not of course give the local density of states directly. He remarked that much of the work done so far referred to adsorbed species in much higher concentrations than those expected for reaction intermediates in catalysis, but that there were experimental reasons for working with adsorbates, like the chalcogens on nickel, which formed well-defined, self-limiting overlayers.

The usefulness of Auger spectroscopy in determining the surface compositions of alloys, and the significance of such work in the understanding of catalysis was noted by Ertl, who also pointed out that, using UPS, one can now directly identify species adsorbed on a surface. For example, if two peaks are seen in the electron energy distribution for CO, then the molecule is undissociated, and it is not necessary to have a complete understanding of the interactions of photons with surfaces, and with adsorbates to use UPS in this way. However, Schrieffer regarded it as absolutely crucial that the theoretical problems posed by UPS and XPS be studied with a view to predicting the angular dependence of energy-resolved photoemission, arguing that if our theoretical models are not good enough to predict the photoemission, we should not believe in their ability to calculate meaningful potential energy surfaces for catalytic reactions. If this view is accepted, theoreticians are committed to the task of unfolding the final-state relaxation processes in UPS, for example, which lead to large, but in a sense, uninteresting, level shifts.

On the question as to whether or not a reaction intermediate stays on the catalyst surface long enough for modern surface spectroscopic techniques to be used to detect it, Somorjai commented that the low turnover numbers of many catalytic reactions (10^{-4} to 1 product molecule/surface atom/second), as well as some molecular beam studies which measure surface residence times, indicate that reaction products can stay on the surface for 10^{-2} second or longer. These residence times are long enough for many spectroscopic techniques to be used to study the reaction intermediates. Somorjai also suggested that electron spectroscopy could be used to study adsorbates present on a surface in a steady-state flow of low-pressure gas, and he thought that any difference between the results for the steady-state case and the static adsorbate/adsorbent system would, by itself, be very interesting.

Yates suggested that optical absorption methods (photons in, photons out) would be among the most useful for detecting intermediates under steady-state conditions at moderate pressures. Thus, in the infrared method, the absorption bands of the adsorbed species are usually shifted from their gas-phase frequencies, and are not therefore occluded in the absorption bands of the reactant gas. He went on to say that, in his study of formaldehyde adsorption using XPS, no trace of the intermediate responsible for the small yield of methane was found, presumably because its steady-state concentration is too low.

The final contribution to the discussion was made by Plummer, who said first that he did not think that theoretical models like those asked for by Boudart and Bond, with new concepts to replace simple correlating parameters like percentage d-band character, could be provided; the phenomena were fundamentally too complicated [this is essentially Ehrenreich's view too, see Section 5 (Chairman)]. Next, on the subject of surface spectroscopy, Plummer noted that the theoreticians had generally been optimistic about their ability to make calculations, and he said that surface spectroscopy could provide essential checks on their theories. He agreed with Schrieffer that our ability both to measure, and to calculate correctly, the electronic structure of a surface complex must be proved if we are ever to construct a microscopic model of chemisorption with predictive powers. Plummer distinguished two categories of spectroscopic techniques: those which look directly at the valency electrons (UPS, INS, FES), and those which look at the response of some other quantity to changes in the valency electron distribution caused by chemisorption (XPS, infrared spectroscopy). The direct valency electron spectroscopic results seem very difficult to interpret, and clearly need immediate theoretical attention, but he thought that the indirect techniques only *seem* to be simple, and that for example, core-level shifts seen in XPS, would be very hard to interpret quantitatively. He also thought that ability to calculate these shifts may not provide a very stringent test of the theory.

Two roles for surface spectroscopy in heterogeneous catalysis emerge clearly from the above discussion: (1) as a "fingerprinting" technique to identify chemisorbed species (but with considerable doubts about our ability to detect chemically interesting intermediates), (2) as a source of microscopic information on electronic structure and bonding with which to confront theoretical models of chemisorption. Those who are concerned with the interpretation of UPS data may be interested to read parts of the Cambridge (England) Discussion of the Faraday Division of the Chemical Society, *Photo Effects in Adsorbed Species* (to be published in 1975) and perhaps the paper by Higginson and others[15] on the UPS and XPS of the gaseous Group VI A hexacarbonyls.

7 CONCLUSION

It is impossible to say at this stage what progress will accrue from the bringing together of theoreticians skilled in solid state and molecular quantum theory, and experimental workers from surface chemistry and physics, but it will be interesting to see how many references to the Colloquium Proceedings are being made in 3 years' time. The Colloquium provided some surprises. One was the small area of common ground seeming to exist among theoreticians, surface chemists, and surface physicists. Generally speaking, those studying experimentally processes involving small molecules interacting with single crystal planes hold the common ground, but quite clearly, those of us interested in heterogeneous catalysis will have to learn to speak a little more in a common language. There may be genuine difficulties here, because, first, there is as yet no demonstration that conventional concepts based on the Born-Oppenheimer separation of electronic and nuclear motions are adequate for the discussion of molecular rate processes on metal surfaces, and, second, even if the Born-Oppenheimer separation is useful, the concept of a reaction coordinate varying only slowly on the time scale of the motions of all other nuclear coordinates needs some consideration in view of the existence, in the catalyst, of phonon modes of arbitrarily low frequencies. These are important problems, and their resolution will enable us to decide how far conventional concepts found useful in discussing chemical kinetics in the homogeneous gas phase are applicable also to heterogeneous reactions. Thus, some fundamental theoretical work must be undertaken in this area. Another area, referred to frequently during the Colloquium, where fundamental theoretical work is required, is in surface electron spectroscopy, and work is already in progress in several Institutions [Liverpool University, Noordwijk (ESRO), University of Pennsylvania, San Jose (IBM), and Warren (General Motors Technical Center)].

REFERENCES

1. For further details see, for example, Levine, R. D., *Quantum Mechanics of Molecular Rate Processes,* Oxford University Press, Oxford (1969).
2. Lippman, B. A., and Schwinger, J., *Phys. Rev.,* **79,** 469 (1950).
3. Lippman, B. A., *Phys. Rev. Letters,* **15,** 11 (1965); **16,** 135 (1966).
4. Ehrenfest, P., *Z. Physik,* **45,** 455 (1927).
5. Solbakken, A., and Reyerson, L. H., *J. Phys. Chem.,* **64,** 1901 (1960).
6. Turnham, B., Ph.D. Dissertation, Stanford University (1974).
7. Toulouse, G., *Proc. Internat. School of Physics Enrico Fermi Course LVIII* (1973), to be published.
8. Clementi, E., *J. Chem. Phys.,* **46,** 3851 (1967).
9. Grimley, T. B., *Proc. Internat. School of Physics Enrico Fermi Course LVIII* (1973), to be published.
10. Grimley, T. B., and Pisani, C., *J. Phys. C: Solid State Phys.,* **7,** 2831 (1974).
11. Danese, J. B., and Connolly, J.W.D., *J. Chem. Phys.,* **61,** 3063 (1974); Danese, J. B., *J. Chem. Phys.,* **61,** 3071 (1974).
12. Wolken, G., Jr., *J. Chem. Phys.,* **58,** 3047 (1973).
13. Wolken, G., Jr., *J. Chem. Phys.,* **59,** 1159 (1973).
14. Somorjai, G. A., and Brumbach, S. B., *Critical Rev. in Solid State Sci.,* **4,** 429 (1974).
15. Higginson, B. R., Lloyd, D. R., Burroughs, P., Gibson, D. M., and Orchard, A. F., *J. Chem. Soc. Faraday Trans. II,* **69,** 1659 (1973).

AUTHOR INDEX

SUBJECT INDEX

593